Resilient Complex Adaptive Systems

Resilient Complex Adaptive Systems

Modeling, Simulation, Visualization and Engineering Approaches

Edited by

Claudia Szabo
Adelaide University, Australia

Rodrigo Castro
Universidad de Buenos Aires and CONICET, Argentina

Viktoriya Semeshenko
Universidad de Buenos Aires and CONICET, Argentina

Saurabh Mittal
FlightSafety International, Ohio, USA

IEEE Press
WILEY

Published by John Wiley & Sons, Inc., Hoboken, New Jersey.

For general information on our other products and services or for technical support, please contact our Customer Care Department within the United States at (800) 762-2974, outside the United States at (317) 572-3993 or fax (317) 572-4002.

Wiley also publishes its books in a variety of electronic formats. Some content that appears in print may not be available in electronic formats. For more information about Wiley products, visit our web site at www.wiley.com.

Library of Congress Cataloging-in-Publication Data:
Names: Szabo, Claudia, editor | Castro, Rodrigo (Professor), editor |
Mittal, Saurabh editor | Semeshenko, Viktoriya, 1979- editor
Title: Resilient complex adaptive systems : modeling, simulation, visualization and engineering approaches / edited by Claudia Szabo, Rodrigo Castro, Saurabh Mittal, Viktoriya Semeshenko.
Description: Hoboken, New Jersey : Wiley-IEEE Press, [2027] | Includes bibliographical references and index.
Identifiers: LCCN 2026021282 (print) | LCCN 2026021283 (ebook) | ISBN 9781394326273 hardback | ISBN 9781394326297 adobe pdf | ISBN 9781394326280 epub | ISBN 9781394326303 obook
Subjects: LCSH: Nonlinear systems | Decision making | Artificial intelligence
Classification: LCC QA402 .R46 2027 (print) | LCC QA402 (ebook)
LC record available at https://lccn.loc.gov/2026021282
LC ebook record available at https://lccn.loc.gov/2026021283

Cover Design: Wiley
Cover Image: Rosi/stock.adobe.com

To all our collaborators and families

Contents

List of Contributors

Zeeshan Ali
IMS UMR 5218
University of Bordeaux
CNRS
Bordeaux
INP
Talence, France

J. Ignacio Alvarez-Hamelin
Facultad de Ingeniería
Universidad de Buenos Aires
Ciudad Autónoma de Buenos Aires,
Argentina

INTECIN
CONICET—Universidad de Buenos Aires,
Ciudad Autónoma de Buenos Aires,
Argentina

Rob Baker
Defence Science and Technology Group
Adelaide, Australia

Souvik Barat
Tata Consultancy Service Research
Tata Consultancy Services
Pune, Maharashtra, India

Emily Barrett
The MITRE Corporation
McLean
VA, USA

Eva Besada-Portas
Department of Computer Architecture and Automation
Universidad Complutense de Madrid
Madrid, Spain

Román Cárdenas
Department of Electronics Engineering
Universidad Politécnica de Madrid
Madrid, Spain

Rodrigo Castro
Facultad de Cs. Exactas y Naturales and Instituto de Ciencias de la Computación
Universidad de Buenos Aires and CONICET
Ciudad Autónoma de Buenos Aires,
Argentina

Kevin S. Chan
U.S. Army Combat Capabilities
Development Command Army
Research Laboratory
DEVCOM ARL
Adelphi, MD, USA

Xuan Chau
The MITRE Corporation
McLean, VA, USA

Nicolás De la Peña
Faculty of Economics, Business, and
Sustainable Development
Universidad De la Salle
Bogotá, D.C., Colombia

Sergio A. De Raco
Facultad de Cs. Económicas
Universidad de Buenos Aires
Ciudad Autónoma de Buenos Aires,
Argentina

Mama Diakité
IMS UMR 5218
University of Bordeaux
CNRS
Bordeaux
INP
Talence, France

Anh Viet Do
School of Computer and Mathematical
Sciences
The University of Adelaide
Adelaide, Australia

Roberto Domínguez
Department of Computer Architecture
and Automation
Universidad Complutense de Madrid
Madrid, Spain

Giovanni Dosi
Institute of Economics
Scuola Superiore Sant'Anna
Pisa, Italy

Segundo Esteban
Department of Computer Architecture
and Automation
Universidad Complutense de Madrid
Madrid, Spain

Theresa Fersch
The MITRE Corporation
McLean
VA, USA

Bradley Fraser
Sensors and Effectors Division
Defence Science and Technology
Group
Adelaide
SA, Australia

Juan I. Giribet
Universidad de San Andrés and
CONICET
Victoria, Buenos Aires, Argentina

Oscar M. Granados
Computational Finance and Digital
Economics Lab
Universidad Jorge Tadeo Lozano
Bogotá, D.C., Colombia

Hazel Griffith
Department of Systems and Computer Engineering
Carleton University
Ottawa, ON, Canada

Ishan Honhaga
School of Computer and Mathematical Sciences
The University of Adelaide
Adelaide, Australia

Rhett Hull
Sensors and Effectors Division
Defence Science and Technology Group
Adelaide, SA, Australia
UTS Tech Lab
University of Technology Sydney
Ultimo, NSW, Australia

Frank T. Johnsen
Norwegian Defence Research Establishment (FFI)
Kjeller
Akershus, Norway

Shankar Kausley
Tata Consultancy Services Research
Tata Consultancy Services
Pune, Maharashtra, India

Vinay Kulkarni
Tata Consultancy Service Research
Tata Consultancy Services
Pune, Maharashtra, India

Andrew Langone
The MITRE Corporation
McLean, VA, USA

Glenn Lematta
The MITRE Corporation
McLean, VA, USA

Ignacio Mas
Universidad de San Andrés and CONICET
Victoria, Buenos Aires, Argentina

Pablo Mira
Facultad de Cs. Económicas
Universidad de Buenos Aires
Ciudad Autónoma de Buenos Aires, Argentina

Saurabh Mittal
Defense Science Division
Modern Technology Solutions, Inc.
Dayton, OH, USA

Kelly J. Neville
The MITRE Corporation
McLean, VA, USA

Emily S. Patterson
The Ohio State University
Columbus, OH, USA

Siddharth Patwardhan
Center for Science of Science and Innovation
Kellogg School of Management
Northwestern University
Evanston, IL, USA

Ezequiel Pecker-Marcosig
Facultad de Ingeniería
Universidad de Buenos Aires
Ciudad Autónoma de Buenos Aires, Argentina
Instituto de Ciencias de la Computación
CONICET—Universidad de Buenos Aires
Ciudad Autónoma de Buenos Aires, Argentina

Marcelo Pereira
Institute of Economics
Scuola Superiore Sant'Anna
Pisa, Italy
Universidade Estadual de Campinas
Campinas, Brazil

Ant Perry
Defence Science and Technology Group
Adelaide, Australia

Gabriel Petrini
Institute of Economics
Scuola Superiore Sant'Anna
Pisa, Italy
Universidade Estadual de Campinas
Campinas, Brazil

J. Francisco Presenza
Facultad de Ingeniería
Universidad de Buenos Aires
Ciudad Autónoma de Buenos Aires, Argentina
INTECIN
CONICET—Universidad de Buenos Aires
Ciudad Autónoma de Buenos Aires, Argentina

Filippo Radicchi
Center for Complex Networks and Systems Research
Luddy School of Informatics, Computing, and Engineering
Indiana University
Bloomington, IN, USA

Beena Rai
Tata Consultancy Services Research
Tata Consultancy Services
Pune, Maharashtra, India

Varun K. Rao
Center for Complex Networks and Systems Research
Luddy School of Informatics, Computing, and Engineering
Indiana University
Bloomington, IN, USA

José L. Risco-Martín
Department of Computer Architecture and Automation
Universidad Complutense de Madrid
Madrid, Spain

Andrea Roventini
Institute of Economics
Scuola Superiore Sant'Anna
Pisa, Italy

Cristina Ruiz-Martin
Department of Systems and Computer Engineering
Carleton University
Ottawa, ON, Canada

Tracy Sanders
The MITRE Corporation
McLean, VA, USA

Viktoriya Semeshenko
Facultad de Cs. Económicas and
Instituto Interdisciplinario de
Economía Política
Universidad de Buenos Aires and
CONICET
Ciudad Autónoma de Buenos Aires,
Argentina

Instituto Interdisciplinario de
Economía Política
CONICET
Ciudad Autónoma de Buenos Aires,
Argentina

Facultad de Cs. Económicas
Universidad de Buenos Aires
Ciudad Autónoma de Buenos Aires,
Argentina

Surabhi Shrivastava
Tata Consultancy Service Research
Tata Consultancy Services
Pune, Maharashtra, India

Claudia Szabo
School of Computer and Mathematical
Sciences
The University of Adelaide
Adelaide, Australia

Eyoel Teffera
Defence Science and Technology
Group
Adelaide, Australia

Puneet Singh Thakur
Tata Consultancy Services Research
Tata Consultancy Services
Pune, Maharashtra, India

Andreas Tolk
The MITRE Corporation
Charlottesville, VA, USA

Mamadou K. Traoré
IMS UMR 5218
University of Bordeaux
CNRS
Bordeaux
INP
Talence, France

Maria Enrica Virgillito
Department of Economic Policy
Università Cattolica del Sacro Cuore
Milan, Italy

Gabriel Wainer
Department of Systems and Computer
Engineering
Carleton University
Ottawa, ON, Canada

Levent Yilmaz
Computer Science and Software
Engineering
Auburn University
Auburn, AL, USA

Foreword

The concept of complex adaptive systems (CAS) has been around for several decades as a means of understanding the remarkable adaptive features of living systems. This book is the first to go well beyond current literature in showing how CAS provides a potent framework to design modern socio-technical systems that are likewise resilient, auditable, and adaptable. In this context, CAS principles such as decentralized control, dynamic learning, and modular interactions become engineering primitives for structuring human–machine ecosystems, policy-driven infrastructures, and collaborative workflows. Pushing the current boundaries of CAS, this book focuses on engineering resilient systems, where resilience is far more than just bouncing back from disruptive events. Rather, it is about how systems absorb, adapt, and transform in response to disruptions, whether these are shocks like cyberattacks, policy shifts, or cascading failures in infrastructure.

As you can imagine, engineering such ability to absorb disruption and recover function raises multiple challenges for current systems engineering and design methods. In my mind, the overriding challenge is the recognition that such challenges must be addressed through both bottom-up and top-down approaches, each dealing with complexities at multiple levels of abstraction from a multiplicity of disciplinary viewpoints. In this vein, the book explicates, critiques, and expands upon current methods and technologies for modeling, simulation, and visualization of resilient CAS across various scientific, social, economic, and multidisciplinary systems engineering fields.

As a prospective reader, you may be reluctant to start into a book with complexity at its core, but rest assured that the book offers a smooth learning experience by introducing resilient complex adaptive system (RCAS) through an explanation of key attributes like adaptability, redundancy, fault tolerance, scalability, self-healing, and diversity. It then proceeds to model RCAS as systems with decentralized components called agents using local information to adapt quickly and reduce localized failures. Finally, it shows how resilience is achieved by self-organization, modular design for easy upgrades, and continuous monitoring and feedback for sustained performance and stability. Of course, the point is that these features are more easily described than achieved. Here is where the book truly breaks new ground. I was happy to see how the authors draw on their expertise in model-based system engineering and simulation modeling methodology to characterize and throw light on the challenging aspects of RCAS. From this perspective, they proceed to lay foundations for RCAS rooted in modeling and simulation concepts, particularly the discrete event system specification (DEVS) formalism. Then on this firm basis, they proceed to examine the engineering of RCAS in terms of iterative, data-driven development of simulation models subject to test and evaluation in live, virtual, and constructive (LVC) environments. This approach is then elaborated in application to exemplary socio-technical systems in the form of digital twins and socio-ecological phenomena. In sum, I heartily urge you, as a curious and discerning reader, to plunge into the intricacies of RCAS. While they may seem intimidating at first, your effort will be amply rewarded with rich insights into how their remarkable properties might be systematically engineered into existence for systems that matter to you.

August 2025

Bernard P. Zeigler
Oakland California

About the Editors

Claudia Szabo is an Associate Professor and leads the Complex Systems Research group in the School of Computer and Mathematical Sciences at the University of Adelaide, Australia. Claudia graduated with a PhD from the School of Computing at the National University of Singapore. Her research work focuses on understanding complex systems and their behavior, and on designing complex systems with desired properties, spanning many fields. Within biology, her work focuses on deriving systemic laws that explain how species evolve in the presence of environmental stressors, such as climate change or mass extinction events. Within defense, her work focuses on systems that operate in contested and dynamic environments, and in particular on designing and implementing self-adaptive and self-organizing autonomous solutions that have to make fast and critical decisions in the presence of various faults. She has authored 130+ papers, over 100+ of which are in the areas of simulation, model-driven engineering, systems of systems, self-adaptive systems, reinforcement learning, multi-agent systems, and their various constraints. She was the chair of the Winter Simulation Conference in 2021 and is an Associate Editor for the ACM Transactions of Modeling and Computer Simulations and one of the four co-editors in Chief of the Journal of Simulation.

Rodrigo Castro is an Associate Professor in the Department of Computer Science at the University of Buenos Aires (UBA). At the Computer Science Institute (ICC) of CONICET (the National Council of Scientific and Technical Research of Argentina), he serves as both Vice Director and an Independent Researcher, where he also leads the Discrete Event Simulation Laboratory. He holds a degree in Electronic Engineering and a PhD in Control Engineering from the National University of Rosario, Argentina, which he complemented with postdoctoral studies at ETH Zurich, Switzerland. With over 25 years of experience in the software and telecommunications industry, he also holds a specialization in University-Business Cooperation for Development (Organization of Ibero-American States). Professor Castro's research focuses on the theoretical foundations, methodologies, and software tools for modeling and simulating complex systems. A central goal of his work is to advance the state of the art in the assisted, systematic construction of simulation models through the rigorous composition of multi-domain, multiscale, and multi-paradigm components. He is an active contributor to the ATLAS and the Geant4 Collaborations at CERN and maintains an active network of collaborators across Europe, America, and Australia. His scholarly contributions include the supervision of 4 completed doctoral theses and the publication of over 100 scientific articles. He serves as an Associate Editor for the journal Simulation: Transactions of the SCS and is an Advisory Editor for the Springer Nature book series, Simulation Foundations, Methods, and Applications.

Saurabh Mittal is Complex Systems Engineering—Subject Matter Expert (SME) at Modern Technology Solutions, Inc., USA. Previously, he was a Principal Scientist and a Project Leader in Air & Space Forces Center at The MITRE Corporation, USA. Earlier at MITRE, he also held roles of Chief Scientist for the Simulation, Experimentation, and Gaming Department, Tech. Advisor and Modeling & Simulation (M&S) SME to various Government sponsors and Principal Investigator for multiple internal research efforts involving cyber-physical systems, digital twins, Internet of Things, and virtual cloud-based simulation system of systems. He received PhD (2007) in Electrical and Computer Engineering with dual minors in Systems and Industrial Engineering and Management and Information Systems from the University of Arizona, Tucson. In his 25 years of technical management and research engineering, he has co-authored over 125 publications as book chapters, journal articles, and conference proceedings, including 5 books, covering topics in the areas of complex systems, system of systems, complex adaptive systems, and M&S-based complex systems engineering. He served as the President for the Society for Modeling and Simulation International (SCS) for 2021 and a member of the Board of Directors (2018–2020). He has also served as General Chair, Vice-General Chair, Program Chair, and Track Chair during the 2013–2023 period in Spring, Summer, and Winter Simulation Conferences. He is a recipient of the US DoD's highest civilian contractor recognition: Golden Eagle award and Outstanding Service and Professional Contribution awards by SCS.

Viktoriya Semeshenko holds a Master's degree in Modeling and Simulation of Complex Realities from the Abdus Salam International Centre for Theoretical Physics, (ICTP)/International School for Advanced Studies (SISSA) in Trieste, Italy, and a PhD in Cognitive Sciences from the Institut National Polytechnique of Grenoble (INPG) in France. She currently works at the Institute of Political Economy (IIEP) of the Economic Science Department at the University of Buenos Aires, where she is head of the Laboratory of Networks and Complex Systems, and is a researcher at the National Council of Scientific and Technical Research of Argentina. Her research interests focus on modeling and analysis of emerging behaviors in social, financial, economic, and urban systems. She specializes in complex systems, complex networks, big data, and the development of new tools for modeling and simulation of complex systems.

Notes on Contributors

Zeeshan Ali is a PhD student at the IMS Lab (Integration from Material to System Laboratory), University of Bordeaux. His research interests include Simulation modeling and Digital Twin engineering with applications in urban context. His email address is zeeshan.ali@u-bordeaux.fr and ORCID is 0009- 0008-8101-9016.

J. Ignacio Alvarez-Hamelin is an Electronic Engineer and Independent CONICET Researcher at the University of Buenos Aires (UBA, Argentina). He leads the Complex Networks and Data Communication Group and is Full Professor in the Engineering School at UBA. His research spans complex systems, Internet topology, routing protocols, and network traffic analysis. A Senior Member of the IEEE and active participant in the IETF and Internet Society, he has authored numerous high-impact publications and collaborates on international grants and research projects.

Rob Baker leads the Advanced Protection Systems Program at DSTG. He holds a degree in Electrical and Electronic Engineering from the University of Adelaide and has 21 years of experience with Defense at the Defence Science and Technology Group. His experience and knowledge include weapons modeling and simulation, probabilistic range safety methodology and implementation, and active protection systems for armored vehicles. Rob has worked closely with Defense during his career, delivering range safety advice for rocket and missile testing activities, deployments to Afghanistan, advising Special Operations Command, and leading research programs focused on protecting Australian soldiers and vehicles.

Souvik Barat is a distinguished Principal Scientist at Tata Consultancy Services Research India, a Visiting Researcher at Middlesex University London, and an Adjunct Professor at Walchand College, Sangli. He has 27 years of experience in industrial research with interests spanning Digital Twin technology, modeling and simulation of complex systems, Artificial Intelligence, Model-Driven Engineering, Software Product Lines, Business Process Management, and Reinforcement Learning. He heads Digital Twin research initiative as part of TCS Research and is actively involved in developing digital twins for complex business and societal systems. His work on digital twins for business systems earned TCS Innovation Award in 2019 and contributed to the conceptualization of the TCS TwinX product. Souvik holds several patents and has authored 50+ journal and international conference papers. He earned his PhD from Middlesex University London, and a Master's degree from the Indian Institute of Technology (IIT), Madras. His email address is souvik.barat@tcs.com.

Emily Barrett is a Cognitive Systems Engineer and data analyst at MITRE. She brings multidisciplinary experience in resilience engineering, aerospace engineering, human–robot interaction, transportation, safety, and customer relations to her work and interests in data analytics and the evaluation and computational modeling of work system dynamics. Emily earned an MS in Industrial and Systems Engineering from The Ohio State University.

Eva Besada-Portas is an Associate Professor in the Department of Computer Architecture and Automation at Universidad Complutense de Madrid (UCM), Spain. She received her PhD in Computer Science from UCM in 2004. She has held pre/postdoctoral research stays at Carnegie Mellon (1999), the Royal Holloway of the University of London (2000), the University of New Mexico (2006, 2007, 2008–2010), and Google (2013). Her research interests focus on Modeling, Simulation and Optimization, often exploiting Artificial Intelligence and Machine Learning approaches. In these fields, she has co-authored more than 210 publications. She has led numerous research projects related to developing technological solutions for monitoring and managing water pollution and has participated in different industry contracts with companies like Airbus and ENAGAS.

Román Cárdenas is an Assistant Professor in the Department of Electronic Engineering at Universidad Politécnica de Madrid (UPM), Spain, where he obtained a PhD in Electronic Systems Engineering in Cotutelle with Carleton University (CU) in 2023. His research interests include modeling and simulation with safety-critical applications in the IoT domain. He participates in the Rust Embedded Working Group, maintaining software for running Rust on RISC-V embedded devices.

Kevin S. Chan is the lead research electronics engineer in the Networks, Cyber, and Computational Sciences Division at the U.S. Army Combat Capabilities Development Command Army Research Laboratory (DEVCOM ARL), Adelphi, USA. His research interests include edge computing, network science, and cybersecurity. Chan received his PhD in electrical and computer engineering from the Georgia Institute of Technology. He received the 2021 IEEE Communications Society Leonard G. Abraham Prize, the 2014 and 2018 NATO Scientific Achievement Awards, and multiple best paper awards. He also served as co-editor of the IEEE Communications Magazine—Military Communications and Networks series from 2017 to 2022. He is a Senior Member of IEEE.

Xuan Chau is a full-stack web designer and developer, translator of abstract client goals into understandable and navigable user interfaces, developer of generative AI applications, and interested in expanding into game development. Xuan is a Senior Software Engineer at MITRE. She earned her BS in Computer Science from the University of Connecticut.

Nicolás De la Peña is Associate Professor at the Department of Business, Politics, and Global Studies, Universidad de La Salle (Colombia). He holds an MA in International Affairs. His research examines the intersection of emerging technologies, international business, and global governance, with recent projects in algorithmic governance and AI supply chains.

Sergio A. De Raco, Universidad de Buenos Aires, Argentina.

Mama Diakité is a Postdoctoral researcher at the University of Bordeaux, within the IMS (Integration from Material to System) Lab. Her research interests include Digital Twin engineering and Applications. Her email address is mama.diakite@u-bordeaux.fr and ORCID is 0009-0005-4538-2975.

Anh Viet Do is a Postdoctoral Research Fellow in the Complex Systems Research Group at the University of Adelaide, Australia. His research focus is on hybridized reinforcement learning algorithms.

Roberto Domínguez is an associate consultant in the Business Security Solutions area at PwC, providing cybersecurity and technology risk services. He is a computer engineer with a bachelor's degree (2023) and a master's degree (2025). He has been a collaborator in the Department of Computer Architecture and Automation (DACyA) at Universidad Complutense de Madrid (UCM), where he developed his master's thesis, "Integration of simulated and real models for the detection of harmful cyanobacteria in aquatic environments," and obtained the highest grade. His research interests focus on modeling and simulation of discrete event systems and cybersecurity risks.

Giovanni Dosi is Professor Emeritus of Economics at Sant'Anna School of Advanced Studies, Pisa, Italy, and editor of the journal Industrial and Corporate Change for the Americas. He is included in the ISI Highly Cited Research list, denoting those who made fundamental contributions to the advancement of science and technology, and is a corresponding member of the Accademia Nazionale dei Lincei. In 2016, he received the Wiley TIM Distinguished Scholar Award from the American Academy of Management. In 2022, he was awarded with Edith Penrose Prize assigned by the European Academy of Management, and in 2023, the International Schumpeter Prize. His major research areas include economics of innovation and technological change, industrial economics, evolutionary theory, economic growth and development, and organizational studies. A major recent contribution is The Foundations of Complex Evolving Economies, Oxford University Press, 2023. https://orcid.org/0000-0002-0126-9349.

Segundo Esteban San Román is an Associate Professor in the Department of Computer Architecture and Automation at Universidad Complutense de Madrid (UCM), Spain. He received his European PhD in Physical Sciences from UCM in 2002. His research interests focus on Modeling, Simulation, and Control of Systems, where he has co-authored 115 publications. He has led the design, implementation, and on fly qualification of the Attitude Control Subsystem for several satellites of INTA: Nanosat-1B, Microsat-1, and Nanosat-2. He is now working on new technologies for the attitude and orbital control of satellites, on floating offshore wind turbines, and on automation of water monitoring.

Theresa Fersch is the creator and lead for Intelligence-Driven Exercises and Solutions (IDEAS) table-top exercise (TTX) capability at MITRE. Her focus is on developing and conducting tailored and scalable solutions that allow experts to explore complex challenges in a meaningful way. Her work contributes to an array of domains, including cyber, health, transportation, climate, defense, and intelligence. Theresa earned her BS in Sociology from Shepherd University and holds an MBA from the University of Maryland.

Bradley Fraser has a deep interest in multi-agent coordination, ranging from swarm intelligence and emergence, to formal methods and learning systems, and their associated modeling and simulation techniques. He obtained his Bachelor of Science (Hons) at Macquarie University in 2008, a Graduate Certificate in Defence Science in 2015, and Doctorate in Computer Science in 2022 from the University of Adelaide. Brad has spent over 15 years with the Australian Defence Science and Technology Group applying multi-agent coordination methods to autonomous uncrewed aerial systems for wireless communications, surveillance, environmental sensing, search and rescue, and force-projection applications. He has participated in numerous field robotics experiments across the globe, demonstrating state-of-the-art capabilities of emerging autonomous systems. Brad is regularly a reviewer for IEEE Access and the Journal of Field Robotics and is an industry advisor of several PhD, MSc, and undergraduate capstone projects. He delighted to contribute to this book.

Juan I. Giribet received his Electrical Engineering degree from the University of Buenos Aires (Argentina), where he also obtained his PhD in Engineering in 2009. He is currently a Researcher at the National Scientific and Technical Research Council (CONICET, Argentina), working on mathematical aspects of control theory with applications to mobile robotics. He is Associate Professor at the University of San Andrés, where he leads the Artificial Intelligence and Robotics Laboratory (LINAR) and directs the PhD program in Applied Sciences. He is also Associate Professor in the Department of Mathematics at the School of Engineering, University of Buenos Aires. Previously, he worked for several years at the National Commission on Space Activities (CONAE) in the Guidance, Navigation, and Control group, as well as in companies in the space sector in Argentina, Europe, and the United States. His research interests include control theory, mobile robotics, and multi-agent systems.

Oscar M. Granados is Professor at the Department of Business, Universidad Jorge Tadeo Lozano. Oscar works in several areas, but primarily in global strategy, business networks, business analytics, machine learning for finance, financial computing, topological data analysis, geometric data analysis, information geometry, and information theory. Oscar leads the Center of Intelligence and Innovation (Ci2) at Tadeo, which includes the Computational Finance and Digital Economics Lab.

Hazel Griffith is a PhD student in the Department of Systems and Computer Engineering at Carleton University in Ottawa, ON, Canada. She holds a MASc in Electrical and Computer Systems Engineering with a concentration in Modeling and Simulation from Carleton University. Her past research focused on DEVS for general-purpose workplace epidemiology models. Currently she is researching model validation and error detection with the National Research Council of Canada's semi-autonomous helicopter project. Her research interests include DEVS, machine learning, model validation, and agent-based modeling.

Ishan Honhaga is a reinforcement learning research engineer who focuses on hybrid approaches to facilitate their application in contested and dynamic environments.

Rhett Hull is a PhD candidate at the University of Technology Sydney with a strong interest in multi-robot coordination, sampling-based planning, and the integration of implicit and explicit coordination modes to achieve more scalable and resilient autonomy. His research focuses on developing planning approaches that enable teams of robots to cooperate effectively under practical communication constraints. Rhett holds a Bachelor of Engineering (Honors) and a Bachelor of Mathematical and Computer Science from the University of Adelaide (2020). Over the past five years with the Australian Defence Science and Technology Group, Rhett has participated in numerous field experiments across Australia and abroad, helping to translate autonomy from fundamental research insights into real-world state-of-the-art autonomous systems.

Frank T. Johnsen is currently a principal scientist in the Strategic Analyses and Joint Systems department at the Norwegian Defence Research Establishment (FFI), Kjeller, Norway. His research interests include edge computing, the Internet of Things, and software architecture for tactical systems. Johnsen received his PhD in informatics from the Faculty of Mathematics and Natural Sciences at the University of Oslo. He is actively engaged in NATO research activities under the Information Systems Technology panel and has been recognized with five NATO Scientific Achievement Awards for his contributions from 2016 to the present. He served as co-editor of the IEEE Communications Magazine—Military Communications and Networks series from 2017 to 2021.

Shankar Kausley is a Senior Scientist at TCS Research. He completed his PhD (Tech) in Chemical Engineering from the Institute of Chemical Technology, Mumbai. His research interests include water and wastewater treatment, food quality prediction, and functional coatings. He is one of the inventors of Tata Swach, the Nanotech water purifier, launched in the Indian market in 2009 as one of the world's most cost-effective water purifiers. He is also a recipient of the "DST-Lockheed Martin-Tata Trusts India Innovation Growth Programme 2018" award for his solution developed for the treatment and reuse of industrial effluents. Shankar is a "Certified Water Specialist" from the Water Quality Association, USA. He is a member of the Global Young Academy (GYA), Germany, and an alumnus of the Indian National Young Academy of Sciences (INYAS). His email address is shankar.kausley@tcs.com.

Vinay Kulkarni is a TCS Fellow, Distinguished Chief Scientist, and Head of Software Systems & Services Research at Tata Consultancy Services. His research interests include learning-native software systems, multi-agent systems, model-driven software engineering, and enterprise modeling. Several of his works have found their way into OMG standards and have also received notable mention in respected international print media. Vinay holds several patents and has authored over 100+ papers in scholarly journals and conferences. He is frequently invited to deliver talks at major conferences and events. Vinay has served as the conference and program chairperson for premier ACM and IEEE international conferences in the area of software engineering and is on the technical program committees of many international conferences. A Fellow of the Indian National Academy of Engineering, Vinay is an alumnus of the Indian Institute of Technology Madras and also serves as a Visiting Professor at Middlesex University, London. His email address is vinay.vkulkarni@tcs.com.

Andrew Langone is Director of Multi-Domain Innovation & Emerging Tech at Tenebrae, where he develops modular adversarial-testing frameworks and multi-agent validation tools to accelerate technology readiness. His current work builds on two decades in systems engineering and operational testing, advancing innovation frameworks that translate research into field-ready capabilities. A former U.S. Army counterintelligence agent, he applies operational rigor and red/blue/white-team methodology to strengthen how organizations evaluate, trust, and deploy emerging systems. He holds a BA in Corporate Systems from Boston College.

Glenn Lematta specializes in the study and application of approaches to and measures of human–machine teaming for high-consequence work operations. He is a Senior Cognitive Engineer in the Human-Centered Engineering and User Experience Department at MITRE. He earned his PhD in Human Systems Engineering from Arizona State University. Ignacio Mas obtained his PhD in Mechanical Engineering from Santa Clara University, Santa Clara, California (USA), in

2011, and his degree in Electrical Engineering from the University of Buenos Aires (Argentina) in 2001. He is currently an Independent CONICET Researcher at the University of San Andrés (Argentina). Dr. Mas has worked as a systems development engineer in aerospace projects at NASA Ames Research Center and at the NASA/Stanford National Biocomputation Center in California, USA. He is an Assistant Professor at the University of Buenos Aires. Currently, his research areas include mobile robotics, robot formation control, and the development of autonomous ground, aerial, and underwater vehicles, as well as their applications.

Pablo Mira is a Researcher at the Interdisciplinary Institute of Political Economy (IIEP, UBA–CONICET) and Coordinator of its Macroeconomics Area. He holds a PhD in Economics from the University of Buenos Aires, where he also teaches Behavioral Economics, Macroeconomics, and Philosophy of Economics. His research examines the intersections between behavioral economics, macroeconomic fluctuations, and epistemology, with a focus on consumption dynamics and expectation formation. He has published in journals such as European Economic Review and Filosofía de la Economía, and co-authored Homo Falsus (2022) and Economía al Diván (2016). Mira has served as Senior Advisor to the Ministry of Economy of Argentina.

Kelly J. Neville contributes to ensuring new technologies improve work operations by enabling human expertise and contributing to operational resilience and agility, including through development of the Transform with Resilience during Upgrades to Socio-Technical Systems (TRUSTS) Framework (https://trusts.mep.mitre.org). She is a Principal Cognitive Systems Engineer in MITRE's Modeling and Analysis Innovation Center. Kelly earned her PhD in Experimental Psychology, specializing in cognitive and human factors psychology, from Rice University.

Emily S. Patterson is a Professor in the School of Health and Rehabilitation Sciences, College of Medicine, at The Ohio State University. She is a Fellow of the Human Factors and Ergonomics Society (HFES) and an associate editor for the Human Factors in Healthcare journal. Emily has served as a scientific advisor on using human factors engineering to improve patient safety for The Joint Commission, the National Board of Medical Examiners, the National Institute of Standards and Technology, and the Society of Hospital Medicine. Dr. Patterson earned her PhD in Industrial and Systems Engineering at The Ohio State University.

Ezequiel Pecker-Marcosig received his Electrical Engineering degree and PhD from the University of Buenos Aires (UBA, Argentina) in 2011 and 2021, respectively. He is currently an Assistant Researcher at the Institute of Computer Science (ICC) of the National Scientific and Technical Research Council of Argentina (CONICET) and an Adjunct Professor at UBA. His research interests include M&S-driven controller design for cyber-physical systems and Model-Based Systems Engineering for Digital Twins.

Marcelo Pereira is Assistant Professor in economics and quantitative methods at the University of Campinas (Brazil), having also acted as Senior VP at large domestic and multinational ICT companies for many years. His main research interests include the application of complex systems theory to macro- and microeconomics and the development of software and methodologies for agent-based modeling. He is a regular Visiting Professor to the Sant'Anna School of Advanced Studies (Italy). https://orcid.org/0000-0002-8069-2734.

Gabriel Petrini is a Associate researcher in the Economics Department at the University of Campinas (Unicamp). He is an associated researcher at the Berlin School of Economics and Law (HWR, Berlin) and the Sant'Anna School of Advanced Studies (Pisa, Italy). His primary research interests encompass demand-led growth models, agent-based macroeconomic models, model selection and validation, and macroeconomic time series analysis. Petrini is the developer and maintainer of the pysolve3 package for numerical model simulation and has contributed to the SFCR package for simulating macroeconomic stock-flow consistent (SFC) models in R. Petrini has experience with economic outlooks for the Brazilian economy and produced policy notes at the Centro de Estudos de Conjuntura e Política Econômica (CECON) and at the Centro de Estudos Sindicais e de Economia do Trabalho (CESIT). https://orcid.org/0000-0002-3523-9826.

J. Francisco Presenza received his Electromechanical Engineering degree in 2017 from the National University of Mar del Plata (Argentina) and his PhD in 2023 from the University of Buenos Aires (Argentina). He is currently a Postdoctoral Research Fellow at the National Scientific and Technical Research Council (CONICET) of Argentina, Teaching Assistant at University of Buenos Aires, and a Professor at University of San Andrés (Argentina). His research interests include collaborative robotics, network control, and graph theory.

Filippo Raddichi, Indiana University, USA.

Beena Rai is Chief Scientist and Head of Physical Sciences Research Area at Tata Consultancy Services (TCS). Her pioneering research in Computational Material Science & Engineering, especially leveraging physics-based models and data analytics tools and techniques in the design, development & deployment of materials and chemicals, has led to the creation of 150+ research papers and 100+ granted patents & patent applications. Beena is a Fellow of the Indian National Academy of Engineering and a Chevening Scholar at Said Business School, University of Oxford. She has earned multiple honors: TCS Distinguished Scientist Award, Distinguished Alumnus of National Chemical Laboratory Pune, IIME Mineral Beneficiation Award, Vasvik Research Award for Women Scientist, Materials Research Society of India Medal, SAIL Gold Medal, Maha-Intrapreneur Award from Praj Industries and Symbiosis International Institute of Business Pune, and WILL Best Mentee Award from WILL-Women in Leadership Forum, India. Her email address is beena.rai@tcs.com.

José L. Risco-Martín is a Full Professor in the Department of Computer Architecture and Automation at Universidad Complutense de Madrid (UCM), Spain. He received his PhD in Physics from UCM in 2004. He has held postdoctoral research stays at the University of Arizona (2006), EPFL (2019), and Carleton University (2022). His research interests focus on Discrete Event Systems, Cyber-Physical Systems, Modeling, Simulation and Optimization, Parallel and Distributed Simulation, and Artificial Intelligence in M&S. In these fields, he has co-authored more than 210 publications and holds 6 Spanish patents. He has participated in numerous research projects and industry contracts with companies like Airbus, Repsol, and ENAGAS. He received the SCS Outstanding Service Award (2017) and the HiPEAC Technology Transfer Award (2018). He is an associate editor of SIMULATION and ACM Computing Surveys, an ACM Senior Member, and an SCS Senior Member.

Andrea Roventini is professor of economics at Scuola Superiore Sant'Anna (Italy), where he is the director of the Institute of Economics. He is research fellow at OFCE, Sciences Po (France), and World Inequality Lab (France). He holds a PhD in Economics and Management from Scuola Superiore Sant'Anna. His main research interests include complex system analysis, agent-based computational economics, business cycles, economic growth, climate change, and inequality. He has been the principal investigator and consortium coordinator of the Horizon 2020 GROWINPRO project, financed by the European Commission, and the unit leader and coordinator of the EEIST project (https://eeist.co.uk) financed by the UK's Department for Business, Energy and Industrial Strategy (BEIS) and the Children's Investment Fund Foundation (CIFF). He is editor of Industrial and Corporate Chance—Macro Economics and Development and advisory editor of the Journal of Evolutionary Economics. His work has been published in many international peer-reviewed journals. A recent contribution is Agent-Based Macroeconomics (with Giovanni Dosi), Cambridge University Press, 2025. https://orcid.org/0000-0001-5518-3084.

Cristina Ruiz-Martin is Associate Professor (Teaching Stream) at Carleton University. She received her MSc in Industrial Engineering (2014) and Project Management (2013) from Universidad de Valladolid. She received her PhD from both Universidad de Valladolid (Spain) and Carleton University (Ottawa, Canada) in cotutelle. Dr. Ruiz's research area includes modeling and simulation of complex systems and resilience. Her research interests include the theory and application of modeling and simulation and the transformations between formalisms and automated model development. In the application domain, she is focused on using Discrete Event Modeling and Simulation, Agent-Based Modeling, and Network Theory to design, analyze, and, hence, complex systems, including their resilience. She received the DEVS Modeling and Simulation PhD Dissertation Award from

SCS (2019) and the Young Simulation Scientist Award from SCS (2020). She has been a member of the Board of Directors of SCS since July 2021.

Tracy Sanders is the Chief Engineer for the Human Centered User Experience Engineering Department at MITRE and holds an Adjunct Lecturer position in the Maryland Institute College of Art (MICA). Tracy earned her PhD in Experimental Psychology from the University of Central Florida.

Surabhi Shrivastava is a Researcher at Tata Consultancy Services (TCS) Research, Pune, India. Her current research interests include modeling, simulation, optimization, and analysis of engineering processes. Her email address is sh rivastava.surabhi2@tcs.com.

Puneet Singh Thakur is a Scientist at Tata Consultancy Services (TCS) Research, India. His current research interests include computer vision, photonics sensing, water and wastewater treatment, smart agriculture, and food quality prediction. He was a MeitY fellow and holds a PhD from IIT Indore. His email address is pun eets.thakur@tcs.com.

Eyoel Teffera, Defence Science and Technology Group, Australia.

Andreas Tolk is Chief Scientist for Complex Systems Modeling at the MITRE Corporation in Charlottesville, Virginia, USA. He holds an MSc and PhD in Computer Science from the University of the Federal Armed Forces in Munich, Germany. Previously, he was a Professor of Systems Engineering at Old Dominion University and worked with the Virginia Modeling, Analysis, and Simulation Center. He also held industry roles and served as an Officer in the German Army. His research focuses on interoperability and composability of simulation systems, including the philosophy of simulation with ontological and epistemological aspects. His applications focus on system of systems engineering and the extension of model-based engineering methods. Andreas has authored 17 books and over 300 publications. He chairs the ACM Special Interest Group for Simulation and Modeling, is a Fellow of the Society for Modeling and Simulation International, and is a Senior Member of IEEE. His ORCID is 0000-0002-4201-8757.

Mamadou K. Traoré is a Full Professor at the University of Bordeaux and Deputy Director of the IMS (Integration from Material to System) Lab, in charge of International Research. His research interests include modeling, simulation and digital twin engineering. His email address is mamadoukaba. traore@u-bordeaux.fr and ORCID is 0000-0001-9464-6416.

Maria Enrica Virgillito is Professor of Economics, Department of Economic Policy, Università Cattolica del Sacro Cuore , Sant'Anna School of Advanced Studies, where she teaches advanced courses in Evolutionary Development, Institutional Labor Economics, Technical change, Income distribution, and Macroeconomic Dynamics and coordinates the Seasonal School on "Agent-based models in Economics: theory, toolkit, and policy laboratories." She is Global Labor Organization

Fellow and serves as Editor for the Macro and Development yearly issue of Industrial and Corporate Change and as Associate Editor for Structural Change and Economic Dynamics and for Review of Evolutionary Political Economy. Her publications have been hosted in a number of international scientific peer-reviewed journals on complexity economics, institutional labor economics, and evolutionary economics. Actively engaged in media and social dissemination activities and action research. https://orcid.org/0000-0001-5958-9647.

Gabriel Wainer is a Full Professor of Systems and Computer Engineering at Carleton University, where he has served since 2000 and directs the Advanced Real-Time Simulation Lab. He is a globally recognized expert in Modeling and Simulation (M&S) methodologies, focusing on the Discrete Event System Specification (DEVS) formalism, real-time, cellular, and distributed modeling systems. Wainer leads interdisciplinary, industry-collaborative research programs that translate theory into deployable simulation tools and platforms. He has held faculty and visiting positions internationally, authored extensive publications and open-source software, supervised numerous graduate students, and served on editorial boards and conference committees. His work is noted for advancing simulation methodologies, promoting reproducible research, and fostering applied education in modeling and real-time systems. A Fellow of the Society for Modeling and Simulation International (SCS) and an ACM Distinguished Speaker, he received various awards in the field recognizing his sustained technical contributions to the profession.

Levent Yilmaz is Alumni Distinguished Professor of Computer Science and Software Engineering at Auburn University. He holds MS and PhD degrees in Computer Science from Virginia Tech. His research interests are in agent-directed simulation, multi-agent systems, and model-driven science and engineering. His email address is yilmaz@auburn.edu.

Preface

The accelerating complexity of contemporary socio-technical systems across the domains of energy, health, infrastructure, the economy, and cyber-physical systems renders resilience an imperative research objective, both in terms of understanding and design. However, engineering resilience is not an upfront activity in such complex and adaptive systems. These socio-technical systems evolve, and resilience arrives organically, which makes these systems sustainable.

We conceived this volume to bring together methods, evidence, and engineering practices that address that imperative for defining a specialized version of complex adaptive systems: resilient complex adaptive systems (RCAS). Our goal has been to assemble contributions that jointly help with our understanding of the—often elusive—theory of complex adaptive systems (CAS) engineering, let alone RCAS, and show possible ways in which such a theory can be operationalized through modeling, simulation, visualization, and engineering approaches.

The resulting collection—nineteen chapters authored by researchers from multiple disciplines and application domains—offers both conceptual foundations and concrete tools for researchers and practitioners working at the interface of computer science, complexity science, and systems engineering.

We believe that the topic of RCAS is timely, given the significant advancements made in the last decade in various fields such as agent-based and multiscale modeling and simulation frameworks, the availability of high-granularity data, and the rise of artificial intelligence and machine learning (AI/ML) methods for analytics and decision-making. These developments have engendered unparalleled opportunities to comprehend and influence system resilience.

The current state of the art in software tools enables us to understand the interconnectedness, which has increased manifold, breaking the hierarchy and making such systems essentially flat with limited hierarchies. This interconnectedness has increased the potential for localized disturbances to propagate and amplify, producing cascades that can overwhelm previously robust systems that were successfully managed through hierarchical control. We assembled this book to map those dual trends: to document methodological advances that enable rigorous study of RCAS and to collect domain studies that expose the practical limits and promises of these methods.

We organized the volume into thematic sections that move from foundations to practice. Early chapters develop conceptual and epistemological foundations for RCAS, framing resilience in systems-theoretic and system-of-systems perspectives and surveying verification and validation challenges for simulation-based studies. Middle sections present methodological advances—modeling formalisms, multilayer network analytics, digital-twin architectures, and AI/ML decision frameworks—alongside metrics and visualization techniques for emergent behavior and resilience assessment. The concluding chapters bring these threads together through applied studies (from epidemiology and water systems to multi-robot control and urban digital twins) and by proposing a forward-leaning research agenda.

In assembling this book, we sought a balance between rigorous formalizations and pragmatic engineering. Some contributions develop theory and formal methods aimed at replicability and interpretability; others present software toolchains, data-integration strategies, and case studies that emphasize operational relevance. Throughout, contributors were encouraged to make their assumptions explicit and to discuss validation strategies so that modelers and decision makers alike can judge where a given approach is most applicable—and where it is not.

We also emphasized limitations and cautions as key traits of any scientific endeavor. Modeling-, simulation-, and data-driven approaches are powerful but not omniscient. The usefulness of any model or decision-support system depends critically on model structure, parameter choices, data quality, and the design of experiments used for validation.

Dealing with heterogeneity, multilayer interactions, and rare extreme events remains particularly challenging; several chapters explicitly address these issues and propose strategies—ensemble simulation, scenario sweeps, and layered validation—to mitigate them. We urge our readers to treat the models collected here as tools for disciplined exploration rather than as oracle-like solution providers.

Our hope is that this book will serve as both a reference and a stimulus: a reference that collects contemporary ideas and methods that enable RCAS engineering and a stimulus for interdisciplinary work that closes the gaps between conceptual advances and deployed resilient systems. We invite readers to use the models, adopt the metrics, replicate the experiments, and—most importantly—bring these ideas into socio-technical domains where resilient behavior is essential, challenging, and even warranted for a sustainable function.

This book is the result of the generous contributions of researchers, scholars, and students who enthusiastically shared their insights and expertise in their respective fields. We are grateful to the chapter authors for their scholarship and collegiality and to the institutions and funding bodies that made their work possible.

Finally, the book is also the result of our shared passion and relentless curiosity about the topic and our belief that the scientific community can help shape a better world by improving its understanding of the complex socio-technical systems in which we live.

With gratitude,
Claudia Szabo, Rodrigo Castro, Saurabh Mittal, and Viktoriya Semeshenko
(Editors)

Part I

RCAS Foundations

1

Modeling and Simulation of Resilient Complex Adaptive Systems

Claudia Szabo,[1] Rodrigo Castro,[2] Saurabh Mittal,[3] and Viktoriya Semeshenko[4]*

[1] School of Computer and Mathematical Sciences, The University of Adelaide, Adelaide, Australia
[2] Facultad de Cs. Exactas y Naturales and Instituto de Ciencias de la Computación, Universidad de Buenos Aires and CONICET, Ciudad Autónoma de Buenos Aires, Argentina
[3] Defense Science Division, Modern Technology Solutions, Inc., Dayton, OH, USA
[4] Facultad de Cs. Económicas and Instituto Interdisciplinario de Economía Política, Universidad de Buenos Aires and CONICET, Ciudad Autónoma de Buenos Aires, Argentina

*Corresponding author: claudia.szabo@adelaide.edu.au

1.1 Introduction

Complex adaptive systems (CAS) are ubiquitous in modern society, permeating domains as diverse as finance [1], social networks [2, 3], medicine [4], and military operations [5]. The engineering of these systems presents numerous challenges, which can be approached from both bottom-up and top-down perspectives, each grappling with complexities at micro-, meso-, and macro-scales. This book provides the state-of-the-art in methods and technologies for the modeling, simulation, and visualization of resilient CAS in various scientific, social, economic, and multidisciplinary systems engineering disciplines that are now defining the new frontiers for mankind.

A CAS is made up of a series of interconnected and interacting components with one or more levels of feedback. The CAS exhibits emergent behavior and self-organization and produces nonlinear dynamic behavior [6]. In Carmichael's perspective, agents are considered self-similar, e.g., largely homogeneous, and within the system a large number of agents have correlated interactions. Some systems may exhibit specific agent hierarchies or role specializations [7] to a specific application domain or task. Other CAS definitions [8] highlight specific properties such as distributed control, connectivity, and coevolution [8, 9].

Examples of CAS that are studied within this perspective are numerous, and many new application domains will be introduced in this book. For example, power grids are often studied from a CAS perspective, as they contain numerous components (generators, transformers, transmission lines) that interact in a decentralized manner [10]. Ecological systems are complex connections of interacting organisms that operate within a closed or open environment, with inter- and intraspecies interactions over time [11]. The strength of the interaction varies from weak and diffuse to strong and direct and is modified by the various feedback loops with the environment and other species [12]. Ecosystems exhibit resilience through diversity and self-organization of species populations, food chains, and nutrient cycling processes. Shocks or disturbances may cause shifts in the balance between different components of the ecosystem; however, these changes can lead to new emergent behaviors that allow it to return to a stable state over time. Many other examples of resilient complex adaptive systems (RCAS) include social networks [2, 3], biological systems [4, 13], and ant-colonies [14], to name a few, and research has identified that long-lasting natural CAS tend to be resilient. However, the resilience of engineered CAS needs to be a focus within the design process of the CAS for the end result to manifest resilience and robustness. We focus in this book on the definition, engineering and modeling, simulation, and visualization of RCAS.

From a research perspective, RCAS is typically approached from two distinct perspectives: bottom-up [15] and top-down [6], with a focus across both perspectives on micro-, meso-, and/or macro-scales and within specific application domains [16, 17]. The design and implementation of RCAS following a bottom-up approach requires the definition, implementation, and interconnection of numerous micro-level components, with a perspective and focus on the holistic behavior that must occur at the macro-level. Further, there is additional complexity due to the multitude of aspects that need to be considered, including component interaction, rules and behaviors within the environment that could ultimately lead to desired properties of the CAS, including resilience.

In contrast, many CAS evolve from a collection of preexisting systems that tend to self-organize to achieve a holistic system-wide behavior. The challenge in this top-down, a posteriori approach is ensuring that the RCAS behaves appropriately at multiple scales, that interactions among components and with the environment are appropriate, and that the system configurations at micro- and macro-levels are such that the CAS is resilient and that observed emergent behavior falls within acceptable operational boundaries.

Resilience is a critical CAS property that ensures that results from CAS modeling, simulation, and visualization efforts find direct application into the engineering of CAS within the real world and that they influence policy and decision-making across several critical aspects of our daily lives [18]. However, a perspective on resilience is not widely considered by existing works, and definitions, formalisms, and modeling, simulation, visualization, and engineering approaches are broadly missing. We advance in this book that the focus should be deeper and that research focus should shift from CAS to RCAS to ensure that we are best placed in addressing the engineering challenges of our ever-changing world.

1.2 Resilient Complex Adaptive Systems

We define *RCAS* as follows:

> A CAS that displays characteristics of resilience, i.e., the presence of perturbation and associated source(s) of perturbation, mechanisms to ascertain the negative impact of perturbation to quantify CAS behavior, inherent or engineered structural and controllable aspects to mitigate the negative effect of perturbation, and recovering the CAS to the same baseline or a new acceptable baseline behavior.

The first use of the concept of resilience in a scientific context was by Sir Francis Bacon in 1625 [19], looking at the strength of echoes, followed by its use to describe the ductility of steel beams in 1858 by Rankine [20]. The broad understanding of resilience within CAS is that it encompasses the behavior of the system before and after a disastrous event and can be viewed, according to Francis and Bekera [21], as three capacities, namely, an adaptive capacity to the disastrous event, an absorbing capacity of the effects and consequences of the event, and a recoverability of the system to a state that was before the event or a better state [22]. The "bouncing back" ability is considered by Gunderson's ecological resilience definition to be measured by a threshold, with resilience defined as "the capacity of a system to absorb impacts before a threshold is reached where the system changes to a different state" [23].

A typical resilient system's effectiveness as function of its dynamical performance in the face of perturbation can be visualized as shown in Figure 1.1.

The figure illustrates the temporal evolution of a RCAS effectiveness when exposed to an external perturbation. The system's effectiveness—its capacity to perform intended functions—is represented through four phases: preparedness, response, recovery, and stabilization.

At the initial stage, the system operates at an *initial performance* level (P_i), reflecting its baseline capability before disturbance. Through adaptive processes such as learning and adaptation, the system may enhance its operational preparedness, reaching a *reference performance* (P_r).

When a perturbation occurs, the system experiences a temporary degradation in effectiveness, dropping to a *degraded performance* level (P_d). The lowest point on

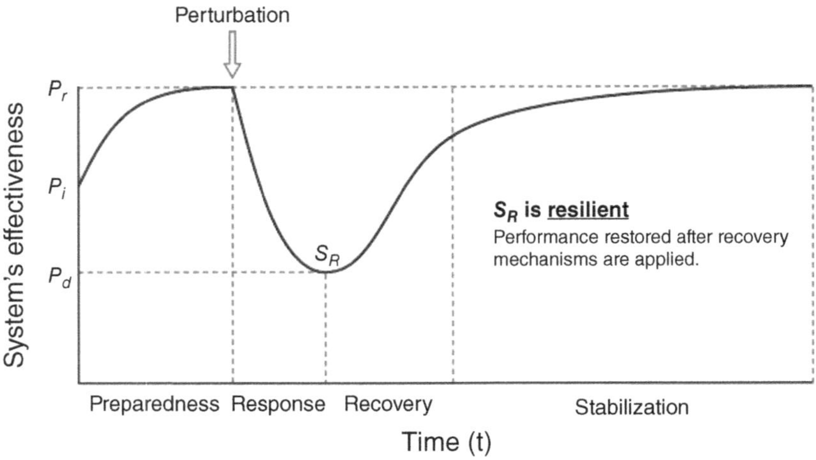

Figure 1.1 RCAS effectiveness in the presence of perturbation over time.

this trajectory (S_R) marks the *system resilience threshold*. At this stage, the interplay of properties such as redundancy, diversity, and modularity enables the system to absorb and redistribute the impact of disruption without systemic collapse.

During the recovery phase, mechanisms gradually restore performance (such as adaptive feedback loops and self-healing). The trajectory ascends toward stabilization, approaching (or eventually surpassing) P_r depending on the degree of learning and adaptation achieved. When the system reaches a new stable state, labeled again S_R on the right, it is considered *resilient*, as its effectiveness has been reestablished following perturbation through endogenous mechanisms rather than external control.

Overall, the figure encapsulates the dynamic resilience of RCAS (the ability to sustain function, reorganize internally, and evolve toward renewed equilibrium despite shocks). The curve emphasizes that resilience is not merely resistance to change but an emergent property itself, arising from continual interaction among adaptive components.

Another perspective considers four aspects of the systems and operating environments that build the case for resilience, according to Tierney and Bruneau: robustness, redundancy, resourcefulness, and rapidity [24, 25]. A system is resilient if the occurrence of undesired incidents is minimized, and, once an incident occurs, its undesirable consequences are reduced, and the system returns to normal operation within a short time [24, 26–28].

From the perspective of the resilience of engineered system of systems (SoS), resilience can be viewed as an inherent property of the system that is influenced by the SoS components and their behaviors, with an understanding that the resilience of a system is dynamic and changes as the environment changes [20]. In a similar approach to the CAS definition above, Enos [29] recognizes that quality, robustness, and agility are components of resilience and identifies repairability, extensibility, flexibility, adaptability, and versatility as how the system achieves robustness after a shock. In addition to these components, Haimes [30] identifies also "the ability to prepare and plan for, absorb or mitigate, recover from, or more successfully adapt to actual or potential adverse events."

One way to incorporate resilience within a system is by incorporating some of the resilience considerations described above within the behaviors and decisions of all individual system components. This could be done by designing node-specific rules or behaviors relying on the partial or complete understanding that a node has of the system context and the role this plays in the system-wide resilience [20, 26, 31]. The node would then be able to reason over the context and prioritize information-sharing activities as it suits the mission plan and the current context. This complex systems design approach subsumes the understanding that individual nodes will make decisions both in cooperation with other nodes and in isolation, caused by various network partitions or specific attacks, and that the local node decisions are cooperative, in the sense that they lead to a complex system with a desired behavior [28, 32].

1.2.1 Properties of RCAS

Several properties of RCAS make them particularly appealing for their wide application in a variety of domains, as discussed below. These include adaptability, redundancy, fault tolerance, scalability, and self-healing, which stem from their decentralized design that encompasses heterogeneity and diversity. In an RCAS, decision-making processes for both managed and managing systems are typically decentralized, with no central authority controlling any aspects of the system's operation. As defined in [33], the *managed system* comprises the application logic that provides the system's domain functionality, while the *managing system* oversees the managed system and comprises the adaptation logic that deals with various concerns. In a RCAS, individual agents make decisions based on local information and interactions with other agents within their immediate environment, allowing rapid adaptation to changing conditions while minimizing potential impacts from failures or disruptions in one area on the rest of the system. Another critical aspect of RCAS is the heterogeneity of system components. A RCAS is characterized by a high degree of diversity among its component parts, which allows for adaptation, redundancy, and resilience. We consider 10 characteristics of RCAS as follows:

1. **Adaptability**: A RCAS is capable of modifying its behavior, structure, or function in response to changes in its environment, learning from its experiences, and adjusting accordingly to maintain optimal performance levels over time.

2. **Redundancy**: A RCAS has multiple pathways for information flow within the system and redundant components that allow it to continue functioning even when some parts fail or are removed. In addition, the system may reorganize itself in response to failures or changes in its environment by redistributing resources or creating new connections between system components.
3. **Fault Tolerance**: A RCAS exhibits fault tolerance through adaptability and self-organization rather than having predefined mechanisms for error detection and correction. The system can evolve strategies for handling failures or errors based on interactions between its components, allowing it to maintain overall performance levels in the face of challenges.
4. **Scalability**: A RCAS is inherently scalable as it grows and changes over time through the addition of new system components (agents) that interact with existing ones, creating complex networks of relationships and emergent behaviors. This allows the system to adapt to changing conditions while remaining robust and resilient.
5. **Self-healing**: A RCAS can exhibit self-healing properties by reorganizing itself following failures or disruptions in order to maintain its overall performance and stability. This may occur through the rearrangement of existing components, the creation of new connections between them, or even the emergence of entirely new subsystems.
6. **Diversity**: A RCAS is characterized by a high degree of diversity among its component parts (agents). This diversity allows for innovation and adaptation as different agents bring unique perspectives and behaviors to the system while ensuring redundancy and resilience through multiple pathways for information flow.
7. **Decentralization**: In RCAS, decision-making processes are typically decentralized with no central authority controlling all aspects of the system's operation. Instead, individual agents make decisions based on local information and interactions with other agents within their immediate environment, allowing rapid adaptation to changing conditions while minimizing potential impacts from failures or disruptions in one area on the rest of the system.
8. **Robustness**: A RCAS is inherently robust due to its ability to self-organize and adapt to changes in its environment through complex interactions between diverse components, creating emergent behaviors that help it cope with stresses, shocks, or disturbances without compromising overall performance or stability.
9. **Modularity**: In RCAS, modularity is often present at multiple levels of the system hierarchy. Individual agents may be organized into modular subsystems that can function independently while still contributing to the overall behavior of the larger system. Additionally, these modules can be

easily replaced or upgraded as needed without affecting the entire system's operation.

10. **Monitoring and Feedback**: Continuous monitoring and feedback mechanisms are essential for understanding RCAS' current state and making adjustments based on changes in its environment or internal interactions among components. This allows CAS to maintain optimal performance levels while adapting to new challenges and opportunities, ensuring long-term stability and resilience.

1.2.2 Research Focus into RCAS

Within this context, we define *RCAS Engineering* as the engineering approach to manage the lifecycle of a RCAS. Six major RCAS defining aspects form the focus of significant research effort, including emergent behavior, adaptation, resilience and robustness, decision-making, visualization, and knowledge engineering. We consider these as follows.

- **Emergent Behavior**: A RCAS is a system where it is likely that microscopic level entities will interact yielding macroscopic level behaviors, structures, and properties which in turn can feed back to the microscopic level. In this book, we present approaches for the modeling, identification and engineering emergent behavior. Provisions for the management of emergent behavior should be part of the RCAS modeling, simulation and engineering approach from the start. This includes among others the representation of the micro- and macro-levels and an understanding of their relationships and their side effects.
- **Adaptive Behavior**: Adaptation refers to the ability of a system to independently identify the need to adapt, plan an appropriate course of action and undertake changes to address the adaptation requirement. This may include reconfiguring components, reallocating resources, establishing new links between systems and services, restarting systems and repurposing systems. It is critical that RCAS are modeled and simulated in such a way that the specific conditions in which these necessary properties are realized are well understood, and the many possible existing approaches are evaluated. We explore in this book adaptation at many levels of a RCAS and in various application domains.
- **Resilience and Robustness**: Resilience refers to the RCAS's ability to continue to operate at acceptable levels of quality of service even when the system is unexpectedly disturbed, through cyberattacks in various forms, full or partial failures of components, and network failures or attacks. We explore in this book the integration of research advances in distributed fault-tolerant

control, cybersecurity, human–machine teaming, digital twins, and the design and implementation of appropriate metrics to measure resilience.

- **Control Theory and Decision-making**: Decision-making in RCAS, at micro and macro levels, dramatically influences the resilience and adaptability of a RCAS. The heterogeneity of RCAS components and their continuous interaction require dynamic decision approaches and appropriate feedback mechanisms that are able to predict and adapt to disruption, to ensure resilience. We include artificial intelligence (AI) and machine learning (ML) approaches for both distributed and centralized control across a variety of domains.
- **Visualization and Evaluation**: Metric definition, data collection, experimental design, experimentation, visualization and analysis are essential components within a CAS study but are challenging as heterogeneous micro-micro, micro-macro interactions need to be captured and evaluated. In this book, we focus on the evaluation of adaptation, resilience and robustness and emergent behavior across many contexts. We include AI/ML design consideration and approaches for ensuring advanced analytics in resilient RCAS.
- **Data Architecture, Knowledge Engineering and Ontological Frameworks**: Data engineering is an integral aspect of modeling, simulation, visualization, and analytics (MSVA) solution. A data-driven MSVA engineering approach must describe the underlying data model, knowledge engineering methodology and structure of the any ontological framework supporting the MSVA Solution Architecture for RCAS.

1.3 Aligning RCAS with Classical Systems Theory and System of Systems Engineering

CAS theory builds upon systems theory by explicitly representing the diversity and heterogeneity that systems theory tended to aggregate within homogeneous stocks and flows. CAS differs from traditional systems theory because it explicitly incorporates the role of adaptation in governing the dynamics and responses of these now heterogeneous reservoirs and therefore extends the definition of traditional systems theory by recognizing that CAS contain adaptive components and capacities. Adaptive components allow systems to change and evolve over time in response to feedback and changes in the system context. This means that CAS have a memory and the capacity to learn from previous responses and configurations and thus, influence and shape current and future system trajectories. Thus, CAS contain "no a priori assumptions about key variables, emphasize nonlinear causal effects between and within systems, and view system equilibrium as multiple, temporary, and moving" [34].

Mittal [35] discussed 18 CAS properties, namely, clustering, scale-free topology, preferential attachment, growth and decay, threshold and affinity, interconnectivity, modularity, hierarchy, agent persistence, environment persistence, interactive transition systems, self-organization, weak emergence, strong emergence, nonlinearity, concurrency, upward causation, and downward causation, and how they could be modeled using discrete systems' perspective, especially through the discrete event systems specification (DEVS) formalism [36].

Succinctly, the DEVS formalism provides a rigorous, modular, and hierarchical framework for modeling and simulating dynamic systems whose behavior evolves through dynamic discrete events. It defines system structure and behavior in a mathematically precise yet implementation-neutral manner, supporting interoperability across heterogeneous DEVS-compliant models. DEVS captures state transitions triggered by events, enabling scalable composition of components and emergent behavior analysis—making it especially suitable for representing CAS where multiple interacting subsystems coevolve over time. Moreover, many other discrete and continuous modeling formalisms can be mapped into DEVS, allowing it to serve as a unifying theoretical foundation for multi-paradigm and hybrid system representations.

Mittal argued that employing a formal framework such as DEVS guarantees the repeatability of weak emergent behavior in a DEVS-enabled modeling and simulation (M&S) environment. Szabo and Teo [37] provided further evidence of formalizing the specification for weak emergent behaviors. Weak emergent behavior, stated simply, is the expected holistic CAS behavior that is non-causal, repeatable, measurable, and controllable in a test bed first, and later assured in the real world. Strong emergent behavior, which is causal, produces novelty that first needs to be semantically labeled and eventually subjected to an engineering process rooted in control theory [38]. These ideas were then further realized in describing both systems and model engineering approaches for Netcentric CAS [39], Cyber CAS [40], and simulation-based CAS [41]. Moreover, acknowledging this potential, Emergent Behavior-DEVS was introduced by Castro and collaborators [42] as a generalization of DEVS specifically designed to deal formally and practically with emergent dynamics in CAS, allowing for live detection and handling of novelty in hierarchical multilayered systems.

Resilient CAS adds an extra focus on resilience within CAS and targets the engineering effort to bring the system back to its identifiable stable state despite negative emergent behaviors. We now look at the identified 10 RCAS properties from the first principles of systems theory and system of systems engineering (SoSE).

1.3.1 RCAS Alignment with Classical Systems Theory

The classical systems theory (CST) decomposes a system into structure and behavior that are closed under composition, i.e., structure defines behavior and vice versa. The CST is represented mathematically using set theory. The closure under the composition principle ensures that the interactions between the subsystems are explicitly defined, such that there is a direct correspondence between structure and behavior at multiple scales. This closed under composition principle is the hallmark of CST that ensures behaviors are robust at a specific level of hierarchy and are reproduced if the associated structure exists. This assurance led to established systems engineering practices, where requirements engineering resulted in a functional (behavioral) system, which subsequently led to the component/subcomponent architecture. Accordingly, one can make sense of the system function by looking at the system architecture. Various key performance indicators (KPIs) can be defined that quantify the performance of the architecture's components and the effectiveness of the system's overall function. The discipline of systems engineering has instituted processes to eliminate emergent behaviors of any kind, ensuring high predictability and robust control within a target environment that hosts the system itself. As Mittal [35] elaborated, the application of DEVS-based systems engineering, which is founded on mathematical systems Theory and the closure under composition principle, is a recommended engineering methodology to model accurate and repeatable emergent behaviors.

As we defined the 10 RCAS properties and research focus areas for RCAS in Section 1.2, there is an assumption of the presence of unknown factors resulting from either the environment or the heterogeneous, diverse nature of RCAS, leading to both positive and negative emergent behaviors that must be accounted for in enabling RCAS engineering. However, limited methodologies exist for applying the closed under composition principle to the RCAS structure, with the exception of DEVS-based systems engineering. Nevertheless, we partitioned the 10 RCAS properties into structural and behavioral aspects. The RCAS structure constitutes properties of redundancy, scalability, and modularity. The RCAS behavior constitutes the remaining properties of adaptability, fault tolerance, self-healing, diversity, decentralization, robustness, and monitoring and feedback.

1.3.2 SoSE Approach to RCAS

A SoS approach to RCAS considers the interactions between multiple interconnected components or subsystems operating within a larger, more encompassing framework. This perspective allows for understanding the dynamic nature of RCAS as they adapt, recover and evolve over time in response to external pressures, challenges, and opportunities. SoS engineering involves developing metrics for ensuring a holistic SoS behavior in a tractable manner [43, 44] and harnessing emergence to guide SoS behavior to the desired holistic behavior and mitigate any negative emergent behavior [38, 45].

The SoSE approach emphasizes five fundamental characteristics, consolidated from [39] as applied to RCAS:

1. **Managing Complexity**: The system consists of multiple components that can be further decomposed into smaller subsystems with their own internal interactions and emergent behaviors. This complexity increases the potential for novel solutions to arise from diverse sources within the SoS, fostering adaptability and resilience. SoSE manages complexity using modular systems approach and composition principles with defined interfaces. Within the 10 RCAS properties, the properties of redundancy, scalability, diversity, decentralization, and modularity facilitate managing complexity.
2. **Assuring Interoperability**: The various components of an SoS must be able to communicate effectively with one another and share resources as needed, allowing them to cooperate in achieving shared objectives efficiently both technically and semantically. This interoperability enables coordinated action across the entire SoS through the incorporation of data standards and community involvement in developing such standards. Within the 10 RCAS properties, the properties of redundancy, scalability, diversity, robustness, modularity, and monitoring and feedback facilitate interoperability assurance.
3. **Allowing Heterogeneity**: Components within an SoS may possess unique capabilities, limitations, or perspectives that contribute to their overall function but may also lead to disparities in performance under certain conditions. Acknowledging this heterogeneity is essential for understanding how components collaborate effectively while ensuring individual strengths are leveraged and potential weaknesses addressed. This is essential because failure at multiple levels is unavoidable, and components need to be replaced in rapid manner to restore RCAS function. Further, configuration management for such components is a huge challenge, and having a configuration management plan will allow rapid integration and minimization of downtime. Within the 10 RCAS properties, the properties of adaptability, diversity, robustness, and modularity facilitate heterogeneity.
4. **Autonomy and Control**: Individual components within an SoS maintain a degree of autonomy, enabling them to adapt independently in response to local challenges or opportunities without requiring explicit direction from a central authority. This autonomy allows the system as a whole to be more resilient by fostering innovation and agility at the component level. This autonomy is implemented through centralized or decentralized control. Developing appropriate fail-safe mechanisms and feedback loops will ensure autonomy boundaries and control pathways. Within the 10 RCAS properties, the properties of adaptability, fault tolerance, self-healing, decentralization, modularity, and monitoring and feedback facilitate autonomy and control features.

5. **Positive and Negative Emergence**: The interactions between components within an SoS can give rise to emergent properties, behaviors, or patterns that were not anticipated during initial design or planning efforts. These emergent phenomena may provide new insights into potential solutions for addressing complex challenges faced by the SoS as a whole. When positive, they aid resilience, and when negative, they erode resilience. Developing metrics to establish positive and negative trajectories and associated instrumentation and collection will ensure guided emergence. Within the 10 RCAS properties, the properties of adaptability, self-healing, decentralization, robustness, and monitoring and feedback facilitate the handling of emergent behaviors for enabling resilience.

The SoS approach is particularly valuable for understanding the resilience and adaptability of RCAS due to its emphasis on collaboration, interoperability, and emergence between multiple interacting components. By acknowledging these key characteristics, we can better design and manage complex systems that respond more effectively to changing circumstances while maintaining a high degree of resilience over time.

Figure 1.2 maps various aspects of CST and SoSE with the 10 RCAS properties. As we mapped various CST and SoSE aspects to the 10 RCAS properties, the property of modularity arrived at the top (score of 5), followed by the properties of adaptability, diversity, decentralization, robustness, and monitoring and feedback, with a score of 4.

		RCAS properties									
		Adaptability	Redundancy	Fault tolerance	Scalability	Self-healing	Diversity	Decentralization	Robustness	Modularity	Monitoring and feedback
Theory/Engineering	**Aspect**	1	2	3	4	5	6	7	8	9	10
Classical systems theory	System structure		x		x					x	
	System behavior	x		x		x	x	x	x		x
System of systems engineering	Managing complexity		x		x		x	x		x	
	Assuring interoperability		x		x		x		x	x	x
	Allowing heterogeneity	x					x		x	x	
	Autonomy and control	x		x		x		x		x	x
	Positive and negative emergence	x				x		x	x		x

Figure 1.2 Mapping RCAS properties with classical systems theory and system of systems engineering aspects.

The property of modularity has been well recognized by industry, defense, academia, and various consortia, and it is not surprising to see this property as the most significant RCAS property that is in alignment with CST and SoSE. The modular open systems approach (MOSA) is incorporated as a law within Title 10 U.S. Code 4401(b) (https://www.dsp.dla.mil/Programs/MOSA/), which states all major defense acquisition programs (MDAP) are to be designed and developed using a MOSA. Modularity, in conjunction with open systems architecture, allows for standardization that supports interoperability, scalability, portability, and resilient supply chains within an SoS environment. Needless to say, any RCAS with defense applications must have a resilient supply chain!

1.4 Challenges in Modeling, Simulation, Analysis, and Visualization of RCAS

The M&S of RCAS sits at the intersection of computer science, systems theory, and domain-specific applications. As RCAS exhibit inherent complexity through their dynamic nonlinear interactions and adaptive behaviors, isolated traditional modeling approaches often prove insufficient for capturing the full spectrum of system behaviors, particularly those related to resilience mechanisms and emergent properties. In this context, rigorous analysis and advanced visualization techniques become essential to interpret, communicate, and derive actionable insights from the complex data structures generated by RCAS models and their simulations.

1.4.1 The Challenge of RCAS Modeling

As mentioned before, the complexity inherent in RCAS stems from several fundamental characteristics that distinguish them from simpler systems. These systems exhibit nonlinear dynamics where small perturbations can cascade into system-wide changes, self-organization capabilities that emerge from local interactions without central control, and adaptive mechanisms that allow components to learn and modify their behavior based on experience and environmental feedback. Such properties make traditional analytical approaches very limited and necessitate tailored computational M&S techniques [5, 46].

The modeling of RCAS requires a multi-faceted approach that can capture both the microscopic behaviors of individual components and the macroscopic patterns that emerge from their interactions. This challenge is compounded by the need to represent the system's ability to maintain functionality while adapting to disruptions—the core of resilience. Current modeling approaches must therefore integrate concepts from complex systems theory, resilience science, and domain-specific knowledge to create representations that are both computationally tractable and scientifically meaningful.

1.4.1.1 Modeling Approaches for RCAS

The modeling of RCAS has evolved, striving to encompass multiple paradigms, each offering distinct advantages for capturing different aspects of system behavior. Agent-based modeling (ABM) has emerged as a particularly suitable approach, allowing researchers to model individual autonomous agents and their interactions to understand how system-level patterns arise. This approach is valuable for exploring the impact of decentralized decision-making and adaptive behavior on system resilience.

Hybrid modeling approaches that combine different methodologies are increasingly recognized as necessary for capturing the full complexity of RCAS. These may integrate system dynamics models to capture aggregate behaviors and feedback loops, discrete event simulation for operational processes, and network-based models to represent structural relationships and their evolution over time. The integration of these approaches requires advanced model composition techniques and standardized interfaces that can handle the semantic consistency required for meaningful results.

The advent of increasingly potent data-driven modeling approaches also presents new opportunities for RCAS representation. ML techniques can be integrated with mechanistic models to capture behaviors that are difficult to specify explicitly while maintaining the interpretability required for understanding resilience mechanisms. This integration represents a significant methodological challenge, necessitating novel methodologies for model validation and verification that can account for both the first principles-driven and the data-driven components.

1.4.2 Simulation Challenges and Opportunities

The simulation of RCAS presents unique computational challenges that extend beyond traditional simulation requirements. These systems often operate across multiple temporal and spatial scales, requiring multiscale simulation approaches that can capture both rapid adaptive responses and long-term evolutionary processes. The computational demands of such simulations necessitate the development of efficient algorithms and the utilization of high-performance computing resources.

Simulation efforts for RCAS must address the challenge of parameter uncertainty and model validation in systems where controlled experimentation is often impossible or impractical. This has led to the development of advanced uncertainty quantification techniques and ensemble simulation approaches that can explore the space of possible system behaviors under different assumptions and conditions. The integration of real-time data streams through digital twin architectures represents a promising approach for maintaining model fidelity and enabling adaptive simulation strategies.

The sustainability of simulation studies has emerged as a critical concern, encompassing both the computational efficiency of simulation execution and the reusability of simulation artifacts. Modern simulation frameworks for RCAS must support model composition and reuse while maintaining provenance information that enables the reproducibility and extension of simulation studies [REF]. This requires the development of standardized modeling languages, metadata systems, and workflow management tools that can support the complex lifecycle of RCAS simulation studies.

1.4.3 Challenges in Analysis, Interpretation, Visualization, and Communication

The analysis of RCAS simulation results also presents specific challenges that stem from the high-dimensional, dynamic nature of the data generated and the need to distinguish between meaningful patterns and simulation artifacts. Traditional statistical analysis methods can often prove inadequate for systems where nonlinear interactions can produce counterintuitive behaviors and where emergent properties may only be observable at specific scales or under particular conditions.

Advanced analytical approaches must incorporate methods from network science to understand structural evolution, time series analysis to capture temporal dynamics, and ML to identify patterns in high-dimensional simulation outputs. The development of analysis pipelines that can automatically detect regime shifts, identify critical transitions, and quantify resilience metrics stands out as key for extracting actionable insights from RCAS simulations.

Furthermore, interpretation of analysis results demands more elaborated visualization tools that can represent multiple dimensions of system behavior simultaneously while maintaining accessibility for diverse stakeholder communities. This challenge is particularly acute in policy-relevant applications where simulation results must be communicated to decision-makers who may not have technical expertise in complex systems or simulation methods.

Traditional visualization approaches often fail to capture the dynamic, multiscale nature of RCAS behavior, necessitating the development of interactive, multi-perspective visualization tools that can adapt to different user needs and analytical objectives.

Exploratory analysis of RCAS simulation results requires interactive tools that allow users to manipulate parameters, explore counterfactual scenarios, and drill down from system-level patterns to component-level behaviors. This requires the integration of real-time simulation capabilities with responsive visualization systems that can maintain interactivity while processing complex simulation data.

The communication of RCAS M&S results to diverse audiences represents a critical challenge for translating research insights into actionable knowledge. This requires the development of adaptive communication strategies that can adjust the level of technical detail and the choice of visualization metaphors based on audience expertise and decision-making requirements.

1.5 Verification and Validation of RCAS with Modeling and Simulation

CAS requires an engineered sandbox or a test bed to experiment with various characteristics, including those of resilient behaviors. M&S enables the development of such a test bed where an RCAS model amenable to large-scale simulation allows simulation-based experimentation [41]. Having a closed-loop CAS/RCAS model will continue to be a challenge due to their inherent complexity. Many piecewise verification and validation (V&V) approaches will need to be brought together for an end-to-end V&V effort. The following are some aspects of V&V for the 10 characteristics identified earlier:

- **Adaptability** (Property 1): Adaptation and evolution demand the creation of models capable of learning from changing external conditions, simulating realistic scenarios to validate predictions against data or expert knowledge while remaining adaptable. Learning models must be designed using appropriate performance metrics, verified against realistic data or expert knowledge, and compared against expected outcomes for accurate prediction.
- **Redundancy and Fault Tolerance** (Properties 2 and 3): Fault tolerance and redundancy models must simulate component failures or disruptions realistically, validate them against realistic data or expert knowledge, and ensure that predictions remain relevant as components adapt over time.
- **Scalability and Modularity** (Properties 4 and 9): Modularity demands the design of scalable and adaptive individual modules without disrupting overall system architecture while validating each module independently for accuracy and compatibility with other modules [35]. Verification of various data/information exchanges must be done through standardized data models and community-developed ontologies.
- **Self-healing and Self-organization** (Properties 5 and 8): Self-organization models require validation with observed data and subject matter expertise, ensuring they accurately represent emergent behavior patterns in real-world systems while remaining adaptable to changing conditions. Emergent behavior necessitates designing scenarios that encourage cooperation among

agents within a CAS, ensuring these emerging behaviors promote overall resilience and adaptability, and verifying model predictions against observed data or subject matter expertise. Gathering real-world data for this property is the hardest unless repeatable emergent phenomena are captured in a test bed [47].

- **Diversity** (Property 6): Distinguishing distinct behaviors among different agents in a CAS, validating simulations under stress conditions with updated components, and maintaining relevance as new elements are introduced. Interconnectedness challenges involve designing CAS responses to external systems or stimuli appropriately without compromising overall resilience and adaptability, validating connections with observed data or expert knowledge, and ensuring accurate predictions under changing conditions. Heterogeneity within the RCAS must be tracked at both syntactic and semantic levels [39].
- **Decentralization** (Property 7): Designing agent behavior models based on local information processing, ensuring they promote resilience and adaptation within the RCAS, and verifying model predictions against realistic data or expert knowledge. Various metrics associated with multiple control strategies need to be developed explicitly [38].
- **Monitoring and Feedback Control** (Property 10): Feedback loops require validation of models responding appropriately to external stimuli while promoting system resilience and adaptation in a realistic manner, using data or subject matter expert (SME) knowledge. Designing of monitoring and instrumentation in alignment with first- and second-order cybernetics [48] is essential.

To summarize, each of the identified 10 RCAS properties has an appropriate digital engineering component within the RCAS test bed. Accordingly, methods from complexity science, statistics, operations research, control theory, systems engineering, SoSE and distribution simulation engineering can be employed to build upon the existing V&V methodologies. Gathering data from real world will only be piecewise continuous, so an end-to-end V&V will continue to be a challenge. There exists limited information in the literature for such end-to-end V&V of CAS at large.

1.6 Book Overview

RCAS have become increasingly pervasive in our daily lives, finding applications in a wide array of domains, including finance, economics, social networks, smart cities, medicine, computer systems, autonomous vehicles, and military operations, among others. With a focus on RCAS design and engineering, this book considers theoretical and applied aspects of critical RCAS properties such as self-organization and data-enabled emergent behavior. With a focus on modeling, simulation, and visualization, it provides insights into analyzing, designing, and deploying RCAS across various domains. From examining and instrumenting emergent behavior using advanced AI/ML analytics in SoS to harnessing novel AI/ML techniques for decision-making, this book serves as an indispensable resource for understanding and mastering RCAS engineering.

To address the above challenges and provide the state-of-the-art in performing CAS engineering, novel approaches are showing up across multiple disciplines due to the pervasive nature of CAS. This book aims to provide a vision of how modeling, simulation, and visualization (MS&V) can facilitate the analysis, design, implementation, and deployment of a resilient CAS. We explore the resiliency aspect of CAS through fundamental first principles and existing literature, and extend it toward engineering approaches for resilient CAS in applicable domains. The book will provide a comprehensive overview of existing research, describe research challenges across various perspectives identified above, including CAS MS&V, planning, modeling and analysis of a resilient CAS, emergent behavior, control and decision-making, utilization of advanced analytics using AI/ML methods, and practical examples.

This book argues for elevating resilience from a by-product of CAS to a primary design objective, defining RCAS as systems that anticipate perturbations, absorb impacts, recover function, and adapt to new equilibria. We formalize RCAS through a set of structural and behavioral properties (e.g., redundancy, modularity, adaptability, decentralization, self-healing, monitoring/feedback), situated them within classical systems theory and SoSE, and highlighted formal modeling foundations for representing and harnessing emergence. We then outlined methodological priorities across modeling, simulation, analysis, visualization, and V&V—emphasizing hybrid and multiscale models, uncertainty management, provenance and traceability, and the role of analytics and interactive visualization in making resilience a critical, measurable focus. This book is structured into four sections that address these concerns, as follows.

1.6.1 RCAS Foundations

This section of the book sets the foundations for RCAS, outlining key challenges, methodologies, and research directions in modeling and simulating CAS, quantifying resilience through measures of entropy, performance, and effectiveness, and situating RCAS research within an epistemological context.

This section presents an overview of the state-of-the-art MS&V strategies for RCAS, identifying challenges in modeling paradigms, formalisms, and toolkits, current solutions, and open problems. Szabo et al. [49] present an overview of existing literature on the M&S of CAS, highlighting their characteristics and discussing existing research focus in the literature.

Mittal et al. [50] discuss emergent behavior within RCAS and focuses on metrics for the quantification of resilient CAS through three categories: measures of entropy, measures of performance, and measures of effectiveness, each category contributing to the measures of resilience in RCAS. The metrics are utilized in a proposed engineering approach for RCAS and describe various challenges and methods to build a M&S-enabled RCAS engineering infrastructure to generate verifiable synthetic emergence in a computational RCAS environment.

Tolk [51] sets the scene for the study of RCAS through the epistemology of using M&S as a tool for gaining knowledge, specifically about RCAS. The chapter discusses related aspects such as modeling, simulation, systems, complexity, adaptivity, resilience, and how to conduct research and experiments in RCAS.

1.6.2 Modeling Frameworks and Their Applications

The next section of the book focuses on modeling of RCAS and the applications of the modeling formalisms, in particular on the use of AI and ML for decision-making in RCAS. Current research favors AI and ML for decision-making in RCAS, from expert systems to optimization heuristics and fuzzy logic to cooperative or competitive multi-agent reinforcement learning (MARL) methods that have been designed, developed, and deployed to address challenges in control, planning, adaptation decision-making, and context and situational awareness. This section of the book covers state-of-the-art approaches for the use of AI in RCAS for various aspects of decision-making and ML for instrumentation and feature assignment.

Dosi et al. [52] starts the M&S section with a focus on the use of ABMs for analyzing economic systems and the challenges they face in integrating empirical data. To address this, Roventini et al. propose a model-agnostic protocol using a new multidimensional similarity index that quantifies alignment between simulated and real-world data, enabling effective model ranking and selection. They demonstrate the approach using the K+S ABM family and US economic data, highlighting its potential for policy-informed scenario analysis and forecasting.

Yilmaz [53] uses the theory of explanatory coherence as a computational framework for learning explanatory cognitive maps during an evolving analysis. The framework and its reference architecture offer strategies for learning coherence-driven explanatory models with implications for simulation model development environments for RCAS analysis.

Honhaga et al. [54] propose the use of MARL for decision-making in RCAS that run in contested and dynamic environments. They show the formulation of the decision-making as a MARL problem and introduce a simulation environment that showcases the contested and dynamic challenges for decision-making in RCAS.

Shemeshenko and De Raco [55] study labor mobility in Argentina's formal private sector as an RCAS evolving across pre-pandemic, pandemic, and post-pandemic periods. Using intersectoral labor flow data, we explore how the labor market adapts to stress, evolves, and recovers. They highlight the importance of analyzing meso-structures within labor flow networks. Their work shows the importance of high-granularity labor flow data for the analysis of labor markets as RCAS.

Castro et al. [56] present a conceptual framework integrating agent-based computational economics (ACE) and MARL, in an attempt to shed new light on crisis events in labor markets. The ACE+MARL framework serves as an organized computational test bed for studying crisis emergence, exploring "what if" scenarios, identifying systemic vulnerabilities, testing potential recovery paths, and designing resiliency policies.

Lastly, Griffith et al. [57] propose the use of DEVS to study RCAS. They exemplify their approach using several epidemiology models, such as epidemics, social network models, city planning models, and segregation models, showcasing the versatility of DEVS as a modeling formalism in many RCAS scenarios.

1.6.3 RCAS Engineering

This section focuses on the engineering of RCAS and on the analysis of the engineered RCAS. The authors present state-of-the-art procedures driven by two perspectives, namely: (1) to create a RCAS with desired emergent properties (for both top-down and bottom-up RCAS engineering) and (2) to validate and verify that a desired RCAS property has materialized.

Chan and Johnsen [58] propose a framework to evaluate the emergent behavior and resilience of edge computing systems in a variety of disruptive scenarios such as disaster responses, smart homes, and environmental sensing. Edge computing enhances responsiveness by processing data near its source, aligning with CAS through decentralized control and local interactions. In scenarios without cloud connectivity, edge systems must be self-sufficient, adaptable, and resilient.

Neville et al. [59] address the challenge of maintaining and enhancing resilience in critical infrastructure as CAS are modernized. The chapter presents a set of tools designed to help engineering teams assess and incorporate resilience impacts during technology development and outlines a framework of resilience enablers.

Granados and De la Pena [60] use a CAS perspective to study the evolution of global AI networks. They propose the use of multilayer network methods and eigenvector centrality to analyze the structure and dynamics of global AI networks and provide empirical insights into how knowledge, hardware, and energy shape the development and evolution of these networks.

Fraser and Hull [61] focus on the importance of considering communication in the engineering of RCAS. RCAS are reliant on it to enable the generation of cooperative emergent behavior and suffer when it is interrupted. Further RCAS components can communicate both explicitly and implicitly at the intersection of mobile ad-hoc networking and network routing, RF communications, graph theory, multi-agent systems, and agent-based and mathematical modeling, highlighting the need to consider all these approaches and perspectives in the study of RCAS.

Presenza et al. [62] present an example of RCAS engineering within the application domain of multirobot coordination. They address the combined problem of control and localization of a multirobot network where some agents have access to their absolute positions, while others do not. They propose the use of subframeworks to express rigidity as a set of local conditions, thus eliminating the need for global information exchange. This has the potential to offer significant improvements in the efficiency and robustness of networked multirobot coordination.

1.6.4 Digital Twins and Socio-ecological Applications

This section illustrates how the principles and methods of RCAS engineering can be applied across diverse domains to address real-world challenges through modeling, simulation, and digital integration. Traoré et al. [63] focus the application of M&S approaches by looking at multi-perspective modeling and holistic simulation for digital twins. Urban areas function as CAS facing increasing pressure from growing urbanization, which strains services like transport, energy, and public health. Existing urban management methodologies are insufficient due to their inability to handle the system's complexity and multi-scale interactions. Urban digital twins (UDTs) have emerged as promising tools to model and manage these complexities by integrating multidisciplinary data and enabling informed decision-making. However, challenges such as lack of standardization and integration barriers hinder the widespread adoption of UDTs. This chapter presents a conceptual UDT framework with cognitive and metaverse integrations to enhance adaptability, urban planning, and citizen engagement.

Kausley et al. [64] view river ecosystems as RCAS and use M&S to model and analyze emergent behavior (both stakeholder behaviors and chemical dynamics) under varying conditions. This hybrid approach supports scenario-based analysis—such as pollution events, climate impacts, and demographic changes—enabling better-informed, long-term policy decisions for a resilient approach to river water quality management.

Risco-Martin et al. [65] presents DEVS-BLOOM, a comprehensive framework designed for the real-time monitoring and management of algae bloom with DEVS. The system integrates both simulated and real-world models to enhance the detection and management of algal blooms in aquatic environments. The authors present a layered approach to the design of RCAs, with each layer having a distinct role in data processing and decision-making. The practical challenges of design and deployment are discussed in detail, and the authors show the benefits and challenges of deploying a multi-role, multi-layer RCAS architecture.

Focusing on a specific emergent behavior, specifically the spread of disease, Raddichi [66] considers synthetic multiplex networks with built-in community structure, where we can vary the within-layer strength and the cross-layer correlation of the community structure, which then makes the community CAS more resilient to disruptions from epidemics.

Lastly, Szabo et al. [67] conclude the book and presents an overview of a research agenda for RCAS engineering.

Disclaimer: Dr. Mittal's affiliation with MTSI is provided for identification purposes only and is not intended to convey or imply concurrence with, or support for, the positions, opinions, or viewpoints expressed by this author.

References

1 George A. Polacek, David A. Gianetto, Khaldoun Khashanah, and Dinesh Verma. On principles and rules in complex adaptive systems: A financial system case study. *Systems Engineering*, 15(4):433–447, 2012.

2 Alon Hasgall. Digital social networks as complex adaptive systems. *Vine*, 43(1):78–95, 2013.

3 Marge Benham-Hutchins and Thomas R. Clancy. Social networks as embedded complex adaptive systems. *JONA: The Journal of Nursing Administration*, 40(9):352–356, 2010.

4 ED Schwab and Kenneth J. Pienta. Cancer as a complex adaptive system. *Medical Hypotheses*, 47(3):235–241, 1996.

5 Claudia Szabo, Robin Baker, Glen Pearce, Eyoel Teffera, and Anthony Perry. Overview and challenges of distributed decision making in resource contested and dynamic environments. *ACM Computing Surveys*, 57(9):1–34, 2025.

6 Ted Carmichael and Mirsad Hadžikadić. *The Fundamentals of Complex Adaptive Systems*. Springer, 2019.

7 Bradley Fraser, Claudia Szabo, Andrew Coyle, and Robert Hunjet. Leveraging emergent specialization in a heterogeneous multi-role swarm control architecture for positional-based UAS missions. In *2020 IEEE Symposium Series on Computational Intelligence (SSCI)*, pages 2584–2591. IEEE, 2020.

8 Rob Dekkers. Complex adaptive systems. In *Applied Systems Theory*, Springer, 2017.

9 Mohammad Fakhrul Alam Onik, Erwin Fielt, and Guy Gable. Complex adaptive systems theory in information systems research: A systematic literature review. In *Proceedings of the 21st Pacific Asia Conference on Information Systems (PACIS) 2017*, pages 1–14. Association for Information Systems (AIS), 2017.

10 David P. Chassin, Joel M. Malard, Christian Posse, Asim Gangopadhyaya, Ning Lu, Srinivas Katipamula, and J. V. Mallow. Modeling power systems as complex adaptive systems. Technical report, Pacific Northwest National Lab.(PNNL), Richland, WA (United States), 2004.

11 Gregg Hartvigsen, Ann Kinzig, and Garry Peterson. Use and analysis of complex adaptive systems in ecosystem science: Overview of special section. *Ecosystems*, 1(5):427–430, 1998

12 Rika Preiser, Reinette Biggs, Alta De Vos, and Carl Folke. Social-ecological systems as complex adaptive systems. *Ecology and Society*, 23(4):1–16, 2018

13 Gérard Weisbuch. The complex adaptative systems approach to biology. In *Self-organization and Emergence in Life Sciences*, pages 7–28. Springer, 2006.

14 Claudia Szabo, Joshua Groot, Thomas McAtee, and Christo Pyromallis. Who did this: Identifying causes of emergent behavior in complex systems. In *2019 IEEE Symposium Series on Computational Intelligence (SSCI)*, pages 262–270. IEEE, 2019.

15 Gordon L. Bonser. Method of systems potential as “top-bottom” technique of the complex adaptive systems modelling. In A. Yang and Y. Shan, editors, *Intelligent Complex Adaptive Systems*, IGI Global, 2008.

16 Johan Bergström and Sidney W. A. Dekker. Bridging the macro and the micro by considering the meso: Reflections on the fractal nature of resilience. *Ecology and Society*, 19(4):22, 2014.

17 Seyedmohsen Hosseini, Kash Barker, and Jose E. Ramirez-Marquez. A review of definitions and measures of system resilience. *Reliability Engineering & System Safety*, 145:47–61, 2016.

18 Adelinde M. Uhrmacher, Peter Frazier, Reiner Hähnle, Franziska Klügl, Fabian Lorig, Bertram Ludäscher, Laura Nenzi, Cristina Ruiz-Martin, Bernhard Rumpe, Claudia Szabo, Gabriel Wainer, and Pia Wilsdorf. Context, composition, automation, and communication: The C2AC roadmap for modeling and simulation. *ACM Transactions on Modeling and Computer Simulation*, 34(4): 1–51, 2024.

19 Ed Glinert. *The London Compendium: A Street-by-Street Exploration of the Hidden Metropolis*. Penguin, UK, 2012.

20 Randy K. Buchanan, Simon R. Goerger, Christina H. Rinaudo, Greg Parnell, Adam Ross, and Valerie Sitterle. Resilience in engineered resilient systems. *The Journal of Defense Modeling and Simulation*, 17(4):435–446, 2020.

21 Royce Francis and Behailu Bekera. A metric and frameworks for resilience analysis of engineered and infrastructure systems. *Reliability Engineering & System Safety*, 121:90–103, 2014.

22 Yossi Sheffi and James B. Rice Jr. A supply chain view of the resilient enterprise. *MIT Sloan Management Review*, 2005.

23 Lance H. Gunderson. Ecological resilience—In theory and application. *Annual Review of Ecology and Systematics*, 31(1):425–439, 2000.

24 Michel Bruneau, Stephanie E. Chang, Ronald T. Eguchi, George C. Lee, Thomas D. O'Rourke, Andrei M. Reinhorn, Masanobu Shinozuka, Kathleen Tierney, William A. Wallace, and Detlof Von Winterfeldt. A framework to quantitatively assess and enhance the seismic resilience of communities. *Earthquake Spectra*, 19(4):733–752, 2003.

25 Kathleen Tierney and Michel Bruneau. Conceptualizing and measuring resilience: A key to disaster loss reduction. *TR News*, (250), 2007.

26 Saurabh Mittal and Andreas Tolk. *Complexity Challenges in Cyber Physical Systems: Using Modeling and Simulation (M&S) to Support Intelligence, Adaptation and Autonomy*. John Wiley & Sons, 2019.

27 Dong Wei and Kun Ji. Resilient industrial control system (RICS): Concepts, formulation, metrics, and insights. In *2010 3rd International Symposium on Resilient Control Systems*, pages 15–22. IEEE, 2010.

28 Ishan Honhaga and Claudia Szabo. A simulation and experimentation architecture for resilient cooperative multiagent reinforcement learning models operating in contested and dynamic environments. *Simulation*, 100(6): 563–579, 2024.

29 James R. Enos. Using systems engineering ilities to better understand resiliency. In *Systems Engineering in Context: Proceedings of the 16th Annual Conference on Systems Engineering Research*, pages 25–36. Springer, 2019.

30 Yacov Y. Haimes. Models for risk management of systems of systems. *International Journal of System of Systems Engineering*, 1(1–2):222–236, 2008.

31 Claudia Szabo, Brendan Sims, Thomas Mcatee, Riley Lodge, and Robert Hunjet. Self-adaptive software systems in contested and resource-constrained environments: Overview and challenges. *IEEE Access*, 9:10711–10728, 2020.

32 Claudia Szabo, Rodrigo Castro, Joachim Denil, and Susan M. Sanchez. Resilience and complexity in socio-cyber-physical systems. In *2023 Winter Simulation Conference (WSC)*, pages 660–670. IEEE, 2023.

33 Danny Weyns, Bradley Schmerl, Vincenzo Grassi, Sam Malek, Raffaela Mirandola, Christian Prehofer, Jochen Wuttke, Jesper Andersson, Holger Giese, and Karl M. Göschka. On patterns for decentralized control in self-adaptive systems. In *Software Engineering for Self-Adaptive Systems II*, pages 76–107. Springer, 2013.

34 Andreas Duit and Victor Galaz. Governance and complexity—Emerging issues for governance theory. *Governance*, 21(3):311–335, 2008.

35 Saurabh Mittal. Emergence in stigmergic and complex adaptive systems: A formal discrete event systems perspective. *Cognitive Systems Research*, 21:22–39, 2013. ISSN 1389-0417. https://doi.org/10.1016/j.cogsys.2012.06.003.

36 Bernard P Zeigler, Alexandre Muzy and Ernesto Kofman. *Theory of Modeling and Simulation: Discrete Event & Iterative System Computational Foundations*. Academic press, 2018.

37 Claudia Szabo and Yong Meng Teo. Formalization of weak emergence in multiagent systems. *ACM Transactions on Modeling and Computer Simulation*, 26(1), 2015. https://doi.org/10.1145/2815502.

38 Saurabh Mittal and Larry Rainey. Harnessing emergence: The control and design of emergent behavior in system of systems engineering. In *Proceedings of the Conference on Summer Computer Simulation*, pages 1–10, 2015.

39 Saurabh Mittal. *Netcentric Complex Adaptive Systems*, Chapter 23, pages 639–661. CRC Press, 2014.

40 Saurabh Mittal. Model engineering for cyber complex adaptive systems. In *Proceedings of the European Modeling and Simulation Symposium*, 2014.

41 Saurabh Mittal and José L. Risco-Martín. *Simulation-Based Complex Adaptive Systems*, pages 127–150. Springer International Publishing, Cham, 2017. ISBN 978-3-319-61264-5. https://doi.org/10.1007/978-3-319-61264-5_6.

42 Daniel Foguelman, Esteban Lanzarotti, Emanuel Ferreyra, and Rodrigo Castro. Simulation of emergence in artificial societies: A practical model-based approach with the EB-DEVS formalism. *arXiv preprint arXiv:2110.08170*, 2021.

43 Saurabh Mittal and Sheila A. Cane. Contextualizing emergent behavior in system of systems engineering using gap analysis. In *Proceedings of the Modeling and Simulation of Complexity in Intelligent, Adaptive and Autonomous Systems 2016 (MSCIAAS 2016) and Space Simulation for Planetary Space Exploration (SPACE 2016)*, MSCIAAS '16, San Diego, CA, USA, 2016. Society for Computer Simulation International. ISBN 9781510823198.

44 Katrina Falkner, Claudia Szabo, and Vanea Chiprianov. Model-driven performance prediction of systems of systems. In *Proceedings of the ACM/IEEE 19th International Conference on Model Driven Engineering Languages and Systems*, pages 44–44, 2016.

45 Claudia Szabo and Yong Meng Teo. Post-mortem analysis of emergent behavior in complex simulation models. In *Proceedings of the 2013 ACM SIGSIM Conference on Principles of Advanced Discrete Simulation*, pages 241–252. ACM, 2013.

46 T. Carreira-Municha, A. Bugalloa, C. Szabo, and R. Castro. Complex adaptive systems simulation modeling: A decade's overview. *Complex System Modeling and Simulation*, 2025.

47 Saurabh Mittal. *New Frontiers in Modeling and Simulation in Complex Systems Engineering: The Case of Synthetic Emergence*, pages 173–194. Springer International Publishing, Cham, 2019. ISBN 978-3-030-17164-3. https://doi.org/10.1007/978-3-030-17164-3_9.

48 F. Heylighen and C. Joslyn. Cybernetics and second order cybernetics. *Encyclopedia of Physical Science & Technology*, 4:155–170, 2001.

49 C. Szabo, R. Castro, S. Mittal, and V. Shemeshenko. Modeling and simulation of resilient complex adaptive systems. In C. Szabo, R. Castro, S. Mittal and V. Shemeshenko, editors, *Resilient Complex Adaptive Systems: Modeling, Simulation, Visualization and Engineering Approaches*. John Wiley & Sons, Hoboken, NJ, 2025.

50 S. Mittal, C. Szabo, R. Castro, and V. Shemeshenko. Emergent behavior in resilient complex adaptive systems: Challenges, metrics and engineering approach. In Claudia Szabo, Rodrigo Castro, Saurabh Mittal, and Viktor Shemeshenko, editors, *Resilient Complex Adaptive Systems: Modeling, Simulation, Visualization and Engineering Approaches*. John Wiley & Sons, Hoboken, NJ, 2025.

51 A. Tolk. Epistemology of resilient complex adaptive system simulations. In C. Szabo, R. Castro, S. Mittal, and V. Shemeshenko, editors, *Resilient Complex Adaptive Systems: Modeling, Simulation, Visualization and Engineering Approaches*. John Wiley & Sons, Hoboken, NJ, 2025.

52 Giovanni Dosi, Marcelo Pereira, Gabriel Petrini, Andrea Roventini, and Maria Enrica Virgillito. From ABM back to real data: time series visualization and model selection in the K+S agent-based model. In Claudia Szabo, Rodrigo Castro, Saurabh Mittal, and Viktor Shemeshenko, editors, *Resilient Complex Adaptive Systems: Modeling, Simulation, Visualization and Engineering Approaches*. John Wiley & Sons, Hoboken, NJ, 2025

53 L. Yilmaz. Self-aware models with explanatory capabilities: Learning explanatory models from simulation experiments. In C. Szabo, R. Castro, S. Mittal, and V. Shemeshenko, editors, *Resilient Complex Adaptive Systems: Modeling, Simulation, Visualization and Engineering Approaches*. John Wiley & Sons, Hoboken, NJ, 2025.

54 I. Honhaga, V. A Do, E. Teffera, A. Perry, R. Baker, and C. Szabo. Multi-agent reinforcement learning for decision making in contested environments. In C. Szabo, R. Castro, S. Mittal, and V. Shemeshenko, editors, *Resilient Complex Adaptive Systems: Modeling, Simulation, Visualization and Engineering Approaches*. John Wiley & Sons, Hoboken, NJ, 2025.

55 V. Shemeshenko and S. A. De Raco. Intersectoral labor mobility in Argentina: Evolution and resilience. In C. Szabo, R. Castro, S. Mittal, and V. Shemeshenko, editors, *Resilient Complex Adaptive Systems: Modeling, Simulation, Visualization and Engineering Approaches*. John Wiley & Sons, Hoboken, NJ, 2025.

56 R. Castro, V. Shemeshenko, and P. Mira. A novel integrated agent-based computational economics and multi-agent reinforcement learning framework: Application to macrocrisis phenomena. In C. Szabo, R. Castro, S. Mittal, and V. Shemeshenko, editors, *Resilient Complex Adaptive Systems: Modeling, Simulation, Visualization and Engineering Approaches*. John Wiley & Sons, Hoboken, NJ, 2025.

57 H. Griffith, C. Ruiz-Martin, and G. Wainer. DEVS as a method to design resilient complex adaptive systems. In C. Szabo, R. Castro, S. Mittal, and V. Shemeshenko, editors, *Resilient Complex Adaptive Systems: Modeling, Simulation, Visualization and Engineering Approaches*. John Wiley & Sons, Hoboken, NJ, 2025.

58 K. Chan and F. T. Johnsen. A framework for characterizing, understanding and predicting behavior in complex edge computing systems. In C. Szabo, R. Castro, S. Mittal, and V. Shemeshenko, editors, *Resilient Complex Adaptive Systems: Modeling, Simulation, Visualization and Engineering Approaches*. John Wiley & Sons, Hoboken, NJ, 2025.

59 K. J. Neville, T. Sanders, E. Barret, T. Fersch, G. Lematta, X. Chau, A. Langone, and E. S. Patterson. Resilience-aware development tools: Translating resilience concepts into engineering practice. In C. Szabo, R. Castro, S. Mittal, and V. Shemeshenko, editors, *Resilient Complex Adaptive Systems: Modeling, Simulation, Visualization and Engineering Approaches*. John Wiley & Sons, Hoboken, NJ, 2025.

60 O. M. Granados and N. De la Pena. Resilience-aware development tools: Translating resilience concepts into engineering practice. In C. Szabo, R. Castro, S. Mittal, and V. Shemeshenko, editors, *Resilient Complex Adaptive Systems: Modeling, Simulation, Visualization and Engineering Approaches*. John Wiley & Sons, Hoboken, NJ, 2025.

61 B. Fraser and R. Hull. Resilient complex adaptive systems: A communications perspective. In C. Szabo, R. Castro, S. Mittal, and V. Shemeshenko, editors, *Resilient Complex Adaptive Systems: Modeling, Simulation, Visualization and Engineering Approaches*. John Wiley & Sons, Hoboken, NJ, 2025.

62 F. J. Presenza, E. Pecker-Marcosig, I. Mas, I. J. Alvarez-Hamelin, and J. I. Giribet. Scalable rigidity maintenance strategies for networked multirobot systems: Overcoming decentralized coordination challenges. In C. Szabo, R. Castro, S. Mittal and V. Shemeshenko, editors, *Resilient Complex Adaptive Systems: Modeling, Simulation, Visualization and Engineering Approaches*. John Wiley & Sons, Hoboken, NJ, 2025.

63 M. K. Traoré, M. Diakité, and Z. Ali. Immersive and cognitive urban digital twins for sustainable and resilient smart territories. In C. Szabo, R. Castro, S. Mittal, and V. Shemeshenko, editors, *Resilient Complex Adaptive Systems: Modeling, Simulation, Visualization and Engineering Approaches*. John Wiley & Sons, Hoboken, NJ, 2025.

64 S. Kausley, S. Barat, P. S Thaku, S. Shrivastava, V. Kulkarni, and B. Rai. The role of digital twins in improving river water quality. In C. Szabo, R. Castro, S. Mittal, and V. Shemeshenko, editors, *Resilient Complex Adaptive Systems: Modeling, Simulation, Visualization and Engineering Approaches*. John Wiley & Sons, Hoboken, NJ, 2025.

65 J. L Risco-Martin, R. Dominguez, R. Cárdenas, E. Besada-Portas, and S. Esteban. Integrating simulation and real-world models for harmful algal and cyanobacterial blooms management. In C. Szabo, R. Castro, S. Mittal, and V. Shemeshenko, editors, *Resilient Complex Adaptive Systems: Modeling, Simulation, Visualization and Engineering Approaches.* John Wiley & Sons, Hoboken, NJ, 2025.

66 F. Raddichi. Optimal strategies to contain and prevent the spread of diseases in group-structured populations. In C. Szabo, R. Castro, S. Mittal, and V. Shemeshenko, editors, *Resilient Complex Adaptive Systems: Modeling, Simulation, Visualization and Engineering Approaches.* John Wiley & Sons, Hoboken, NJ, 2025.

67 C. Szabo, R. Castro, S. Mittal, and V. Shemeshenko. Research agenda for resilient complex adaptive systems. In C. Szabo, R. Castro, S. Mittal, and V. Shemeshenko, editors, *Resilient Complex Adaptive Systems: Modeling, Simulation, Visualization and Engineering Approaches.* John Wiley & Sons, Hoboken, NJ, 2025.

2

Emergent Behavior in Resilient Complex Adaptive Systems: Challenges, Metrics, and Engineering Approach

Saurabh Mittal,[1] Claudia Szabo,[2] Rodrigo Castro,[3] and Viktoriya Semeshenko[4]*

[1] *Defense Science Division, Modern Technology Solutions, Inc., Dayton, OH, USA*
[2] *School of Computer and Mathematical Sciences, The University of Adelaide, Adelaide, Australia*
[3] *Facultad de Cs. Exactas y Naturales and Instituto de Ciencias de la Computación, Universidad de Buenos Aires and CONICET, Ciudad Autónoma de Buenos Aires, Argentina*
[4] *Facultad de Cs. Económicas and Instituto Interdisciplinario de Economía Política, Universidad de Buenos Aires and CONICET, Ciudad Autónoma de Buenos Aires, Argentina*

*Corresponding author: saurabh.mittal@mtsi-va.org

2.1 Introduction

In this chapter, we shall focus on the modeling, simulation, analysis, and engineering of emergent behavior in resilient complex adaptive systems (RCASs). We first present an overview of emergent behavior definitions and examples within RCAS. We then introduce a systems-of-systems engineering approach to RCAS, which considers the interactions between the connected systems in obtaining a deeper understanding of the dynamic nature of the RCAS. We present an overview of metrics for quantifying RCAS emergent behaviors, focusing on three main perspectives: considering the amount of *information*, the *operational* performance of the system, and the overall *effectiveness* of the RCAS emergent behavior.

2.1.1 Theoretical Foundations of Emergent Phenomena

Emergent behavior is an inherent property of complex adaptive systems (CASs), together with self-organization and feedback loops [1]. An extensive body of work has focused on the definition, identification, and validation of unexpected behaviors in complex systems [2–5]. Holland [6] and Bedau [7] define *weak* and *strong* emergence as well as defining these terms using different degrees [8]. We consider *weak* emergence as being the *macrolevel* behavior that is a result of *microlevel* component interactions, and *strong* emergence as the *macrolevel* feedback or causation on the *microlevel*. The *microlevel*, in the context of a CAS, is defined by all the individual system components and their interactions [9]. In contrast, the *macrolevel* captures the collective, system-wide patterns of behaviors that are usually the result of microlevel interactions.

An emergent property is defined by Bedau [7] as "a property of assemblage that could not be predicted by examining the components individually." This is similar to Carmichael's and Hadzikadic's definition, who consider emergent behavior as "the macrolevel behavior that is not defined at the macrolevel but rather depends on the rules and interactions of agents defined at the microlevel" [1]. Many examples of emergent properties exist, including social networks [10, 11], power grids [12], and economics [13, 14], but three well-known models are the Flock of Birds [15], the cellular automata Game of Life [16], and the ant-colony models [17]. The flock of birds model achieves its emergent properties through each bird flocking around a perceived flock center. In the Game of Life model, the emergent properties are achieved by the patterns that are formed by the cells' transitions between states. Lastly, ants in an ant colony self-organize to create resilient pathways to food sources. In the flock of birds model, independent birds aggregate around an invisible center and fly at similar speeds, forming flocks. Flocks of birds are resilient to obstacles, re-forming upon encountering them [18, 19]; however, the study of their resilience when being targeted by a large number of predators has not yet been studied from a complex systems' perspective [20].

Many other examples of emergent behavior have been studied using a CAS perspective, including social networks [10, 21] and power grids [12]. The structure of *social networks*, such as friendships on social networks or professional relationships in a business setting, can exhibit resilience by maintaining connections even when some nodes (individuals) drop out or disappear. New nodes enter the network and create new links, while existing ones adapt their interactions to changing circumstances. [21] proposes a series of models that can be used to study social networks from a CAS perspective, with a focus on adaptability and the dynamic evolution over time. These include statistical physics models, evolutionary models, threshold models, neural network models, adaptive voter models, and social percolation models. The resilience of social networks to disruptions has been studied from the point of view of analyzing the number of connections [22].

Power grids are another example of application domains where emergent phenomena are inherent [12], as the interaction of all system components (generators, transformers, and transmission lines) can lead to cascading failures that propagate through the system. Often modeled as complex networks in which generators and loads can be considered as nodes connected over power cables or communication links, resilience analysis of power grids focuses on identifying small-world characteristics [23] and centrality components that disintegrate considerably faster when nodes are removed under attacks or failures. Studies show that power grids are resilient and can be restored after faults, attacks, or blackouts using black-start strategies [24], relying for success on optimal start-up, reconfiguration, and distribution strategies.

In swarms operating in search-and-rescue missions or data ferrying operations, individual drones operate autonomously and collaborate with other drones to coordinate actions in response to local environmental changes [25–27]. Swarms adapt and specialize under environmental disruptions [28], showcasing resilient and robust behaviors.

2.2 Challenges in Measurement, Modeling, Simulation, Analytics, and Visualization

To understand, model, and measure emergent behaviors in complex adaptive resilient systems, we count on a holistic analytical framework that captures micro- and macrodynamics and provides an effective and comprehensive analytical groundwork. This framework relies on four fundamental pillars: data collection, modeling, analytical techniques, and visualization. Data collection provides the granular observations, while modeling offers a coherent representation of how microlevel rules may lead to emergent behaviors. Analytical techniques then help to distinguish between weak and strong emergence and examine how resilience is fostered through feedback loops. Visualization ties the three efforts together, making the dynamic interplay between micro and macro levels clear and communicable. This integrated approach allows for a comprehensive understanding of how microbehaviors coalesce into macrophenomena. However, many challenges exist in bringing these four pillars together.

2.2.1 Data Collection

The foundational step of data collection, however, presents a significant challenge. This difficulty stems from the fact that these systems operate on two distinct, yet interconnected, scales. To fully grasp their dynamics, one must simultaneously consider both the microlevel, where individual agents interact, and the macrolevel, where large-scale phenomena emerge from those interactions. Only then can we accurately understand and model how individual actions scale up and reshape the structure of the entire system and collect high-quality data.

The sheer volume and speed of these microlevel interactions make it very difficult to isolate and track individual behaviors within a massive volume of data. Yet, it is this very activity that gives rise to broader system-wide behaviors, such as a stock market crash or a sudden change in public opinion. These are patterns that cannot be explained by a single event but rather by the aggregation, often nonlinear effects, of a large number of microlevel actions.

Given this complex, dual-scale reality, the process of data collection and preparation must overcome the following challenges: volume, complexity, quality, and relevance. A large-scale system poses an immediate challenge due to its volume and inherent complexity. With countless components that interact and produce diverse data streams in real time, the system quickly overwhelms conventional storage and processing methods. This makes it incredibly difficult to accurately monitor local interactions while simultaneously grasping the overall, emergent dynamics. Another significant challenge is data quality. Real-world data is inherently noisy and inconsistent. Incomplete or contradictory readings, if not carefully managed within a relevant context, can easily distort our understanding of emergent patterns at both the micro and macro levels.

2.2.2 Modeling and Simulation

Developing an RCAS model requires an understanding of the hybrid modeling paradigm [29] that includes models with multiple paradigms, especially, agent-based, discrete, and continuous modeling paradigms. Another challenge is the selection of tools for hybrid modeling. For example, AnyLogic (www.anylogic.com) provides such a hybrid model engineering environment. Other options include a formal universal modeling formalism, such as discrete event systems (DEVSs) that serve as a common integrator for hybrid models [30, 31]. Hybrid models must also be defined according to different levels of fidelity and timescales for behavior and phase transitions. Another challenge would be selecting real-time or faster-than-real-time model execution. While the real-time model is cost prohibitive for simulation experimentation, there exist technical challenges in running hybrid models faster than real-time. Developing a simulation environment to conduct simulation-based experimentation is a nontrivial undertaking [32] and, accordingly, the ensuing V&V considerations. Ultimately, the simulation infrastructure must interface with real word for hardware-in-the-loop or software-in-the-loop testing to capture real-world data from a cyber-physical instantiation of the RCAS [33, 34].

2.2.3 Analysis

Analysis of structural and behavioral properties of RCAS will involve accessing methods from statistics, network science, time-series pattern matching employing machine learning (ML), and operations research. Coordinating data persistence infrastructure to enable an analytics infrastructure will require targeted effort beginning with a unified data model, or lack thereof. Use of advanced AI and/or generative artificial intelligence (AI) may speed up ML-enabled analysis and will require a cloud-enabled persistent data infrastructure. Developing a cohesive analytical framework that enables the quantification of weak emergent behaviors and allows the discovery of strong emergent behaviors using a plethora of metrics at multiple RCAS specification levels is a critical aspect of RCAS engineering [35]. No such metrics are available for RCAS analytics.

2.2.4 Visualization

Real-time interaction allows analysts to respond quickly to changing conditions within the environment and make informed decisions based on real-time data and insights. Conveying multiple interconnected variables or metrics, such as agent behaviors, interactions, and environmental factors within a CAS in an easily digestible manner, is a huge challenge. Representing spatio-temporal dynamics of emergent behavior in real-time requires advanced computational infrastructure, which requires expensive graphical processing units (GPUs)-enabled hardware. Developing interactive visualization tools capable of allowing analysts to explore RCAS at multiple scales, from individual components up through system-wide behavior patterns, is a significant challenge that must balance complexity with usability for effective analysis and decision-making purposes. Advanced software architectures are needed that decouple the simulation infrastructure from visualization constraints for multi-scale, multi-dimensional, multi-temporal, multi-fidelity, and multi-resolution data.

2.3 Metrics for Quantification of RCAS Emergent Behaviors

The engineering of a system requires instrumentation and control of the variability of various factors that influence or affect the system. RCAS engineering must deal with the quantification of emergent behaviors, so adequate instrumentation can be built into the system, leading to the establishment of various control mechanisms to control both the weak and strong emergent behaviors. While weak emergent behavior is a desired emergent behavior because it can be studied in a laboratory setting [30], the strong emergent behavior must be limited due to its potential negative impact [36] on the RCAS emergent behavior. From a safety perspective, strong emergent behavior, once detected, must involve human-in-the-loop to be deemed causal for RCAS.

We discuss different types of metrics that can be used to quantify emergent behaviors in RCAS. We consider in this setting both *a posteriori* perspectives and *a priori* perspectives. *A posteriori* perspectives imply knowledge about the system, the connections between micro and macro behaviors, and use metrics after an emergent behavior is observed, in order to identify its causes or predict future outcomes. *A priori* perspectives do not use any insight about the system, micro–macro connections, and tend to identify emergent behavior live, as the system is running.

We consider the following measures of entropy (MoENs), which measure the information content of a system [37, 38], measures of performance (MoPs), which use the underlying system or formalism metrics to quantify threshold violations [39–41], and measures of effectiveness (MoEs) [42, 43], which provide a nonfunctional, application-based view on the behavior of a CAS.

We categorize measures in three broad categories, namely, (1) MoENs; (2) MoPs; and (3) MoEs.

MoENs quantify the level of uncertainty, randomness, or unpredictability within a system, reflecting its ability to resist control and maintain complexity. Entropy is a concept used to describe the measure of disorder, randomness, or uncertainty in a system. In the context of RCAS, entropy can be thought of as a way to quantify the level of unpredictability or uncontrollability. MoENs aim to capture the amount of information that is not predictable or explainable by current knowledge about the system. These measures are essential for understanding the behavior and performance of RCAS, which often exhibits emergent properties and adapts in response to changing conditions. This category reflects the measurement of information content to identify the emergent behavior to begin with.

MoPs capture the internal view of a system [42]. In systems engineering theory, MoPs capture a necessary but not sufficient condition for a system, in that a system must be precise, accurate, and utilized in an optimal manner, and meet specific performance bounds, but its overall behavior must also be effective.

MoEs evaluate the actual outcome or result of a system's performance. They answer questions like "What happened?" and provide insight into how well a system is achieving its objectives. MoEs focus on the "why" behind the operational characteristics described by MoPs. MoEs evaluate the actual outcomes or results of those performances, quantified by MoPs, helping understand whether SoSs are achieving intended objectives.

Together, MoENs measure information content, MoPs measure constituent systems' performance, and MoEs measure the overall outcome of the system synthesizing the MoENs and MoEs. This allows quantifying RCAS behavior across multiple levels.

2.3.1 Measures of Entropy

Entropy can be regarded as a quantitative lens for emergence and resilience in CAS.

Resilience arises from the dynamic interplay between order and disorder, where emergent behaviors enable adaptation to perturbations, thereby leading to the preservation of core system functions. Entropy—fundamentally a measure of uncertainty, information, or disorder—provides a well-established mathematical framework borrowed from statistical mechanics for quantifying such emergent dynamics.

Its domain-agnostic nature makes it suitable to gain insights into the dynamics of systems across ecological, economic, social, engineered, and physical domains, just to name a few.

Given entropy's rich and extensive history across scientific disciplines, the list below offers a selection of fundamental entropy-based measures—representing a core subset that serves as the foundation for a continually expanding array of derived metrics adapted to specialized domains and applications.

- **Shannon Entropy**: Measures average uncertainty or information content in a probability distribution. Foundational for assessing system-wide disorder. In RCAS, it helps to monitor how uncertainty fluctuates, which is a proxy to the system's capacity to absorb disturbances while maintaining key functions.
- **Boltzmann Entropy**: Quantifies disorder via microscopic configurations in physical systems. Adapted for spatial/dynamic data patterns. It can reveal how the microscale diversity within a system enables it to reorganize and adapt following shocks, contributing to resilience.
- **Rényi Entropy**: Generalizes Shannon entropy with a parameter to weight rare/common events. Enables multiscale complexity analysis. By capturing different scales of system behavior, Rényi entropy can detect subtle shifts that precede critical transitions or adaptations within complex systems.
- **Tsallis Entropy**: Extends entropy to nonextensive systems (e.g., long-range interactions). Captures deviations from classical statistics. It can be useful to portray how resilience emerges from correlations and memory effects in systems with complex internal interactions.
- **Permutation Entropy**: Evaluates complexity through ordinal patterns in time series. Distribution-free and robust for temporal dynamics. When RCAS can exhibit characteristic temporal patterns, permutation entropy helps identify changes in these patterns signaling loss or gain of resilience.
- **Sample Entropy**: Estimates regularity in time series by measuring pattern repetition. This measure tracks the predictability of system dynamics; a stable sample entropy over time can suggest persistent adaptive capabilities under stress.
- **Transfer Entropy**: Quantifies directional information flow between system components. Identifies causal relationships. It can hint at how information and influence propagate in RCAS to study how sub-systems coordinate to maintain resilience.

- **Spectral Entropy**: Assesses power distribution across frequencies. Higher entropy indicates broader frequency dispersion. Spectral entropy highlights shifts in dynamical regimes, such as moving from ordered to chaotic behavior, which may reflect changes in system resilience.
- **Statistical Complexity (e.g., LMC Measure)**: Combines entropy with disequilibrium. Peaks at intermediate order–disorder balances. This balance embodies the "edge of chaos" where resilient systems often operate, maximizing adaptability and robustness—avoiding extremes of rigidity (total order) or chaos (total randomness).
- **Relative/Conditional Entropy**: Compares observed vs. reference distributions or measures residual uncertainty given another variable. These measures quantify how much new information or uncertainty arises under changing conditions, providing insights into a system's capacity to reorganize.

In RCAS, the evolution of entropy can be used to capture the system's capacity to balance exploration (creative innovation) and exploitation (conservative stability) across scales. Figure 2.1 depicts various metrics associated with various entropy categories.

As articulated in [44], self-organizing systems exhibit emergent properties through micro-level interactions among agents, and these properties can be captured and evaluated using entropy-based metrics. Robot swarms use local rules (e.g., collision avoidance) for global tasks. Entropy metrics guide exploration (high entropy, dispersed search) and exploitation (low entropy, formation control), while feedback loops adjust entropy to maintain resilience.

In the case of biological systems, the authors in [45] stress that the choice of any specific entropy-based estimator is determined by the goal type, as there can be a range of such measures. Underlying is the idea of a dynamical system under study whose state variables (at different scales) can be represented as time series. In this context, various entropy measures were developed to compare time series (distinguish their periodic, chaotic, and random nature), such as sample entropy, control entropy, spectral entropy, wavelet entropy, singular-value decomposition entropy, permutation entropy, base-scale entropy, and entropy based on the Lempel-Ziv algorithm, each offering different performance depending on the base data and the explanatory goal.

Entropy has also been addressed in the context of social systems as a measure of disorder and complexity often linked to the ability to predict a system's next state. Theoretical frameworks like Social Entropy Theory (SET) have been developed to quantify entropy as an indicator of social systems' internal states [46, 47].

In [48], entropy is used to approximate energy, which in many systems is natural to estimate. This approximation is particularly relevant in high-dimensional data, where the entropy of individual elements is considered. In latent variable models with high-dimensional data, the covariance of element-wise (microlevel) entropies can lead to significant variance in the system's macrolevel entropy. This

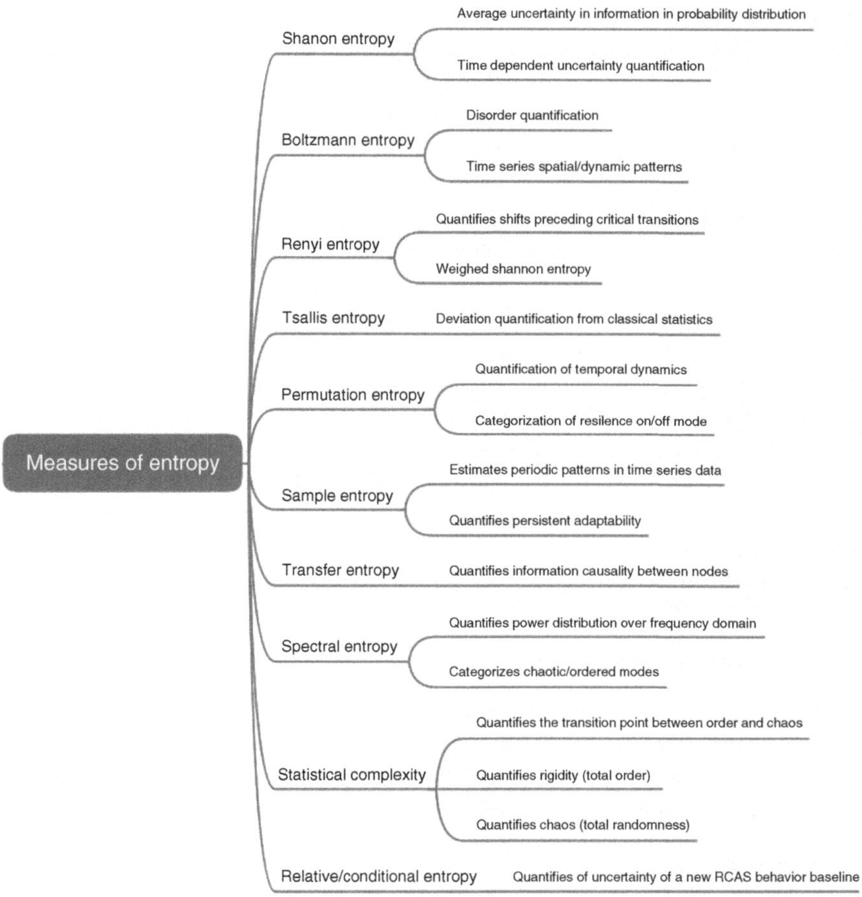

Figure 2.1 A taxonomy of entropy metrics for RCAS.

variance scales as $O(n^2)$, indicating a relationship between microlevel entropy changes and macro-level variance. In some cases, the entropy of a system can be expressed as a sum of terms, each corresponding to individual elements, making this approach helpful in measuring to what extent microlevel interactions contribute to macrolevel entropy.

Moreover, high-dimensional latent variables can induce Zipf's law (a signature of self-organizing dynamics in CAS, where the frequency of events—or components—follows a power-law relationship with rank), showing that covariation in individual entropies can lead to macro-level patterns.

This demonstrates a link between micro- and macroscales through entropy. In this context, Zipf's law arises when systems balance energy minimization and entropy maximization. This powerful concept links directly to the resilience capability of a CAS, when viewed as a balance between stability (through robustness via redundancy) and adaptability (through the capability to self-organize).

In the context of Power Grids, a key entropy-based metric is the entropic degree [23], which is used for quantifying the vulnerability of nodes (busbars). Grounded in entropy concepts from complex network theory, it captures the distribution and strength of connections within the grid. This measure enables the identification of critical components, supporting resilience and reliability analyses by highlighting the key elements for protection and maintenance. A higher entropic degree indicates greater importance or criticality of a busbar, as its failure would substantially reduce the grid's total line margin. Compared to other node centrality measures, the entropic degree demonstrates superior precision in identifying the most vulnerable nodes in large-scale networks.

It is worth noting that the effectiveness of entropy measures depends heavily on appropriate parameterization and the quality and availability of data; therefore, overinterpretation must be avoided, and employing multiple complementary entropy measures in parallel is usually recommended to achieve more robust and reliable insights into RCAS.

2.3.2 Measures of Performance

In Chapter 1, we discussed the Classical Systems Theory perspective on RCAS. SoSE provides us a foundation for developing MoPs based on Systems Theory. MoPs reflect a system's functional health. MoPs for RCAS are built upon the SoSE MoPs and are categorized into six broad categories:

- **Structural Performance**: Measures the state of structural integrity contributing to RCAS properties of redundancy, modularity, and scalability.
- **Dynamic Performance**: Measures the time-dependent and transient behavioral integrity for RCAS properties of adaptability, robustness, self-healing, and fault tolerance.
- **Emergent Behavior Performance**: Measures the novelty or time-dependent deviation from the expected holistic RCAS behavior for properties of adaptability, diversity, and self-organization.
- **Control and Autonomy Performance**: Measures the centralized and decentralized control effectiveness and stability measures reflecting the degree of autonomy (from manual, to semi-autonomous, to fully autonomous) for RCAS properties of decentralization, monitoring, and feedback.
- **Interoperability Performance**: Measuring the capacity to exchange information and means across the RCAS structure for RCAS properties of modularity, scalability, and diversity.

- **Reliability and Maintainability Performance**: Measures the failure rates and sustenance of RCAS behaviors for properties of self-healing, robustness, and fault tolerance.

Structural performance is measured through system's:

1. Degree of Connectivity: Measures the percentage of nodes and the relative strength of connection based on network traffic or data exchange between adjacent nodes to identify hubs and influencer nodes [30].
2. Degree of Cohesion and Modularity: Measures the percent reduction of RCAS system after structural removal that continues to retain the overall RCAS function. A cohesive and modular system gets reduced to a hub-and-spoke network graph that demonstrates resilience.
3. Degree of Scalability: Measures the percentage of increased nodes to an RCAS that strengthens the overall RCAS function.

Dynamic performance is measured through system's:

1. Degree of Sensitivity: Measures the deviation of system's output function as a function of inputs.
2. Degree of Reconfiguration/Self-organization: Measures the percentage of times the system maintains function through structural reconfiguration as new system functions are added/removed.
3. Degree of Robustness: Measures the degree to which RCAS maintains dynamic behavior as a function of degree of resilience.

Emergent behavior performance is measured through system's:

1. Degree of Successful Weak Emergent Behaviors: Measures successful weak emergent behavior as a percentage of planned weak emergent behaviors.
2. Degree of Detection of Strong Emergent Behaviors: Measures the percentage of emergent behaviors that deviated from planned emergent behaviors and subsequent identification and labeling of novel emergent behavior by a subject matter expert (SME). The newly labeled emergent behavior has the potential to impact overall RCAS function, but controls have not been established yet [36].
3. Degree of Resilience: Measures percentage RCAS weak emergent behaviors that resume their planned weak emergent behavior function in the presence of perturbation.
4. Degree of Evolution of RCAS Function: Measures the percentage of positive deviation from the planned weak emergent behavior that contributes to the overall RCAS function.

Control and autonomy performances are measured through system's:

1. Degree of Instrumentation: Measures the percentage of data instrumentation available for establishing control variables.
2. Degree of Autonomy: Measures the number of systems (nodes) according to their varying levels of autonomy. This metric utilizes a scaled autonomy framework with n-levels, starting from no autonomy (manual mode) to full autonomy.
3. Degree of Decentralization: Measures the centrality of the RCAS architecture based on the number of system nodes that issue directives for controlling RCAS constituent systems.
4. Degree of Indirect Influencers: Measures the modularity of the RCAS structure by a percentage of system nodes that are influenced by nodes outside their a single control structure. This also measures the associated feedback structures.

Interoperability performance is measured through system's:

1. Degree of Reliable Communication Networks: Measures the percentage of communication networks reliably connecting various system nodes.
2. Degree of Usage of Data Standards: Measures the percentage of communication links using data standards (e.g., MIL-STD, JSON, XML, etc.) Lack of use of data standards for syntactic interoperability makes RCAS a closed system with proprietary data interfaces, reducing the interoperability in general.
3. Degree of a Unified Data Taxonomy: Measures the percentage of inclusion of data model across various communication links into a unified data taxonomy.
4. Degree of Semantic Layering: Measures the inclusion of data taxonomy elements in a knowledge graph or an ontology that transforms data into synthesized knowledge.

Reliability and maintainability performance is measured through system's:

1. Degree of Availability: Measures the percentage of uptime of various constituent systems.
2. Degree of Maintainability: Measures the percentage of downtime of various constituent systems.
3. Degree of Fault Tolerance: Measures the percentage of redundant system nodes (with respect to their contributing function) before they transition from uptime to downtime. Figure 2.2 summarizes the metrics with associated MoPs.

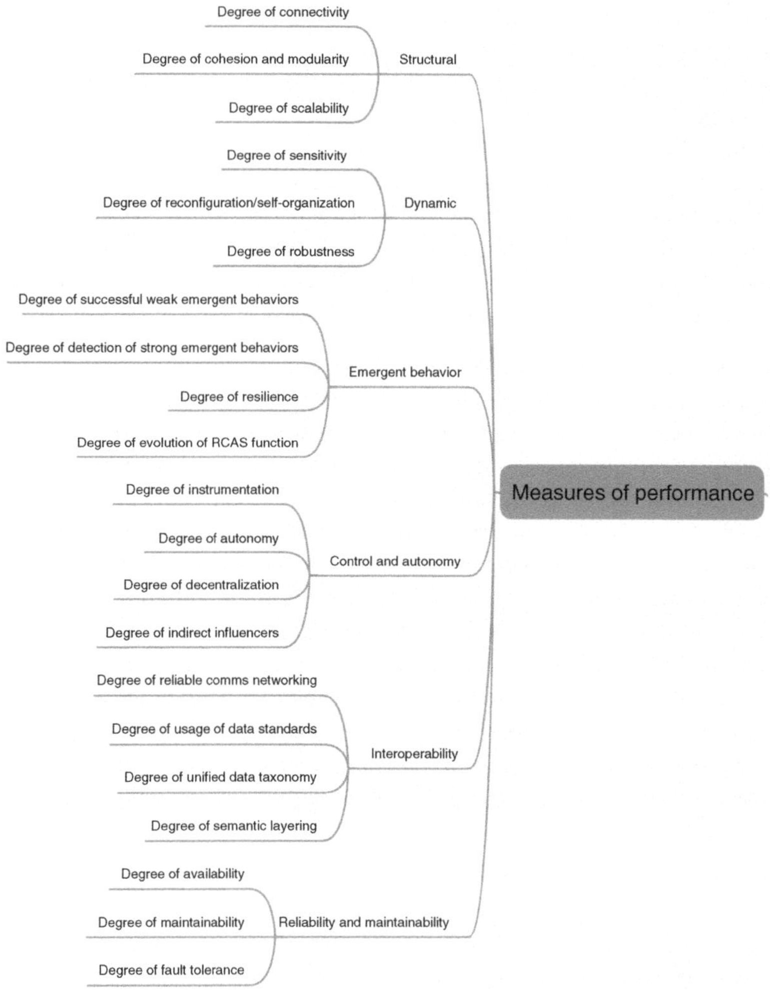

Figure 2.2 A taxonomy of measures of performance for RCAS.

2.3.3 Measures of Effectiveness

The MoEs are largely domain dependent, as the RCAS purpose is aligned to the context. However, some objective outcomes can be defined that are universal and can be applied to multiple domains. To evaluate the actual outcomes resulting from RCAS emergent behavior, the following are some of the MoEs:

- Survivability: Measures the deviation from planned RCAS behavior in the presence of varying degrees of perturbation.

- Restoration: Measures the time it takes to restore the RCAS behavior baseline for varying degrees of perturbation.
- Adaptability: Measures the deviation from planned system (node) function for the entire RCAS to restore the RCAS function.
- Robustness: Measures the success rate of demonstrating resilience over broad RCAS structure for varying degrees of perturbation.
- Scalability: Measures the evolution and increase of constituent system nodes to strengthen the RCAS function
- Economic Value: Measures the efficiency in RCAS behavior as a function of survivability, adaptability, robustness, and scalability.

Figure 2.3 depicts the metrics associated with MoEs. Demonstrating these by the ant-colony example described in Section 1.2, The ant colony must *survive* as a whole. There may be clusters formed when a new food source is found, but that clustering eventually leads to new trail formation, thereby, increasing the ant-traffic to the food source. If the trail gets broken due to external perturbation, the individual ants reestablish and *restore* the trail either through removing the physical obstacle or moving around the obstacle. The colony is highly adaptable when it comes to assignment of tasks and duties to specific ant-groups for foraging, transfer, storage, etc. New food sources are integrated into existing plans, and resources are allocated. The colony is resilient in terms of perturbation and establishes deep pheromone trails for ants to find their way back to the colony to deal with perturbations. This structural aspect becomes an inherent property of the ant-colony and results in the resilient emergent behavior of decentralized ant navigation through the process labeled as "stigmergy." Once an ant colony gets

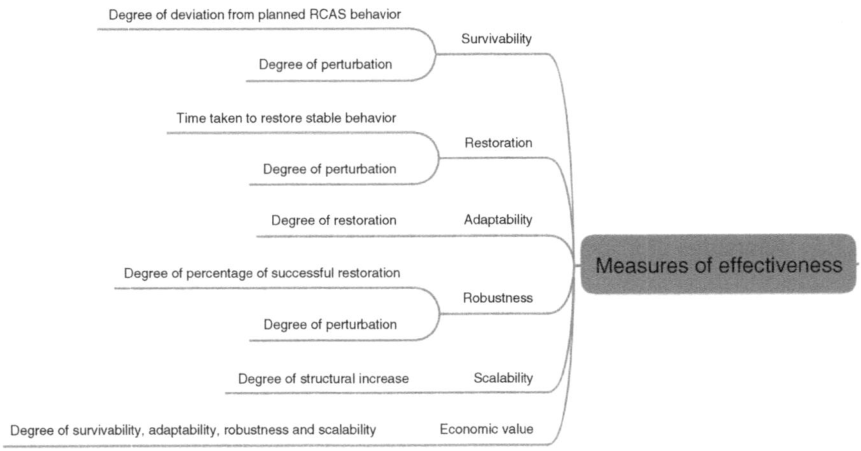

Figure 2.3 A taxonomy of measures of effectiveness for RCAS.

established, the stigmergic behavior is strengthened further via pheromones, leading to structural integrity and a perceived economic value of "food availability" for the entire colony.

2.4 Measures of Resilience

Measuring resilience in CAS presents a significant challenge, primarily attributed to their dynamic and context-dependent characteristics. Consequently, there is no single universally applicable metric. Instead, a suite of indicators is typically employed, each focusing on distinct aspects of system behavior. These need to be understood from the fundamental system's performance in the face of perturbation caused by either the environment or a negative emergent behavior that may have a potential to collapse the entire RCAS. A typical resilient system's effectiveness as function of its dynamical performance in the face of perturbation can be visualized as it performs over various phases such as preparedness, response to perturbation, recovery, and stabilization/transformation. Figure 2.4 describes a typical graph of system's effectiveness for four systems: S_A, S_B, S_C, and S_D.

S_A denotes a default system that is stable in the face of perturbation and negative emergent behaviors. S_B denotes a system that takes a performance hit but has enough restoration mechanisms to bring back the system to its original, planned effectiveness. This is a resilient system that has engineered resilience measures into its architecture. S_C denotes a system that is fragile and never recovers. It withers away. S_D is an anti-fragile system that capitalizes on the restoration

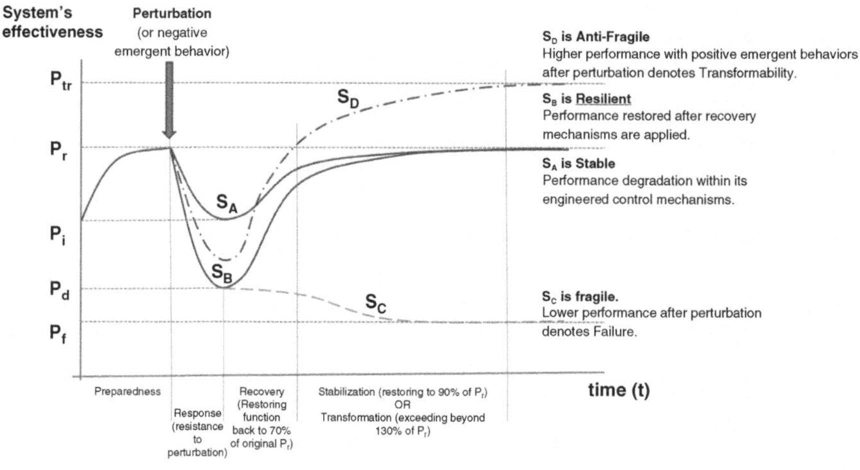

Figure 2.4 System's resilience behavior for notional values of 70% of performance recovery, 90% of original performance levels for stability, and 130% of original performance for transformation.

mechanisms and manifests positive emergent behaviors that strengthen the overall performance and increase the effectiveness. S_D gets transformed into a new system as a result of positive emergent behaviors contributing to the overall effectiveness.

The four phases identified in 2.4, viz., preparedness, response or resistance to perturbation, recovery (restoration to 70% of original performance), and stabilization and/or transformation, can each be associated with MoENs, MoPs, and MoEs as follows. The use of 70% is only notional here.

Resistance to perturbation and timely response can be measured through:

- Duration of Stress on RCAS Structural MoPs: The structural MoPs defined in Section 2.3.2 can be utilized for their dynamic behavior. These need to be monitored for time-dependent effects.
- Duration of Stress on RCAS Emergent Behavior MoPs: The emergent behavior MoPs defined in Section 2.3.2 need to be understood from negative effects perspective and need to be monitored for time-dependent analytics.
- Degree of Percent Failure Under Stress: This measures the percentage degree in RCAS effectiveness as a function of perturbation impact.

The restoration or recovery phase can be measured through:

- Time Taken to Restore Relevant MoPs: All the MoPs defined in Section 2.3.2 may not be applicable to all the RCAS. Consequently, applicable MoP operating ranges need to be identified and, through the recovery mechanisms, restored to original function.
- Time Taken to Restore All MoEs: While MoPs are contextual, the MoEs are context-independent (Section 2.3.3). Accordingly, this metric relates to tracking time restoring them.
- Time Taken to Restore Appropriate MoENs: Like MoPs, the MoEns are context dependent (Section 2.3.1). This metric tracks the restoration of appropriate MoENs.

The stabilization phase can be measured through:

- Time taken to reduce MoEs.
- Time taken to reduce stressors impacting MoPs.
- Time taken to achieve restoration through the restoration measures as indicated above.

The transformability phase can be measured through:

- Time taken to mitigate negative emergent behavior MoPs.
- Time taken to establish control and autonomy MoPs.
- Time taken to establish economic value MoE.

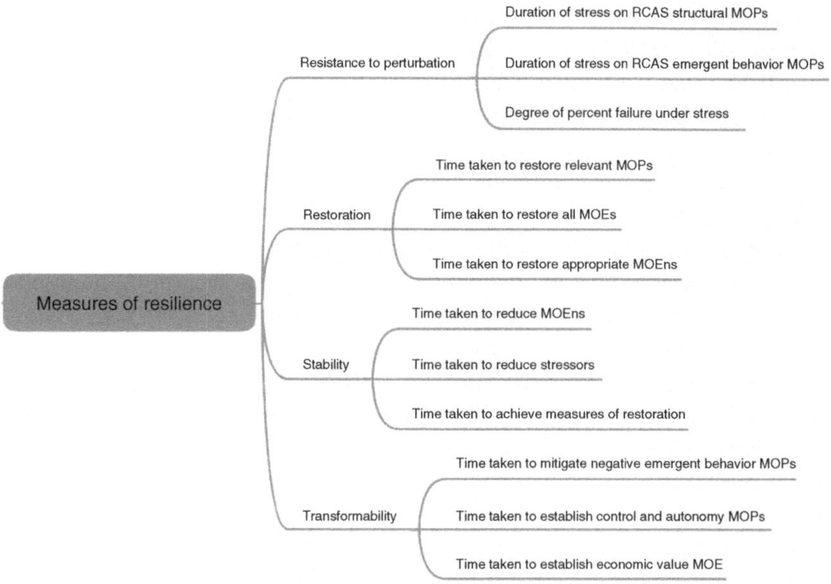

Figure 2.5 A taxonomy of resilience metrics derived from MoENs, MoPs, and MoEs for RCAS.

Figure 2.5 depicts various metrics associated with measures of resilience. Selecting appropriate metrics for resilience assessment depends heavily on the specific system and the type of resilience being evaluated. There's no one-size-fits-all solution; what works for one system might be irrelevant for another. Often, qualitative assessments and participatory approaches are just as valuable as quantitative data. By involving actors in identifying key vulnerabilities and capacities, we gain crucial insights that numbers alone might miss. This collaborative process of measurement can also serve as a powerful learning opportunity, helping everyone involved collectively understand the system's intricate dynamics. The measures of resilience discussed above are synthesized from MoENs, MoPs, and MoE described in Section 2.3. These synthesized measures are time-dependent measures that assure a timely RCAS and provide guidance to establish instrumentation and control of RCAS.

2.5 M&S-based RCAS Engineering

RCAS is a specialized class of CAS with added features/constraints of resilient behaviors. There are three broad classes of systems we must clearly articulate: (1) Class I: the RCAS as a system that exists in the real world; (2) Class II: the RCAS test bed built using the latest in modeling and simulation (M&S), analytics, and visualization technologies; and (3) Class III: the artificial RCAS, which we intend to engineer using the RCAS test bed.

Class I: The first class of RCAS inspires us to identify resilience in CAS and various interconnections between the perceived players/agents in a contextual environment. It gives us an observation medium to study natural CAS, where examples include financial markets, brains, ant colonies, population stability, ecological systems, power grids, and many others.

A deep understanding of the system's purpose and specific research questions is very important. This insight provides the necessary framework for guiding the entire data selection and processing pipeline, allowing us to effectively trace how microbehaviors translate into macropatterns and vice versa. Therefore, the first and most critical action is always to understand the overall SoS goals through Class I RCAS, expected emergent behaviors, and associated data elements. This may be broken down into part-SoS structure with associated emergent behaviors and data elements. While the associated data is used in a twofold manner—to inform model engineering and to validate the simulation model—the expected emergent behavior must be understood from a "synthetic" perspective. This is important because it is one thing to observe emergent behavior in the real world, and it's a different thing to reproduce it in a computational environment with a set of interacting agents. Mittal [35] defined synthetic emergence as:

> Emergent Behavior is created in artificial environments that include both computational and physical manifestations in controlled settings.

Class II: The natural RCAS is then brought into the M&S laboratory within a dedicated computational environment geared to reproduce the natural RCAS behaviors within a controlled setting in a computational way. This requires leveraging modeling, simulation, analytics, visualization, verification, validation, and engineering technology to create a test bed that can "accurately" reproduce the observed natural Class I RCAS behavior.

Mittal [35] also laid out principles of synthetic emergence with 3 primary goals, necessary pre- and postconditions, and a 10-step process to achieve synthetic emergence. The goals for synthetic emergence for RCAS in a computational environment remain the same:

1. There is no emergent behavior as a result of the simulation platform.
2. There is explicit communication protocol between the component systems.
3. The observed emergent behavior as useful (validated) or eliminated (pursuant to further investigation and resolution).

The preconditions are enumerated as:

1. There are at least two systems engaged in communication.
2. The systems are autonomous and maintain their individuality.

3. The systems need to be on the same time base, i.e., their clocks need to be synchronized if they are on a different time base.
4. The propagation delays in communication are accounted for during abstract time execution.
5. They share a common ontology or data dictionary.
6. The systems do not exhibit zenoness property, i.e., not advancing time or running an infinite loop calculation in zero virtual time with real-time consequences.
7. The systems are intended to be connected and share semantic context.

The postconditions, after the simulation is conducted and the observed computational emergent behavior deemed novel and consistent in the presence of various experimental conditions, are enumerated as:

1. The new semantic spatio-temporal signature can be recorded for a finite duration in a controlled manner.
2. The system components share a semantic context.
3. The system falls out of emergent behavior as context is switched.
4. The granularity of time advancement in hybrid systems is explicitly handled, i.e., the emergent behavior should be reproducible at higher granularity, but the system may fall out of emergence at coarse granularity.

Developing an M&S-enabled process for synthetic emergence for Class II RCAS requires understanding the emergent behavior from a layered perspective, i.e., decomposing an emergent behavior to its minimalist set of component systems and interactions first and then incrementally adding components to understand the effect and impact of the emergent behavior for resilience considerations. This requires us to decompose the top-level SoS to part-SoS and implement the process to achieve synthetic emergence. Adding this aspect resulted in a 15-step Class II RCAS engineering process:

- **Step 1**: Identify top-level SoS requirements and objectives
- **Step 2**: Identify part-requirements and part-objectives
- **Step 3**: Develop part-functional architecture
- **Step 4**: Develop part-metrics and data acquisition methods
- **Step 5**: Develop part-synthetic emergence preconditions with associated modeling paradigms
- **Step 6**: Develop design of experiments (DoE) configuration to conduct simulation experimentation
- **Step 7**: Conduct simulation experimentation on RCAS computational infrastructure to include: Hybrid models, multiscale data, multilevel analytics, and multidimensional visualization on an abstract time simulation infrastructure, supported by various persistent stores for models, data, tools, and software libraries

- **Step 8**: Verify postconditions for expected part-synthetic emergent behavior
- **Step 9**: Develop heuristics to refine DoE for limiting negative or undesired part-synthetic emergent behaviors
- **Step 10**: Develop control measures and resilience metrics/thresholds for part-synthetic emergent behavior
- **Step 11**: Develop data and communication protocols for critical spatio-temporal interactions causing part-synthetic emergent behavior
- **Step 12**: Upgrade part-functional architecture and consolidate with Step 3. You may repeat the cycle Step 3 -> Step 12 to develop confidence in instrumenting and controlling the part-synthetic emergent behavior
- **Step 13**: Validate the part-synthetic emergent behavior with available data from the real world
- **Step 14**: Develop epistemology and/or data/functional standards
- **Step 15**: Integrate and validate part-SoS synthetic emergent behaviors

The process is shown in Figure 2.6. It shows the top-level SoS decomposed into three parts, viz., Parts X, Y, and Z. Each part executes Steps 2–13, and then aggregated through Steps 14 and 15. Within each part cycle, Step 9 provides initial feedback on developing heuristics to improve the DoE focused on repeatable part-synthetic emergent behavior, and Step 12 provides part-System level improvements to include control and resilience measures into the part-functional architecture. These two feedback loops ensure the part-Functional architecture can generate part-synthetic emergence in a reliable manner verifiable through DoEs. Class II RCAS computational infrastructure for simulation experimentation in Step 7 needs further description. It is composed of six elements:

1. Hybrid models
2. Multiscale data
3. Multilevel analytics
4. Multidimensional visualization
5. Abstract time simulation infrastructure
6. Computational hardware

Hybrid Models: Models represent a system's components and their interactions in varying levels of fidelity. These models should capture the dynamic and often unpredictable nature of emergent behaviors. By simplifying real-world systems, models serve as controlled simulations, enabling experimentation and analysis in a structured environment. This is especially useful when it's too hard, expensive, or impossible to experiment directly with the real system. On one side, these systems exhibit profound intricacy, characterized by numerous constituent units engaging in complex, nonlinear interactions. Consequently, even minor perturbations can precipitate substantial and unpredictable systemic effects. Accurately encapsulating these nuanced interdependencies within a computational

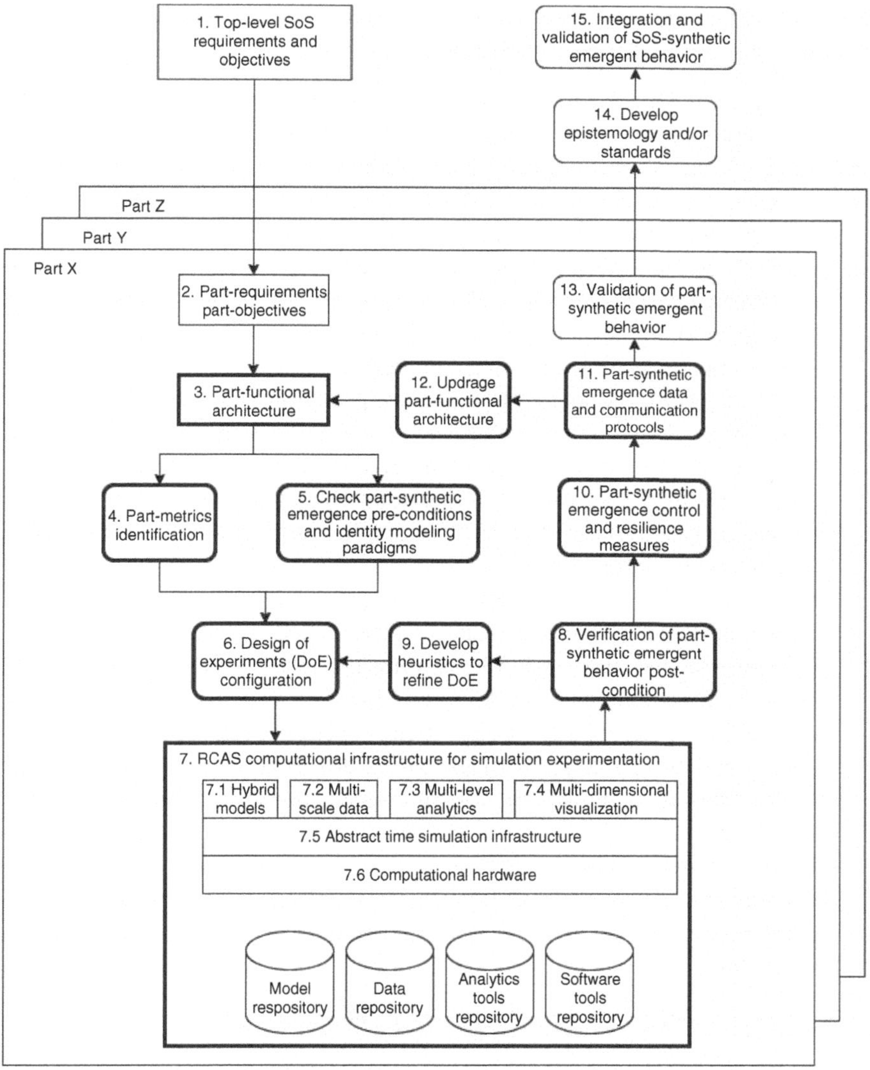

Figure 2.6 RCAS engineering process flow.

model presents a significant methodological hurdle. On the other hand, given this complexity, it's important to carefully consider the system's setup and validate assumptions about its behavior and components when building models and associated abstractions, avoiding oversimplification that could lead to misrepresentation. Analyzing resilience in CAS demands sophisticated analytical tools capable of capturing their dynamic, nonlinear, and interconnected nature. Different types of models are considered for complex systems:

- Agent-based Models (ABMs): These models involve the simulation of discrete, autonomous agents within a defined system boundary. These agents can represent diverse entities such as individuals, organizations, biological organisms, or cellular units. ABMs illustrate how these agents exhibit behaviors and engage in interactions governed by predefined rules or decision-making heuristics. Through the observation of these interactions, it becomes possible to detect how microlevel behaviors collectively generate macro-level patterns or unanticipated system-wide emergent phenomena. This approach is instrumental in understanding complex dynamics.
- System Dynamics Models: These models primarily emphasize the identification and analysis of causal feedback loops and the dynamics of resource or information flow within a system over temporal sequences. Their efficacy stems from their capacity to illustrate how endogenous actions within the system propagate, leading to either amplifying (reinforcing) or stabilizing (balancing) feedback effects.
- Network Models: These models are conceptualized as graphical representations that depict system interactions through nodes (representing individual components) and edges (delineating the relationships or connections between these components). Analyzing network structure (e.g., connectivity, centrality, modularity) helps identify critical nodes, potential failure points, and the pathways through which disturbances propagate. Metrics like network robustness to random or targeted attacks can inform resilience strategies.
- Hybrid Models: These models are an ensemble of discrete, continuous dynamical, and network models that act in a unified way through formal integration mechanisms, specified at the same time base across multiple abstraction levels.

Typical methodology for model reuse and evolution includes:

1. Developing mathematical models (Step 5) that represent the behavior of individual agents within the system based on observed patterns or theoretical principles from relevant disciplines such as physics, biology, economics, or social sciences. These ABMs can account for interactions between agents and their environment, as well as emergent behaviors arising from these interactions (e.g., flocking behavior in birds or fish).
2. Implementing stochastic simulations (Step 6) to explore the system's dynamics over different time scales and conditions while incorporating uncertainty due to random factors affecting agent behavior or environmental variables

3. Develop modular structures for the models (Step 7), allowing easy modification of individual agents or components without disrupting the overall architecture of the model (enabling scalability).
4. Use techniques like heuristics and AI/ML algorithms (Steps 8 and 9) to improve the accuracy of predictions and identify underlying patterns in the data that might not be immediately apparent through traditional analytical methods, however, reinforcing the expected emergent behaviors and mitigating the unexpected emergent behaviors.

Multi-scale Data: To effectively bridge the gap between individual actions and system-wide outcomes, data collection must be a methodical process designed to capture both micro-level and macro-level dynamics. This process begins with multi-modal and multi-scale data collection, where diverse data is acquired to quantify system attributes. At the micro-level, the data is collected on the behaviors and interactions of individual agents, while simultaneously data is gathered at the macro-level on external environmental factors and system-wide performance metrics. This comprehensive approach ensures to have the necessary information to connect individual actions to global patterns.

Then, a strategic sampling is required. To capture the fast-paced nature of micro-interactions, we need to collect high-frequency data, whereas observing long-term macro-level trends requires sampling over extended periods. This step demands a deliberate approach to temporal scales and an awareness of potential biases and data quality limitations.

Finally, robust data preparation is important for making multi-scale data useful. This involves cleaning the data by filtering out noise and handling missing information to accurately represent individual agent behaviors. The prepared micro-level data can then be properly aggregated to analyze macrolevel patterns, ensuring the generated insights are both accurate at the agent level and meaningful at the system level.

Multi-level Analytics: Once the appropriate model is selected, we can then employ various analytical techniques to dissect the system's complexity, uncover patterns, identify relationships, and gain insights into why certain behaviors emerge across different parts. Each technique offers unique strengths, making it suitable for different data types and system dynamics. The key analysis techniques include:

1. Statistical analysis methods are fundamental for identifying patterns and relationships between variables. Regression analysis, for instance, models how changes in one system aspect affect an outcome. Clustering groups similar data points to identify subgroups. Principal Component Analysis (PCA)

simplifies complex, high-dimensional datasets by finding the main factors that explain most of the data's variance, highlighting significant trends without being overwhelmed by complexity.

2. Employ time-series analysis techniques for understanding the temporal dynamics of CASs by analyzing trends over time and identifying periods of stability or instability within the system (e.g., autocorrelation functions, Fourier transforms).
3. Implement clustering algorithms to group agents based on similarities in their behavior patterns or performance metrics, enabling a better understanding of emergent behaviors arising from interactions between different types of agents within the system (e.g., K-means clustering, hierarchical clustering).
4. Network analysis techniques focus on interconnected systems, examining the connectivity and structure among agents within a CAS. This involves using centrality measures and community detection algorithms to understand how information flows through the network and to identify potential vulnerabilities or redundancies in its architecture.
5. ML techniques, encompassing both supervised and unsupervised learning, offer powerful ways to analyze complex systems, often without requiring predefined labels or explicit models.
6. Optimization techniques, like gradient descent or evolutionary algorithms, to optimize agent behavior patterns or overall system performance under specific constraints (e.g., minimizing energy consumption while maintaining a desired level of service).

While powerful tools like statistical analysis, network analysis, and ML are invaluable, they grapple with challenges such as high-dimensional data and inherent uncertainty. Overcoming these issues demands a careful selection of methods, appropriate modeling assumptions, and a deep understanding of the system being studied.

Multi-dimensional Visualization: Finally, effective visualization is crucial, playing a critical role both during model simulation and in the subsequent communication of research findings. A well-designed, multidimensional, and accurate visualization of data frequently facilitates an initial comprehension of the problem's inherent complexity, thereby guiding the selection of the appropriate modeling and analysis techniques. By turning raw data into visual formats across different semantic dimensions, we can identify patterns, relationships, and trends that would otherwise be hard to discern. Consequently, presenting complex data and model outputs with enhanced clarity significantly improves the understanding of a system's underlying dynamics and the broader implications of its emergent behaviors, empowering both researchers and decision-makers to identify important areas and formulate better choices.

Different visualization techniques are available, each suited to specific system types and offering unique insights:

1. Interactive visualizations allow easy exploration of the CAS and its underlying dynamics, enabling users to gain insights into the emergent behaviors arising from agent interactions or environmental conditions.
2. Animations or simulations of the modeled system to showcase key patterns in behavior over time or under different scenarios (e.g., sensitivity analysis) inform fidelity requirements for temporal analysis.
3. Time-series plots and dashboards are fundamental for observing system behavior over time and integrating various metrics. Time-series plots show how one or more variables change over a continuous period. They're essential for identifying trends, seasonality, and sudden changes in system performance or agent behavior. The interactive dashboards integrate multiple visualizations into a single interface, providing a comprehensive overview of key performance indicators and system states. Dashboards often allow users to filter data, drill down into specifics, and track real-time changes, making it easy to monitor system health and pinpoint areas for intervention or improvement.
4. Agent trajectories and spatial visualizations are particularly useful for ABMs, which help track and understand agent behavior in a spatial context. Agent trajectories animate the movement paths of individual agents over time within a simulated environment. One can often see their positions, velocities, and how they interact dynamically. Spatial heatmaps/density plots show the concentration or density of agents in different areas of the environment, revealing emergent spatial patterns like clustering or dispersion. 3D visualizations can depict agent locations and connectivity patterns, offering a more immersive and intuitive understanding of complex relationships (e.g., Netlogo; Vensim, Stella, Anylogic for System Dynamics; Gephi, Cytoscape for Networks, etc.).
5. Network graphs are powerful tools for visualizing interactions within connected systems. These graphs make it easy to identify key players, spot isolated parts, and understand overall connection patterns. More advanced versions of network graphs can even show the strength of interactions (using weighted edges) or the direction of influence (with directional edges). This added detail makes it much simpler to analyze how specific connections affect a system's overall dynamics.
6. System dynamics diagrams (causal loop diagrams and stock-and-flow diagrams) designed for system dynamics models, which focus on feedback loops and flows over time.

Abstract Time Simulation Infrastructure: Simulation tools provide a computational means to execute a model over time. Many simulation tools are available that run models faster than real-time (wall-clock time), at real-time, and slower than real-time. The simulation architecture should be engineered in such a manner that it allows tuning the relationship factor between the simulation clock and the real-time clock. Faster than real-time simulation environment fosters simulation experimentation and massively parallel concurrent executions for Monte Carlo analysis, etc. Real-time simulation environment fosters interactive simulations that include human-in-the-loop or Test and Training use cases. Slower than real-time simulation environment is less of a feature but more of a necessity where the model fidelity is so high that the sheer computational load of doing calculations takes longer than the simulation step advance time; for example, a simulation time step of 0.1 seconds may need 10 seconds of computational real-time to complete the calculation needed to advance 0.1 seconds. So, the simulation is running 10 times slower than real-time.

Hybrid model execution using formal DEVS allows integration of discrete and continuous systems in an abstract time manner [49].

Computational Hardware Infrastructure: A simulation infrastructure is an SoS itself. This virtual test bed for hybrid models, software tools, their dependencies, and execution platforms such as Windows and Linux is necessary to support iterative, continuous testing and iterative development. The test bed may be extended to include emulators and live-data interfaces and may need to be deployed in target cloud environments. This warrants the development of a virtual environment that provides the ground truth in a simulated scenario, interfaces with real-world data, and uses various domain-specific and domain-agnostic simulation systems for development, testing, and evaluation. These required features present a nontrivial challenge wherein constructive models and systems at different levels of fidelity need to interoperate to advance the testing, evaluation, and integration of CAS. Such a virtual and constructive SoS architecture should be independent of the underlying platform infrastructure. Mittal et al. [50] showcased a modular Simulation, Experimentation, Analytics, and Test (SEAT) layered Architecture Framework and a 10-step methodology to construct a service-oriented architecture that employs service facades over various software components. The SEAT approach describes the organization of the M&S infrastructure in various layers:

- Hardware Infrastructure as a Service (IaaS)
- Cloud Platform as a Service (PaaS)
- Simulation Infrastructure as a Service (SIMaaS)
- Software Utilities as a Service (SaaS)
- Modeling Services (MaaS)

- Experimentation and Gaming Services (EXaaS)
- Visualization Services (VISaaS)
- Data Repository Layer
- User Application Layer
- Security Layer

Class III: The third RCAS is an artificial RCAS that is an engineered one, which replicates some of the characteristics of natural CAS and is reproducible in a Class II RCAS test bed. This version of RCAS is a live, virtual, and constructive (LVC) system [35], an open system that leverages the virtual constructive Class II RCAS test bed to inform, influence, and control the live systems, which are active participants in the RCAS. Such RCAS incorporates Digital Twins into the RCAS architecture and integrates a robust computational infrastructure of Class II RCAS test bed for real-time experimentation and heuristics evaluation. An engineering process for Class III RCAS is left for future work.

2.6 Summary

Emergent behavior taxonomy consists of simple, weak, strong, and spooky emergent behaviors. Of these four, the simple and weak ones are well understood and reproducible in a computational environment [36]. The strong emergent behavior needs a verifiable M&S environment to produce credible synthetic emergent behavior in a computational environment [35]. There exist many challenges in data collection, modeling paradigms, analytics, and visualization techniques that limit our understanding of the CAS phenomenon in a synthetic environment. Engineering of CAS, rightly so, is a grand challenge, let alone RCAS.

Measurement, instrumentation, and control are integral to any engineering process. What cannot be measured and instrumented ultimately cannot be controlled. With CAS, the issue is of identifying the "what," especially in strong emergence. Consequently, the CAS engineering process must include identification of the novel emergent behavior from the computational test bed. This must be verifiable and repeatable. This chapter proposed a set of metrics in three broad categories: MoENs, MoPs, and MoEs, which align CAS with SoS engineering practices. We further synthesize Measures of Resilience from these three categories to arrive at measures that may be able to quantify resilience in CAS. These measures are currently a proposal and need to be tested and implemented in an example, something left for the future work.

Building on the case for synthetic emergence, this chapter further defined a 15-step engineering process that incorporates metrics and proposed two feedback loops: (1) to refine the design of experiments, strengthening repeatability and verifiability; and (2) to label the novel emergent behavior and develop resilience measures and data and communication protocols, strengthening the functional architecture of the SoS, thereby addressing novelty and making it repeatable and controllable through the functional architecture refinement. We also provided an overview of challenges and methodologies for hybrid modeling, multi-scale data collection, multi-level analysis, and multi-dimensional visualization, and how the tools, techniques, and procedures need to be brought together in a service-oriented manner through an abstract time simulation infrastructure.

Indeed, RCAS engineering through a virtual test bed is a hard problem, but that's the only option we have. We are never going to find a real-world CAS to experiment upon in a controllable manner.

Disclaimer: Dr. Mittal's affiliation with Modern Technology Solutions, Incorporated (MTSI) is provided for identification purposes only and is not intended to convey or imply concurrence with, or support for, the positions, opinions, or viewpoints expressed by this author.

References

1 Ted Carmichael and Mirsad Hadžikadić. *The Fundamentals of Complex Adaptive Systems*. Springer, 2019.

2 C. W. Johnson. What are emergent properties and how do they affect the engineering of complex systems? *Reliability Engineering and System Safety*, 12:1475–1481, 2006.

3 J. C. Mogul. Emergent (mis)behavior vs. complex software systems. In *Proceedings of the 1st ACM SIGOPS/EuroSys European Conference on Computer Systems*, pages 293–304, New York, USA, 2006.

4 James Odell. Agents and emergence. *Distributed Computing*, pages 51–53, 1998.

5 Jochen Fromm. On engineering and emergence. *arXiv preprint arXiv:nlin/0601002*, 2006.

6 J. Holland. *Emergence, From Chaos to Order*. Basic Books, 1999.

7 M. Bedau. Weak emergence. *Philosophical Perspectives*, 11:375–399, 1997.

8 O. T. Holland. Taxonomy for the modeling and simulation of emergent behavior systems. In *Proceedings of the 2007 Spring Simulation Multiconference*, pages 28–35, 2007.

9 Melanie Mitchell. *Complexity: A Guided Tour*. Oxford University Press, 2009.

10 Alon Hasgall. Digital social networks as complex adaptive systems. *Vine*, 43(1):78–95, 2013.

11 Marge Benham-Hutchins and Thomas R. Clancy. Social networks as embedded complex adaptive systems. *JONA: The Journal of Nursing Administration*, 40(9):352–356, 2010.

12 David P. Chassin, Joel M. Malard, Christian Posse, Asim Gangopadhyaya, Ning Lu, Srinivas Katipamula, and J. V. Mallow. Modeling power systems as complex adaptive systems. Technical report, Pacific Northwest National Lab. (PNNL), Richland, WA, 2004.

13 Herbert Gintis. The economy as a complex adaptive system, 2006.

14 Roberta De Angelis. Circular economy business models as resilient complex adaptive systems. *Business Strategy and the Environment*, 31(5):2245–2255, 2022.

15 C. Reynolds. Flocks, herds, and schools: A distributed behavioral model. In *Proceedings of ACM SIGGRAPH*, pages 25–34, 1987.

16 M. Gardner. *The Fantastic Combinations of John Conway's New Solitaire Games*. Mathematical Games, 1970.

17 Claudia Szabo, Joshua Groot, Thomas McAtee, and Christo Pyromallis. Who did this: Identifying causes of emergent behavior in complex systems. In *2019 IEEE Symposium Series on Computational Intelligence (SSCI)*, pages 262–270. IEEE, 2019.

18 Simranjeet Singh Dahia and Claudia Szabo. Detecting emergent behavior in complex systems: A machine learning approach. In *Proceedings of the 38th ACM SIGSIM Conference on Principles of Advanced Discrete Simulation*, pages 81–87, 2024.

19 Claudia Szabo and Yong Meng Teo. Post-mortem analysis of emergent behavior in complex simulation models. In *Proceedings of the 1st ACM SIGSIM Conference on Principles of Advanced Discrete Simulation*, pages 241–252, 2013.

20 Jure Demšar and Iztok Lebar Bajec. Simulated predator attacks on flocks: A comparison of tactics. *Artificial Life*, 20(3):343–359, 2014.

21 Ilias Lymperopoulos and George Lekakos. Analysis of social network dynamics with models from the theory of complex adaptive systems. In *Collaborative, Trusted and Privacy-Aware e/m-Services: 12th IFIP WG 6.11 Conference on e-Business, e-Services, and e-Society, I3E 2013*, Athens, Greece, (April 25-26, 2013), *Proceedings 12*, pages 124–140, Springer, 2013.

22 Albert-László Barabási. *Linked: How Everything Is Connected to Everything Else and What It Means for Business, Science, and Everyday Life*. Basic Books, 2014.

23 Ali Moradi Amani and Mahdi Jalili. Power grids as complex networks: Resilience and reliability analysis. *IEEE Access*, 9:119010–119031, 2021.

24 En Lu, Ning Wang, Zhijun Qin, Haoming Liu, and Yunhe Hou. Black-start strategy for power grids including fast cut thermal power units. In *2013 IEEE Power & Energy Society General Meeting*, pages 1–5. IEEE, 2013.

25 Ross Arnold, Jonathan Jablonski, Benjamin Abruzzo, and Elizabeth Mezzacappa. Heterogeneous UAV multi-role swarming behaviors for search and rescue. In *2020 IEEE Conference on Cognitive and Computational Aspects of Situation Management (CogSIMA)*, pages 122–128. IEEE, 2020.

26 Dariusz Marek, Marcin Paszkuta, Jakub Szyguła, Piotr Biernacki, Adam Domański, Marta Szczygieł, Marcel Król, and Konrad Wojciechowski. Swarm of drones in a simulation environment—efficiency and adaptation. *Applied Sciences*, 14(9):3703, 2024.

27 Bradley Fraser, Andrew Coyle, Robert Hunjet, and Claudia Szabo. An analytic latency model for a next-hop data-ferrying swarm on random geometric graphs. *IEEE Access*, 8:48929–48942, 2020.

28 Bradley Fraser, Claudia Szabo, Andrew Coyle, and Robert Hunjet. Leveraging emergent specialization in a heterogeneous multi-role swarm control architecture for positional-based UAS missions. In *2020 IEEE Symposium Series on Computational Intelligence (SSCI)*, pages 2584–2591. IEEE, 2020.

29 Saurabh Mittal. Model engineering for cyber complex adaptive systems. In *Proceedings of the European Modeling and Simulation Symposium*, 2014.

30 Saurabh Mittal. Emergence in stigmergic and complex adaptive systems: A formal discrete event systems perspective. *Cognitive Systems Research*, 21:22–39, 2013. https://doi.org/10.1016/j.cogsys.2012.06.003.

31 Ezequiel Pecker-Marcosig, Sebastian Zudaire, Martin Garrett, Sebastian Uchitel, and Rodrigo Castro. Unified DEVS-based platform for modeling and simulation of hybrid control systems. In *Proceedings - Winter Simulation Conference*, 1051–1062, Institute of Electrical and Electronics Engineers Inc, 2020. https://doi.org/10.1109/WSC48552.2020.9384025.

32 Saurabh Mittal and José L. Risco-Martín. *Simulation-Based Complex Adaptive Systems*, pages 127–150. Springer International Publishing, Cham, 2017. https:doi.org/10.1007/978-3-319-61264-5_6.

33 Saurabh Mittal and Andreas Tolk. *Complexity Challenges in Cyber Physical Systems: Using Modeling and Simulation (M&S) to Support Intelligence, Adaptation and Autonomy*. John Wiley & Sons, 2019.

34 Ezequiel Pecker-Marcosig, Juan I. Giribet, and Rodrigo Castro. DEVS-over-ROS (DoveR): A framework for simulation-driven embedded control of robotic systems based on model continuity. In *Proceedings - Winter Simulation Conference*, pages 1250–1261, Institute of Electrical and Electronics Engineers Inc, 2018. https://doi.org/10.1109/WSC.2018.8632504.

35 Saurabh Mittal. *New Frontiers in Modeling and Simulation in Complex Systems Engineering: The Case of Synthetic Emergence*, pages 173–194. Springer International Publishing, Cham, 2019.https:doi.org/10.1007/978-3-030-17164-3_9.

36 Saurabh Mittal and Larry Rainey. Harnessing emergence: The control and design of emergent behavior in system of systems engineering. In *Proceedings of the Conference on Summer Computer Simulation*, pages 1–10, 2015.

37 Mikhail Prokopenko, Fabio Boschetti, and Alex J. Ryan. An information-theoretic primer on complexity, self-organization, and emergence. *Complexity*, 15(1):11–28, 2009.

38 Carlos Gershenson and Nelson Fernández. Complexity and information: Measuring emergence, self-organization, and homeostasis at multiple scales. *Complexity*, 18(2):29–44, 2012.

39 Gloria Pumpuni-Lenss, Timothy Blackburn, and Andreas Garstenauer. Resilience in complex systems: An agent-based approach. *Systems Engineering*, 20(2):158–172, 2017.

40 Sree Ram Boyapati and Claudia Szabo. Self-adaptation in microservice architectures: A case study. In *2022 26th International Conference on Engineering of Complex Computer Systems (ICECCS)*, pages 42–51. IEEE, 2022.

41 Hussain Ahmad, Christoph Treude, Markus Wagner, and Claudia Szabo. Towards resource-efficient reactive and proactive auto-scaling for microservice architectures. *Journal of Systems and Software*, 225:112390, 2025.

42 Noel Sproles. Formulating measures of effectiveness. *Systems Engineering*, 5(4):253–263, 2002.

43 G. Eoyang and T. Berkas. *Evaluating performance in a CAS*. Human Systems Dynamics Institute, Circle Pines, MN, 1998.

44 Somayeh Kalantari, Eslam Nazemi, and Behrooz Masoumi. Entropy-based goal-oriented emergence management in self-organizing systems through feedback control loop: A case study in NASA ANTS mission. *Reliability Engineering & System Safety*, 210:107506, 2021.

45 Kazimieras Pukėnas, Jonas Poderys, and Remigijus Gulbinas. Measuring the complexity of a physiological time series: A review. *Baltic Journal of Sport and Health Sciences*, 1(84), 2012.

46 Thomas Mavrofides, Achilleas Kameas, Dimitris Papageorgiou, and Antonios Los. On the entropy of social systems: A revision of the concepts of entropy and energy in the social context. *Systems Research and Behavioral Science*, 28(4):353–368, 2011.

47 Kenneth D. Bailey. Social entropy theory: An overview. *Systems Practice*, 3:365–382, 1990.

48 Laurence Aitchison, Nicola Corradi, and Peter E. Latham. Zipf's law arises naturally when there are underlying, unobserved variables. *PLoS Computational Biology*, 12(12):e1005110, 2016.

49 Bernard P. Zeigler, Alexandre Muzy, and Ernesto Kofman. *Theory of Modeling and Simulation: Discrete Event & Iterative System Computational Foundations*. Academic Press, 2018.

50 Saurabh Mittal, Robert L. Wittman, John Gibson, Josh Huffman, and Hans Miller. Providing a user extensible service-enabled multi-fidelity hybrid cloud-deployable SoS test and evaluation (T&E) infrastructure: Application of modeling and simulation (M&S) as a service (MSaaS). *Information*, 14(10), 2023. https://doi.org/10.3390/info14100528.

3

Epistemology of Resilient Complex Adaptive System Simulations

Andreas Tolk

The MITRE Corporation, Charlottesville, VA, USA

Corresponding author: atolk@mitre.org

3.1 What Can We Learn from Resilient Complex Adaptive System Simulations? A Prologue

Our focus in this chapter is epistemic: how can we use simulations of resilient complex adaptive systems to gain understanding and increase our knowledge? What is the nature of such gained knowledge, and where are its limits?

This is not the first time I have personally addressed this central question of philosophy of science applied to simulation. I had the pleasure of participating in the first American incarnation of the Epistemological Perspectives on Simulation (EPOS) conference series in 2012, a series that started in Europe in 2004, driven by the growing interest to use simulation, particularly in social sciences [1].

The simulation community I was engaged with at that time was shaped by the need for training and representing high-fidelity physics-based models to create a realistic training environment. As the laws and processes of physics were well understood, the role of simulation was predominantly to simulate real physical objects, and empirical data was used to support their validation. The challenges of validation were still significant, and the work of Oberkampf and Roy [2] as well as Winsberg [3] shows how numerical approximation, heuristics, and other computational practices to improve simulation efficiency introduce errors and uncertainties that are artificial and not characteristic of the simulated system. Thus, the mechanical engineering community emphasizes the need to quantify uncertainty in their efforts [4].

In the operations research community, uncertainty quantification is less emphasized in the foundational contributions, such as [5–7]. Their work focuses more on the procedures and methods, as well as practical limits, and how these can constrain the validity of solutions. However, while computer science and simulation started to think about the applicability of simulation to gain insight into social challenges, exemplified by Epstein and Axtell [8], the community of computational social sciences was still in its early stages then, and the use of simulation to express social theories had only just been broached [9].

Computational social science has a scientific focus on understanding complex socio-technical systems. Its goal is to uncover the underlying mechanisms of social systems and provide insights that can inform policymaking, business strategies, and the design of technology and infrastructure. The application of computer science, data science, and social sciences to analyze and model complex social systems, often using interdisciplinary teams, is of deep interest to the question of this chapter.

As the need grew to add human aspects to the training environment and the analytic toolbox [10], the broader simulation community became interested in extending beyond the physical realm to address socio-technical systems. Computational social science and behavioral representations had to be addressed. In parallel, ethics and philosophy of simulation established their place in the Body of Knowledge [11], and, with it, the quest to understand epistemology better.

In social science, the Johari window is often used to explore questions of epistemology [12], see Figure 3.1. This model is divided into four sections that address understanding and awareness. The first section, the "known knowns," includes facts we are aware of and understand, which we use in decision-making. The second section, the "unknown knowns," consists of hidden facts or untapped knowledge that we have not yet utilized. The third section, the "known unknowns," represents areas where we recognize that solutions exist, but the facts are not yet identified. The final section, of "unknown unknowns," describes complete ignorance, where neither knowledge nor awareness of that knowledge exists.

There is a deep connection between the Johari window and the epistemological contribution of simulation: simulation is deeply rooted in the known knowns. Only in this quarter of the Johari window is validation possible, as we always need to have a valid model at hand to start from. This valid model may be the laws of physics, operations research insights, or perhaps established social theories.

We can then use simulation to tap into the area of the unknown knowns by deriving what we should know from applying the simulation to what we know. We can also assume possible facts that we don't know about, but that may be derivable from simulation results. The unknown unknowns, however, are hardly reachable. Nonetheless, simulations can always help to test hypotheses in all unknown areas and help to gain new understanding.

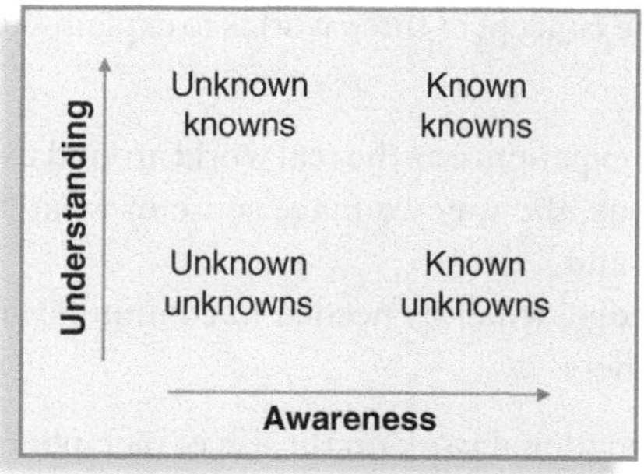

Figure 3.1 Application of the Johari Window.

What simulations cannot do is create new knowledge. Simulations are transformative by their very nature: they use their algorithms to transform input data into output. These outputs may spawn new ideas or insights in the user, but most likely this results in uncovering an unknown known.

In the following sections, we will look in more detail at these important concepts that address the question of what and how we can learn something from Resilient Complex Adaptive System Simulations and how we apply this knowledge. While books have been written about each individual topic, I hope to provide an overview of important ideas and research in all these relevant domains.

To show that the discussion about the epistemology of resilient complex adaptive system simulations has far-reaching consequences for practical applications, we use the example of how to facilitate the design of resilient urban transportation systems with such simulations, focusing on a practical problem known to all of us. At the end of each section, a short paragraph addresses the practical application of the theoretical concepts discussed.

3.2 The Nature of Resilient Complex Adaptive System Simulations

3.2.1 The Nature of Modeling

In his essays on life itself, Rosen [13] states: "I have been, and remain, entirely committed to the idea that modeling is the essence of science and the habitat of all epistemology." It is therefore fitting to start the quest for an understanding of the epistemology of resilient complex adaptive system simulations with a question about the nature of modeling.

As Rosen's quote illustrates, modeling is anything but unique to simulation. Humans are used to using models to understand and explain things. The philosophy of science has long dealt with modeling and the use of models to capture and gain knowledge. Popper [14] introduced the concept of three worlds to explain the logic of scientific discovery:

- the physical world of objects, what we experience as the real world around us;
- the mental world of conceptualizations, the way we make sense of what is going on around us in the real world; and
- the formal world of objective knowledge, which is needed to communicate our sensemaking in usable form to others.

Another example is Vygotsky's [15] foundational work on the use of metaphors in scientific concept formation, very similar to the ideas discussed by Popper. In summary, there is a strong foundation for the use of models in the philosophical community that goes further back than the foundations for computer simulations.

When we address the real world, we use modeling to develop conceptualizations to make arguments. To gain insights, we need to build a formal model to capture and code our knowledge with others. Models are the cognitive part of sensemaking, the result of our conceptualization and formalization of our observations, or even of our ideas of what could be.

Models have a long history in the scientific community as mediators, contributors, and enablers of scientific knowledge, as observed by Gelfert [16] (p. 127):

> Whereas the heterogeneity of models in science and the diversity of their uses and functions are nowadays widely acknowledged, what has perhaps been overlooked is that not only do models come in various forms and shapes and may be used for all sorts of purposes, but they also give unity to this diversity by mediating not just between theory and data, but also between the different kinds of relations into which we enter with the world. Models, then, are not simply neutral tools that we use at will to represent aspects of the world; they both constrain and enable our knowledge and experience of the world around us: models are mediators, contributors, and enablers of scientific knowledge, all at the same time.

Such models are conceptualizations. They are born in a process of creativity out of observations, experience, and data, but also out of ideas. These conceptualizations are focused on the object of interest, and this interest is usually shaped by a question or a task to be supported. Things not important for this question or task are eliminated, and an abstraction level is chosen that is best suited to the answer or support sought. As a result, the model becomes a task-driven simplification and abstraction of the object of interest. Like in a scientific experiment, where isolation of the object is used to eliminate unwanted side effects that would only blur the insights that can be gained, the simplification of a model focuses on the capabilities needed to accomplish the task. The hypothesis to be tested and the metrics used to collect results are defined by the question or task at hand.

This is true for models used to implement simulations as well. In this chapter, the focus lies on digital computer simulations that, in the ideal world, are derived verifiably from the model specification. One of the most rigorous forms to define models is the use of ontologies. Interestingly, Gruber [17] originally defined the notion of an ontology as the "explicit specification of a conceptualization," which is a fitting description in our context.

Of particular importance is that the *model becomes the reality of the resulting simulation*. Only things that are part of the conceptualization will become elements in the simulation. There is no representation outside of the model. As discussed in the first part of this chapter, a modeler can imagine something unknown, but only what is imagined can make it into the conceptualization and become part of the simulation. This makes the model crucially important. The conceptualization is as important as the implementation, as discussed in detail by Tolk [18]; a machine-readable version, such as an ontological representation, becomes very important [19], particularly when artificial intelligence (AI) is used to enrich the process. The use of generative AI to capture the concepts of a simulation retrospectively is the subject of current research; some successful first applications are described by Akhavan and Jalali [20]. However, it should be noted that the initial development of conceptual models is a creative process and will remain the domain of human creativity for the foreseeable future. How AI in general—and generative AI in particular—can support the modeling process deserves its own treatment and goes beyond the scope of this chapter.

For our resilient urban transportation system example, the modeling process is very important, as it defines the simplifications and abstractions of the socio-technical urban transportation system and its components, attributes, relations, and processes. We must define the different transportation modes, such as buses, trains, private vehicles, bicycles, and pedestrians, and we have to conceptualize their interactions within the urban infrastructure. We must decide who our actors are, what their choices will be, what strategies they may follow, etc. Will they be cooperative in their choices? How much do they know about the choices of others? What are the options, and how well are they aware of them? Will they learn from their choices? All these are choices that will define what and how much we will be able to learn from our simulation to address the real-world problem. In general, selecting the task-appropriate simplifications and abstractions is a creative process that requires simulation engineers with a good understanding of the system of interest as well as the task to support and the mix of modeling paradigms that are most appropriate.

3.2.2 The Nature of Simulation

As mentioned in the last section, we are interested in digital computer simulations that implement underlying models. As computer simulations are computer programs, all the limits of computability apply. A computer program consists of algorithms implemented by computable functions that use a finite and discrete set of input data and produce a finite and discrete set of outputs over a finite amount of time.

These characteristics of computable functions lead to the distinction between tractable problems, intractable problems, and incomputable problems, as discussed in multiple textbooks, such as [21] or [22]. Tractable problems can be solved efficiently, intractable problems can be solved in theory but are too complex to be solved in a reasonable time with reasonable resources, and incomputable problems simply have no solution, such as Turing's Halting Problem [23]. Computer simulations are no exception to these rules. If a problem is incomputable, representing it as a simulation will not solve it either.

Another challenge is discretization. By discretizing our problem space, we are artificially introducing thresholds to continuous spaces. No matter how dense our digitalization mesh is, there will always be points that are not part of it. Although Wolfram [24] makes the argument that at extremely small scales space is made up of discrete units, for practical purposes, the resolution of our models appears far coarser than the resolution of discrete space or spacetime.

We will revisit nonlinearity in our section about the nature of complexity, but we can note here already that introducing discrete representations and the concomitant truncation of parameter values can significantly affect the solutions provided by the simulation, driving them away from the real systems. These and additional examples are given by Oberkampf et al. [25].

In addition to these general challenges of using discrete computer simulation, Winsberg [26] states that simulation is not a mere number-crunching technique, but rather involves a complex chain of inferences that transform conceptual structures into specific algorithms representing the systems of interest. They often must deviate from the conceptual model derived from theory to avoid the challenge of intractability. Heuristics and approximations are needed that add additional constraints to the validity of the simulation. As Winsberg states [3] (p. 9):

> Simulationists deal with the problem of computational intractability by deviating from the most principles model suggested by theory. This can work in one of two ways. They can develop a model that is simplified and hence computationally cheaper than the theoretically principled one. ... But in many cases a fine enough computational approximation is vastly out of reach or may not even exist in principle. In such a case they can supplement the model with features that have nothing to do with theory at all but are

designed to compensate for the errors that the coarseness of the approximation is found to create. Either way, depending on what aspect of the solution the simulationist is interested in resolving, it is often advantageous to trade any theoretical rigor for expediency. The deciding factor is not which approach is most true to theory, but which approach will produce the best solution set as the outcome of the simulation. The best solution set is the one that will best uncover or reveal the features of the system that are important for understanding it.

Computer simulations provide dependable results when we can validate them against real-world observations or compare them with other proven modeling approaches. This verification process ensures that our simulations accurately represent the theories they're based on. However, when we push these simulations beyond the realm of observable phenomena, we encounter a fundamental shift: we're no longer testing the original theory, but rather the simplified computational version we've created to make the simulation possible.

This distinction doesn't invalidate simulation results that extend beyond observable limits. These results can still offer valuable insights and predictions about potential outcomes. However, we must be transparent about an important limitation: such results cannot be empirically validated against real-world data. This means we should treat these findings with appropriate caution and clearly acknowledge their speculative nature.

While Winsberg focuses on scientific applications, similar observations are true in many engineering-driven applications, such as healthcare, traffic, or defense applications. Although the elicitation of knowledge from domain experts is part of the conceptualization process, those building the simulation are often not domain experts but nonetheless have to fill in gaps and assume certain relations and transformations that make sense for the practicalities of application despite not being rooted in generally accepted theory or doctrine. In other words, in the lack of existence, the lack of access, or in ignorance of underlying theoretical principles, the simulationist implements a solution that represents the ideas of the implementer, not of the theory or doctrine. Winsberg [3] provides several examples in his book. It is good practice to capture such decisions in the conceptual model as well, as the as-is of the implementation is more important for conceptual alignment than the theoretical to-be of the original concept.

Epistemologically, that can lead to the problem of simulationist's regress [27]. In philosophy, regress is a series of statements in which a procedure is continually reapplied to its own result without taking us closer to a useful conclusion. Regress often leads to concluding that an assumption is valid by assuming that the assumption is valid. Such circularity may look to be trivially avoidable but, following the observations above, a simulation can embed the non-theory-based assumptions of a simulationist, which the simulationist does driven by the best intentions to make the simulation run and to provide the results as expected by the specification. A user of this simulation who now uses the simulation to gain insight into the system may discover the bias of the implementation as a system feature, hence the term simulationist's regress: *there is a danger of proving something with the simulation system that is actually an assumption encoded into the simulation system.*

As simulation systems grow to millions of lines of code or hundreds of thousands of agents interacting with each other, we need methods to establish trustworthiness in the simulation results [28]. At the least, whenever we endeavor to learn something about the nature of resilient complex adaptive systems by simulating them, we must understand that the validity of the simulation is constrained by the observations made in this section.

For our example of the resilient urban transportation system, it is of interest that there are many examples of traffic simulations in the transportation literature, such as summarized in [29] or [30], including how to model behaviors, such as described by Suo et al. [31], and addressing computational challenges [32]. As we want to use our simulation to learn about the resilience of our urban transportation systems, besides the realistic behavior of our agents, the realistic representation of the situated environments is a topic of particular interest. As the simulation of complex adaptive systems requires the use of heuristics and numerical approximations due to their nonlinear and multimodal relations, the insights of this section are of particular interest.

3.2.3 The Nature of Systems

Our understanding of systems is constantly evolving. The International Council on Systems Engineering (INCOSE) provides some definitions for their members that are applicable for our challenge of understanding systems as well [33] (p. 3):

> A system is an arrangement of parts or elements that together exhibit behavior or meaning that the individual constituents do not. More specifically, an engineered system is a system designed or adapted to interact with an anticipated operational environment to achieve one or more intended purposes while complying with applicable constraints.

The INCOSE definitions include people, products, services, information, and processes as possible parts of the systems. When such a system is simulated, all these components, relations, and capabilities must be captured in the conceptualization and implemented in the simulation. The report also addresses conceptual systems and emphasizes the role that modeling plays for that category, stating that they "consider the structural ontology of physical and conceptual systems to be the same, at the fundamental level of parts and relationships—this is what allows us to model physical systems using conceptual systems" [33] (p. 8).

Clear boundaries separate the system from the environment that it interacts with. This clarity is beneficial when, during simulation design, the time comes to determine which environmental elements need to be included. For simulationist, the Experimental Frame, introduced by Zeigler in his work on the Discrete Event Simulation formalism [34], is a well-known representation of external conceptualization that interacts with the simulated system. This frame provides a formal way to specify the conditions under which a model is to be observed or experimented with, defines inputs and observed outputs, and enables systematic and repeatable simulation studies.

Buede and Miller [35] propose defining external systems that interact with the system of interest across these boundaries, exchanging inputs. Additional inputs can be drawn from the context, which, unlike external systems, does not receive inputs from the system. This distinction can be illustrated by flight simulators: the aircraft being simulated is "the" system of interest; other aircraft close enough to interact with it are considered external systems, while all other aircraft and environmental factors are part of the context. This example is used by Menke [36] to recommend a novel architecture for this family of simulators. As it will be revisited in the section on the nature of resilience, the environment and external systems play pivotal roles in understanding possible effects or even attacks that the system of interest may be exposed to, which motivates the need for resilience.

The ever-increasing use of information technology (IT) driven by societal requests for quick reaction to new requirements presents a challenge to our understanding of systems as well. Traditionally, system components offer well-defined capabilities through well-defined interfaces. However, with IT integration, updates that change capabilities and modernize interfaces can occur almost instantaneously. Just by downloading new software, the system's behavior can be significantly modified. This allows a system to adapt without the need for hardware changes or cumbersome manual maintenance processes. However, at the same time, it also necessitates keeping simulations of the system and its environment constantly up to date. The conceptual description of the system, essential for evaluating whether a system in its current configuration can collaborate meaningfully with other systems to form a system-of-systems, remains a crucial element in the pursuit of learning and understanding. As one example, the work of Damm and Vincentelli [37] describes how to use such conceptual models to plan collaborations of systems as a swarm.

In our example, the urban transportation system is already a system of systems, combining multiple transportation modes from pedestrians or people using private cars to mass-transportation options. This system of systems itself is embedded into an even bigger system of systems describing the city with all its local capabilities and constraints. Understanding what parts of the system and its context we are interested in is pivotal, and the simulation engineer must guide the decision makers using it to keep any recommendations within its validity constraints.

3.2.4 The Nature of Complexity

The term "complexity" is used in many disciplines, such as biology, social science, or mathematics. In this section, we are mainly interested in complexity in systems theory, where it refers to the interconnectivity and interdependence of components within a system.

As discussed in Tolk [38], system science has evolved over the years. Initially, it dealt with linear approximations of relationships between system components, which were predictable and easy to engineer. Even nonlinear relationships could be approximated linearly for short-term predictions. Systems were understood through reductionism: by understanding the components, the whole system could be grasped. As systems became more intricate, the nonlinearity of relationships was increasingly recognized to embody important information. Although these complicated systems could still be approached with reductionist principles, a holistic view became necessary. While behavior remained predictable, experts needed more advanced tools to manage these systems. Eventually, a new class of systems emerged: complex systems, which could not be understood through reductionism alone. Complexity science showed that new properties could emerge, making behavior generally unpredictable, though often explainable after the fact, using holistic methods. System thinking now differentiates between simple systems with consistent behavior, complicated systems with predictable behavior, and complex systems with unpredictable but explainable behavior.

It is worth evaluating emergence in complex systems a little bit more. The idea of emergence was introduced to a wider community by the philosopher [39]. He observed that, in natural complex systems, something new can emerge that cannot be associated with anything less than the complex system in its entirety. While Lewes was interested in life, mind, and consciousness emerging from biochemical processes, the idea was subsequently adapted to characterize complex system behavior in general as the emerging of something new or unexpected. This implies that emergence needs to be evaluated in the context of the system as well as its observer [40]. We must understand emergence in its relationship with the observer, the model, and the observations. Philosophers therefore distinguish between ontological and epistemological emergence, as explained by Silberstein and McGeever [41].

Ontological emergence occurs in natural complex systems, leading to the creation of something entirely new—such as life emerging from organic chemistry, or the wetness of water. Something that did not exist before emerges in the process. It is not transformed, but rather is something that was neither buried in the input nor in the processes and therefore cannot be mimicked by simulations. In contrast, epistemological emergence involves phenomena that are in principle explainable, but their interconnections and processes are too complex for us to predict the system behavior. We need to gain new knowledge by applying models and meta-models to understand a system. As Crutchfield [42] observes in his work, this interplay of is often observed.

"The emergence ... is the product of both the complicated behavior of nonlinear dynamical systems and the limitations of the observer. ... The newness ... is in the eye of an observer: the observer whose predictions fail or the analyst who notes that the feature of statistical self-similarity captures a commonality across length scales."

In other words, the model user gains additional understanding of how something emerges by working with the simulation, like understanding how weather patterns emerge, or discovering how social behavior can be explained using examples provided by Epstein and Axtell [8]. In terms of the Johari Window classification mentioned in the introduction, we can uncover an "unknown known" by using a simulation that highlights what is already known, making it clear to an observer who was previously unaware of this knowledge until they used the simulation.

In our example, the simulation of urban transportation can be extremely helpful to better understand the behavior and capacities of this system, particularly under stress. How can the system adapt to disruptions, such as accidents and traffic jams? What if public transportation is overloaded due to a big sport event with limited parking? Are there patterns that emerge based on traffic participants' behavior? Using such insights from epistemological emergences can help to make recommendations regarding this behavior or technical improvements that help to optimize the system, e.g., providing a better insight of effects and side effects for city planning.

The insight that ontological emergence cannot be recreated by computers relates to the nature of modeling as well as to the nature of simulation: they can only transform information; they cannot create new information! Everything a simulation presents is either in the implementing algorithms or the input data used to produce the output. However, if our concepts capture knowledge in one form, its rearrangement in the simulation can result in new insights for the observer. This implies that simulations of complex systems surely can lead to epistemological emergence, but not to ontological emergence.

3.2.5 The Nature of Adaptivity

As observed for the term "complexity," the term "adaptivity" is also used in several disciplines. In engineering, adaptive systems are designed to adjust their operations in response to changes in their environment or system parameters. This understanding is connected to that in computer science, where adaptive systems are implemented by algorithms that can modify their behavior based on input from the environment.

One of the earlier publications on the theory behind adaptivity is described by Holland [43], who uses generation procedures and generated populations to provide a theory of adaptivity. He also observes that the modification of the generated procedures can only be carried out in terms of available information about the environment. In other words, it entails collecting sufficient information about the environment, the effects of activities of the system on the environment, and a set of metrics that allow us to understand if the effects are contributing to the intended results.

Kauffman [44] introduced the idea of coevolution, which allows elements in a system to change based on their interactions with one another and with the environment. He introduced the fitness landscape, which maps all possible fitness strategies to the expected outcome; this map is then converted into a fitness metric that reflects how well this outcome contributes to the intended outcome, very similar to Holland's ideas. These ideas of capturing the response surface as a reflection of the metrics for success opened the door for the application of many heuristics to look for good solutions, such as the use of genetic algorithms, simulated annealing, and various other methods [45]. The increasing use of IT in modern systems directly contributes to adaptivity as well, as the modification of software and IT interfaces facilitates the adaptation to a new and changing environment, which includes new systems to react to or to collaborate with.

Interestingly, the task of identifying metrics to assess the effectiveness of a strategy or the impact of an effect on a desired outcome is often not addressed in sufficient detail. These metrics must incorporate a model of the system within its environment to determine what changes are possible and how the system can adapt to these changes, especially when focusing on proactive rather than reactive adaptation. This model should include the essential concepts, and calculating effects requires accurate implementation with adequate detail and precision. All the assumptions and constraints and foundational principles covered so far must be well understood before these metrics can be identified and implemented in a trustworthy manner.

A related challenge when simulating adaptive systems is the use of surrogate models. Particularly when simulation systems become computationally very expensive, such as agent-based models with hundreds of thousands of agents and complicated rules, machine learning is applied to create a surrogate model. The surrogate recreates the response surface as learned from the original model, as described by, among others, Angione et al. [46]. This response surface provides the same result for a set of input parameters like the original system, but without the need of invoking computationally expensive simulation runs. However, because the system is adaptive, it changes as soon as the change is triggered by changes in the environment or, potentially, by the unwanted results of actions of the system. If such changes alter the response surface of the system, the surrogate model needs to be recreated as well. This effect is similar to concept drift effects in other machine learning applications, so the lessons learned should be applicable also to adaptive system models using surrogate models; see, among others, Suárez-Cetrulo et al. [47].

In our resilient urban transportation system example, the ability of agents to observe and react to the outcome of their decisions becomes pivotal. If they cannot change their behavior and decision rules, the simulation will not be flexible enough to support adaptivity studies. On the other hand, the new behaviors and rules need to remain realistic, or the results will become invalid. Possible ways to adapt to a new situation are changing the preferred mode of transportation, changing routes, adapting transportation schedules, etc.

3.2.6 The Nature of Resilience

Resilience and adaptivity are connected by distinct concepts. Adaptivity emphasizes altering a system to deliver necessary capabilities in a changed environment, whereas resilience refers to the ability to bounce back to normal conditions—or even to bounce back to become a better system—following a disruptive event. Closely linked to resilience are the notions of robustness [48], the ability to function effectively across a broad range of conditions and to withstand or swiftly recover from unexpected disruptions or failures, and the concept of anti-fragility [49], the ability to improve under stress. Overall, the concept of resilience is complex, multidimensional, and comprises many heterogeneous, dynamic, and interrelated parts.

Hosseini et al. [50] compiled an overview of various definitions of resilience and metrics that allow system resilience to be measured. While traditional approaches focus on reducing the likelihood of disruptive events rather than their consequences, the resilience approach accepts that protection and prevention are necessary but not sufficient, as many disruptive events cannot be foreseen or prepared for. As discussed earlier, such events may emerge from the system's complexity, so that they are only explainable after the fact. Other events can be of such a magnitude that protection and prevention is hardly possible, such as hurricanes, earthquakes, or a pandemic. In such cases, disruption is unavoidable. Instead, the system should remain sufficiently intact to return to a functioning state and recover from there, potentially evolving to be less vulnerable in the future, which overlaps to some degree with the ideas of adaptivity.

Several definitions provided by Hosseini et al. [50] emphasize the importance of minimal or vital functions, along with measurements for both the actual and desired capabilities of a system. These elements are crucial for understanding resilience. Firstly, there is a set of core capabilities that are essential; without them, the system collapses and cannot recover. These core capabilities differ depending on the application area. Secondly, much as for adaptivity, there are metrics that measure capabilities and effects. As in that case, defining these metrics requires a comprehensive model of the system and its environment. Lastly, some definitions introduce a temporal aspect, specifying how much time is reasonably needed for recovery. This involves understanding how much capability loss a system can endure during disruption before losing its vital functions, and how quickly and effectively it can regain its original capabilities afterward. This process may include concepts such as prevention, protection, absorption, adaptation, and other supporting capabilities. Dorton et al. [51] discuss resilience engineering as good practice when dealing with complex socio-technical systems.

There are numerous examples in the literature that use all forms of simulation paradigms—discrete event models, agent-based models, system dynamics, Monte Carlo, and hybrid approaches. Hosseini et al. [50] enumerate several examples of studies conducted before 2016. Examples for more recent ones are Epiphaniou et al. [52], Zhang and Wang [51], and Saisridhar et al. [52, 53]. They all show that simulation is an applicable quantitative method to better understand the nature of resilience but, so far, no general conceptual framework has been developed that is widely accepted.

Our example of the resilience of the urban transportation system allows us to evaluate the applicability of various metrics for minimal or vital functions, along with measurements for both the actual and desired capabilities of a system within the simulation and utilize the most promising ones to observe the real systems. Examples are the number of people who reach their goal in the tolerated time, even if accidents lead to traffic jams on their way. This evaluation should also include policy guidelines, different states of cooperation among the agents, and other means to bounce back better from a disruptive event, like the flooding of major routes, significant accidents that hamper the transportation system over a longer period, etc. It can support the design of infrastructure improvements, optimizing traffic management systems, or developing contingency plans for emergencies.

3.3 Epistemological Insights from Simulation-based Experiments

Thus far, we have observed that modeling involves simplifying and abstracting the system of interest to create a formal specification of its conceptualization. This process forms the foundation for all subsequent artifacts. Computer simulations bring this conceptualization to life, often requiring the use of heuristics and alternative perspectives, as the original concepts may be inefficient or even intractable. These necessary modifications must be documented in the as-is conceptualization, providing information essential for reuse and composition. Computer simulations are inherently transformative, allowing them to model both simple and complex systems. In many cases, simulations offer the quantitative methods needed to understand highly complicated systems. However, if a system is truly complex and exhibits ontological emergence, it falls outside the realm of computable systems. In practice, many systems labeled "complex" are actually just highly complicated. In these instances, simulations can address epistemological emergence, enhancing our understanding of the complicated system. To address adaptivity and resilience, a model of the system of interest in its situated environment is needed to define the measurements for system capabilities, the effects of disruptive events, and system activities. While adaptivity allows a system to evolve in changing environments, resilience enables a system to recover after a disruptive event. Both require modeling and simulation of the system and are not inherently new challenges but rather special applications of the content of the earlier sections on modeling, simulation, systems, and complexity.

How can simulation, despite all the observed constraints, help to support research and experiments whose nature is systematic and methodical, aiming at generating new knowledge, validating existing theories, or exploring new phenomena? Since [54] fundamentally shaped how philosophers and scientists understand the role of models and analogies in science, simulation has matured to a point where it is widely accepted as an analysis and design tool complementary to theoretical considerations and experimental investigations, becoming the third pillar of gaining scientific knowledge. Ihrig [55] proposed a research architecture to address how simulation can be used as the third pillar of science, complementary to theory building and empirical observations, including conducting experiments (see Figure 3.2).

In the original article, Ihrig uses the term "model" instead of "simulation." To avoid confusion, I use the term "simulation" in the center of the figure instead. The focus lies on the role that simulation-based experimentation plays when it is theory-driven while grounded in the real world. The well-established relationship between theory building and empirical observations is not captured explicitly but can easily be extracted: the existing theory is an abstraction of the real-world issue or is embedding the abstraction of the real-world issue, like physics in physics-based systems. Propositions resulting from theorizing need to be confirmed by empirical data, which are usually gathered systematically in experiments or observational studies.

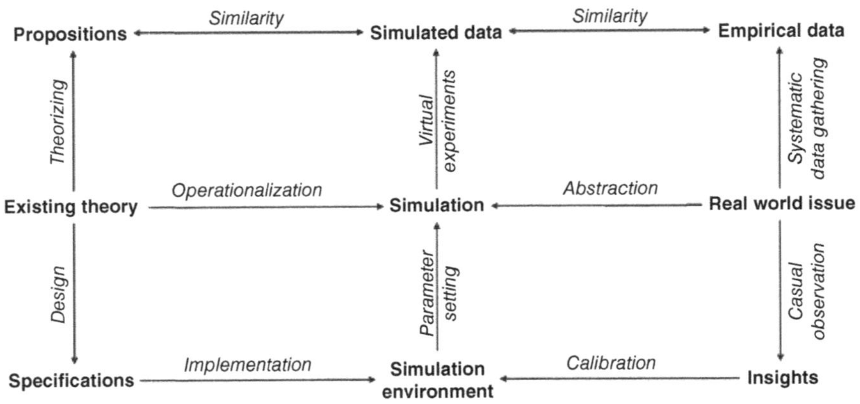

Figure 3.2 Research architecture. Adapted from [55].

Ihrig [55] suggests using existing theory as the foundation for simulations to better integrate simulation-based experiments into the framework of scientific discovery. His model's upper two quadrants deal with conducting experiments, while the lower two quadrants concentrate on creating a simulation. He uses a few terms differently, which will be explained in more detail below.

In order to guarantee that the model is grounded on accepted knowledge, the conceptualization specification is first created using current theory. The resulting concepts are then implemented in a simulation environment, which Ihrig defines as a simulation system not yet populated with parameter values. Once the parameters are set, the simulation, based on theory, is ready for experimentation. The right side of the model emphasizes grounding the simulation in reality, using parameters derived from insights gained by observing the real system, which is represented as an abstraction and simplification within the simulation. This approach ensures that both the system of interest and existing theory form the foundation.

The upper part of the model provides the structure for running simulation-based experiments to test hypotheses or to derive new knowledge. Conducting experiments with the simulation generates simulated data, often referred to as synthetic data. These data are realistic without being real, making them useful for studies that would otherwise require protected data, such as health or other personal information. Because the simulation is rooted in existing theory and grounded in reality, the results are reliable. The simulated data can lead to new propositions, which can be further evaluated through real-world experiments or observations. Scientists can hence experiment with and better understand complex, nonlinear, dynamic systems represented by the simulation.

The explicit integration of simulated systems with existing theory is not a novel concept for developers and users of physics-based simulations. The primary distinction between these physics-based simulations and those in social sciences, which include socio-technical systems, lies in the availability of well-established theories. In physics, theories like Newtonian physics are well-known and serve as a solid foundation. In contrast, social science lacks a universally accepted theory equivalent to Newtonian physics. However, there is an increasing number of examples of how social science theories can be used to ground simulations in both theory and reality. There are discussions about to what degree social theories shall be used as the basis for validation, but it is good practice using theories accepted by social scientists, as they are more useful to predict social and societal processes than assumptions by computer scientists and simulationists alone. As an example, Bullock et al. [56] describe how they employ identity fusion theory and moral foundations theory to implement agent behavior rooted in these theories, which is then applied to evaluate real-world behaviors. Clemen [57] provides a general framework for policy support utilizing these ideas for artificial societies.

In our case study, complex adaptive system principles can guide simulation experiments for resilient infrastructure, such as testing adaptive responses to traffic jams or failure of one or more transportation modes. As long as our simulations are grounded in the identified system theory, they can reveal robust strategies for urban transportation adaptation and recovery. This must include the comparison of simulation data with empirical data from real-world transportation systems, as exemplified in [58].

3.4 Selected Implications of the Nature of Resilient Complex Adaptive System Simulations

3.4.1 The End of the One-model Era

The section above highlights the absence of well-established theories, resulting in several complementary yet competing theories. Initially, the decision-making science community recognized the challenges of decision-making under deep uncertainty, a topic discussed by Lempert et al. [59]. This challenge is addressed by providing methods to support solutions, as outlined by Marchau et al. [60]. Focusing on simulation challenges, Tolk et al. [61] utilize ontological representations of underlying knowledge as existing theories. From this complete but possibly inconsistent knowledge representation, they derive sets of consistent simulations that represent the different aspects or facets of a problem and then use the ensemble of the resulting simulations to compute possible effects of all recognized aspects.

All these cases exemplify deep uncertainty, characterized by the absence of a well-established and consistent theory that is accepted and applicable to all stakeholders in the simulated system. In practice, this results in ignorance or disagreement regarding an appropriate conceptual representation of the driving forces and their relationships, key variables and parameters, their mathematical representations, and the metrics that provide value functions for comparing outcomes. To better examine a wide range of possible futures and outcomes when making decisions, Lempert et al.[59] recommend exploratory analysis. By creating a diverse set of plausible scenarios that capture a wide range of possibilities and applying sensitivity analyses to all possible constellations, they identify a set of robust recommendations that work well across various potential developments. These robust solutions replace highly optimized single-point solutions that rely on strict assumptions and constraints, which are rarely achievable. Having a robust solution applicable to a set of possible developments is preferable to an optimized solution that only works for one possible development.

As discussed above in the section on the nature of simulation, every computer simulation is constrained by the laws of computability. Lempert et al. [61] elaborate on this, emphasizing that implementations must be consistent to avoid faulty executions. While many alternatives can be generated through parametric variations, some alternatives may necessitate conceptual changes that lead to a fundamentally different simulation. This observation suggests that relying on a single simulation is inadequate, as it can only model developments captured by parametric variations of one conceptualization. Therefore, instead of depending on one simulation, we should employ an ensemble of simulations to explore a range of alternative developments. A well-known example is meteorology, in which ensembles of models are run in parallel, and the outcomes are compared for their agreement, like in the various forecasts of hurricane paths and their likelihood. Similarly, Page [62] observes that diverse perspectives and models can lead to better decision-making and problem-solving and highlights the effectiveness of ensemble approaches as well.

Regardless of how complex and extensive a single simulation may be, it remains limited to the theory it is based on. To truly understand resilient complex adaptive systems, relying solely on one simulation is unlikely to be sufficient.

To be able to compare the alternatives and variations, we make two assumptions, namely that the simulation systems of the ensemble can be initialized from the same starting point, and that the results can be mapped to the same solution surface, so they can be compared to each other.

3.4.2 The End of Single-objective Optimization

The insights summarized above suggest that just one simulation is insufficient. In this section, we observe that evaluating a system based on a single objective is also inappropriate. This observation, that single-objective optimizations often fall short for complex systems, has emerged from the broader fields of systems theory and operations research. The limitations of single-objective optimization have been discussed by various researchers, including [63–65], and many others. Instead, complex systems require multiobjective optimization approaches that can handle interdependencies, balance competing goals, and find robust solutions across multiple performance criteria simultaneously.

The main arguments for multiobjective evaluation of complex systems behavior are that they allow for addressing crucial system interactions and emergent behaviors that arise from multiple interrelated components. They also capture real-world constraints and requirements that often conflict and require balancing rather than a singular solution, which allows us to understand system-wide performance and stability implications of all possible developments. To communicate the effects on the resulting multidimensional utility or value function, effective visualization techniques are under development [66]. Rouse [67] recommends increasing the immersion for the decision maker, providing a similar experience to a flight simulator for decision-making that provides immediate feedback on all dimensions. Some ideas about possible implementations are described by Haberlin and Page [68]. In addition, the work on immersive analytics will be helpful in addressing these new requirements [69].

The need for multiobjective evaluation results from the tendency of decision makers to focus mainly on the main issue that drives the need to decide. If effects of effects or unintended effects, as well as the associated uncertainties, are not presented, they quickly are overlooked. Some examples and their neuroscientific background are provided by Starcke and Brand [70]. In their article, they review several research examples that stress and decision-making are intricately connected on the behavioral and on the neural level, and that immersive representations are beneficial in particular in stressful situations. An example from the recent past is provided by Tolk et al. [71], who describe simulation-based decision support in the COVID-19 Healthcare Coalition.

3.4.3 The Need for New Metrics

Our above considerations on adaptivity and resilience both emphasize the importance of capturing the value of a system change in relation to the desired outcome. The resilient complex adaptive systems of interest are made up of multiple interconnected and interrelated components that not only constantly adapt their behavior but may also produce emergence through these interactions. The effects of system actions may not occur immediately and can unfold in a series of cascading effects, not all of which are directly observable. Nonlinear interactions create a high sensitivity to initial conditions, where even small changes can lead to significantly different outcomes. While traditional utility functions, as famously discussed by Keeney and Raiffa [72], remain valuable for many use cases, additional metrics are needed to better capture dependency and dynamics.

There are many open research questions, such as how to connect an effect that results from cascading effects initialized several decision cycles ago with this decision? How can we capture that measurable effects may only be enabled by another effect of a system activity that is not measured explicitly? In general, whenever we need to understand causal as well as temporal dependencies, traditional metrics fall short in our quest to gain a better understanding of the system. We must replace metrics focusing exclusively on effects and system status changes with metrics that capture dynamics and causality over time and space.

Complexity theory may provide a good foundation, as it is the home for several metrics of interest to us, like the use of trajectories instead of instantaneous events, the understanding of singularities in the event trace, small-scale and large-scale processes, alternative cause-and-effect models, and more. These metrics are quickly becoming models by themselves. However, one of the objectives behind developing them is that the new metrics will be applicable not only to the simulations we use for gaining insights, but also to the real-world resilient complex adaptive systems as well.

3.4.4 A New View at System Categories

As published in Tolk [38], the discussion of the various categories of emergence implies also the need to revisit our understanding of systems and what can be simulated in order to gain the desired epistemological insights. Maier [73], who is known for his contributions to the discussion of systems of systems, defines four categories of emergence from a systems science perspective, depending upon how well the emergence observed in the natural system can be reproduced and explained through a simulation system.

- Simple emergence: The emergent property or behavior is congruent with the expected behavior and can be observed by simple systems.
- Weak emergence: The emergent property is readily and consistently reproduced in the simulation of the system but not in reduced complexity nonsimulation models of the system; that is, simulation is necessary to reproduce it. It is predictable but requires complicated systems.

- Strong emergence: The emergent property is consistent with the known properties, but is inconsistently reproduced without any justification of its manifestation. It is explainable but no longer predictable, leading to complex systems beyond computability.
- Spooky emergence: The emergent property is inconsistent with the known properties of the system and is impossible to reproduce in a simulation of a model of equal complexity as the real system. The system is beyond complexity, and the behavior is not even explainable after the fact.

In nature, all four classes can be observed. However, as observed in the section on simulation and systems, computers can only simulate simple and complicated systems. As all knowledge is either in the algorithm or in the input data, they can only produce simple and weak emergence. Figure 3.3 shows the classes of emergence with their characteristic attribute on the left and the resulting system category on the bottom.

To what degree the trans-complex systems with their spooky emergence resulting in systems beyond systems thinking are related to wicked problems, as introduced by Rittel and Webber [74] and recently revisited in the context of complexity and uncertainty by Head [75], is an open research question that deserves formal evaluation.

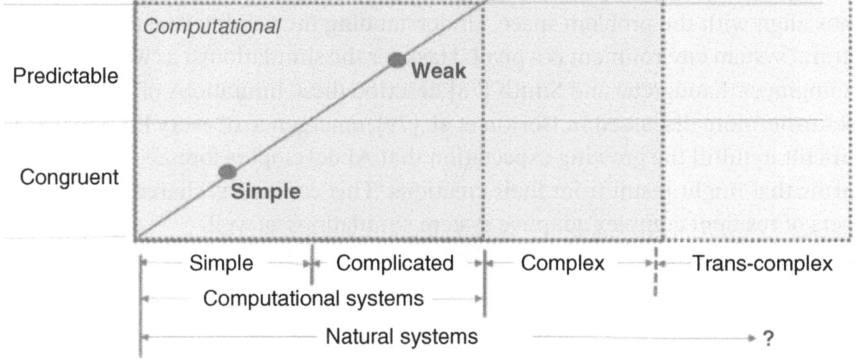

Figure 3.3 Emergence categories and corresponding systems categories [38].

Given these observations about emergence, let's revisit the definition of systems provided earlier: "A system is an arrangement of parts or elements that together exhibit behavior or meaning that the individual constituents do not." [33]. In engineering contexts, systems arise through the purposeful arrangement of parts by engineers, who ensemble components to provide specific functions that implement capabilities fulfilling the system's designed purpose. However, parts can also coalesce into collectives either through coincidence or necessity, as frequently observed in biological systems [76] and increasingly recognized in military and defense applications [77].

The categories of designed systems/engineered ensembles and emergent systems/natural collectives represent endpoints of a spectrum rather than mutually exclusive classifications. In engineered ensembles, parts are deliberately arranged for specific functions, while in natural collectives, order and function emerge from interactions between parts without preconceived top-down design, but mixed forms are likely and maybe even necessary to enable adaptivity and resilience. Understanding that resilient complex adaptive systems can exist anywhere along this spectrum, while recognizing the limitations of their computational representation through simulation, is crucial for deriving meaningful epistemological insights.

An interesting corollary is that computational AI faces the same fundamental constraints as simulations when dealing with complexity and trans-complexity. When AI systems are deployed in complex or trans-complex environments, their effectiveness becomes restricted to the computational domain. Beyond this domain, the reliability of AI-generated as well as simulation-provided output cannot be guaranteed, and the results may prove misleading or invalid when applied to real-world situations. Therefore, it is prudent to carefully bound these applications to domains where their computational limitations and underlying assumptions align with the problem space. Understanding the validity constraints in the natural system environment is a pivotal task for the simulationist as well as for the AI engineer. Landgrebe and Smith [78] describe these limitations of AI in detail. As furthermore discussed in Dorton et al. [79], emergence of every kind makes it difficult to fulfill the growing expectation that AI developers foresee and mitigate harms that might result from their creations. This concern is shared with developers of resilient complex adaptive system simulations as well.

3.5 Summary

Resilient complex adaptive system simulations are primarily computer simulations that implement an underlying conceptual framework. While simulations can aid in understanding complexity, they inherently cannot represent characteristics beyond what computer solutions can handle [38]. Adaptivity and resilience introduce additional challenges, particularly as they demand new metrics to capture dynamics and cascading effects, though these do not pose systemic epistemological challenges. Especially when simulations are used to create immersive experiences, they can enhance decision-making.

Ensembles of complex adaptive system-based simulations can be used to explore diverse scenarios for resilient infrastructure, such as evaluating multiple designs for transportation networks under varying environmental as well as behavioral conditions. They support multiobjective optimization in balancing competing goals like cost, efficiency, and resilience, which can provide actionable insights for designing infrastructure that adapts to disruptions and recovers effectively, making complex adaptive systems theory applicable via simulation results.

Humphreys [80] compares the use of computers to extend our capabilities to the use of microscopes and telescopes. These tools do not fundamentally change the observable phenomena, but they practically allow us to see things that are inaccessible to the naked eye. Similarly, computer simulations enable us to solve mathematical challenges related to complex adaptive systems and to simulate them within feasible constraints, provided we adhere to the principles of being theory-rooted and reality-grounded, as discussed by Ihrig [55]. Violating this principle risks creating a simulationist's regress, where we might inadvertently "prove" something within the simulation that was actually encoded as an assumption [27]. Additional challenges include the use of heuristics and deviations from theory to manage computational complexities [3]. Despite these challenges, simulations remain the best tool available for enhancing our understanding of resilient complex adaptive systems.

3.6 What Can We Learn from Resilient Complex Adaptive System Simulations? An Epilogue

Simulation is increasingly recognized as a valuable quantitative tool for facilitating communication and collaboration among cross-disciplinary teams. By providing a common formal framework for understanding complex systems and processes, simulation allows team members from different disciplines to visualize scenarios, test hypotheses, and analyze outcomes in a shared context. Immersive environments even allow us to experience such solutions on an empathetic level. Using common simulations can help bridge the gaps in terminology, perspectives, and methodologies that often occur in cross-disciplinary work [81].

By using simulation models, teams can create a "lingua franca" that transcends disciplinary boundaries, making it easier to integrate knowledge and expertise from various fields [9]. This has been particularly evident in areas such as healthcare, engineering, environmental science, and urban planning, where multidisciplinary collaboration is essential for addressing complex challenges. Simulation allows team members to engage in discussions based on concrete data and visualizations, fostering a more inclusive and collaborative environment. As a result, simulation has been advocated in academic literature and practical applications to improve communication and decision-making in cross-disciplinary settings. Creating such compositions of simulations created by the experts of multiple disciplines will furthermore allow for reducing epistemological emergence, as new facets on problem-solving are introduced and result in new insights among all participating experts, such as discussed by Page [62].

Although there are epistemological limits to simulation helping to learn about resilient complex adaptive systems, we should end this chapter on a positive note. Simulation disciplines continue to push the boundaries of feasibility every day and, as such, follow the philosophical principles of practicability [80] (pp. 153, 154)

> "Much of philosophy, and of philosophy of science, is concerned with what can be done in principle. ... What cannot be done in principle, can't be done in practice either. ... Yet when one is concerned with positive rather than negative results, the situation is reversed—what can be done in principle is frequently impossible to carry out in practice. Of course, philosophy should not ignore these 'in principle' results, but if we are concerned with, among other things, how science progresses, then the issue of how science pushes back the boundaries of what can be known in practice should be a primary concern."

As scientists and engineers, we will continue to take advantage of new developments and methods, such as data science to provide broader and better access to empirical data or generative AI to support and extend simulation activities, or even our modeling approaches, as envisioned by Frydenlund et al. [82]. However, all such solutions must be applied through an awareness of the foundations as they are exemplified in this chapter, which again emphasizes the need for a broad education of simulationists far beyond the scope of good software engineering, as captured recently in our body of knowledge [11]. Extending this body is the task of every simulationist to enable to learn as much as possible—but not more—from resilient complex adaptive system simulations and their represented systems. There are many open research questions in the context of complex socio-technical systems that require a closer collaboration of cross-disciplinary teams [83]. These tasks require simulationists to know the foundations of modeling and simulation as a discipline, rather than limiting their knowledge to simulation tools and their applications. Understanding the nature of modeling as well as simulation and being aware of the limits of computability [22] is essential for every epistemological support.

Acknowledgments: I am grateful to my friends and colleagues who provided valuable feedback on this chapter, particularly Stephen L. Dorton, Dr. Chris Glazner, Dr. Daniel T. Maxwell, Anandhi Sutti, and Dr. David E. Willmes. I would also like to express my deepest appreciation to my lifelong mentors, Dr. Tuncer Ören, Dr. William Rouse, and Dr. Bernard Zeigler for their enduring guidance and contributions.

Disclaimer: Dr. Tolk's affiliation with MITRE is provided for identification purposes only and is not intended to convey or imply concurrence with, or support for, the positions, opinions, or viewpoints expressed by this author. This article has been approved for Public Release, Distribution Unlimited. Public Release Case Number 24-01888-18.

References

1 U. Frank and K. G. Troitzsch. Epistemological perspectives on simulation. *Journal of Artificial Societies and Social Simulation*, 8(4), 2005.

2 W. L. Oberkampf and C. J. Roy. *Verification and Validation in Scientific Computing*. Cambridge University Press, Cambridge, MA, 2010. https://doi.org/10.1017/CBO9780511760396.

3 E. Winsberg. *Science in the Age of Computer Simulation*. University of Chicago Press, Chicago, IL, 2010.

4 American Society of Mechanical Engineers. *Verification, Validation, and Uncertainty Quantification (VV&UQ) in Scientific Computing*. American Society of Mechanical Engineers Press, New York, NY, 2006.

5 O. Balci and R. G. Sargent. Bibliography on validation of simulation models. *Newsletter of TIMS College on Simulation and Gaming*, 4(2):11–15, 1980.

6 O. Balci. Validation, verification, and testing techniques throughout the life cycle of a simulation study. *Annals of Operations Research*, 53:121–173, 1994. https://doi.org/10.1007/BF02136828.

7 R. G. Sargent. Verification and validation of simulation models. *Journal of Simulation*, 7:12–24, 2013. https://doi.org/10.1057/jos.2012.20.

8 J. M. Epstein and R. Axtell. *Growing Artificial Societies: Social Science from the Bottom Up*. The MIT Press, Cambridge, MA, 1996.

9 A. Tolk, W. J. Wildman, F. L. Shults, and S. Y. Diallo. Human simulation as the lingua franca for computational social sciences and humanities: Potential and pitfalls. *Journal of Cognition and Culture*, 18(5):462–482, 2018. https://doi.org/10.1163/15685373-12341418.

10 P. K. Davis, A. O'Mahony, and J. Pfautz. *Social-Behavioral Modeling for Complex Systems*. John Wiley & Sons, Hoboken, NJ, 2019. https://doi.org/10.1002/9781119485019.

11 T. Ören, B. P. Zeigler, and A. Tolk. *Body of Knowledge for Modeling and Simulation: A Handbook by the Society for Modeling and Simulation International*. Springer Nature, Cham, 2023. https://doi.org/10.1007/978-3-031-26706-9.

12 J. Luft and H. Ingham. The ohari indow, a raphic odel of interpersonal awareness. In *Proceedings of the Western Training Laboratory in Group Development*. University of California, Los Angeles, CA1955.

13 R. H. Rosen. *Essays on Life Itself*. Columbia University Press, New York, NY, 1998.

14 K. R. Popper. *Objective Knowledge: An Evolutionary Approach*. Clarendon Press, Oxford, 1979.

15 L. Vygotsky. *Thought and Language*. The MIT Press, Cambridge, MA, 1934.

16 A. Gelfert. *How to Do Science with Models: A Philosophical Primer*. Springer International Publishing, Cham, Switzerland, 2016. https://doi.org/10.1007/978-3-319-27954-1.

17 T. R. Gruber. A translation approach to portable ontologies. *Knowledge Acquisition*, 5(2):199–220, 1993. https://doi.org/10.1006/knac.1993.1008.

18 A. Tolk. Conceptual alignment for simulation interoperability: Lessons learned from 30 years of interoperability research. *Simulation: Transactions of the Society for Modeling and Simulation International*, 100(7):709–726, 2024. https://doi.org/10.1177/00375497231216471.

19 S. Robinson, G. ArbezL. G. BirtaA. Tolk, and G. Wagner. Conceptual modeling: Definition, purpose and benefits. Huntington Beach, CA, IEEE. 2015.

20 A. Akhavan and M. S. Jalali. Generative AI and simulation modeling: How should you (not) use large language models like ChatGPT. *System Dynamics Review*, 40(3):e1773, 2024. https://doi.org/10.1002/sdr.1773.

21 M. R. Garey and D. S. Johnson. *Computers and Intractability: A Guide to the Theory of NP-Completeness*. Freeman, San Francisco, CA, 1979.

22 J. MacCormick. *What Can Be Computed? A Practical Guide to the Theory of Computation*. Princeton University Press, Princeton, NJ, 2018. https:doi.org/10.1515/9781400889846.

23 M. Davis. *Computability and Unsolvability*. McGraw-Hill, New York, NY, USA, 1958.

24 S. Wolfram. *A New Kind of Science*. Wolfram Media, Champaign, IL, 1997.

25 W. L. Oberkampf, S. M. DeLand, B. M. Rutherford, K. V. Diegert, and K. F.Alvin.Error and uncertainty in modeling and simulation. *Reliability Engineering & System Safety*, 75(3):333–357, 2002. https://doi.org/10.1016/S0951-8320(01)00120-X.

26 E. Winsberg. Sanctioning models: The epistemology of simulation. *Science in Context*, 12(2):275–292, 1999. https://doi.org/10.1017/S0269889700002239.

27 A. Tolk. Bias ex silico: Observations on simulationist's regress. In *Proceedings of the 50th Annual Simulation Symposium*, pages 1–9, 2017. https://doi.org/10.5555/3106388.3106403.

28 A. Harper, N. Mustafee, and M. Yearworth. Facets of trust in simulation studies. *European Journal of Operational Research*, 289(1):197–213, 2021. https://doi.org/10.1016/j.ejor.2019.07.064.

29 J. Nguyen, S. T. Powers, N. Urquhart, T. Farrenkopf, and M.Guckert. An overview of agent-based traffic simulators. *Transportation Research Interdisciplinary Perspectives*, 12:100486, 2021. https://doi.org/10.1016/j.trip.2021.100486.

30 J. Huang, Y. Cui, L. Zhang, W. Tong, Y. Shi, and Z.Liu. An overview of agent-based models for transport simulation and analysis. *Journal of Advanced Transportation*, 2022(1):1252534, 2022. https://doi.org/10.1155/2022/1252534.

31 S. Suo, S. Regalado, S. Casas, and R. Urtasun. Trafficsim: Learning to simulate realistic multi-agent behaviors. In *IEEE/CVF Conference on Computer Vision and Pattern Recognition (CVPR)*, pages 10400–10409, 2021. https://doi.org/10.1109/CVPR46437.2021.01032.

32 H. Naing, W. Cai, H. Nan, W. Tiantian, Y. Liang. Dynamic data-driven microscopic traffic simulation using jointly trained physics-guided long short-term memory. *ACM Transactions on Modeling and Computer Simulation*,34(4):28, 2022. https://doi.org/10.1145/3514255.

33 H.Sillitto, J. Martin, D. McKinney, R. Griego, D. Dori, D. Krob, P. Godfrey, E. Arnold, and S. Jackson. *Systems engineering and system definitions.* In *INCOSE – International Council on Systems Engineering*, San Diego, CA, 2019

34 B. P. Zeigler. *Theory of Modeling and Simulation*. Wiley & Sons, New York, NY, 1976.

35 D. M. Buede and W. D. Miller. *The Engineering Design of Systems: Models and Methods* 4 edition, John Wiley & Sons, Hoboken, NJ, 2024.

36 T. Menke. Joint simulation environment for united states air force test support. In *NATO Science and Technology Organization Meeting Proceedings*, pages 17.1–17.14, Vienna, Austria, 2019. https://doi.org/10.14339/STO-MP-MSG-171-17-PDF.

37 W. Damm and A. S. Vincentelli. A conceptual model of system of systems. In *Proceedings of the 15th International Conference on Embedded Software (EMSOFT '15)*, pages 19–27. ACM, Seattle, WA, 2015. https://doi.org/10.1109/EMSOFT.2015.7318257.

38 A. Tolk. Limitations and usefulness of computer simulations for complex adaptive systems research. In *Summer of Simulation: 50 Years of Seminal Computer Simulation Research*, pages 77–96. Springer Nature, Cham, 2019. https://doi.org/10.1007/978-3-030-17164-3_5.

39 G. H. Lewes. *Problems of Life and Mind*. Trübner & Company, London, 1877.

40 E. Pessa. What is emergence? In *Emergence in Complex, Cognitive, Social, and Biological Systems*, pages 379–382. Kluwer Academic Publishers, New York, NY, 2002. https:doi.org/10.1007/978-1-4615-0835-8_35.

41 M. Silberstein and J. McGeever. The search for ontological emergence. *The Philosophical Quarterly*, 49(195):201–214, 1999. https://doi.org/10.1111/1467-9213.00138.

42 J. P. Crutchfield. The calculi of emergence: Computation, dynamics and induction. *Physica D: Nonlinear Phenomena*, 75(1–3):11–54, 1994. https://doi.org/10.1016/0167-2789(94)90273-9.

43 J. H. Holland. Outline for a logical theory of adaptive systems. *Journal of the ACM (JACM)*, 9(3):297–314, 1962. https://doi.org/10.1145/321127.321128.

44 S. A. Kauffman. *The Origins of Order: Self-organization and Selection in Evolution*. Oxford University Press, Oxford, 1993.

45 R. Salomon. Increasing adaptivity through evolution strategies. In *Parallel Problem Solving from Nature—PPSN IV*, pages 411–420. MIT Press, Cambridge, MA, 1996.

46 C. Angione, E. Silverman, and E. Yaneske. Using machine learning as a surrogate model for agent-based simulations. *PLoS One*, 17(2):e0263150, 2022. https://doi.org/10.1371/journal.pone.0263150.

47 A. L. Suárez-Cetrulo, D. Quintana, and A. Cervantes. A survey on machine learning for recurring concept drifting data streams. *Expert Systems with Applications*, 213:118934, 2023. https://doi.org/10.1016/j.eswa.2023.118934.

48 A. Zakarian, J. W. Knight, and L. Baghdasaryan. Modelling and analysis of system robustness. *Journal of Engineering Design*, 18(3):243–263, 2007. https://doi.org/10.1080/09544820701211932.

49 N. N. Taleb. *Antifragile: Things that Gain from Disorder*. Random House, New York, NY, 2014.

50 S. Hosseini, K. Barker, and J. E. Ramirez-Marquez. A review of definitions and measures of system resilience. *Reliability Engineering & System Safety*, 145:47–61, 2016. https://doi.org/10.1016/j.ress.2015.08.006.

51 S. L. Dorton, G. J. Lematta, and K. J. Neville. The tough sell of resilience engineering. *IEEE Transactions on Technology and Society*, 6(1):47–53, 2025. https://doi.org/10.1109/TTS.2025.3533091.

52 G. Epiphaniou, M. Hammoudeh, H. Yuan, C. Maple, and U. Ani. Digital twins in cyber effects modelling of IoT/CPS points of low resilience. *Simulation Modelling Practice and Theory*, 125:102744, 2023. https://doi.org/10.1016/j.simpat.2023.102744.

53 P. Saisridhar, M. Thuerer, and B. Avittathur. Assessing supply chain responsiveness, resilience and robustness (triple-R) by computer simulation: A systematic review of the literature. *International Journal of Production Research*, 62(4):1458–1488, 2024. https://doi.org/10.1080/00207543.2023.2177369.

54 M. Hesse. *Models and Analogies in Science*. Sheed & Ward, London, 1963.

55 M. Ihrig. A new research architecture for the simulation era. In A. Tolk, S. Y. Diallo, I. O. Ryzhov, L. Yilmaz, S. Buckley, and J. A. Miller, editors, *Seminal Contributions to Modelling and Simulation: 30 Years of the European Council of*

Modelling and Simulation, pages 47–55. Springer Nature, Cham, 2016. https://doi.org/10.1007/978-3-319-30061-0_5.

56 J. Bullock, J. E. Lane, I. Miklousić, and F. L. Shults. Modeling nationalism, religiosity, and threat perception: During the COVID-19 pandemic. *PLoS One*, 18(4):e0281002, 2023. https://doi.org/10.1371/journal.pone.0281002.

57 T. Clemen, A. Tolk, U. A.Clemen, D. Glake, G. Günther. Creating artificial societies for policy decision support: A research agenda and call to action. *Simulation: Transactions of the Society for Modeling and Simulation International*, 101(6):645–660, 2025. https.//doi.org/10.1177/00375497241295765.

58 T. Clemen, N. Ahmady-Moghaddam, U. A. Lenfers, F. Ocker, D. Osterholz DJ. Ströbele, D. Glake, Multi-agent systems and digital twins for smarter cities. In *Proceedings of the Principles of Discrete Advanced Simulation Conference*, pages 45–55, 2021. https://doi.org/10.1145/3437959.3459254.

59 R. J. Lempert, S. W. Popper, and S. C. Bankes. *Shaping the Next One Hundred Years: New Methods for Quantitative Long-term Policy Analysis*. RAND Report MR-1626. RAND Corporation, Santa Monica, CA, 2003.

60 V. A. W. J. Marchau, W. E. Walker, P. J. T. M. Bloemen, and S. W. Popper. *Decision Making under Deep Uncertainty: From Theory to Practice*. Springer Nature, Cham, 2019. https://doi.org/10.1007/978-3-030-05252-2.

61 A. Tolk, S. Y. Diallo, J. J. Padilla, and H. Herencia-Zapana. Reference modelling in support of M&S—foundations and applications. *Journal of Simulation*, 7:69–82, 2013. https://doi.org/10.1057/jos.2012.28.

62 S. Page. *The Difference*. Princeton University Press, Princeton, NJ, 2008.

63 H. A. Simon. Rational decision making in business organizations. *The American Economic Review*, 69(4):493–513, 1979.

64 J. H. Holland. Complex adaptive systems. *Daedalus*, 121(1):17–30, 1992.

65 D. Savic. Single-objective vs. multiobjective optimisation for integrated decision support. In *Proceedings of an International Conference*119, Lugano, Switzerland, 2002.

66 H. Chen, S. Zhang, W. Chen, H. Mei, J. Zhang, A. Mercer, R. Liang, and H. Qu. Uncertainty-aware multidimensional ensemble data visualization and exploration. *IEEE Transactions on Visualization and Computer Graphics*, 21(9):1072–1086, 2015. https://doi.org/10.1109/TVCG.2015.2410278.

67 W. B. Rouse. Understanding the complexity of health. *Systems Research & Behavioral Science*, 38(2):197–203, 2021. https://doi.org/10.1002/sres.2701.

68 R. J. and E. H. Page. Visualization support to strategic decision-making. In C. Turnitsa,C. Blais, and A. Tolk editors. *Simulation and Wargaming*, John Wiley & Sons, Hoboken, NJ, 2021.

69 K Marriott, F. Schreiber, T. Dwyer, K. Klein, N. H. Riche, T. Itoh, W. Stuerzlinger, and B. H. Thomas. *Immersive Analytics*. Springer Nature, Cham, 2018. https://doi.org/10.1007/978-3-030-01388-2.

70 K. Starcke and M. Brand. Decision making under stress: A selective review. *Neuroscience & Biobehavioral Reviews*, 36(4):1228–1248, 2012. https://doi.org/10.1016/j.neubiorev.2012.02.003.

71 A. Tolk, C. Glazner, and J. Ungerleider. Computational decision support for the COVID-19 healthcare coalition. *Computing in Science & Engineering*, 23(1):17–24, 2020. https://doi.org/10.1109/MCSE.2020.2968541.

72 R. L. Keeney and H. Raiffa. *Decisions with Multiple Objectives: Preferences and Value Tradeoffs*. John Wiley & Sons, New York, NY, 1976.

73 M. W. Maier. The role of modeling and simulation in system of systems development. In *Modeling and Simulation Support for System of Systems Engineering Applications*, pages 11–41. John Wiley & Sons, Hoboken, NJ, 2015. https://doi.org/10.1002/9781119176627.ch2.

74 H. W. J. Rittel and M. M. Webber. Dilemmas in a general theory of planning. *Policy Sciences*, 4(2): 155–169, 1973. doi: https://doi.org/10.1007/BF01405730

75 B. W. Head. *Wicked Problems in Public Policy: Understanding and Responding to Complex Challenges*. Springer Nature, Cham, Switzerland, 2022. doi: https://doi.org/10.1007/978-3-030-94580-0

76 K. Kelly. *Out of Control: The Rise of Neo-biological Civilization*. Addison-Wesley Longman Publishing Co., Inc., Reading, MA, 1994.

77 M. T. Koehler, J. L. Bricio-Neto, E. H. Page, and A. Tolk. Applying complex adaptive systems research results to combat simulations of the generation-after-next. *Journal of Defense Modeling and Simulation*, 15485129241233608, 2024. https://doi.org/10.1177/15485129241233608.

78 J. Landgrebe and B. Smith. *Why Machines Will Never Rule the World: Artificial Intelligence Without Fear*. Routledge, New York and London, 2022. https:doi.org/10.4324/9781003244757.

79 S. L. Dorton, L. M. Ministero, B. Alaybek, and D. J. Bryant. Foresight for ethical AI. *Frontiers in Artificial Intelligence*, 6:1143907, 2023. https://doi.org/10.3389/frai.2023.1143907.

80 P. Humphreys. *Extending Ourselves: Computational Science, Empiricism, and Scientific Method*. Oxford University Press, Oxford, 2004.

81 A. Tolk, A. Harper, and N. Mustafee. Hybrid models as transdisciplinary research enablers. *European Journal of Operational Research*, 291(3):1075–1090, 2021. https://doi.org/10.1016/j.ejor.2020.11.018.

82 E. Frydenlund, J. J. Padilla, K. Palacio, and D. Shuttleworth. Modeler in a box: How can large language models aid in the simulation modeling process? *Simulation: Transactions of the Society for Modeling and Simulation International*, 100(7):727–749, 2024. https://doi.org/10.1177/00375497241243509.

83 S. Elsawah, T. Filatova, A. J. Jakeman, A. J. Kettner, M. L. Zellner, I. N. Athanasiadis, S. H. Hamilton, R. L. Axtell, D. G. Brown, J. M. Gilligan, and M. A. Janssen. Eight grand challenges in socio-environmental systems modeling. *Socio-Environmental Systems Modelling*, 2:16226, 2020. https://doi.org/10.18174/sesmo.2020a16226.

84 J. Zhang and T. Wang. Urban resilience under the COVID-19 pandemic: A quantitative assessment framework based on system dynamics. *Cities*, 136:104265, 2023. https://doi.org/10.1016/j.cities.2023.104265.

Part II

Modeling Formalisms and Their Applications

4

From ABM Back to Real Data: Time Series Visualization and Model Selection in the K+S Agent-based Model[1]

Giovanni Dosi,[1] Marcelo Pereira,[1,2] Gabriel Petrini,[1,2] Andrea Roventini,[1] and Maria Enrica Virgillito[3]*

[1] *Institute of Economics, Scuola Superiore Sant'Anna, Pisa, Italy*
[2] *Institute of Economics, Universidade Estadual de Campinas, Brazil*
[3] *Department of Economic Policy, Università Cattolica del Sacro Cuore, Milan, Italy*

*Corresponding author: gpetrinidasilveira@gmail.com

4.1 Introduction

Understanding and evaluating how much a model is similar/dissimilar vis-à-vis the empirical evidence is certainly a challenge for model development. In this chapter, we respond to this challenge by proposing a systematic protocol for model selection. The method relies on a composite similarity index that synthesizes multiple dimensions of comparison between simulated and empirical time series. This contribution fits within the broader complex systems approach to socioeconomic phenomena [1], which seeks to explain macroeconomic dynamics as emergent from micro- and meso-level interactions. Within the complex system approach, agent-based models (ABMs) stand out as a growingly important technique to study evolving economies. One of the aims of building an ABM is to understand mechanisms and processes that guide the real-world dynamics, and they often represent thought experiments or metaphorical models [2], that is, simulated complex maps that can be used to build scenario analyses and test for the relevance of parameters/interactions, providing diagnosis. The incipient [3] May's model of population ecology, where high ecological diversity leads to unstable rather than stable outcomes, serves as a classic example of a metaphorical model. Another common

1 The authors wish to thank participants at the workshop "Agent-based models for political economy", Pisa 2025, for their helpful comments and suggestions.

usage of ABMs is comprehending the mechanisms of a social phenomenon, as [4] referred to as Generative Social Science.

Such metaphorical or diagnostic models are not primarily designed to replicate any specific empirical dataset but to investigate how the proposed laws of motion yield coherent qualitative results. With this virtual laboratory, the modeler can generate scenarios and data that are typically unobservable or unavailable for empirical studies. ABMs applied to economic phenomena are medium-/large-scale simulated models that cover micro, industry, and macro aggregate variables, providing a powerful tool for scenario analysis and policy experiments [5–9]. One of the substantial advancements of ABMs in economic modeling is the ability to create simulated environments for diagnostic exercises, focusing on qualitative empirical patterns. Overall, ABMs are designed to generate stylized facts (SFs) resulting from the interactions of numerous disaggregated units connected by various forms of structures and interactions, allowing the emergence of aggregated properties, which are tested against robust qualitative patterns.

While ABMs have proven helpful in capturing emergent properties and complex interactions, their integration with empirical data remains an open issue. The relationship between empirical data and models in economics generally goes in the realm of four different domains, which are estimation, calibration, validation, and selection. While the first two problems mainly pertain to parameter definitions, validation, and selection pertain instead to model mechanisms, which depend on the inner structure of interactions and the structure of variables.

The literature has addressed such issues in the last years, distinguishing between input and output validation and highlighting the importance of external validity [10–13]. External validity is a dimension that ABMs particularly require because of their attempt to be empirically disciplined models [14]. Recently, there has been growing interest in applying ABMs to forecasting exercises [15] and to inform specific policy decisions [16], stressing the shift toward the quantitative dimension.[2] Overall, this indicates a rich research venue for data-friendly ABMs.

However, this involves closely examining and utilizing data. Still, today there is a lack of a comprehensive approach able to confront with the inner complexity of such models. Reconciling simulated models with real-world data is nontrivial, especially in socioeconomic systems impacted by interacting processes and environmental changes like crises, catching-up, economic development, and stagnation. Developers of socioeconomic models encounter numerous challenges, starting with determining the appropriate data selection methods, including aggregation levels, historical periods, and types of series to be analyzed. In addition, there is still a lack of a protocol able to synthetically rank models in their

2 It is important to note that this quantitative dimension should not be seen as an alternative test of model validity but rather as a complementary step to the diagnostic capabilities of the models.

performance ability to be close to the data. Metrics of distance between simulated and real time series, such as simulated minimum distance, root mean squared errors, or correlation coefficients, fail to capture the complete range of information and structure encoded in the data. Indeed, unidimensional distance measures, such as Euclidean distance, may fail to capture the global model dynamics and typical attributes of time series produced by ABMs, such as persistent deviations from average behavior, path dependency, and structural breaks. In addition, information regarding the signal domain of the series is not usually embraced by unidimensional distance measures.

Signal processing and the frequency domain of economic variables are crucial information that should be embraced when comparing data with a time-series structure. Economic time series are, in fact, characterized by some distinctive features. Indeed, a common characteristic of this data type is a smoothly decreasing spectral shape, which indicates the significance of low-frequency phenomena, such as long-term fluctuations [17]. Long waves, for example, have been associated with low-frequency events in the literature [18], particularly in the context of technological change [19, 20] and structural change [21]. In this specific context, for example, large-scale adoption of new technological paradigms can also be driven by short-term euphoria [22]. Neglecting the frequency domain would overlook the distinctions associated with short versus medium versus long run cycles. Indeed, a similarity measure of time series should be able to embrace the time spectra, including time intervals, lag structures, and misalignments. There is still a lack of a single indicator able to account for different dimensions, therefore the quest for a multidimensional approach.

Responding to the increasing demand for data-friendly ABMs and the challenges of integrating data back into the models, we propose a step to complement existing techniques in the literature. This chapter introduces an alternative approach to similarity measurement, focusing on model selection through a composite similarity index aggregating multiple dimensions of comparison. The primary purpose of this approach is to provide a systematic method for model selection, which can assist other methods in integrating data back into models.

The contribution of our approach lies in constructing a composite similarity index that aggregates multiple statistical and dynamic measures into a single ranking system. Instead of relying on a single indicator, we combine several metrics that capture different aspects of similarity, ensuring a more detailed assessment. The ranking system is designed to iteratively refine model selection by penalizing those models that achieve the highest aggregate dissimilarity score. This iterative selection process ensures a continuous refinement of model selection, enabling a structured analysis that can assist in integrating data back into models.

We apply the protocol to different versions of the Schumpeter meeting Keynes ABMs [5, 21, 23–26]. We use US data from 1948Q1 to 2019Q1 for our empirical dataset throughout the analysis. Our protocol is designed to be easily adaptable to different empirical cases and sets of simulated models, with the flexibility to apply to models beyond ABMs.

This protocol consists of two main phases. During the first phase, we systematically analyze the empirical-simulated pair for each variable across different K+S model versions. Specifically, we conduct a battery of tests to dynamically compare time-series dissimilarities, taking into account different alignments in the time spectra. This comparison examines statistical moments, different frequencies, and paths of the signal as a whole. We obtain a variable-wise multidimensional index by applying this procedure to every variable across all models under scrutiny. In the next phase, we use the information from the multidimensional index to perform an iterative model selection. In this way, we advance the analysis of time series similarity and model selection.

We finally present an extension to this protocol by assigning different weights (*ex ante*) to fine-tune the model selection for specific purposes. We apply the experiment to variables typically associated with policymaking decisions and with technological change-related variables. The analysis reveals that the finance-augmented K+S model demonstrates robust performance compared to recent developments, particularly when examining different frequencies. Likewise, the labor-augmented model, adjusted for competitive institutional settings, yields satisfactory results across most configurations.

The remainder of this chapter is structured as follows. Section 4.2 briefly introduces the key aspects of the K+S models. The information about the structure of the models grounds the rationale behind pairing decisions between the simulated and empirical data. Section 4.3 describes the data selection used to build the empirical dataset for comparison with the simulated models. Section 4.4 outlines the first phase of the protocol, which consists of creating a comprehensive evaluation of model similarity to empirical data. Next, Section 4.5 presents the model selection routine to identify the least dissimilar model. This section also presents a purpose-guided model selection to evaluate models' performance with a subset of target variables. Finally, Section 4.6 concludes the chapter and offers insights into other potential uses of this protocol and possible improvements.

4.2 K+S Family Model

This section briefly presents the evaluated models and their underlying structural assumptions. These assumptions will guide the data selection described in Section 4.3. Table 4.1 lists the different model versions. The K+S model [5, 27, 28] is a macroeconomic ABM where Schumpeterian and Keynesian mechanisms influence business cycle fluctuations and long-run economic growth. The original version encompasses two industries, each populated by heterogeneous firms. The households, banking, and government sectors are stylized in a parsimonious way and refined in other versions. In this artificial economy, all agents make decisions

Table 4.1 Collection of model versions considered.

Main reference	Contribution	Label
[5]	First version of the K+S model	`K+S-original`
[23]	Detailed financial sector	`K+S-finance`
[29, 30]	Decentralized labor market and institutional growth regimes	`[K+S-Ford, K+S-Comp]`
[21]	Disruptive technological change and multiple sectors	`K+S-multi`

using bounded-rational rules and simple heuristics. This model was designed on purpose to capture a rich ensemble of micro-macroeconomic regularities (SFs). As a consequence, this model (and its later extensions) was not intended to precisely replicate the empirical time series of any specific economy or time period. Therefore, some degree of dissimilarity with empirical data is expected.

Before comparing the models with empirical data, it is important to understand their general structure. The models feature two distinct sectors. The first sector consists of capital goods firms that produce heterogeneous, durable machine tools using only labor. These firms operate on a made-to-order basis, producing capital goods in response to the investment plans of the other industry, underscoring the interdependence between them. The labor productivity growth rate is endogenous and evolves according to the level of R&D expenditures, solely comprised of labor. Thus, this sector is the primary locus of endogenous innovation embedded in capital goods of different vintages. Consumption-good firms employ a stock of machine tools whose technologies evolve over time. Consumption goods are demanded by households and can be stored if unsold. Both industries have access to bank loans (up to a limit) and operate under conditions of imperfect information. The capital goods sector, in particular, faces Schumpeterian competition. Firms in both sectors set prices based on a mark-up rule over production costs.

The model includes a single bank, representing a stylized financial sector. At each time step, the bank collects deposits from the private sector and offers interest-bearing loans to finance firms' production and investment activities. The bank does not employ any workers for its operations, and only firms have access to credit. If a firm is credit-constrained, it cannot carry out its production or investment decisions as planned, requiring further revisions. The central bank sets the policy rate, completing the financial circuit of the model. The prime rate anchors deposit and loan rates, which are determined using markdown and mark-up rules.

The household sector is also stylized and populated by homogeneous workers facing a centralized labor market, receiving a homogeneous wage if employed. The government sets both the tax rate and unemployment benefits. In this setup, all firms pay the same wage, and unemployed households receive uniform unemployment benefits. The government is responsible for setting and collecting taxes, issuing bonds to finance the public deficit, and maintaining a sustainable public debt trajectory over the long term. Notably, the government does not provide goods or services to the private sector. Like the banking sector, the government does not employ any workers.

Despite its stylized representation, the original K+S model successfully captures a wide range of micro- and macroeconomic regularities. Table 4.2, adapted from [21], lists the SFs replicated by this model. At the macroeconomic level, the `K+S-original` model generates endogenous, self-sustained growth with persistent fluctuations. It also produces fat-tailed growth rate distributions, cross-correlations among macroeconomic variables, and cyclicality in GDP components, among other features. At the firm level, the model exhibits lumpy investment patterns and persistent productivity heterogeneity across firms. The table also highlights the SFs associated with major extensions of the original model. Most of these extensions build on the core principles outlined above, but are tailored to address specific research questions.

The first major modification to the model was introduced by Dosi et al. [23], who added a more complex banking sector. This extension aimed to explore the relationship between finance and economic growth. Subsequently, Dosi et al. [24] introduced bank heterogeneity. In this version, banks decide how much credit to extend to firms and prioritize borrowers based on a pecking-order list determined by the firms' liquidity ratios. The amount of credit granted also depends on the bank's net worth, subject to Basel-type regulatory capital adequacy constraints. These refinements provide a more detailed representation of banks' balance sheets and lending channels. This enables the model to reproduce systemic financial phenomena, particularly those associated with banking crises and their macroeconomic consequences [31]. Notably, the model demonstrates that banking crises can engender fiscal burdens for the government [32]. The incorporation of diverse transmission channels enriches the model's descriptive capacity and expands the scope for policy experiments, enabling revisiting the implications of both fiscal and monetary rules [33].

Later, the centralized labor market was replaced with a decentralized one, inaugurating the labor-augmented versions [14, 25, 26, 29, 30]. In these versions, the economy is populated by workers with heterogeneous skills. This means labor market conditions now influence productivity growth. When the decentralized labor market opens, firms and workers engage in a search-and-matching process. In addition to imperfect information, this market has other forms of imperfect competition. For example, larger firms receive more job applications than smaller ones. Unlike in earlier versions, the assumption of a single nominal wage no

Table 4.2 Stylized facts replicated by the K+S family models.

Microeconomic SFs	**Macroeconomic SFs**
Skewed firm size distribution[a]	Endogenous self-sustained growth with persistent fluctuations and crises[a]
Fat-tailed firm growth rates distribution[a]	Fat-tailed GDP growth rate distribution[a]
Heterogeneous productivity across firms[a]	Endogenous volatility of GDP, consumption and investment[a]
Persistent productivity differentials[a]	Cross-correlation of macro variables[a]
Lumpy investment rates of firms[a]	Pro-cyclical aggregate R&D investment and net entry of firms in the market[a]
Firm bankruptcies are counter-cyclical[b]	Cross-correlations of credit-related variables[b]
Firm bad-debt distribution fits a power-law[b]	Cross-correlation between firm debt and loan losses[b]
	Banking crises duration is right skewed[b]
	Fiscal costs of banking crises to GDP distribution is fat-tailed[b]
Heterogeneous skills distribution[c]	Persistent and counter-cyclical unemployment[c]
Fat-tailed unemployment time distribution[c]	Endogenous volatility of productivity, unemployment, vacancy, separation and hiring rates[c]
Fat-tailed wage growth rates distribution[c]	Unemployment and inequality correlation[c]
Cross-sectional Engel Law[d]	Pro-cyclical workers skills accumulation[c]
Heterogeneous propensity to save and consume[d]	Beveridge curve[c]
	Okun curve[c]
	Wage curve[c]
	Matching function[c]
	Engel Law[d]
	Non-satiation in luxury goods[d]
Technology-level SFs[d]	**Sectoral-level SFs[d]**
Stepwise increase in technological frontier	Product life cycle
Lower rate of radical versus incremental innovation	Exponential age distribution
Fast diffusion of dominant techniques	Sectoral wage and productivity differentials

Notes: *Source:* [21] with permission of Elsevier.

[a] Since [5, 27].

[b] Since [23, 24].

[c] Since [25, 26, 30].

[d] Since [21].

longer holds. The evolution of the labor market will be contingent upon the institutional arrangements within the economy.

Dosi et al. [25] examine a setting that resembles a Fordist labor market in the spirit of [2] Jean-Philippe Bouchaud's metaphorical model. In this regime, wage determination is insensitive to the tightness of the labor market, while the productivity gains are passed on to wages. On the behavioral side, firms and workers maintain longer-lasting contractual relationships. This includes conditional firing schemes and less frequent job switching. While unemployed, households receive government unemployment benefits provided by a welfare state, which also guarantees a minimum wage. The authors also evaluate a "competitive regime" that reflects post-1980s changes. Unlike the Fordist regime, wages respond more rapidly and markedly to changes in unemployment levels. Firms have greater flexibility to fire workers in response to production plans, facilitating reductions in excess workforce. Both minimum wage and its indexation to productivity are no longer guaranteed. Workers also exhibit increased job-seeking behavior, even while employed. Regardless of the institutional setting, the labor-augment versions can replicate an additional collection of SFs, also listed in Table 4.2. These include wage, Beveridge, and Okun curves, unemployment, and vacancy rate volatility. This chapter applies the proposed protocol to both regimes, labeling the first as `K+S-Ford` and the second as `K+S-Comp`.[3]

While earlier model versions featured only two sectors, each with a single industry, [21] introduced a multi-sector version (`K+S-multi`). The key contribution is the introduction of disruptive technological change and the inclusion of multiple consumption goods sectors. One sector represents basic nondurable consumption goods, as in previous versions. The second simulates a luxury goods sector, produced through multiple stages. This extension, combined with a decentralized labor market, enables the exploration of relationships between income distribution, consumption patterns, and the impact of technological paradigms on job creation and displacement. These advancements enable the reproduction of a new batch of SFs, such as Engel's law, non-satiation in luxury goods, heterogeneous propensity to save, paradigm changes, and product life cycle.

One of the advantages of this family of macroeconomic ABMs is its flexibility in generating different metaphorical economic systems. While each model has been empirically validated by comparing its outputs with real-world regularities, the extent to which they resemble/mirror empirical data remains an open question. This chapter aims to address this question tentatively. We simulate each model version discussed above using their benchmark configurations for 600 time steps without changing the source code. All simulations were conducted using the

3 In the original papers [25, 29], the authors examine the effect of an institutional change during the simulations. Here, we compare the two regimes, ignoring the institutional transition phase.

Laboratory for Simulation Development [34]. The results were integrated into R using the LSD interface library [35]. The first 350 periods are excluded from the analysis to ensure that the effects of institutional changes, which occur during simulations in some model versions, do not skew the results. The following section outlines the process of selecting empirical data, guided by the theoretical framework of the models presented here.

4.3 Data Selection

The first step of the analysis consists of the collection of data, which will be later contrasted with the simulated models presented in Section 4.2. The analysis is restricted to the variables common to the most stylized model version presented in Dosi et al. [5], due to the different levels of disaggregation among the model versions. As those models are calibrated for quarterly data, we downsample the empirical time series with higher frequency (e.g., monthly) by aggregating the interquarter data using the mean, a method chosen for its simplicity and ability to preserve the overall trend of the series. When available, we make use of seasonally adjusted series. For each variable considered, we estimate the trend (`TREND`) and the cycle (`CYCLE`) for both empirical and simulated series using the [36] filter to the series in levels. This enables us to evaluate whether the models under consideration are similar to empirical data in other frequency domains alongside series without filtering (`unfiltered`).

A critical aspect of the analysis is developing a mapping procedure to associate each simulated series with its corresponding empirical counterpart. The mapping strategy needs to be robust to what we refer to as *ex ante dissimilarity*. This concept refers to the necessary degree of divergence between simulated and empirical data that ensures the model does not artificially reproduce the observed series due to structural simplifications or missing real-world complexities.[4] We illustrate the issue of *ex ante* dissimilarity with an example. Suppose a model produces a simulated GDP series identical to the observed one. Given that the model lacks crucial real-world elements such as the external sector or more granularity, this perfect match suggests that the model might be similar through artifacts rather than correctly reproducing the economic dynamics of the series. In this case, a slight dissimilarity might indicate a more realistic representation of the core economic mechanisms, given how stylized the models are. Still, determining the tolerance level of dissimilarity is not straightforward.

One way to reduce the occurrence of spurious matches is by selecting time series that have a similar structure to the simulated ones, that is, selecting the empirical series whose generative mechanisms are included in the model. Therefore, for

---

4 Although this concept is not documented in the literature, it illustrates the challenges of comparing simulated models with real data, which is not restricted to the K+S family or ABMs.

this exercise, the theoretical aspects of the model and its underlying assumptions serve as the selection criteria—even though it implies using shorter series—so the empirical data possess a structure most akin to the model. This increases the likelihood that models closely resembling the data are the best candidates for accurately representing a higher similarity. However, it is important to note that controlling for *ex ante* dissimilarity is particularly challenging for GDP series, as simply aggregating demand subcomponents consistent with the model does not fully capture the indirect effects necessary to replicate GDP dynamics coherently. Since this is out of the scope of this chapter, we chose to select this variable in real terms (GDPC1) without further modification.

Table 4.3 shows the empirical-simulated mapping for the remaining evaluated variables.[5] We use US data from the FRED database and, where possible, we select time series spanning from 1948Q01 to 2018Q04. The decision to restrict the dataset before 2019 removes any potential effects of the COVID-19 crisis—a phenomenon not intended to be captured by the models under scrutiny.

As Section 4.2 describes, capital-good firms produce durable machine tools, which are subsequently supplied to firms producing consumer goods. The aggregate production of these firms is mapped to empirical data on durable industrial goods (IPG333S), reflecting their role in producing long-lasting capital equipment. The real series is obtained using the implicit deflator of industrial equipments (Y033RD3Q086SBEA). Additionally, we are repurposing this series to compare with the producer price inflation rate. Furthermore, firms in this sector also invest in R&D (Y006RC1Q027SBEA), allocating a portion of their workforce to innovative activities, which must be considered. Specifically for the R&D series, we use the empirical R&D series as it is, deflating it by the intellectual products implicit deflator (Y001RD3Q086SBEA), which lacks an equivalent in the model. This particular implicit deflator is not utilized further.

Firms producing consumer goods combine capital and labor to produce goods consumed by households. In the model, the consumption decisions exhibit characteristics akin to nondurable personal consumption expenditures (PCND), as purchasing such goods in one period does not affect decisions in the following periods. In obtaining the real series, we used the implicit price deflator for nondurable PCE goods (DNDGRD3Q086SBEA) instead of the implicit GDP deflator and repurposed this implicit deflator for the basic goods inflation rate. Because the production of consumer goods requires industrial equipment, it can be mapped to the production of manufactured nondurable (and non-energy) goods (IPB51210S). Likewise, we are using the observed capacity utilization rate for the nondurable goods sector (CAPUTLGMFNS) to contrast it with the simulated series. As for aggregate investment decisions, we focus on industrial equipment

5 We delay the explanation of its last column to Section 4.4.

(A680RC1Q027SBEA), the closest sub-component to capital goods in the simulated models. This approach, mirroring that used for the capital goods sector, employs the implicit price deflator for industrial equipment at the finest level of disaggregation available. Additionally, the pairing basic goods/nondurable goods is used to compare labor productivity in the consumption goods sector by contrasting the simulated series with labor productivity in the manufacturing sector (OPHMFG). Although this decision has certain limitations, it is primarily driven by data availability.

For government expenditures, federal nondefense government consumption (A542RC1Q027SBEA) is used for comparison with the corresponding model variable, as public investment and defense expenditures are excluded from the models. The real government consumption series is derived by applying the federal nondefense implicit deflator (A825RD3Q086SBEA); however, since no equivalent exists in the models, it is not used further. Additionally, total federal debt as a share of GDP (GFDEGDQ188S) is directly contrasted with the model counterpart.

Labor-market variables are examined by comparing model-generated data with the general unemployment rate. For real wages, hourly compensation for all workers in the nonfarm business sector (COMPNFB) is used and deflated using the GDP implicit deflator (GDPDEF). Although the model presents specific wages and unemployment by sector, we employ aggregate variables because the weight of the labor market in the capital-goods sector is very tiny. All data selected and corresponding variables in the model are listed in Table 4.3.

Private inventories are not considered due to the high level of aggregation of empirical series. Similarly, variables related to the financial sector are omitted primarily due to data availability constraints, which limit establishing a credit/debit relationship in the analysis. Future stages of this comparative exercise may incorporate additional data sources to address these limitations.

4.4 Multidimensional Similarity Index

4.4.1 Motivation and Data Pre-processing

This section presents a systematic approach to capturing various aspects of similarity between a set of simulated models and their empirical counterparts. Our point of departure is that an analysis limited to only one aspect of similarity—as the Euclidean distance, or statistical moments—fails to properly capture all information embedded into signals. For this reason, we propose a multidimensional composite index of similarity. The key principle guiding the development of this protocol is to create a versatile and scalable procedure that can be applied to different pairs of objects across the time domain, regardless of whether they originate from ABMs. This multidimensional approach is necessary because of the inherent

Table 4.3 Empirical time series and corresponding equivalents in the models. Real series are estimated using the deflators indicated in the text.

FRED Code	Empirical series	Simulated variables	Transformation
GDPC1	Real GDP	GDP	Growth rate
PCND	Nominal personal nondurable goods expenditure	ScBas	Growth rate
DNDGRD3Q086SBEA	Nondurable goods implicit price index	CPIbas	Growth rate
A542RC1Q027SBEA	Nominal federal non-defense government expenditure	Gnom	Growth rate
A680RC1Q027SBEA	Private investment on industrial equipment	Inom	Growth rate
Y033RD3Q086SBEA	Private equipment implicit price index	PPI	Growth rate
Y006RC1Q027SBEA	R&D	RDnom	Growth rate
Y001RD3Q086SBEA	Intellectual property products implicit price index	–	–
PRS32006093	Labor productivity	A	Growth rate
UNRATE	General Unemployment Rate	U	Divided by 100
IPG333S	Industrial production of machinery	Q1	Growth rate
IPB51210S	Nondurable and nonenergy industrial production	QcBas	Growth rate
CAPUTLGMFNS	Capacity utilization rate in the nondurable manufacturing sector	QcUbas	Divided by 100
GFDEGDQ188S	Total public debt as percent of GDP	DebGDP	Divided by 100

complexity of time series data and the absence of a single measurement capable of capturing all relevant aspects of similarity at once.

In order to motivate the importance of building a multidimensional similarity index, we start with a counter-example, showing the case of a unidimensional similarity metric and the weakness of such an approach. Figure 4.1 presents the trajectories of two time series. Suppose that the solid gray one corresponds to the simulated series produced by a computational model, while the solid black line is the empirical counterpart that this series was designed to represent. One immediate way to measure how similar these two series are is to compute the point-wise

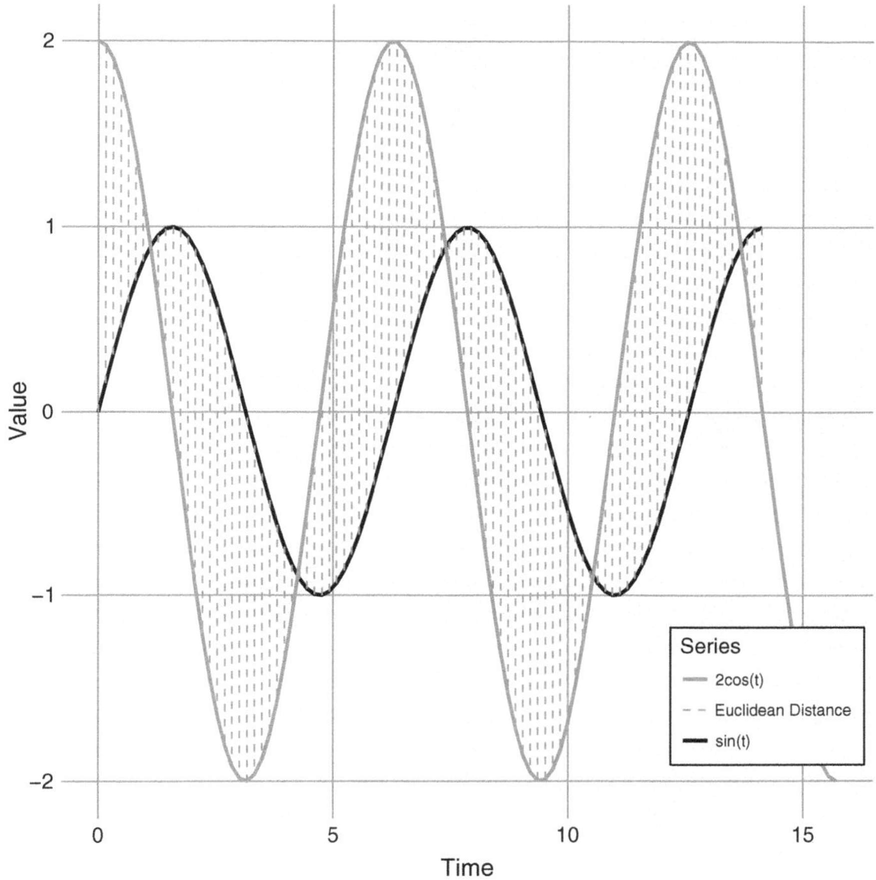

Figure 4.1 Example of measuring similarity of two generic and misaligned time series.

Euclidean distance between them, indicated by the dashed lines. A naive interpretation of this metric might suggest that the model lacks closeness to empirical data because the two series are substantially distant. This conclusion changes when we reveal that the data-generating processes of the two series are $\sin(t)$ and $2\cos(t)$, implying that although the data-generating processes are different, the two series might overlap by means of a time shift transformation. This simple example highlights the consequences of disregarding time shifts when measuring similarity.

Time shift is not the only drawback of the use of a Euclidean distance. Because computational models can generate as many data points as desired, series lengths may differ. Additionally, models may produce series with mismatches in scale and frequency. Labeling a model as similar/dissimilar without properly addressing these issues might lead to wrongly ranking model distance vis-à-vis the data structure.

Series with similar temporal structures may be deemed dissimilar solely due to differences in scale. Typically, nonstationary series may initially share a similar scale but quickly diverge. While taking the first difference is a standard method to transform a nonstationary series into a stationary one, it does not address the scaling issue. To mitigate this issue, we calculate growth rates when comparing two non-stationary series, thereby minimizing the problem of mismatched scales. The last column of Table 4.3 denotes whether the series is being considered in terms of growth rates. Whenever further normalization is necessary, we utilize the z-score recommended by Paparrizos and Gravano [37].[6] This normalization is applied to the cyclical components of the series while computing growth rates to control scale mismatches in trends if the original series is nonstationary.

Another source of incomparability is length mismatches. Although some similarity measurements can handle differing lengths, others cannot. A common approach is to interpolate the shorter series to match the length of the longer one or to pad zeros to the shorter series. However, this requires another layer of decisions, such as the interpolation method, which can introduce artifacts into the signals. Instead, we truncate the longer series by removing excess time steps to ensure that both series have the same length. This procedure is not harmful in the case of ergodic series. The next decision concerns the placement of truncation. We opt for a centered downsampling, ensuring that an equal number of observations are dropped from the beginning and end of the series.

To provide an intuitive understanding of our protocol, Figure 4.2 visually illustrates the concept of multidimensional similarity. We rely on multiple indices to measure different aspects of similarity between time series, as indicated by the inputs on the left side of the figure. Each index will be thoroughly explained in

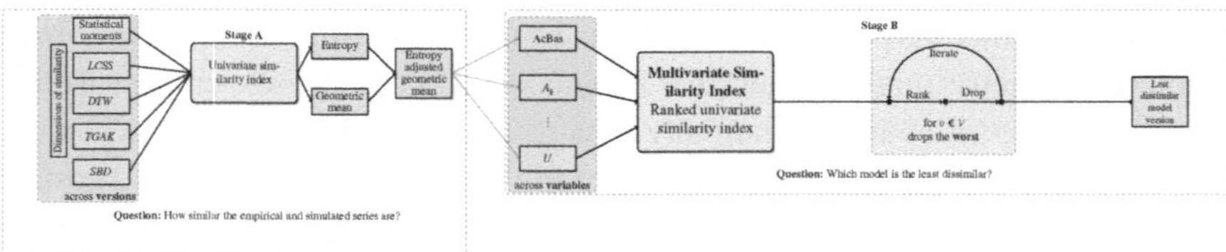

Figure 4.2 Flowchart representation of the protocol.

6 See Keogh and Kasetty [38] for a discussion about the necessity of normalization before measuring the distances.

the following subsections, and we will revisit the motivation example when appropriate. Each measurement is computed for all models for a single variable. This process is repeated for each variable across different model versions represented by the inputs on the right side. This results in a univariate multidimensional similarity index, which will later be used to develop a ranking system for model selection. For the sake of clarity, we defer the details of the computation of our multidimensional index to Section 4.4.7.

We now introduce a general notation to ensure broad applicability across different model classes. Let $\mathbf{x}$ represent the empirical time series vector (reference) with a length of m. The models discussed in Section 4.2, denoted as $\mathcal{V}$, will be considered. A generic model version is indexed by v, enabling us to represent the simulated (query) time series vector for model v as $\mathbf{y}_v$, with a time-length of n data points. It is important to note that the reference/query pair may not have the same length, with $m \ll n$ being a typical case. Some similarity measures require all possible trajectory alignments between two series, denoted as $\mathcal{A}(m, n)$, while $\pi \in \mathcal{A}(m, n)$ represents one possible trajectory that belongs to this space. Specific alignments have a special property referred to as $\pi^\star$, while some measurements impose restrictions on the alignment space, denoted as ω. Using $\mathcal{M}$ to represent a generic index, we define the normalized distance as $\mathcal{D}_\mathcal{M}$, and the corresponding similarity as $\mathcal{S}_\mathcal{M} = 1 - \mathcal{D}_\mathcal{M}$ when the respective distance measurement is a metric and limited to closed range.

In the following subsections, we describe the individual measurements that compose the multidimensional similarity index. The index shows how closely different K+S model versions resemble their empirical counterparts. All similarity and distance measurements are applied to trend, cyclical, and unfiltered components of the series. When presenting each measurement, we use the capacity utilization rate as a variable of reference, as this series does not require any further transformation.

4.4.2 Statistical Moments

The first measurement, likely one of the most intuitive, evaluates the difference between selected summary statistics of the simulated and empirical series. This index aims to measure dissimilarity in terms of the probability distribution of the series. We consider the mean, standard deviation, skewness, kurtosis, and the number of structural breaks. This selection is guided by the significance of these moments in capturing key patterns in time series data, particularly for economic series.

Each moment provides distinct insights into the distributional characteristics of the data. The mean μ conveys both the central tendency and the scale of the distribution. The standard deviation (SD) measures volatility, a fundamental aspect of business cycles [39, 40, and references therein]. Skewness ($Skew$) and kurtosis ($Kurt$) further describe the shape of the distribution. Skewness measures asymmetry and is relevant for analyzing the magnitude and duration of fluctuations in economic time series [31, 41, 42]. Kurtosis evaluates tailedness and the presence of fat tails, which are empirically observed features of output growth rates [32, 43–45], for instance. Finally, the inclusion of structural breaks (SB) accounts for sudden changes in both simulated and empirical series using the [46] method.

Let $\mu_{\mathbf{x}}$ denote the mean of empirical series and $\mu_{\mathbf{y}_v}$, denote the mean of the simulated series v. The relative deviation of the mean for model v, denoted as ($\text{rel}_{\text{mean},v}$) is computed as follows:

$$\text{rel}_{\text{mean},v} = \left| \frac{\mu_{\mathbf{y}_v} - \mu_{\mathbf{x}}}{\mu_{\mathbf{x}}} \right| \cdot \gamma \qquad \forall v \in \mathcal{V}.$$

Normalization using the empirical mean mitigates scale mismatches. This procedure is applied consistently to standard deviation, skewness, and kurtosis, each normalized by its empirical counterpart. Specifically, for skewness and kurtosis, we introduce a penalty of $\gamma = 2$ when their sign differs from the empirical series and set $\gamma = 1$ otherwise. The computation of the distance in terms of the number of structural breaks is slightly different, as shown below:

$$\text{rel}_{\text{SB},v} = \left| \frac{SB_{\mathbf{y}_v} - SB_{\mathbf{x}}}{\max(1, SB_{\mathbf{x}})} \right| \cdot \gamma \qquad \forall v \in \mathcal{V}.$$

The modification in the denominator ensures that the metric remains well-defined even when the empirical series has no structural breaks.

After computing the relative deviations for the selected moments, we take their geometric mean to form a composite index. This measure indicates how much a given simulated series deviates from the empirical counterpart. A value closer to zero suggests higher similarity. Note that this index is nonmetric, ranging from $[0, \infty)$ meaning there are no strict boundaries. Thus, we define only the distance measure $\mathcal{D}_{\text{statistical}}$ without an absolute similarity scale.

Table 4.4 presents the computed statistical moments for the capacity utilization rate and their relative deviations from the empirical series (indicated in bold). This allows for a direct assessment of how different K+S model versions replicate key distributional properties. Examining each moment individually, we observe that the original version (`K+S-original`) performs best in terms of mean and skewness, aligning more closely with the empirical values, while the multi-sector model (`K+S-multi`) exhibits the highest deviation in both moments. Besides the good relative performance in terms of the mean, the original version does not perform well in terms of standard deviation, displaced by the finance-augmented

Table 4.4 Statistical moments similarity index contrasting simulated and empirical capacity utilization series across different K+S model versions.

	K+S-Comp	K+S-finance	K+S-Ford	K+S-multi	K+S-original	Empirical
Mean	0.9887	0.7099	0.9839	0.3972	0.7028	0.8061
Mean diff	0.2265	0.1193	0.2205	0.5072	0.1282	0.0000
SD	0.0002	0.0130	0.0016	0.0107	0.0040	0.0416
SD diff	0.9953	0.6879	0.9610	0.7414	0.9047	0.0000
Skewness	−0.1934	−0.0516	−0.8938	0.6245	−0.4619	−0.4109
Pen. Skewness diff	0.5292	0.8743	1.1753	5.0396	0.1241	0.0000
Kurtosis	−0.3428	−0.8460	0.2072	−0.3846	−0.4024	−0.5809
Pen. Kurtosis diff	0.4099	0.4565	2.7134	0.3379	0.3072	0.0000
Breaks	2.0000	5.0000	3.0000	4.0000	4.0000	4.0000
Breaks diff	0.5000	0.2500	0.2500	0.0000	0.0000	0.0000
Statistical moments similarity index	0.5125	0.4517	0.8908	0.8421	0.2586	0.0000

(`K+S-finance`). Regarding structural breaks, most models display similar occurrences, except for `K+S-Comp`, which presents a lower frequency. In terms of skewness, the multi-sector version (`K+S-multi`) stands out for having an opposite sign compared to the empirical series, whereas the Fordist version (`K+S-Ford`) shows a similar discrepancy in kurtosis. Aggregating these aspects—using the geometric mean—the statistical moments similarity index identifies the original version as the closest to the empirical data. This suggests that, among the given alternatives, it better captures the distributional attributes of the observed capacity utilization series.

4.4.3 Longest Common Subsequence

Time series data are characterized by their dynamic nature, often representing evolving entities that do not necessarily progress at the same pace or share the same length. As a consequence, an appropriate metric is needed to measure the similarity of the trajectories between the two series. Wagner and Fischer [47] introduced a method later called edit-based distance, which measures the cost of transforming one sequence into another through a series of insertions, deletions, or substitutions. This approach has been widely employed in diverse fields, including writing recognition and biological pattern identification [48, 49]. Based on this method, [50] proposed the longest common subsequence (LCSS), a modification of the edit-based framework explicitly tailored for time series analysis.

The LCSS distance is computed by first constructing a distance matrix that classifies pairs of points from time series $\mathbf{x}$ and $\mathbf{y}_v$ as either matching or nonmatching. A pair (x_i, y_j) is considered a match if its Euclidean distance is smaller than or equal to a given threshold ϵ; in this case, their distance is set to 0; otherwise, it is set to 1. Using this matrix, a dynamic programming (DP) algorithm iteratively builds a cost matrix $C(i, j)$, where each entry represents the length of the LCSS found up to that point. The cost matrix is updated using the recurrence relation:

$$C(i, j) = \begin{cases} C(i-1, j-1) + 1, & \text{if } |x_i - y_j| \leq \epsilon \quad \text{(match)} \\ \max(C(i-1, j), C(i, j-1)), & \text{otherwise (gap)} \end{cases}.$$

If x_i and y_j match, the LCSS count increases based on previous matches; otherwise, the algorithm propagates the maximum value from the adjacent entries, effectively allowing gaps without imposing an explicit penalty. This flexibility enables LCSS to handle time shifts and noise by permitting unmatched regions while focusing on the longest sequence of similar values. Another notable feature of this index is the ability to accommodate cases where the length of the series diverges ($m \neq n$). The final entry of the cost matrix, $C(m, n)$, contains the length of the LCSS distance measurement between the two series:

$$\text{LCSS} = C(m, n). \tag{4.1}$$

By definition, a larger value of LCSS indicates a greater degree of similarity, implying that the two sequences share a longer contiguous subsequence within the given tolerance ϵ. To ensure comparability across time series of different lengths, the similarity index is normalized by the length of the shorter series:

$$\mathcal{S}_{\text{LCSS}} = \frac{\text{LCSS}(\mathbf{x}, \mathbf{y}_v, \epsilon)}{\min(m, n)},$$

which allows us to compute the *normalized* LCSS distance measurement as:

$$\mathcal{D}_{\text{LCSS}} = 1 - \mathcal{S}_{\text{LCSS}}$$

Figure 4.3 revisits the prototypical example and presents the length of LCSS under different ϵ values. The occurrence in the matching condition is indicated with gray lines. This example illustrates the relevance of the choice of ϵ in determining the sensitivity of the LCSS measure: a larger ϵ increases the likelihood

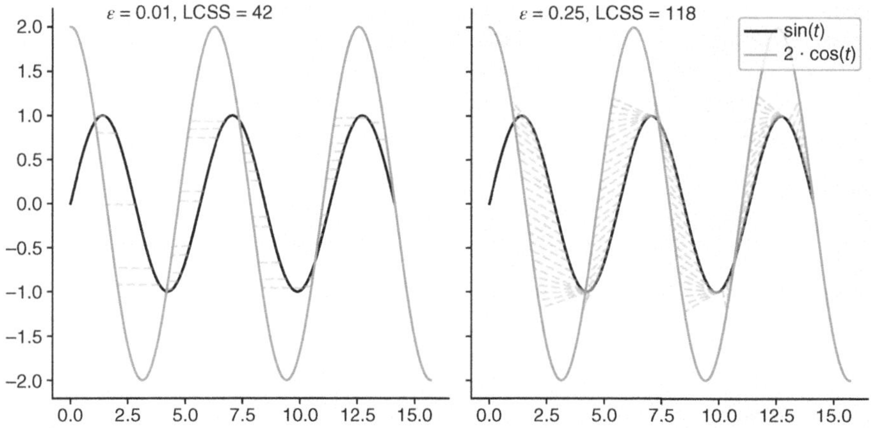

Figure 4.3 Time series similarity using LCSS for the prototypical example under different ϵ values.

of matching between observations, thereby producing a longer common subsequence. To mitigate the discretion in selecting ϵ, an alternative approach is to compute an overall distance measure by averaging across multiple distance computations up to a predefined maximum threshold, $\epsilon_{max} = 0.005$.[7] The interpretation of ϵ is meaningful only when applied to time series that are consistently scaled. For example, directly comparing the GDP levels across different series may yield few common subsequences due to differences in scale. To address this, it is often necessary to preprocess data. This is why we transformed variables into growth rates where relevant, as indicated in the last column of Table 4.3.

Table 4.5 presents the length of the LCSS for the simulated and observed capacity utilization rate according to Eq. (4.1) under different values of ϵ. As expected, relaxing the restriction on ϵ leads to increased matches. A closer inspection of the index reveals that the original (`K+S-original`) and finance-augmented (`K+S-finance`) versions outperform the others, regardless of the threshold value selected. Notably, the other model versions fail to have any common subsequence with the observed data, suggesting the need for finer calibration. Still, this should not be interpreted as a justification for discarding such models outright; instead, it highlights the importance of evaluating similarity across multiple dimensions.

4.4.4 Dynamic Time Warping

When using a pointwise method to compare two sequences, there is a risk of falsely identifying them as dissimilar when they are simply out of phase. This issue, as illustrated in the opening example of this section, can lead to the exclusion of

7 In the context of growth rates, the value of $\epsilon_{max} = 0.005$ is not overly stringent.

Table 4.5 LCSS index contrasting simulated and empirical capacity utilization series across different K+S model versions.

	K+S-comp	K+S-finance	K+S-Ford	K+S-multi	K+S-original
0.001	0	6	0	0	2
0.005	0	7	0	0	4
0.01	0	8	0	0	5
0.015	0	13	0	0	6
0.025	0	29	0	0	8
0.05	0	60	0	0	33
0.075	0	171	0	0	64
0.1	0	272	0	0	249

relevant candidates. The dynamic time warping (DTW) algorithm addresses this problem by providing a distance measure robust to time shifts, making it well-suited for comparing nonaligned time series [51, 52].[8] Due to this property, DTW has been widely applied across various fields, including speech recognition [55, 56], gene expression analysis in bioinformatics [57, 58], handwriting comparison [59], and, to a lesser extent, economics [60–62]. As an alternative to conventional L_p norms, such as Euclidean distance (L_2 norm), DTW is widely regarded as a state-of-the-art approach for sequential pattern matching [52, 63].

The DTW algorithm aims to compute the optimal alignment between two sequences by minimizing the distance between corresponding data points, even when they are not temporally aligned (as Euclidean distance requires). Instead, DTW allows for temporal shifts by aligning points along the time axis [64]. The alignment cost is computed iteratively by recursively updating the distortion function d, which measures discrepancies between elements of the two sequences. The distortion cost is updated for $i = 2, 3, \dots, m$ and $j = 2, 3, \dots, n$, while the direction of alignment follows a predefined step function. Following [63], we adopt the symmetrical step pattern with no slop restrictions, initially proposed by Sakoe and Chiba [55], which remains a standard choice in the literature [65]. The cost of aligning two sequences using DTW is formally defined as follows:

$$D(i,j) = d(x_i, y_{v,j}) + \min \begin{cases} D(i-1,j) & (\rightarrow) \\ D(i,j-1) & (\uparrow) \\ D(i-1,j-1) & (\nearrow) \end{cases} . \tag{4.2}$$

8 We refer to Kruska and Liberman [53] for a more detailed discussion about DTW and [54] for its implementation.

Equation (4.2) represents the recurrence relation for computing the alignment cost along the step pattern. The alignment proceeds monotonically by either moving right (skipping an element in the observed sequence), upward (skipping an element in the simulated sequence), or diagonally (aligning both). Consistent with Sardá-Espinosa [54], we use the L_1 (Manhattan distance) as the discrepancy function for computing the distortion costs.

After computing the distortion costs, the algorithm constructs an $m \times n$ local cost matrix (LCM), which accumulates alignment costs along the alignment space $\mathcal{A}(m, n)$.[9] Starting from the first element of each sequence $(x_1, y_{v,1})$ and proceeding to the last $(x_m, y_{v,n})$, the LCM enables the identification of multiple possible warping paths (π) that traverse the alignment space. One limitation of DTW is that it requires the first and last elements of the query/reference pair to be aligned as a boundary condition. In other words, it needs to start at x_1, y_1 and end at x_m, y_n. Some modifications can relax this restriction by allowing the algorithm to choose the starting/ending point with the lowest alignment costs. However, these modifications can lead to matches with little significance. For instance, the algorithm might suggest a warping path in which only a small portion $m' \ll m$ of the sequences are aligned. In other words, the resulting alignment could show that the series are alike, but only for a short time window. Thus, the decision was made to maintain the first/last alignment constraint. These paths define how indices (i.e., time steps in the case of time series) are remapped between two sequences, effectively capturing distortions in the time axis. DTW aims to determine the optimal warping path $\pi^\star$ that minimizes the cumulative distortion over time. This is expressed as:

$$\mathcal{D}_{DTW} = DTW(\mathbf{x}, \mathbf{y}_v) = \min_{\pi \in \mathcal{A}(m,n)} D_{\mathbf{x},\mathbf{y}}(\pi).$$

At this stage, it is important to emphasize that our goal is not to align the sequences directly but rather to measure their similarity. Hence, our primary focus is on computing distortion costs rather than optimizing the alignment.

Before delving into how DTW is applied in our comparative exercise, we revisit the motivating example presented at the beginning of this section. This example highlights two common issues in time series similarity measurement: misalignment (time shifts) and scale mismatches. Figure 4.4 exemplifies how the DTW algorithm works. In all subfigures, the horizontal axis represents the time indices of the $\sin(t)$ series, while the vertical axis corresponds to the time indices of $2 \cdot \cos(t)$. The cumulative local cost matrix for our prototypical example, computed using Eq. (4.2), is in the main plot of Figure 4.4a. The solid line represents the

9 One drawback of this algorithm is the computational cost of obtaining the matrix, which is of the order of $\mathcal{O}(m \cdot n)$. These aspects also motivate the departures from it, which will be discussed in the Section 4.4.5.

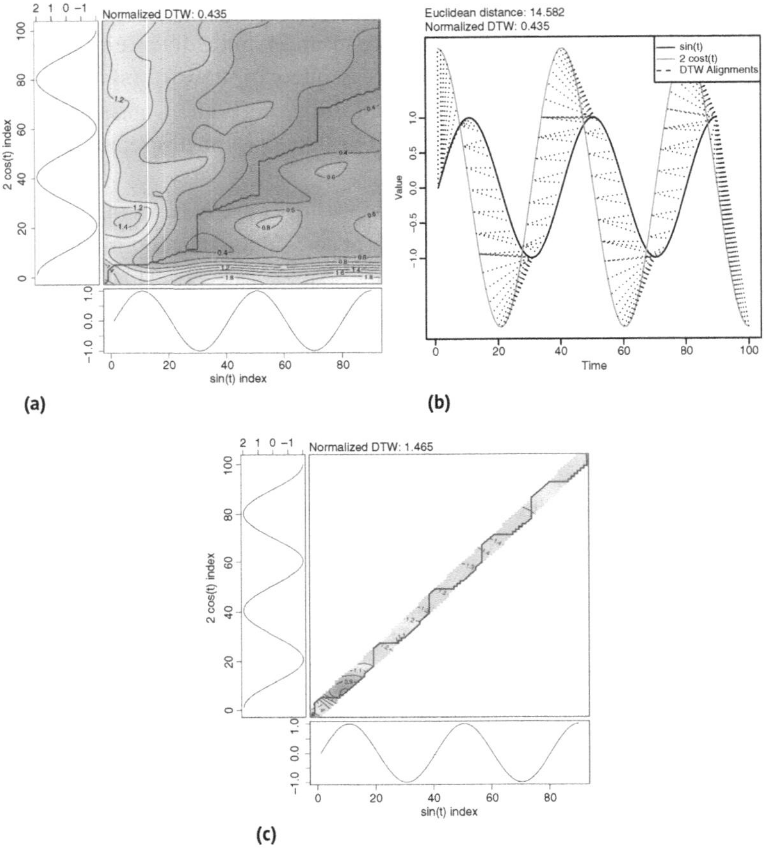

Figure 4.4 Computing similarity using DTW: Revisiting the prototypical example: (a) cumulative cost matrix and sequence alignments (no constraints); (b) time series alignment using DTW; (c) cumulative cost matrix and sequence alignment (with global constraints).

warping path, which traverses regions of minimal distortion cost within the alignment space. This warping path provides an optimal alignment of the two series, as shown in Figure 4.4b. As demonstrated in this figure, the Euclidean distance substantially overestimates dissimilarity compared to DTW, underscoring DTW's robustness to time shifts.

DTW ensures that the warping path follows the trajectory with the least temporal distortion. Still, this alignment may not always be economically meaningful. To address this issue, we impose global window constraints (ω), which restrict the warping path to align points within ω time periods, regardless of the distortion costs involved.[10] Paths falling outside this range are excluded from the analysis, even when producing a costless alignment. By imposing these constraints, we ensure that the computed alignments remain within an economically relevant time frame.[11] Figure 4.4c illustrates the distortion cost between the two series, restricted to a defined window. The resulting distance measurement increases as the algorithm is no longer allowed to traverse paths with minimal temporal distortion.

Having established the operationalization of DTW, we now outline its application in our analysis. Our approach involves computing the normalized DTW distance to align various model versions with a common empirical reference.[12] A lower DTW distance indicates greater similarity, with a zero value representing perfect alignment.[13] Table 4.6 presents the results for the simulated and observed capacity utilization rate time series. The distance computation is performed under different window sizes. As expected, increasing the window size reduces alignment costs, as the algorithm can explore a broader range of warping paths. To summarize these results, we compute the average distance up to a maximum window size ($\omega_{max} = 8$), which is further utilized for constructing the multidimensional similarity index.

Table 4.6 DTW index contrasting simulated and empirical capacity utilization series across different K+S model versions.

	K+S-Comp	K+S-finance	K+S-Ford	K+S-multi	K+S-original
2	0.177	0.090	0.172	0.402	0.097
4	0.172	0.084	0.167	0.396	0.092
8	0.166	0.075	0.161	0.387	0.083
20	0.158	0.060	0.153	0.372	0.068
40	0.153	0.047	0.148	0.357	0.054
60	0.149	0.042	0.145	0.352	0.050

10 In signal processing, various windowing procedures have been described in the literature. Sakoe and Chiba [55] introduced one of the most well-known methods. However, this technique applies when the time series dimensions are equal ($m = n$). To address cases where $m \neq n$, rather than employing downsampling, we have chosen to utilize a slanted band window function [65].
11 Another alternative would be to adopt a lower-bound DTW proposed by Refs. [64, 66]. However, this algorithm only applies in cases where $m = n$.
12 Normalization is important to yield comparable results in cases where $m \neq n$.
13 It is worth noting that this distance is nonmetric—not defined in a closed range—so only $\mathcal{D}_{\text{DTW}}$ is defined. Hence, additional re-scaling procedures are necessary to enable comparison with other measurements. Further details on this topic will be discussed in Section 4.4.7.

The similarity index for these specific sequence pairs reveals that the finance-augmented model version (`K+S-finance`) shows the lowest dissimilarity, indicating minimal temporal distortion relative to the empirical series. The multi-sector version (`K+S-multi`), on the other hand, has the highest time distortion. Notably, the original model version (`K+S-original`) ranks second in similarity. The two labor-augmented institutional settings produce relatively close results. As these two models share the same structure, this outcome is expected.

4.4.5 Triangular Global Alignment Kernel

A key limitation of DTW is that it only explores a subset of possible alignments between two sequences, constrained by a predefined step pattern. This restriction prevents DTW from considering the full spectrum of alignments in the alignment space $\mathcal{A}(m, n)$. Cuturi et al. [67] introduced the Global Alignment Kernel (GAK), a kernel-based method that measures similarity by aggregating contributions from all possible alignments to address this restriction. Unlike DTW, which focuses on the minimum-cost alignment, GAK computes the exponentiated soft-minimum of all alignment costs, as defined in Eq. (4.3).[14] This approach provides a more comprehensive assessment of dissimilarity, capturing both strong and weak alignments. As a result, GAK offers a richer similarity measure compared to DTW, particularly when sequences have significant divergence.

The GAK is formally defined as:

$$k_{\mathrm{GA}}(\mathbf{x}, \mathbf{y}_v) = \sum_{\pi \in \mathcal{A}(m,n)} \prod_{i=1}^{|\pi|} \kappa\left(\pi_x(i), \pi_{y,v}(i)\right), \tag{4.3}$$

where $\kappa(\mathbf{x}, \mathbf{y}_v) = \exp \phi_\sigma(\mathbf{x}, \mathbf{y}_v)$ is a local kernel computed over the aligned points in $\mathbf{x}$ and $\mathbf{y}_v$, while ϕ_σ is a negative definite kernel given by:

$$\phi_\sigma(\mathbf{x}, \mathbf{y}_v) = \frac{1}{2\sigma^2}\|\mathbf{x} - \mathbf{y}_v\|^2 + \log\left(2 - \exp - \frac{\|\mathbf{x} - \mathbf{y}\|^2}{2\sigma^2}\right). \tag{4.4}$$

Building on the GAK framework, [63] introduced the triangular global alignment kernel (TGAK), which incorporates a triangular local kernel $T(i, j)$ to constrain the alignment space—as the global constraint in DTW.[15] Unlike DTW, where a larger constraint window typically reduces the distortion cost, TGAK does not guarantee that relaxing the window constraint will lead to lower dissimilarity.

14 The computation of the kernel requires estimating a hyperparameter (σ), and we adopt the suggestion of Cuturi [63] implemented by Sardá-Espinosa [54], which relies on a subsample of both series. Unlike previous methods, this sampling introduces randomness into the distance estimation, which can be controlled by setting a random seed.

15 According to Cuturi [63], one technical advantage of using the triangular kernel is reduce the complexity from $\mathcal{O}(mn)$ to $\mathcal{O}(\min(m, n))$.

A larger ω window may result in higher dissimilarity due to the inclusion of sub-optimal alignments. The TGAK is defined as:

$$T(i,j) = \left(1 - \frac{|i-j|}{\omega}\right)_+ \tag{4.5a}$$

$$\text{TGAK}(\mathbf{x}, \mathbf{y}_\upsilon, \sigma, \omega) = \tau^{-1}\left(T \otimes \frac{1}{2}\kappa\right)(i, \mathbf{x}; j, \mathbf{y}_\upsilon) = \frac{T(i,j)\kappa(\mathbf{x}, \mathbf{y}_\upsilon)}{2 - T(i,j)\kappa(\mathbf{x}, \mathbf{y}_\upsilon)}, \tag{4.5b}$$

where τ^{-1} is the inverse mapping function, and ω controls the width of the triangular kernel.[16]

Figure 4.5 exemplifies the local similarity matrix (LSM) computed with the local kernel using Eq. (4.4) for the prototypical $\sin(t)$ and $2\cos(t)$ example. Each entry of this matrix contains pairwise similarities between individual elements of the two time series being compared. The difference between the two subfigures is the presence of a global constraint over the alignment space. In the case paired with the triangular kernel (Figure 4.5b), the similarity is reduced if the compared points are far apart from the time restriction modulated by ω. This ensures that only temporally close points contribute substantially to the overall similarity measurement.

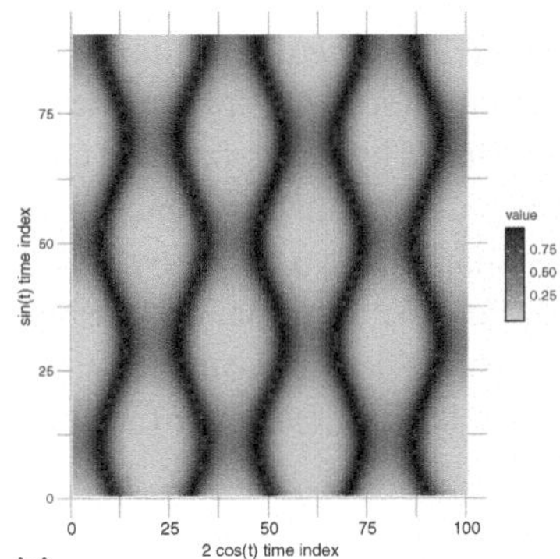

(a)

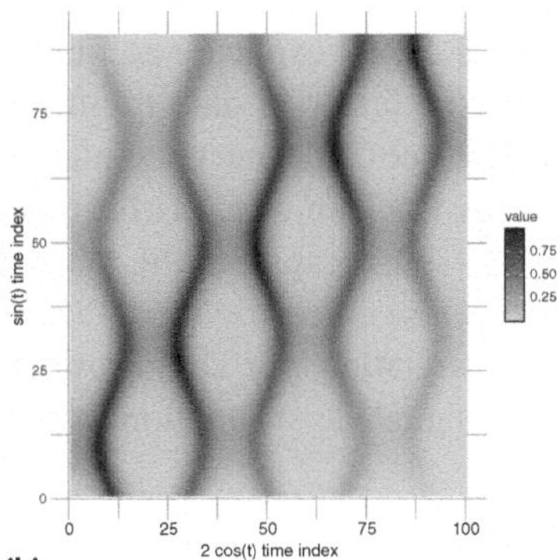

(b)

Figure 4.5 Local similarity matrix (LSM) for the prototypical example: (a) LSM without global constraint; (b) LSM paired with triangular kernel.

16 A practical challenge with GAK is diagonal dominance, which occurs when the lengths of the sequences being compared differ considerably ($2 \cdot m \ll n$ or $m \gg 2 \cdot n$). In such cases, the kernel values $\kappa(\mathbf{x}, \mathbf{y}_\upsilon)$ may be biased toward self-alignments, distorting the similarity analysis [67]. This is a probable scenario as we encounter cases where $m \ll n$. To mitigate this issue, we apply centered downsampling, truncating the longer sequence by removing an equal number of elements from both the beginning and the end. This ensures that the lengths of the sequences are comparable, reducing the risk of diagonal dominance.

The next step consists of accumulating the similarity scores for all possible alignments. By doing this, TGAK returns similarity scores ($\mathcal{S}_{\text{TGAK}} = \text{TGAK}$) ranging from zero (perfect similarity) to one (maximum dissimilarity), providing a proper metric distance $\mathcal{D}_{\text{TGAK}} = 1 - \mathcal{S}_{\text{TGAK}}$. Similar to the implementation of DTW, described in Section 4.4.4, we explore different global constraints ω, averaging results up to $\omega_{\text{max}} = 8$.

Table 4.7 presents the results of applying TGAK to compare empirical capacity utilization across different model versions. The results show that the similarity scores are numerically close across all model versions, regardless of the window constraint ω. The finance-augmented model (`K+S-finance`) produces the most similar series, closely followed by the original model (`K+S-original`), while other models fall within a narrow range.[17] This suggests that TGAK alone may not isolate the most similar model. To refine the analysis, we use a min-max normalized distance index, which provides a finer-grained assessment to compare similarity. This highlights the need for a multidimensional similarity approach to distinguish between closely related models, discussed in Section 4.4.7.

4.4.6 Shape-based Distance

The previous indices do not explicitly account for the lag structure of the series, a commonly neglected factor in data mining literature (see [68] for a review). The shape-based distance (SBD) measurement, introduced by Paparrizos and Gravano [37], addresses this gap and has been successfully applied in time series clustering [69]. This index captures the temporal shape of two series by utilizing the normalized cross-correlation ($\text{NCC}(\mathbf{x}, \mathbf{y}_\upsilon)$, Eq. 4.6). In signal processing, normalized cross-correlation—or sliding inner-product—is a measure of similarity between two signals as a function of the time lag slid over one of them.

Table 4.7 TGAK index contrasting simulated and empirical capacity utilization series across different K+S model versions.

	K+S-Comp	K+S-finance	K+S-Ford	K+S-multi	K+S-original
2	0.023	0.020	0.023	0.023	0.021
4	0.026	0.022	0.025	0.026	0.023
8	0.027	0.024	0.027	0.028	0.024
20	0.029	0.025	0.029	0.030	0.026
40	0.030	0.026	0.030	0.030	0.027
60	0.030	0.026	0.030	0.031	0.027

17 It is worth noting that these findings are consistent with the ones produced by DTW.

The result is scaled to a range of [−1, 1], facilitating interpretable comparisons.[18] For two sequences $\mathbf{x}$ and $\mathbf{y}_v$ the discrete form normalized cross-correlation is given as:

$$\text{NCC}(\mathbf{x}, \mathbf{y}_v, k) = \frac{\text{CC}(\mathbf{x}, \mathbf{y}_v, k)}{\sqrt{\sum_{n=1}^{n}(x[n] - \bar{x})^2 \cdot \sum_{n=1}^{n}(y[n+k]_v - \bar{y}_v)^2}} \tag{4.6}$$

$$\text{CC}(\mathbf{x}, \mathbf{y}_v, k) = \sum_{n=1}^{n}(x[n] - \bar{x})(y[m+k]_v - \bar{y}_v), \tag{4.7}$$

where $CC(\mathbf{x}, \mathbf{y}_v)$ is the cross-correlation between $\mathbf{x}$ and the shifted version of $\mathbf{y}_v$, and k represents the time index used to compute the shift.

The cross-correlation is computed over $m + n - 1$ time steps. The algorithm identifies the lag i at which the two series have the highest cross-correlation, effectively capturing the similarity of out-of-phase signals (see Eq. 4.8). Like dynamic time warping (DTW) and the TGAK, SBD is robust to temporal shifts, making it particularly suitable for comparing time series with phase differences.

$$\text{SBD}(\mathbf{x}, \mathbf{y}_v) = 1 - \max\{(\text{NCC}(\mathbf{x}, \mathbf{y}_v, k))\}_{k=1}^{k=m+n-1}. \tag{4.8}$$

The SBD metric is constrained to the range [0, 2], where zero indicates identical temporal shapes between the two series. For comparability with other measurements, we normalize the results by dividing them by two (details in Section 4.4.7). As SBD provides a closed-form distance measurement, we can represent the results in terms of the distance $\mathcal{D}_{\text{SBD}} = \text{SBD}(\mathbf{x}, \mathbf{y}_v)$ and similarity $\mathcal{S}_{\text{SBD}} = 1 - \mathcal{D}_{\text{SBD}}$.

Before applying SBD, several practical issues must be addressed. First, the metric is sensitive to the scale of the input series. Following [37], we standardize each series by computing its z-score, ensuring scale invariance. Second, length mismatches between series can affect results as the SBD implementation utilizes the convolution theorem to reduce the computational complexity. In cases of length mismatch, zero-padding is applied during convolution to equalize series lengths. As this operation may introduce artifacts into the series, potentially distorting the signals, we adopt the centered downsampling approach, resizing both series to match the length of the shorter one.

Figure 4.6 illustrates the SBD alignment process for the prototypical example. Figure 4.6a shows the normalized cross-correlation (after z-normalization) between two series, while Figure 4.6b displays the same series after shifting one of

18 While the signal processing definition emphasizes time-domain analysis and signal detection, the statistical and econometric interpretation focuses on assessing the linear relationship and lead-lag dynamics between variables, often using a similar normalization to quantify correlation strength.

them by the lag that maximizes cross-correlation. Visual inspection confirms the robustness of SBD to phase differences, as it correctly identifies the shift required to realign the sequences. The distance measure is computed at this optimal lag, reflecting the shape similarity of the series.

Table 4.8 presents the similarity analysis of simulated series compared to the empirical capacity utilization, using the $\mathcal{D}_{\text{SBD}}$ distance measure. Because SBD requires no additional parameterization, we apply it directly to the raw (unfiltered), trend, and cyclical components of the series. For the unfiltered series, the results align with earlier findings: labor-augmented versions perform relatively worse, while the original (`K+S-original`) model has the least dissimilarity. The finance-augmented (`K+S-finance`) and multi-sector (`K+S-multi`) models show comparable performance. This finding persists in the cyclical component, where the original model remains among the most similar while the other models have smaller relative differences. For the trend, the finance-augmented is considerably dissimilar compared to the other models, while the original version—indicated as the least dissimilar at other frequencies—no longer stands out.

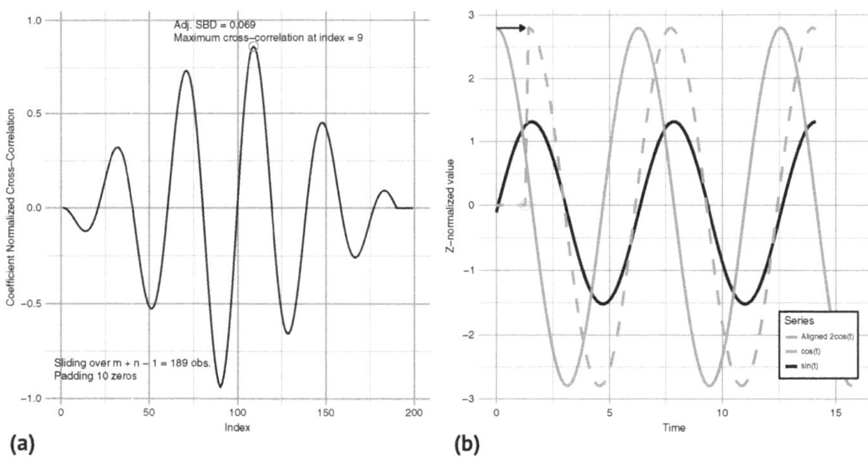

Figure 4.6 Computing similarity using SBD: Revisiting the prototypical example (z-normalized): (a) normalized cross-correlation; (b) series alignment based on maximum NCC.

Table 4.8 SBD index contrasting simulated and empirical capacity utilization series across different K+S model versions.

	K+S-Comp	K+S-finance	K+S-Ford	K+S-multi	K+S-original
UNFILTERED	0.385	0.233	0.311	0.267	0.192
TREND	0.349	0.307	0.352	0.342	0.350
CYCLE	0.370	0.378	0.370	0.374	0.367

4.4.7 Building the Multidimensional Index

The preceding sections introduced various measurements designed to capture distinct aspects of similarity between two series along different domains. Section 4.4.2, presents an index incorporating information on the probability distribution of the series. Section 4.4.3 utilized the LCSS to account for the similarity of the series' trajectories. Sections 4.4.4 and 4.4.5 describe dynamic time warping (DTW) and triangular global alignment (TGAK), which are elastic measurements that handle temporal misalignments between sequences. Finally, Section 4.4.6 discussed the SBD, which is robust to time shifts and provides insights into the lag structure of the series.

These measurements highlight a critical insight: no single similarity metric can fully capture the complexity embedded in the signals produced by the models. To address this limitation, we propose a multidimensional index that combines these diverse measurements, thereby achieving a more comprehensive approximation of similarity that embraces both the sequential and distributional domains of the data structure.

Certain adjustments are necessary to create our multidimensional index. First, we address the issue of differing ranges across measurements. Except for TGAK and SBD, the other criteria operate on non-closed ranges. Additionally, the LCSS measurement, as implemented, interprets similarity in the opposite direction compared to the others—higher values indicate greater similarity.[19] To ensure consistency, we transform all measurements to convey the same information.

As we have non-closed and nonmetric indexes, we present the results in terms of dissimilarity. Thus, we use the normalized LCSS distance ($\mathcal{D}_{LCSS}$) so that higher values indicate greater dissimilarity. Next, we rescale all measurements using min-max normalization, where one represents the most dissimilar and zero represents the least dissimilar. This normalization ensures that all measurements are on a comparable scale. We then compress the information from all measurements into a single index using the geometric average.[20] This approach is chosen because it is less sensitive to extreme values than the arithmetic mean, making it more robust for combining dissimilarity metrics. Equation (4.9) presents the geometric mean computation.

$$\text{Geometric Average}_{v} = \sqrt[5]{\mathcal{D}_{\text{Statistical},v} \cdot \mathcal{D}_{\text{NLCSS},v} \cdot \mathcal{D}_{\text{DTW},v} \cdot \mathcal{D}_{\text{TGAK},v}, \mathcal{D}_{\text{SBD},v}}. \tag{4.9}$$

19 This is the case because implementing the LCSS in the TSdist package [70] returns the occurrence of contiguous subsequences that respect the matching condition, so higher values indicate greater similarity.

20 We use the geometric average extension robust to the presence of zeroes, as proposed by Cruz and Kreft [71].

We introduce a penalizing factor based on [72] entropy, to refine the index further. The entropy is computed to quantify the consistency of a model's performance across all measurements. Specifically, we calculate the number of times it occupies each position in the ranking system for each model across all measurements. This is achieved by tabulating the positions of each model across the rankings and computing the entropy of the resulting distribution. A high entropy value indicates that a model performs inconsistently, appearing in multiple positions across the indices. Conversely, a low entropy value suggests that the model consistently occupies a similar position across all measurements.[21] To ensure comparability, we normalize the entropy by dividing it by the maximum possible entropy, which occurs when a model occupies a different position for each measurement. The normalized entropy, ranging from zero to one, is then used as a penalizing factor in the geometric average, which we refer to as the final multidimensional index. This formulation ensures that models with high entropy (inconsistent performance) are appropriately down-weighted in the final index. The penalized index is computed as follows:

$$\text{Entropy Adj. Geometric Average}_{\upsilon} = \text{Geometric Average} \cdot (1 + \text{Entropy}). \tag{4.10}$$

To showcase our approach, we present results for the capacity utilization rate as a representative variable in Table 4.9. The numbers in parentheses represent the rank across measurements, with one indicating the least dissimilar model and five indicating the most dissimilar.[22] We will use this example to discuss the multidimensional index in more detail and how to interpret it. According to the multidimensional index, the original model `K+S-original` outperforms the others, while the multi-sector version is the most dissimilar. This result is driven by the original model's good performance in replicating similar statistical moments and lag structures as observed in the empirical data (captured by SBD). Although the finance-augmented version performs well in terms of trajectory similarity (captured by LCSS) and alignment paths (captured by DTW and TGAK), the original model's consistent performance across all measurements explains its position as the relatively most similar model.

Our findings also provide valuable insights into domains where specific models require improvement. For instance, the `K+S-multi` model has poor performance in terms of statistical moments and shows notable dissimilarity regardless

21 It is worth noting that a low entropy does not necessarily mean that the model under consideration has a relatively good performance. In other words, a model with a consistently poor performance will present a low entropy.

22 Due to space limitations, we present the results for the unfiltered series, although the same process is repeated for the filtered variables.

Table 4.9 Example of the computation of the multidimensional index for the unfiltered capacity utilization index across different K+S model versions.

	K+S-Comp	K+S-finance	K+S-Ford	K+S-multi	K+S-original
Statistical	0.512 (3)	0.452 (2)	0.891 (5)	0.842 (4)	0.259 (1)
LCSS	1 (3)	0.155 (1)	1 (3)	1 (3)	0.375 (2)
DTW	0.171 (4)	0.083 (1)	0.167 (3)	0.395 (5)	0.091 (2)
TGAK	0.025 (4)	0.022 (1)	0.025 (3)	0.026 (5)	0.023 (2)
SBD	0.385 (5)	0.233 (2)	0.311 (4)	0.267 (3)	0.192 (1)
Geometric Average	6e-01 (3)	3e-03 (2)	7e-01 (4)	8e-01 (5)	1e-03 (1)
Entropy	7e-01 (4)	4e-01 (1)	6e-01 (3)	7e-01 (4)	4e-01 (1)
Entropy adj Geom. Avg.	1e+00 (3)	4e-03 (2)	1e+00 (4)	1e+00 (5)	1e-03 (1)

of temporal elasticity (as indicated by DTW and TGAK). In particular, the labor-augment and multi-sector versions perform poorly in capturing the trajectories of the series, with no common subsequences identified.

Figure 4.7 provides a comprehensive overview of the relative similarity across all variables, models, and filters using the multidimensional index.[23] The heatmap is divided into three subplots—Unfiltered, Trend, and Cycle—each corresponding to a different frequency component. The Y-axis represents the model versions, while the X-axis denotes the variables listed in Table 4.3.[24] Each cell displays the ranking (one to five) within the same variable, where lower values indicate superior relative performance.

In the unfiltered dataset, model rankings appear more dispersed, suggesting that when all frequency components are considered, differences in model performance become more pronounced. This could be attributed to unfiltered data incorporating both short-term fluctuations and long-term trends, requiring models to balance these dynamics effectively. Conversely, in the trend and cycle decompositions, the rankings exhibit more structured patterns, indicating that specific models (notably `K+S-Comp` and `K+S-finance`) better capture long-term structural movements or cyclical dynamics. The systematic shifts in rankings across these panels underscore the importance of aligning model selection with the specific frequency components of interest in economic analysis.

Overall, these findings emphasize the sensitivity of model performance to the frequency domain and variable characteristics. For instance, while the

23 The Table in the online appendix offers a more detailed breakdown for each model, variable, and filter across all measures.

24 Variables transformed into growth rates are prefixed with "d" to reflect this adjustment. For instance, dGDP represents the growth rate of real GDP.

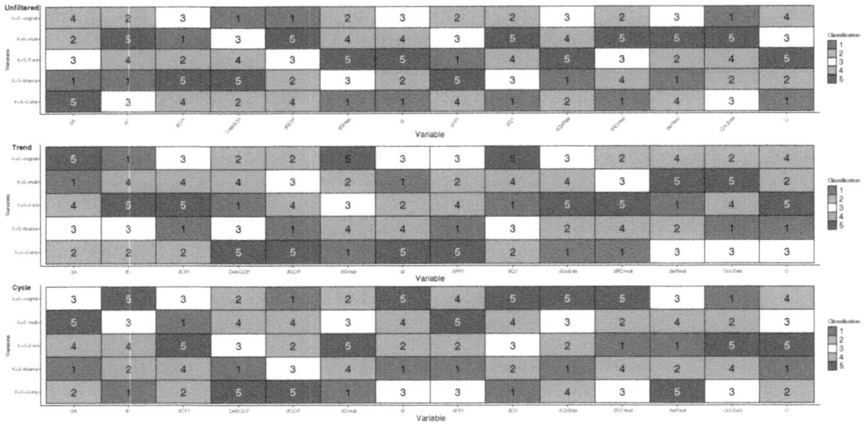

Figure 4.7 Relative performance of all models according to the multidimensional index across all variables and filters.

K+S-original and K+S-Comp models perform well in the unfiltered dataset, their rankings deteriorate under different frequency filters. Specific economic indicators, such as GDP growth rate (dGDP) and general price inflation (dCPI), show relatively stable rankings across filtering methods, suggesting that model performance for these variables is more robust to data transformations. In contrast, variables like government consumption (dGreal) and consumer goods production (dQcUbas) growth rates show substantial ranking shifts depending on the filtering approach. This underscores the necessity of selecting a model tailored to the specific characteristics of the dataset and research objectives.

This section has demonstrated how to capture different aspects of similarity between two signals. Several key insights emerge from this approach. First, relying on a single similarity index may overlook critical information embedded in the signals, reinforcing the importance of considering multiple dimensions of similarity. Second, our method identifies specific aspects where a model requires refinement (indicated with higher-ranking positions), enabling targeted improvements. Finally, the multidimensional index makes constructing a ranking system that sorts models based on their similarity to empirical data possible. In the next section, we leverage the ranking system derived from our proposed index to conduct model selection exercises.

4.5 Model Selection

Following the workflow illustrated in Figure 4.2, we have computed a multidimensional index for each variable across all models being compared. This index quantifies the degree of similarity between each model and the empirical data, providing a comprehensive basis for comparison, discussed in the previous section. A striking result from this analysis is the absence of a single model version that consistently outperforms the others across all variables. Performance rankings vary notably depending on the specific variable and the frequency component analyzed, highlighting the need for a nuanced approach to model selection.

A wide array of complementary approaches exists in model selection, each tailored to specific modeling contexts and objectives. A critical aspect of model selection is its coexistence with validation and calibration exercises, which ensure that the chosen model replicates key empirical patterns. One well-known set of techniques, rooted in statistical and econometric traditions, relies on information criteria and forecasting accuracy [11, 15, 73]. Another group prioritizes the ability of models to generate realistic qualitative outcomes, such as SFs, through history-friendly calibration [13, 74] or pattern-oriented modeling [75]. Despite this diversity, the predominant focus has been on predictive performance or analyzing impulse response functions rather than measuring the direct similarity between models and empirical data.

This chapter introduces an alternative selection scheme based on our multidimensional index, departing from traditional practices. Starting from the ranking system described in the previous section, we remove models with a higher occurrence of the most dissimilar variables rather than selecting those with the highest frequency of similar cases. Algorithm 1 outlines the pseudocode for this iterative process. The first selection exercise does not attribute a distinct importance to any variables. Later, we imposed different weights to adopt a purpose-driven selection framework, similar to what [76] did.

The first group of Table 4.10 provides the number of relative "worst" cases for each variable using this algorithm. Although this analysis is applied to both cyclical and trend components, we illustrate our approach using the unfiltered series for clarity. In the first iteration, the `K+S-multi` version is removed, as it is the most dissimilar for the highest number of variables. The ranking may change after each removal because dissimilarity is measured in relative terms. We then sort the remaining candidates, assigning a value of one to the least dissimilar and four (the number of kept models, $\mathcal{V} - 1$) to the most distant one. This iterative process stops when only one model remains. Continuing this procedure, the next removed models are `K+S-Ford`, `K+S-finance`, and `K+S-Comp`. Based on this, we find that the `K+S-original` is the least dissimilar for the unfiltered series, while the `K+S-Comp` ranks as the second best across all frequencies. When analyzing other

Algorithm 1 Pseudocode of the Iterative Process

```
Require: Dimension(MultidimensionalIndex) = Variables x V        ▷ Dataframe
    produced in phase 1
Require: length(TieBreaker) = length(V)        ▷ TieBreaker is a vector of strings
Require: length(Weights) = length(Variables)        ▷ Weights is numeric vector
Output: Least dissimilar model configuration
Versions = V
repeat
    RankedTable = matrix(Variables, Versions)
    for var in Variables do
        for ver in Versions do
            if RankedTable[var, ver] = WorstLevel then
                RankedTable[var, ver] = 1
            else
                RankedTable[var, ver] = 0
            end if
        end for
    end for
    RankedTable = RankedTable * Weights
until length(Versions) = 1
return Versions
```

components of the frequency domain, the `K+S-finance` outperforms the other alternative versions.

Although the iteration ends here, this should not be seen as a final step. We suggest revisiting the performance of the selected candidates to refine the analysis. We refer to Figure 4.7, which provides a visual representation of the model performance across variables. A quick inspection of this heatmap reveals variables where the models could be improved. For example, the `K+S-original` shows relatively higher dissimilarity for productivity and unemployment series, indicating variables requiring further attention. Not only is the `K+S-original` the least dissimilar model for the unfiltered series, but it also has no occurrences of relatively worst variables before the iterative removal process begins. The same applies to the `K+S-finance` model for other frequencies. This indicates that the selection criteria successfully identify a good candidate as the least dissimilar model version.

Table 4.10 Number of relatively worse variables across different K+S versions, filters, and purpose-driven experiments.

	Unfiltered					Trend					Cycle				
Iterations	K+S-Comp	K+S-finance	K+S-Ford	K+S-multi	K+S-original	K+S-Comp	K+S-finance	K+S-Ford	K+S-multi	K+S-original	K+S-Comp	K+S-finance	K+S-Ford	K+S-multi	K+S-original
Baseline															
1	1	3	4	6	0	4	0	5	2	3	3	0	4	2	5
2	3	4	7	-	0	4	1	-	5	4	4	1	5	4	-
3	5	6	-	-	3	5	2	-	-	7	4	4	-	6	-
4	8	-	-	-	6	8	6	-	-	-	8	6	-	-	-
Policy-oriented															
1	1	6	4	12	0	10	0	5	5	3	6	0	7	5	5
2	6	7	10	-	0	-	1	13	6	3	6	6	-	5	6
3	11	9	-	-	3	-	2	-	13	8	7	7	-	9	-
4	-	17	-	-	6	-	4	-	-	19	14	9	-	-	-
Research-oriented															
1	4	6	7	12	0	10	0	8	2	9	3	0	4	8	14
2	6	10	13	-	0	-	4	13	3	9	4	4	5	16	-
3	8	15	-	-	6	-	8	-	7	14	12	4	13	-	-
4	14	-	-	-	15	-	13	-	16	-	17	12	-	-	-

4.5.1 Targeting Variables: A Policy-oriented Exercise

It is worth noting that the selection procedure discussed so far is variable-agnostic. However, model design often targets specific research questions or policy-oriented debates. To address this, we propose an alternative approach that incorporates external criteria to evaluate the performance of any simulated model collection. At this stage, it is important to emphasize that the similarity of any two series does not depend on these criteria. Thus, we assign penalization weights during the iterative selection process rather than adjusting the multidimensional similarity index. Table 4.11 indicates the penalization weights associated with each variable in every experiment, in which a higher value forces the algorithm to demote the variable if the model has a relatively worse performance. Weights might be due to higher importance assigned to some specific variable, such as inflation in a phase of inflation spiral, unemployment in case of recessions, or productivity growth in case of prolonged stagnation.

Once the priorities are set, we apply the iterative selection process with slight changes. Our goal is to identify the most similar model, so we use these priority weights as a penalization factor during the elimination step. As before, we first count the number of occurrences of the worst relative performance for each model under scrutiny. Unlike the previous exercise, a model is penalized by these weights when it is more dissimilar with respect to the target variables. This approach allows us to guide the iterative procedure toward a desired purpose.

Table 4.11 Penalization weights for purpose-guided model selection.

	Policy-oriented	Research-oriented
dA	1	4
dC	2	1
dCPI	4	2
DebGDP	1	1
dGDP	4	4
dG	2	1
dI	2	2
dPPI	2	2
dQ1	2	2
dQcBas	2	1
dRDreal	1	4
dwReal	1	1
QcUbas	2	1
U	4	4

Policy-oriented considerations inspire the first experiment without aiming to replicate any specific policy decision. Through this experimental setup, we employ arbitrary weights to demonstrate the refined model selection mechanism.[25] For instance, suppose policymakers are interested in targeting specific macroeconomic variables, such as real GDP growth rate, inflation, and unemployment rate. Labor productivity and R&D variables are assigned lower priority, as they are indirectly affected by government actions in the short run. The iterative process then begins.[26] The second group of Table 4.10 displays the models removed at each iteration. Interestingly, the same model selection emerges in this experiment, with the `K+S-original` producing similar results for the unfiltered series, while the `K+S-finance` remains the least dissimilar for other frequencies. Except for the trend component, the `K+S-Comp` ranks as the second-best model in this targeted exercise.

As previously discussed, the models evaluated here were designed to reflect general regularities rather than a specific economy or time period. Therefore, it is more appropriate to contrast the model collections with their theoretical aims. For this reason, we also evaluate their relative performance in terms of similarity to variables associated with technological change, such as labor productivity, R&D expenditures, and their effects on real GDP and unemployment rates. The procedure remains the same as described earlier, with the only difference being the weights assigned to each variable, as shown in the second column of Table 4.11.

The third group of Table 4.10 presents the models removed at each iteration. Unlike previous exercises, the `K+S-original` is no longer the best candidate for the unfiltered series, displaced by the finance-augmented version, which also ranks as the best model for other frequencies. This result is primarily driven by the fact that the original version does not generate similar series for innovation-related variables and the unemployment rate. As a consequence, the finance-augmented version is more suitable for analyses involving these variables. The `K+S-Comp` maintains its relatively good performance, again ranking as the second-best model. Overall, this model performs well regarding labor productivity and unemployment rates, which is consistent with its theoretical design.

25 The ranging of the values does not have an explicit reasoning, but rather imposes an ordinal hierarchy on the variables. The fine-tuned selection can be performed as long as there is an unambiguous hierarchy among variables.

26 It is important to note that there are two models with the same number of worst cases on the second iteration of the cyclical component. This requires a tie-break rule. We remove the model with the higher mean value according to our proposed multidimensional index (see table in the online appendix). While this tiebreaker criterion is not the only possible option, the number of tied cases is small, as shown in Table 4.10.

Considering all this, model performance and selection must be evaluated in light of the specific elements and objectives the model was designed to address. The `K+S-original` and `K+S-finance` models emerge as the least dissimilar in different contexts, with their performance varying substantially depending on the variables and frequency components analyzed. These findings underscore the importance of aligning model selection with theoretical and practical objectives, offering a flexible framework for identifying the most suitable model for specific applications.

4.6 Concluding Remarks

The primary contribution of this chapter lies in the development of a protocol that enables modelers to evaluate the similarity between their models and empirical data, with specific reference to the K+S macroeconomic ABM. We argue that the similarity of a model vis-à-vis the empirical evidence should be measured from a multidimensional perspective, considering various aspects when comparing two objects along the time domain. Building on this premise, we have created an index to capture these diverse dimensions, providing a detailed understanding of the relative performance of each model.

Our work is part of a larger effort to bridge the gap between simulated models and their real-world counterparts, focusing on a specific category of macroeconomic ABMs. The motivation behind this exercise is the increasing demand for data-friendly models that can enhance forecasting capabilities for policymakers. While our application is rooted in economic literature, the protocol's utility is not limited to this field. With broader applications in mind, the design of the protocol aims to be as adaptable and scalable as possible, making it suitable for numerous research and policy challenges.

In addition to the protocol, we have proposed two potential uses. First, an iterative procedure identifies the most similar model. This procedure is further complemented by a refinement that considers an external criterion in the selection process, assigning weights to specific target variables. The analysis reveals that the finance-augmented K+S model demonstrates robust performance compared to recent developments, particularly when examining different frequencies. Likewise, the labor-augmented model, adjusted for competitive institutional settings, yields satisfactory results across most configurations. Second, we have performed a policy analysis exercise, targeting specific variables and showing that the protocol can be employed to assign specific weights and relevance to variables of interest. The iterative selection process can address several research questions and policy challenges by integrating domain-specific weights and criteria.

Another potential application, not explored due to space limitations, is the interactive calibration of models based on their similarity to empirical data. This approach could assist modelers in determining parameters that are not directly observable. For example, the modeler can compare different parametric settings of the same model against empirical data and choose the one that is more similar, according to the iterative model selection.

While the selection protocol and multidimensional index improve the understanding of the models' similarity with empirical data, they are not without limitations and offer opportunities for further refinement. At a higher level of abstraction, the lack of interaction among the variables under consideration is a notable limitation. Therefore, although the index covers multiple dimensions of similarity, it should not be interpreted as a global measure. Nonetheless, the protocol serves as a first effective step to identify a set of alternative configurations and/or models before resorting to more computationally intensive methods. Finally, the applicability of the proposed method goes beyond ABM, enlarging the understanding of model performance under different theoretical assumptions.

Supplementary Material

Supplementary material related to this chapter can be found online at https://doi.org/10.5281/zenodo.17427980.

References

1 Giovanni Dosi. *The Foundations of Complex Evolving Economies: Part One: Innovation, Organization, and Industrial Dynamics*. Oxford University Press, 2023.

2 Jean-Philippe Bouchaud. From statistical physics to social sciences: The pitfalls of multi-disciplinarity. *Journal of Physics: Complexity*, 4(4):041001, 2023. https://doi.org/10.1088/2632-072X/ad104a.

3 Robert M. May. Will a large complex system be stable? *Nature*, 238(5364):413–414, 1972. https://doi.org/10.1038/238413a0.

4 Joshua M Epstein. *Generative Social Science*. Princeton Studies in Complexity. Princeton University Press, 2007.

5 Giovanni Dosi, Giorgio Fagiolo, and Andrea Roventini. Schumpeter meeting Keynes: A policy-friendly model of endogenous growth and business cycles. *Journal of Economic Dynamics and Control*, 34(9):1748–1767, 2010. https://doi.org/10.1016/j.jedc.2010.06.018.

6 Herbert Dawid, Simon Gemkow, Philipp Harting, and Michael Neugart. Labor market integration policies and the convergence of regions: The role of skills and technology diffusion. *Journal of Evolutionary Economics*, 22(3):543–562, 2012. https://doi.org/10.1007/s00191-011-0245-1.

7 Herbert Dawid, Simon Gemkow, Philipp Harting, Sander van der Hoog, and Michael Neugart. Agent-based macroeconomic modeling and policy analysis: The Eurace@ Unibi Model. In Shu-Heng Chen, Mak Kaboudan, and Ye-Rong Du, editors, *The Oxford Handbook of Computational Economics and Finance*, pages 490–519. Oxford University Press, 2018. https://doi.org/10.1093/oxfordhb/9780199844371.013.19.

8 Silvano Cincotti, Marco Raberto, and Andrea Teglio. Credit money and macroeconomic instability in the agent-based model and simulator eurace. *Economics*, 4(1):20100026, 2010. https://doi.org/10.5018/economics-ejournal.ja.2010-26.

9 Domenic Delli Gatti, Edoardo Gaffeo, and Mauro Gallegati. Complex agent-based macroeconomics: A manifesto for a new paradigm. *Journal of Economic Interaction and Coordination*, 5(2):111–135, 2010. https://doi.org/10.1007/s11403-010-0064-8.

10 Giorgio Fagiolo, Alessio Moneta, and Paul Windrum. A critical guide to empirical validation of agent-based models in economics: Methodologies, procedures, and open problems. *Computational Economics*, 30(3):195–226, 2007. https://doi.org/10.1007/s10614-007-9104-4.

11 Francesco Lamperti. Empirical validation of simulated models through the GSL-div: An illustrative application. *Journal of Economic Interaction and Coordination*, 13(1):143–171, 2018. https://doi.org/10.1007/s11403-017-0206-3.

12 Robert Ernest Marks. Validating simulation models: A general framework and four applied examples. *Computational Economics*, 30(3):265–290, 2007. https://doi.org/10.1007/s10614-007-9101-7.

13 Paul Windrum, Giorgio Fagiolo, and Alessio Moneta. Empirical validation of agent-based models: Alternatives and prospects. *Journal of Artificial Societies and Social Simulation*, 10(2):8, 2007. https://www.jasss.org/10/2/8.html.

14 Giovanni Dosi and Andrea Roventini. More is different... and complex! the case for agent-based macroeconomics. *Journal of Evolutionary Economics*, 29(1):1–37, 2019. https://doi.org/10.1007/s00191-019-00609-y.

15 Sebastian Poledna, Michael Gregor Miess, Cars Hommes, and Katrin Rabitsch. Economic forecasting with an agent-based model. *European Economic Review*, 151:104306, 2023. https://doi.org/10.1016/j.euroecorev.2022.104306.

16 Domenico Delli Gatti and Severin Reissl. Agent-based Covid economics (ABC): Assessing non-pharmaceutical interventions and macro-stabilization policies. *Industrial and Corporate Change*, 31(2):410–447, 2022. https://ideas.repec.org//a/oup/indcch/v31y2022i2p410-447..html.

17 C. W. J. Granger. The typical spectral shape of an economic variable. *Econometrica*, 34(1):150–161, 1966. https://doi.org/10.2307/1909859.

18 N. D. Kondratieff and W. F. Stolper. The long waves in economic life. *The Review of Economics and Statistics*, 17(6):105–115, 1935. https://doi.org/10.2307/1928486.

19 John Clark, Christopher Freeman, and Luc Soete. Long waves, inventions, and innovations. *Futures*, 13(4):308–322, 1981. https://doi.org/10.1016/0016-3287(81)90146-4.

20 G. Silverberg. Long waves: Conceptual, empirical and modelling issues. In Horst Hanusch and Andreas Pyka, editors, *Elgar Companion to Neo-Schumpeterian Economics*. Edward Elgar Publishing, 2007. https://doi.org/10.4337/9781847207012.00060.

21 Giovanni Dosi, Marcelo C. Pereira, Andrea Roventini, and Maria Enrica Virgillito. Technological paradigms, labour creation and destruction in a multi-sector agent-based model. *Research Policy*, 51(10):104565, 2022. https://doi.org/10.1016/j.respol.2022.104565.

22 Carlota Perez. The double bubble at the turn of the century: Technological roots and structural implications. *Cambridge Journal of Economics*, 33(4):779–805, 2009. https://doi.org/10.1093/cje/bep028.

23 Giovanni Dosi, Giorgio Fagiolo, Mauro Napoletano, and Andrea Roventini. Income distribution, credit and fiscal policies in an agent-based Keynesian model. *Journal of Economic Dynamics and Control*, 37(8):1598–1625, 2013. https://doi.org/10.1016/j.jedc.2012.11.008.

24 Giovanni Dosi, Giorgio Fagiolo, Mauro Napoletano, Andrea Roventini, and Tania Treibich. Fiscal and monetary policies in complex evolving economies. *Journal of Economic Dynamics and Control*, 52:166–189, 2015.https://doi.org/10.1016/j.jedc.2014.11.014.

25 Giovanni Dosi, Marcelo C. Pereira, Andrea Roventini, and Maria Enrica Virgillito. When more flexibility yields more fragility: The microfoundations of Keynesian aggregate unemployment. *Journal of Economic Dynamics and Control*, 81:162–186, 2017. https://doi.org/10.1016/j.jedc.2017.02.005.

26 Giovanni Dosi, Marcelo C. Pereira, Andrea Roventini, and Maria Enrica Virgillito. The labour-augmented K+S model: A laboratory for the analysis of institutional and policy regimes. *EconomiA*, 21(2):160–184, 2020. https://doi.org/10.1016/j.econ.2019.03.002.

27 Giovanni Dosi, Giorgio Fagiolo, and Andrea Roventini. An evolutionary model of endogenous business cycles. *Computational Economics*, 27(1):3–34, 2006. https://doi.org/10.1007/s10614-005-9014-2.

28 Giovanni Dosi, Giorgio Fagiolo, and Andrea Roventini. The microfoundations of business cycles: An evolutionary, multi-agent model. *Journal of Evolutionary Economics*, 18(3):413–432, 2008. https://doi.org/10.1007/s00191-008-0094-8.

29 Giovanni Dosi, Marcelo C. Pereira, Andrea Roventini, and Maria Enrica Virgillito. The effects of labour market reforms upon unemployment and income inequalities: An agent-based model. *Socio-Economic Review*, 16(4):687–720, 2018. https://doi.org/10.1093/ser/mwx054.

30 Giovanni Dosi, Marcelo C. Pereira, Andrea Roventini, and Maria Enrica Virgillito. Causes and consequences of hysteresis: Aggregate demand, productivity, and employment. *Industrial and Corporate Change*, 27(6): 1015–1044, 2018. https://doi.org/10.1093/icc/dty010.

31 C. Reinhart and K. Rogoff. The aftermath of financial crises. *American Economic Review*, 99:466–472, 2009.

32 Laeven and L. Valencia. F. Systemic banking crises: A new database, *IMF Working Paper, WP/08/224*, 2008.

33 Marco Amendola and Marcelo C. Pereira. Linear and state-dependent impulse responses in agent-based models: A new methodology and an economic application. 2024. (available at SSRN 4740360). https://papers.ssrn.com/sol3/papers.cfm?abstract_id=4740360.

34 Marco Valente and Marcelo C. Pereira. *Laboratory for Simulation Development - LSD*, 2023. https://labsimdev.org.

35 Marcelo C. Pereira. *LSDinterface: Interface tools for LSD simulation results files. Comprehensive R Archive Network*, 2022. https://CRAN.R-project.org/package=LSDinterface.

36 Lawrence J. Christiano and Terry J. Fitzgerald. The band pass filter. *International Economic Review*, 44(2):435–465, 2003. https://www.jstor.org/stable/3663474.

37 John Paparrizos and Luis Gravano. K-shape: Efficient and accurate clustering of time series. In *Proceedings of the 2015 ACM SIGMOD International Conference on Management of Data*, pages 1855–1870. ACM, 2015. https://doi.org/10.1145/2723372.2737793.

38 Eamonn Keogh and Shruti Kasetty. On the need for time series data mining benchmarks: A survey and empirical demonstration. *Data Mining and Knowledge Discovery*, 7(4):349–371, 2003. https://doi.org/10.1023/a:1024988512476.

39 J. H. Stock and Mark Watson. Chapter 1 Business cycle fluctuations in us macroeconomic time series. In *Handbook of Macroeconomics*, Vol. 1, pages 3–64. Elsevier, 1999. https://doi.org/10.1016/S1574-0048(99)01004-6.

40 Mauro Napoletano, Andrea Roventini, and S. Sapio. Are business cycles all alike? A bandpass filter analysis of the Italian and U.S. cycles. *Rivista Italiana degli Economisti*, 1:87–118, 2006.

41 Alisdair McKay and Ricardo Reis. The brevity and violence of contractions and expansions. *Journal of Monetary Economics*, 55(4):738–751, 2008. https://doi.org/10.1016/j.jmoneco.2008.05.009.

42 Giovanni Dosi. Statistical regularities in the evolution of industries: A guide through some evidence and challenges for the theory. In Franco Malerba and Stefano Brusoni, editors, *Perspectives on Innovation*, pages 153–186. Cambridge University Press, 2007. https://doi.org/10.1017/CBO9780511618390.009.

43 Giorgio Fagiolo, Mauro Napoletano, and Andrea Roventini. Are output growth-rate distributions fat-tailed? Some evidence from OECD countries. *Journal of Applied Econometrics*, 23:639–669, 2008.

44 Giulio Bottazzi and Angelo Secchi. Explaining the distribution of firm growth rates. *The RAND Journal of Economics*, 37(2):235–256, 2006. https://doi.org/10.1111/j.1756-2171.2006.tb00014.x.

45 Giulio Bottazzi, Le Li, and Angelo Secchi. Aggregate fluctuations and the distribution of firm growth rates. *Industrial and Corporate Change*, 28(3):635–656, 2019. https://doi.org/10.1093/icc/dtz016.

46 Jushan Bai and Pierre Perron. Computation and analysis of multiple structural change models. *Journal of Applied Econometrics*, 18(1):1–22, 2003. https://doi.org/10.1002/jae.659.

47 R. A. Wagner and M. J. Fischer. The string-to-string correction problem. *Journal of the ACM (JACM)*, 21(1):168–173, 1974. https://doi.org/10.1145/321796.321811.

48 C. Gruber, T. Gruber, S. Krinninger, and B. Sick. Online signature verification with support vector machines based on LCSS kernel functions. *IEEE Transactions on Systems, Man, and Cybernetics, Part B: Cybernetics*, 40(4):1088–1100, 2010. https://doi.org/10.1109/TSMCB.2009.2034382.

49 Shyong Jian Shyu and Chun-Yuan Tsai. Finding the longest common subsequence for multiple biological sequences by ant colony optimization. *Computers & Operations Research*, 36(1):73–91, 2009. https://doi.org/10.1016/j.cor.2007.07.006.

50 M. Vlachos, G. Kollios, and D. Gunopulos. Discovering similar multidimensional trajectories. In *Proceedings 18th International Conference on Data Engineering*, pages 673–684. IEEE Computer Society, 2002. https://doi.org/10.1109/ICDE.2002.994784.

51 Donald J. Berndt and James Clifford. Using dynamic time warping to find patterns in time series. In *Proceedings of the 3rd International Conference on Knowledge Discovery and Data Mining*, AAAIWS'94, pages 359–370. AAAI Press, 1994.

52 Hansheng Lei and Bingyu Sun. A study on the dynamic time warping in kernel machines. In *2007 Third International IEEE Conference on Signal-Image Technologies and Internet-Based System*, pages 839–845. IEEE, 2007. https://doi.org/10.1109/SITIS.2007.112.

53 Joseph B. Kruskal and Mark Liberman. The symmetric time warping algorithm: From continuous to discrete. In David Sankoff and Joseph B. Kruskal, editors, *Time Warps, String Edits, and Macromolecules*, David Hume Series. Center for the Study of Language and Information, reissue edition, 1999.

54 Alexis Sardá-Espinosa. Time-series clustering in R using the dtwclust package. *The R Journal*, 11(1):22–43, 2019. https://journal.r-project.org/archive/2019/RJ-2019-023/index.html.

55 H. Sakoe and S. Chiba. Dynamic programming algorithm optimization for spoken word recognition. *IEEE Transactions on Acoustics, Speech, and Signal Processing*, 26(1):43–49, 1978. https://doi.org/10.1109/TASSP.1978.1163055.

56 F. Itakura. Minimum prediction residual principle applied to speech recognition. *IEEE Transactions on Acoustics, Speech, and Signal Processing*, 23(1):67–72, 1975. https://doi.org/10.1109/TASSP.1975.1162641.

57 John Aach and George M. Church. Aligning gene expression time series with time warping algorithms. *Bioinformatics*, 17(6):495–508, 2001. https://doi.org/10.1093/bioinformatics/17.6.495.

58 Ziv Bar-Joseph, Georg Gerber, David K. Gifford, Tommi S. Jaakkola, and Itamar Simon. A new approach to analyzing gene expression time series data. In *Proceedings of the Sixth Annual International Conference on Computational Biology*, pages 39–48. ACM, 2002. https://doi.org/10.1145/565196.565202.

59 T. M. Rath and R. Manmatha. Word image matching using dynamic time warping. In *2003 IEEE Computer Society Conference on Computer Vision and Pattern Recognition, 2003. Proceedings*, Vol. 2, pages II-521–II-527. IEEE Comput. Soc, 2003. https://doi.org/10.1109/CVPR.2003.1211511.

60 Philip Hans Franses and Thomas Wiemann. Intertemporal similarity of economic time series: An application of dynamic time warping. *Computational Economics*, 56(1):59–75, 2020. https://doi.org/10.1007/s10614-020-09986-0.

61 Gang-Jin Wang, Chi Xie, Feng Han, and Bo Sun. Similarity measure and topology evolution of foreign exchange markets using dynamic time warping method: Evidence from minimal spanning tree. *Physica A: Statistical Mechanics and Its Applications*, 391(16):4136–4146, 2012. https://doi.org/10.1016/j.physa.2012.03.036.

62 Tasneem Raihan. Predicting US recessions: A dynamic time warping exercise in economics. Unpublished, 2017.

63 Marco Cuturi. Fast global alignment kernels. In *Proceedings of the 28th International Conference on Machine Learning*, 2011.

64 Eamonn Keogh and Chotirat Ann Ratanamahatana. Exact indexing of dynamic time warping. *Knowledge and Information Systems*, 7(3):358–386, 2005. https://doi.org/10.1007/s10115-004-0154-9.

65 Toni Giorgino. Computing and visualizing dynamic time warping alignments in R: The dtw package. *Journal of Statistical Software*, 31:1–24, 2009. https://doi.org/10.18637/jss.v031.i07.

66 Daniel Lemire. Faster retrieval with a two-pass dynamic-time-warping lower bound. *Pattern Recognition*, 42(9):2169–2180, 2009. https://doi.org/10.1016/j.patcog.2008.11.030.

67 Marco Cuturi, Jean-Philippe Vert, Oystein Birkenes, and Tomoko Matsui. A kernel for time series based on global alignments. In *2007 IEEE International Conference on Acoustics, Speech and Signal Processing - ICASSP '07*, pages II-413–II-416, 2007. https://doi.org/10.1109/ICASSP.2007.366260.

68 Xiaoyue Wang, Abdullah Mueen, Hui Ding, Goce Trajcevski, Peter Scheuermann, and Eamonn Keogh. Experimental comparison of representation methods and distance measures for time series data. *Data Mining and Knowledge Discovery*, 26(2):275–309, 2013. https://doi.org/10.1007/s10618-012-0250-5.

69 Fateme Fahiman, James C. Bezdek, Sarah M. Erfani, Marimuthu Palaniswami, and Christopher Leckie. Fuzzy c-shape: A new algorithm for clustering finite time series waveforms. In *2017 IEEE International Conference on Fuzzy Systems (FUZZ-IEEE)*, pages 1–8. IEEE, 2017. https://doi.org/10.1109/FUZZ-IEEE.2017.8015525.

70 Usue Mori, Alexander Mendiburu, and Jose A. Lozano. *TSdist: Distance Measures for Time Series Data*. Comprehensive R Archive Network, 2014. https://CRAN.R-project.org/package=TSdist.

71 Roberto de la Cruz and Jan-Ulrich Kreft. Geometric mean extension for data sets with zeros. 2019. 10.48550/arXiv.1806.06403.

72 Claude E. Shannon. A mathematical theory of communication. *Bell System Technical Journal*, 27(3):379–423, 1948.

73 Mario Martinoli, Alessio Moneta, and Gianluca Pallante. Calibration and validation of macroeconomic simulation models by statistical causal search. *Journal of Economic Behavior & Organization*, 228:106786, 2024. https://doi.org/10.1016/j.jebo.2024.106786.

74 Giorgio Fagiolo, Mattia Guerini, Francesco Lamperti, Alessio Moneta, and Andrea Roventini. Validation of agent-based models in economics and finance. In Claus Beisbart and Nicole J. Saam, editors, *Computer Simulation Validation: Fundamental Concepts, Methodological Frameworks, and Philosophical Perspectives*, Simulation Foundations, Methods and Applications, pages 763–787. Springer International Publishing, 2019. https://doi.org/10.1007/978-3-319-70766-2_31.

75 Volker Grimm, Eloy Revilla, Uta Berger, Florian Jeltsch, Wolf M. Mooij, Steven F. Railsback, Hans-Hermann Thulke, Jacob Weiner, Thorsten Wiegand, and Donald L. DeAngelis. Pattern-oriented modeling of agent-based complex systems: Lessons from ecology. *Science*, 310(5750):987–991, 2005. https://doi.org/10.1126/science.1116681.

76 Roberto Leombruni and Matteo Richiardi. Why are economists sceptical about agent-based simulations? *Physica A: Statistical Mechanics and Its Applications*, 355(1):103–109, 2005. https://doi.org/10.1016/j.physa.2005.02.072.

5

Self-aware Models with Explanatory Capabilities: Learning Explanatory Models from Simulation Experiments

Levent Yilmaz

Computer Science and Software Engineering, Auburn University, Auburn, AL, USA

Corresponding author: yilmaz@auburn.edu

5.1 Introduction

The application of computational models within systems engineering is both widespread and foundational. Nevertheless, while a range of sophisticated tools exist to support engineers in routine aspects of systems modeling, significant challenges remain in advancing causal and evidential reasoning, hence constraining explanation and transparency in systems science [1–3]. These challenges encompass the systematic generation, prioritization, and articulation of modeling assumptions; the design and execution of hypothesis-driven automated experiments; and the iterative refinement of competing behavioral mechanisms through simulated experience within closed-loop analytical processes.

The principal challenges in advancing these methodological steps lie in the persistent disjunction between model discovery and experimentation, as well as the tendency to address justification only subsequent to model construction. Although exploratory analysis has been recognized as highly valuable [4, 5], prevailing computational modeling frameworks are not inherently structured to support such a symbiotic and adaptive process. Overcoming these limitations necessitates a tighter integration of model development and experimental design, facilitating iterative feedback across their respective technical domains. At present, state-of-the-art approaches to coupling remain largely confined to the revision of system specifications, without providing explanatory accounts of plausible mechanisms, coherence relations, or the conditions under which models are applicable.

The limitations outlined above frequently result in premature convergence toward a singular, authoritative model that lacks sufficient robustness to adequately account for variability and uncertainty, particularly within the domain of self-organizing and adaptive systems [6]. In contrast, the retention, systematic experimentation with, and integration of alternative and complementary modeling mechanisms are essential for the development of self-aware and resilient solutions. This concern parallels longstanding debates in the philosophy of science [7], wherein the pursuit of robust theories necessitates balancing the competing demands of generality, precision, and accuracy in explaining observed phenomena. A similar challenge arises in the context of artificial intelligence, where the so-called explainability crisis [8, 9] underscores the need for interpretable systems capable of articulating their reasoning processes [3, 10, 11]. These issues extend more broadly to the modeling and simulation community, where concerns regarding reproducibility and transparency in model use [12, 13], as well as the imperative of establishing credibility and trust in simulations [14, 15], remain central to advancing the field.

Motivating Scenario: To demonstrate the necessity and potential of integrating scientific thinking into the overall fabric of systems thinking, consider the design and development of a packet delivery system. This system enables customers to submit orders for pickup and transportation to specific destinations or delivery locations. The system faces two primary requirements: (1) *Dispatching*: Each new order must be assigned to a truck responsible for handling the requested transport. (2) *Routing*: Trucks must adaptively navigate through a street map to reach pickup locations for assigned orders and delivery locations for orders already in the truck. This problem is highly dynamic, as the information required for routing or dispatching is inherently decentralized across multiple trucks, customers, and streets.

Scientists often seek to uncover the underlying mechanisms of observed phenomena. Similarly, engineers of self-organizing and self-adaptive systems must experiment and discover which competing and complementary mechanisms or features, such as *market-based*, *tag-based*, or *token-based* coordination, can be used to implement the dispatching requirement. What are the alternative options, and how do they interact with the *gradient field* and *infochemical* communication mechanisms that can be employed to implement the routing requirement? Which mechanisms synergize and operate under specific contexts? How can we search, discover, and retain information about the combination of mechanistic features that provides a reasonable and robust explanation for the desired or observed behavior? Answering these questions facilitates the gradual development of a cluster of models. The emergent knowledge then informs the revision of the study's objectives and requirements for the system.

The central premise of this chapter is that the design of self-aware and self-explaining models constitutes a critical advancement in systems engineering, enabling the representation of dynamism frequently observed when scientific objectives are reported as emergent. The chapter seeks to establish a conceptual and methodological foundation for the development of a coherence-driven cognitive framework, which facilitates three key capabilities: (1) the implementation of systems models as adaptive agents that mediate among engineering domain theories, empirical data, requirements, principles, and analogical reasoning; (2) the enhancement of cognitive assistance in model discovery, experimentation, and evidential evaluation, thereby supporting the differentiation among competing system models; and (3) the systematic evaluation of alternative strategies for explanatory justification of hypothesized models through an evolving runtime cognitive architecture that renders coherence judgments explicit.

The overarching thesis advanced herein is that model construction necessitates the coherent integration of diverse elements—including domain theory, data, requirements, analogies, and principles—such that the resulting model satisfies established quality criteria. This thesis is further articulated through two interrelated sub-theses: the context of discovery, which emphasizes the processes by which models are generated, and the context of justification, which concerns the explanatory and evidential grounds upon which models are validated.

Within the context of discovery, the integration of mechanisms assumes a pivotal role. This integration encompasses both the causal mechanistic representation of system processes and the calibration of models to empirical or simulated data. By contrast, the context of justification pertains to embedded justification, wherein the integrated ensemble of elements collectively satisfies system requirements. To address these dual sub-theses, we propose a novel coherence-driven reflective equilibrium framework for systems modeling. Building upon a critical review of the philosophical and psychological literature on explanatory justification, we advance a systematic strategy grounded in the theory of explanatory coherence and the methodological principles of reflective equilibrium. This strategy enables the structured generation of explanations and the iterative learning of explanatory models through simulation-based experimentation. Furthermore, we outline the requirements and methodological strategies necessary for explanatory modeling, thereby contributing to the advancement of simulation infrastructures that support the development of self-aware and self-explaining models.

5.2 Background

Both scientific inquiry and engineering practice are fundamentally characterized by activities of model-based discovery and explanation. Accordingly, questions of explanation have long occupied a central position within the philosophy of science, serving as a framework for articulating normative criteria governing causal reasoning. Normative accounts frequently rely upon idealized strategies, including logic-based deduction, inductive generalization, and abductive inference. Yet, because explanation entails significant cognitive effort and is inherently tied to the construction and manipulation of mental models, cognitive approaches to explanation provide critical value in advancing pragmatic, model-based accounts. In this section, we examine the philosophical, cognitive, and theoretical foundations of model-based explanation, with particular emphasis on the properties and behaviors of idealized abstract models employed to account for observed regularities.

Explanations are fundamentally concerned with causal claims that seek to account for why or how particular phenomena occur. Such accounts generally assume two principal forms: process-centric representations, which emphasize the organization of entities and activities, and theory- or data-driven approaches, which rely on formalized knowledge and empirical evidence. Among the earliest systematic models of explanation is the deductive-nomological (DN) framework [16], which conceptualizes explanation as a deductive argument derived from general laws and initial conditions. Subsequent developments extended the DN model through probabilistic and inductive methods, thereby enabling statistical forms of explanation under conditions of uncertainty. Deductive-statistical explanations derive regularities from statistical laws, whereas inductive-statistical explanations assign likelihoods to outcomes based on probabilistic inference. The statistical relevance (SR) model [17] further advances explanatory theory by formalizing conditional dependence relations among events, thereby situating explanation within the broader context of probabilistic causation. Specifically, given features A, B, and C, the feature C is statistically relevant to attribute B if the probability of event B conditional on A and C is different from the probability of B conditional on A alone; that is, if the probability of event B given A and C is not the same as the probability of B given A.

Building upon the DN and argumentative reconstruction (AR) models, the simulacrum strategy of explanation [18] entails the identification of a model that coheres with the overarching theoretical framework and functions as an analogical representation of the system under study. In parallel, the theory of explanatory coherence [19] offers a computational framework for systematically establishing connections among competing or complementary explanations, guided by principles of local coherence. Within this context, the method of inference to the best explanation serves as a critical evaluative mechanism, assessing the degree to which candidate explanations correspond with empirical observations and align with established theoretical principles and accepted explanatory accounts.

The process of abductive reasoning is initiated by a triggering event. Within agent-based models (ABMs) of complex adaptive systems, such triggers may manifest either as unexpected phenomena that necessitate explanation or as anticipated outcomes that require justification. As noted by Peirce [20], the observed event must possess sufficient significance to warrant explanatory inquiry. Following this initiation, one or more candidate explanations are generated. Although this stage often involves creative reasoning, the selection of relevant explanations is guided and constrained by contextual factors, prior domain knowledge, and heuristic strategies. The resulting explanations are subsequently evaluated to determine their plausibility and relative probability in accounting for the observed event. The strategy of manipulative abduction highlights the critical role of contextual framing and its deliberate manipulation in fostering the emergence of novel explanatory hypotheses [21].

Aligned with the AR model, the framework of explanatory coherence [22] conceptualizes scientific explanation from a cognitive standpoint and delineates it into three principal processes supported by four critical methodological approaches. The processes consist of (1) the selection of an explanation based on the information available, (2) the generation of novel hypotheses to account for observed phenomena, and (3) the evaluation of competing plausible explanations in order to infer the most satisfactory account. The methodological approaches employed within these processes are typically classified as deductive strategies (e.g., rule- or logic-based reasoning), schematic models that draw upon explanatory patterns and analogical structures, probabilistic approaches (e.g., Bayesian inference), and connectionist network models of explanation.

The process of selecting among multiple competing explanations is not exclusively determined by utility considerations or expected value calculations. Within the framework of explanatory coherence [19], explanatory adequacy is assessed through coherence judgments operationalized via constraint satisfaction mechanisms. Tentative explanations identified through this process are subject to ongoing revision and refinement as additional evidence becomes available. The role of coherence as a guiding principle in explanation is further exemplified in the unificationist account [23], which posits that a theory's explanatory relevance derives from its capacity to unify a diverse range of phenomena. From this perspective, understanding is achieved by situating individual phenomena within broader explanatory patterns, with explanatory power increasing proportionally to the scope of unification. Such pattern-oriented explanations are inherently analogical, insofar as they enable the interpretation of distinct phenomena through the subsumption of these phenomena under common, schematic structures of explanation.

The causal mechanical (CM) model of explanation [24] constitutes a process-oriented theoretical framework that emphasizes spatiotemporal dynamics in the formulation of explanatory accounts. Owing to this emphasis, the CM model has been particularly influential in the study and explication of biological mechanisms [25]. Within the causal mechanistic perspective, entities, their activities, and their organizational structures are regarded as the fundamental components of explanation. The interactions and activities occurring within and among entities generate, sustain, and regulate the regularities observed in complex systems [26]. Consistent with the CM framework, explanation is conceived as information relevant to manipulation and control, thereby underscoring the intrinsic connection between explanation and exploration [27]. Central to this view is the role of causal manipulation in enabling counterfactual reasoning, which evaluates how outcomes would vary under hypothetical alterations of the causal factors referenced in explanatory accounts.

5.3 Conceptual Framework: Models as Cognitive Mediators

The vision presented in this chapter is based on the observation that the absence of explicit and dynamic interaction between the experimental and model design spaces impedes the modeling process, leading to premature convergence to a single authoritative model. By integrating the search processes and strategies across the experimentation and modeling domains, it is possible to represent systems as an evolving set of alternative or complementary ensembles of model mechanisms (e.g., a configurable product line of features). Additionally, a cognitive model can be developed to explain how these mechanisms relate to and cohere with each other, subject to contextual constraints.

Figure 5.1 depicts the mediating function of models in bridging the divide between theoretical constructs and empirical data. Theoretical principles provide the foundation for constructing mechanistic hypotheses based on observed behavioral features. A dynamic feedback loop between the model and the data, characterized by processes of adaptive learning, enables the progressive evolution of the

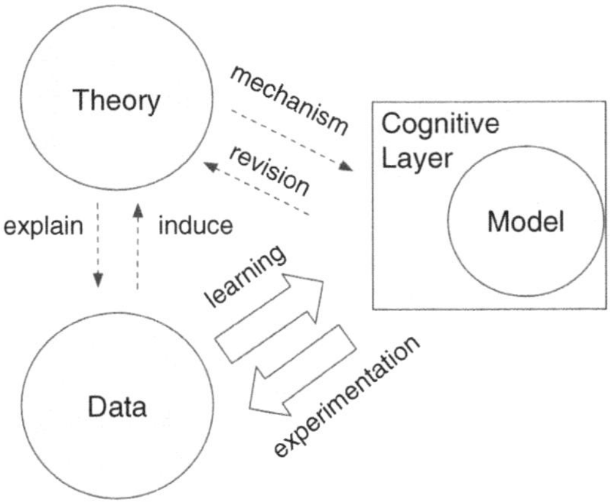

Figure 5.1 Bridging theory and data via cognitive mediator models.

model in response to newly acquired evidence. Through systematic experimentation, data are analyzed to refine and enhance the fidelity of the model representation. When consensus emerges within the scientific community, such revisions may extend beyond the model itself, ultimately contributing to the refinement and potential revision of the underlying theoretical framework.

Consequently, theoretical principles derived from limited datasets progressively attain greater accuracy and reliability in their explanatory capacity. Conceptualizing the coupled dynamics of models and experiments as a cognitive system underscores the necessity for advancements in several domains: the systematic management of variability, the automation of instrumentation to support goal-directed experimentation and analysis, and the establishment of infrastructures capable of integrating data-driven inferences into the representational feature space of models. The provision of run-time explanatory cognitive models, together with mechanisms for dynamic model updating during execution, is essential for closing the iterative loop between theory, experimentation, and model refinement.

5.3.1 Challenges

We view the challenges and opportunities in *accelerating model-driven systems research* in two categories: coupling the technical spaces of experiments and models and the provision of explainable cognitive assistance for model discovery and evaluation:

- **Managing Coupled Model and Experiment Spaces**
  - Engineers often lack specialized expertise in the use of general-purpose programming environments. Consequently, the development of high-level generative architectures becomes essential, as such architectures enable the representation of models through domain-specific abstractions and the execution of appropriate operations upon them. This approach parallels the principles of many-sorted algebra [28], wherein distinct sorts and operations are systematically organized to support rigorous formal reasoning across heterogeneous domains.
  - The demand for mechanistic explanations of engineering phenomena necessitates the construction of models that explicitly embody the causal mechanisms responsible for generating and sustaining targeted system behaviors. To facilitate this process, automated support for the exploratory search of mechanistic hypotheses can be employed. Such support may utilize heuristic search strategies in conjunction with abductive and analogical reasoning methods to systematically postulate, evaluate, and recommend revisions to existing models.
  - The execution and orchestration of simulation experiments designed to evaluate hypothesized models necessitate the variable instrumentation

of configurable model architectures. Such instrumentation is essential to ensure the systematic collection of data that are directly relevant to the research questions under investigation.

- **Cognitive Assistance for Model-driven Systems Science**
  - A cognitive model management system must facilitate learning derived from experimentation in order to effectively distinguish among competing hypothesized models. Such functionality enables online learning of causal-probabilistic structures and cognitive maps that capture relationships between model features and targeted behaviors. In turn, this capability provides a rigorous analytical framework for assessing coherence across alternative explanatory accounts.
  - Evaluating the consequences of hypothesized models necessitates the careful selection of variables to be measured, as well as the systematic updating and prioritization of experimental designs. Such refinement, conceptualized as the formulation of more precise and targeted research questions, serves to maximize information gain by reducing uncertainty across competing model features. In this context, cognitive assistance plays a critical role by facilitating the assessment of the marginal utility of individual experiments and by enabling the systematic comparison and potential reuse of alternative experimental designs.
  - The outputs generated by existing learning techniques often present challenges in terms of effective communication with domain-specific engineers. To address this limitation, appropriate forms of cognitive assistance are required to facilitate the acquisition of explanatory models and to ensure that results are conveyed through a dedicated explanation interface. Such an interface must be informed by psychological theories of practical explanation in order to enhance interpretability, usability, and disciplinary relevance.
  - Following the analysis of experimental results, causal hypotheses may require revision, with such modifications, including the incorporation of novel behavioral mechanisms, being transferred into the model space to initiate a subsequent cycle of experimentation. This process necessitates the systematic refinement of experimental models in order to sharpen analytical focus and to generate new investigatory plans, a task often operationalized through symbolic planning methodologies.

5.3.2 Methodological Foundation

To address the challenges outlined above, we advance a comprehensive strategy consisting of three interrelated components. Collectively, these components are intended to facilitate the development of a theoretical framework, a methodological approach, and an exploratory proof-of-concept aimed at enabling coherence-driven model discovery and providing explainable, built-in justification.

The explanatory learning process functions as a critical mediating mechanism, directing the progression from the initial formulation of hypotheses (e.g., causal mechanistic model features) to their systematic experimental evaluation. This iterative process culminates in the determination of whether one or more hypothesized models should be accepted, based on the weight of accumulated empirical evidence. The strategy underpinning this process comprises three interrelated components, outlined as follows:

- The *Search Feature Space* component is responsible for generating a fully specified set of hypotheses or features, which serve as the foundational input for the subsequent *Hypothesis Testing* phase.
- The *Search Experiment Space* component encompasses the design and execution of simulated experiments, which serve to generate empirical evidence supporting or refuting the hypothesized models.
- The *Evaluate Evidence* component systematically examines the cumulative body of evidence to determine whether it justifies the acceptance, rejection, or continued consideration of the current model variations.

Figure 5.2 illustrates the elements of the goal structure and the computational tasks required to support this strategy.

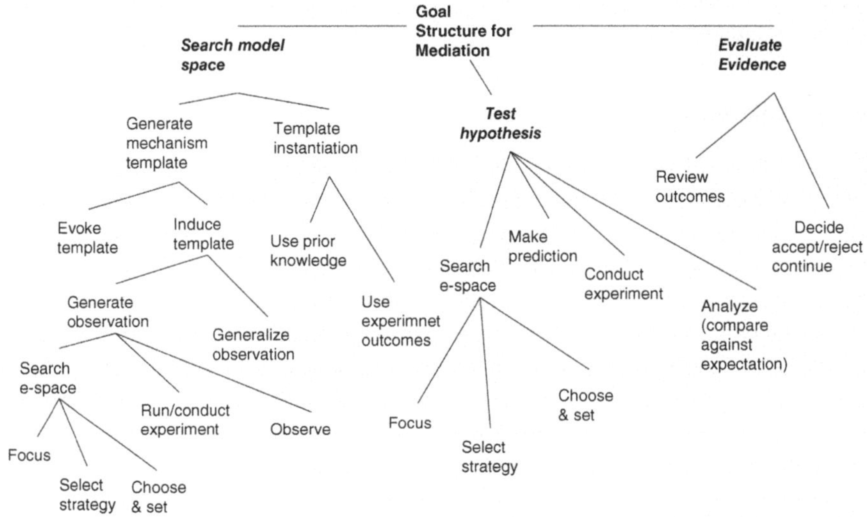

Figure 5.2 Sketch of the goal structure for model-based mediation.

Searching the experiment space entails the systematic design of experiments that correspond to the current set of hypotheses, the execution of simulation-based investigations to generate predictions, and the subsequent comparison of these predictions with expected behaviors or empirical evidence. The initial phase of experimentation is directed toward differentiating among competing hypothesized models by prioritizing experimental designs that maximize the expected information gain metric, thereby reducing entropy across alternative mechanistic accounts. The analyze (compare) objective involves a rigorous examination of discrepancies between the observed evidence, the anticipated behavior, and the actual outcomes produced under the hypothesized model.

Evaluating evidence constitutes the component responsible for determining whether the cumulative body of evidence, assessed within the framework of the explanatory coherence model, justifies the acceptance or rejection of competing hypothesized models. Evidence and hypotheses may be evaluated according to multiple criteria, including coherence maximization. In the absence of well-specified hypotheses, experiments can be generated through systematic exploration of the experiment space. Within the evaluate-evidence phase, three general outcomes are possible: coherent hypotheses may be accepted, rejected, or retained for further investigation. Acceptance of a hypothesis concludes the model discovery process, whereas rejection initiates a return to the search hypothesis space. This reengagement can trigger two potential activities: if only specific hypotheses are rejected, alternative hypotheses may be generated; however, if the entire mechanism template (or frame) is deemed inadequate, the system must attempt to construct a new mechanism through reinforcement learning or schema revision strategies.

5.3.3 Reference Architecture

The goal-directed process delineated above necessitates the establishment of a reference architecture that conceptualizes models as cognitive, autonomous, and adaptive agents. As active entities, cognitively enabled models extend the capabilities of the foundational model, thereby facilitating the mediation process. Within this perspective, an autonomous model is understood as an ensemble of interrelated models that progressively evolves through iterative learning and experimentation. Accordingly, we propose that models should be designed with explicit consideration of variability management, thereby enabling systematic customization and ensuring alignment with experimental objectives, particularly in contexts where the target system encompasses multiple dimensions or aspects.

Figure 5.3 illustrates the core components of the *active model* framework, which is integrated within an experiment management environment to foster a reciprocal and adaptive relationship between domain theory and empirical data. The variability management layer (VML) organizes theoretical constructs and

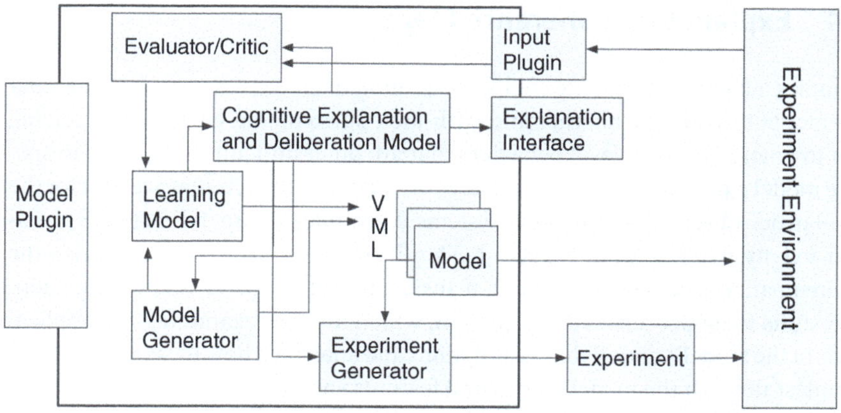

Figure 5.3 Models as mediators between theory and data.

models that embody the mechanistic hypotheses underpinning the domain theory. Within this framework, a family of models is delineated according to features that facilitate the synthesis of alternative model configurations. These features define both mandatory and optional structural building blocks, as well as their interconnections, thereby enabling systematic variation across a family of systems.

Variability management through feature models bears strong resemblance to product-line engineering, wherein the initial construction of products is automated [29]. Within the present context, however, the variability management subsystem must extend beyond the specification of a variability model to encompass runtime adaptation, thereby enabling the incorporation of model updates directly into the system's simulation language. The selection of variants, represented as features, initiates actions that dynamically customize the target model by adding, removing, or modifying functional units during execution.

The accumulation of evidence for hypothesized models necessitates rigorous and objective scrutiny conducted in a transparent and explainable manner. Such practices are essential for fostering trust in the cognitive assistance system and the validity of its recommendations. Within this framework, the explanation model fulfills two principal functions. First, it provides users with insights into the relative utility of competing mechanistic accounts. Second, it operates as a run-time model that guides the iterative revision and selection of hypotheses. In doing so, it facilitates the identification of mechanisms warranting further investigation and informs the design of experiments aimed at differentiating among alternative models. To establish a theoretically robust foundation for the cognitive dimensions of the reference architecture, this approach draws upon the theory of explanatory coherence and the interactive activation dynamics model that underpins its connectionist representation.

5.4 Explanatory Coherence Maps

Complex adaptive systems are composed of numerous interconnected entities that interact both with one another and with their surrounding environment, giving rise to emergent macrolevel behaviors that are often difficult to attribute to specific model parameters or features. The construction of models that account for the regularities observed within such systems constitutes a target-directed explanation activity, wherein scientists seek to identify models capable of reproducing the desired patterns of behavior. Once a plausible model has been established, attention shifts to model-directed explanation, which involves examining the implications of the model's underlying assumptions and analyzing how these assumptions manifest through the model's structural features and dynamic properties.

5.4.1 Theory of Explanatory Coherence

In the evaluation of alternative explanations, the provision of a coherent justification is indispensable for establishing their validity. Drawing on the theory of *reflective equilibrium* [30], conflicts among competing explanations necessitate iterative adjustments until a state of equilibrium is achieved. More broadly, *reflective equilibrium* may be conceptualized as an attractor state within a complex adaptive system, wherein stability emerges through the dynamic reconciliation of competing justificatory elements.

The attractor state arises at the culmination of a deliberative process that entails the systematic revision of explanations and goals within a defined domain of inquiry. Conceptualizing equilibrium as a stable condition that resolves conflicts, the equilibrium state functions as a coherent account of justification. An optimal equilibrium is attained when further revision becomes unnecessary, as the prevailing explanations exhibit the highest degree of acceptability [30, 31]. The principles and judgments established at this equilibrium state provide a contextually grounded and situationally appropriate basis for evaluation. The theory of explanatory coherence offers a principled foundation for this process, furnishing a computational strategy—specifically, constraint satisfaction for determining the state of reflective equilibrium [31].

The constraint satisfaction strategy can be conceptualized as the representation of a state within an N-dimensional space. Within this framework, the activation levels of explanatory hypotheses—represented as nodes in a coherence graph—correspond to the degree of acceptability assigned to the respective propositions. Each node receives input from, and is reinforced by, other nodes to which it is explanatorily connected. These inputs are modulated by the weights associated with the connecting links, thereby determining the relative influence of each connection. The activation value of a node is iteratively updated as a function of the weighted sum of its inputs. This dynamic process continues until the activation values across all units converge to a stable equilibrium. Formally, if we define the activation level of each node j as a_j, where a_j ranges from -1 (rejected) to $+1$ (accepted), the update function can be expressed as follows:

$$a_j(t+1) = \begin{cases} a_j(t)(1-\theta) + \text{net}_j(M - a_j(t)), & \text{if net}_j > 0 \\ a_j(t)(1-\theta) + \text{net}_j(a_j(t) - m), & \text{otherwise} \end{cases}.$$

In this formulation, the variable θ functions as a decay parameter that progressively diminishes the activation level of each unit across successive cycles. In the absence of external input, the activation level of a unit decreases over time, bounded by m, representing the minimum activation threshold, and M, denoting the maximum activation threshold. The net input to a unit, expressed as net_j, is computed as the weighted sum of inputs received from other units, $\sum_i w_{ij} a_i(t)$. These iterative computations are applied to each node until the activation levels converge, thereby enabling the network to achieve equilibrium through self-organizing dynamics. At equilibrium, nodes exhibiting positive activation levels are interpreted as maximally coherent propositional explanations. As an illustrative example, the prey–predator model, widely employed in biological research, simulates interspecies interactions in which prey and predator populations fluctuate as a result of competitive and cooperative dynamics.

Figure 5.4 depicts two properties of central interest: lagged oscillation and the Volterra principle. To account for these phenomena, two hypothesized equation-based mechanisms are introduced: `H1` (prey dynamics) and `H2` (predator dynamics). Together, these mechanisms successfully reproduce the lagged oscillation behavior but fail to capture the Volterra principle. Although equation-based

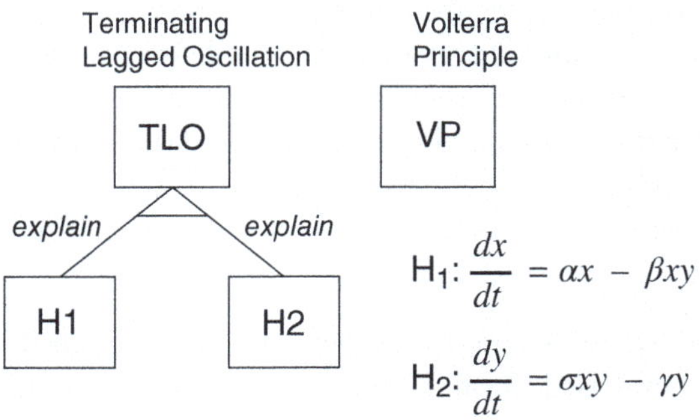

Figure 5.4 Explanatory relations. The Lotka–Volterra equations provide a formal framework for characterizing population dynamics in ecological systems involving two interacting species. Within this formulation, the rate of change in population sizes is expressed as a function of parameters that capture the nature and intensity of interspecies interactions, thereby offering a mathematical representation of predator–prey or competitive relationships.

models offer a highly idealized and simplified representation of system dynamics, they become increasingly intractable as contextual complexity and the heterogeneous behaviors of individual population members are incorporated. Such models typically yield population-level averages, thereby neglecting critical aspects of ecological systems, including community structure, environmental heterogeneity, food chain stability, and the diversity of individual behavioral strategies.

5.4.2 Abductive Reasoning with Explanatory Cognitive Maps

To improve the accuracy and realism of explanatory accounts, increasingly detailed and refined hypotheses are introduced to elaborate upon generic formulations and to incorporate the influence of specific factors such as community structure and individual-level activities. ABMs address limitations inherent in equation-based approaches by reformulating idealized mathematical models in terms of causal mechanisms involving discrete entities, their behaviors, and their interactions. Nevertheless, ABMs often incorporate programmatic simplifications and utilities that may not be directly grounded in the underlying theoretical framework, thereby complicating the task of discerning whether observed outcomes derive from the essential features of the model or from its externalities. As an illustrative example, consider the rules proposed by Weisberg [32] that operationalize hypotheses `H1` and `H2` through the dynamics of two interacting species—prey and predator.

- **Movement Rule**: Agents relocate by selecting a random direction of movement.
- **Consumption Rule**: At each time step, agents verify whether prey is present at their current location. If prey is available, one individual is randomly selected and consumed, thereby increasing the agent's energy level.
- **Reproduction Rule**: When an agent's energy exceeds a predefined threshold, a random number between 0 and 1 is generated. If this value is less than or equal to the specified predator reproduction rate, a new predator agent is instantiated at the same location.
- **Removal Rule**: Agents whose energy level reaches zero are eliminated from the simulation environment.

It is important to note that these rules approximate the constraints embedded within the original equation-based formulation of the Lotka–Volterra model. Translating an equation-based representation into an ABM requires the specification of explicit assumptions regarding individual-level behaviors, many of which are either implicit or undefined in the mathematical formulation. As a result, multiple ABM instantiations can be constructed to reproduce particular system behaviors, such as stabilized oscillations in population sizes. Nevertheless, a significant challenge arises in determining which among the numerous alternative ABM realizations most accurately captures the dynamics of the system for the intended objectives of the study.

For example, the equation-based formulation of the Lotka–Volterra model does not incorporate assumptions regarding the movement of prey or predator populations. In contrast, the movement rules within an ABM may range from purely stochastic trajectories to more sophisticated representations that integrate collective behavior and risk-averse strategies, wherein agents avoid hazardous regions based on experiential learning. The specification of such representational choices requires explicit decisions by the ABM developer. Likewise, to complete the model specification and to evaluate its capacity to generate stabilized oscillations as an emergent property of population dynamics, analogous rules governing prey behavior must also be defined.

- **Movement Rule**: Agents relocate by selecting a random direction of movement within the simulation environment.
- **Reproduction Rule**: A random number is sampled from a uniform distribution. If the sampled value is less than or equal to the predefined prey reproduction rate, the agent reproduces, thereby generating a new agent at the current location.
- **Removal Rule:** If an agent is captured by a predator, it is eliminated from the simulation and no longer participates in subsequent interactions.

Figure 5.5 presents the refined explanatory mechanisms corresponding to each high-level idealized equation model. In the case of `H1`, the birth, death, and movement rules collectively characterize the prey growth rate, while predator behaviors—specifically movement and consumption—are incorporated to represent the interaction rate between prey and predators, which may ultimately result in prey mortality. Similarly, for `H2`, the predator rules of movement, consumption, and death, together with the prey movement behavior, are employed to construct a causal mechanistic explanation of predator dynamics. Substituting the idealized equation models with their respective ABM specifications reveals the phenomenon of terminating lagged oscillation. This occurs when the system exhibits lagged oscillatory dynamics for a finite duration, after which the population of either prey or predators converges to extinction or reaches its maximum capacity, thereby terminating the simulation. As a result, the hypothesized behavioral rules fail to provide an accurate representation of the original theoretical assumptions.

The original hypothesized equation-based model, which does not explicitly account for food availability or resource density, assumes that the prey population grows exponentially in the absence of predator interactions. In practice, however, population growth is constrained by the degree of resource availability. To incorporate this counterbalancing mechanism, the reproduction rule for

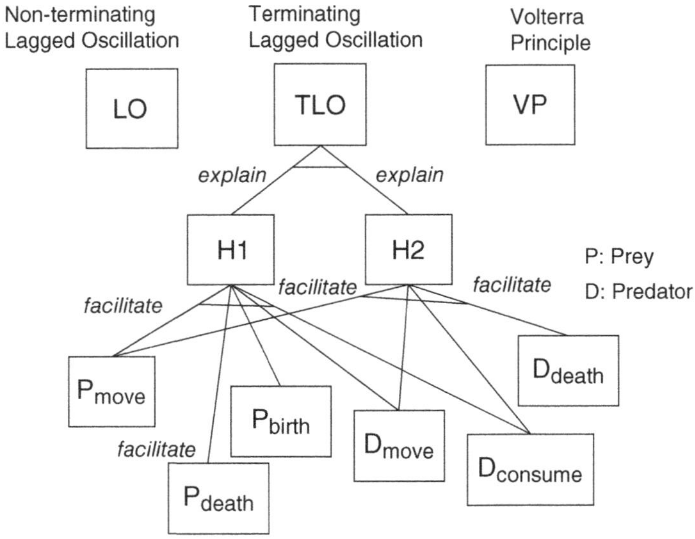

Figure 5.5 Refinement of explanatory hypotheses. Idealized equation-based models are supplanted by their corresponding agent-based model (ABM) specifications, thereby enabling a more granular representation of system dynamics and heterogeneity.

prey is defined as contingent upon the energy obtained from environmental food resources. Accordingly, an environmental component—grass—is introduced as the primary food source for prey agents (e.g., sheep). The density of this resource is regulated by the parameter `grass-growth-rate`. When prey agents consume grass, they acquire energy that facilitates reproduction. Yet, unchecked increases in prey populations are mitigated by declining resource levels, thereby imposing a negative feedback loop that counteracts the positive feedback generated through reproduction.

The prey consumption rule is enabled when the environment incorporates resource elements that account for the emergence of lagged oscillatory behavior. Given that the environment encompasses both resource-based and alternative features, these elements are mutually incompatible and logically contradictory, thereby precluding their simultaneous use as explanatory mechanisms. Nevertheless, the resource-based configuration of the environment supports the manifestation of lagged oscillations, whereas the alternative configuration accounts for the termination of such oscillatory dynamics, consistent with the outcomes observed in the preceding version of the model.

The cognitive explanatory map depicted in Figure 5.6 delineates the hypotheses and corresponding behavioral mechanisms that may be selected to account for a specific targeted behavior. The explanatory strength of hypothesized causal

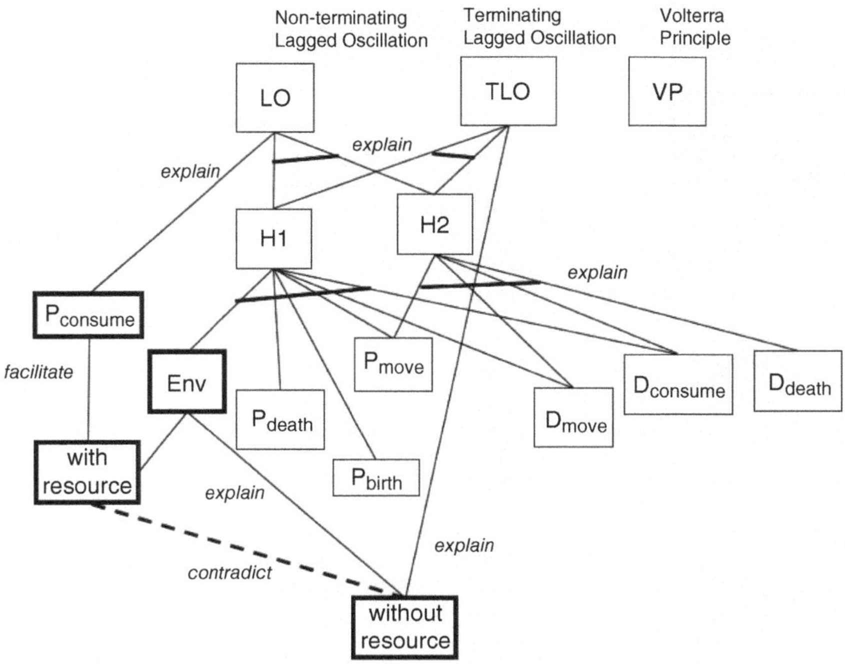

Figure 5.6 Revising explanatory hypotheses. An environmental feature comprising two alternative configurations—one incorporating resources and the other excluding them—is introduced to further refine hypothesis `H1`.

accounts is enhanced when they (1) extend their scope to encompass an increasing range of targeted behaviors and (2) deepen their explanatory capacity by engaging with lower-level fundamental mechanisms that clarify how and why higher-level explanations operate. A particularly noteworthy outcome of the Lotka–Volterra prey–predator dynamics model is the Volterra principle, an empirically observed phenomenon in fishery statistics.

According to this principle, as articulated in `H1` and `H2`, negatively coupled prey–predator dynamics imply that the introduction of biocide (i.e., a toxic substance) into the environment results in an increase in the prey population and a corresponding decrease in the predator population. To extend the applicability of these principles, the model is refined through the incorporation of a new causal mechanism involving the introduction of biocide into the system. Furthermore, the consumption patterns of both prey and predator agents are elaborated by specifying two alternative explanatory mechanisms: one in which agents consume biocide, and another in which consumption is restricted solely to food resources. Figure 5.7 illustrates how these principles, in conjunction with resource

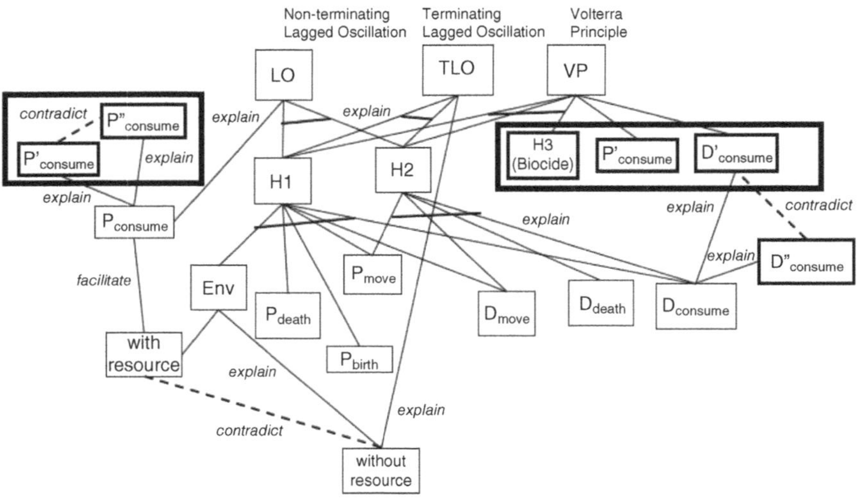

Figure 5.7 Specific configurations of H1 and H2, in conjunction with refined resource-consumption rules and environmental biocide features, provide a coherent explanatory framework for accounting for the Volterra principle.

consumption and environmental biocide features, collectively contribute to the explanation of the Volterra principle.

The preceding discussion highlights three essential components of target-directed explanation. First, it is necessary to select a coherent set of explanatory features from among available alternatives. When existing features prove insufficient to account for the targeted behavior, additional features are introduced and incorporated into the set of plausible explanations. Second, the explanatory accounts must be systematically evaluated to identify the most compelling explanation. This process involves ranking alternative features according to their relevance and their effectiveness in generating the desired or expected behavior. Third, the scope and granularity of explanatory features are progressively expanded to strengthen confidence in their explanatory adequacy. In particular, explanations are broadened to encompass additional behaviors while simultaneously being refined into increasingly detailed and high-resolution causal mechanisms that elucidate the underlying processes.

5.5 Explanatory Modeling Process

The illustrative example outlined above indicates that the construction of explanatory models constitutes a dynamic process encompassing multiple activities and distinct inferential procedures. This example, together with its underlying coherence-driven strategy, provides the basis for the formulation of the following guidelines.

- **Explanation as an Iterative and Incremental Process:** The formulation of a model-based explanation initiates subsequent inquiry aimed at both deepening and broadening the range of plausible accounts. Preliminary explanations frequently function as provisional templates that guide the ongoing search process, allowing model developers to iteratively refine causal mechanisms by incorporating additional detail and complexity. As this process advances, the focus of inquiry may shift in response to the evaluation and comparison of alternative explanatory mechanisms, thereby fostering the progressive development of more comprehensive and robust explanatory frameworks.
- **Explanation as a Symbiotic Adaptive Process:** The pursuit of explanation unfolds through a reciprocal and adaptive search process that develops between the hypothesis space (e.g., model structures) and the experimental domain. Within the architectural space of models, experimental conditions are deliberately constructed to elicit novel information that would otherwise remain inaccessible. This newly generated information subsequently shapes and redirects the search within the hypothesis space, thereby reinforcing the iterative interplay between modeling and experimentation, as articulated by Magnani [21].
- **Emergence as a Catalyst for Explanation:** ABMs of complex systems frequently uncover emergent properties that arise from the interactions among heterogeneous agents situated within spatial environments. Such emergent behaviors are often unexpected and may yield novel insights that cannot be derived as a simple aggregation of individual agent attributes. The recognition of these emergent phenomena necessitates explanatory accounts that articulate the underlying causal mechanisms responsible for the observed dynamics.
- **Explanation Through Self-awareness:** The capacity to introspectively evaluate the consequences of actions in relation to their underlying premises is essential for reflective inquiry. Explanatory reasoning entails examining the causal factors underlying observed behaviors, thereby enabling their interpretation in terms of the beliefs, objectives, and intentions that motivate observable actions. A model endowed with self-awareness is capable of comparing its simulated behaviors against its predefined objectives, thereby attaining an understanding of its structural features and the conditions under which these features contribute to the realization of the desired behavior.
- **Explanation as an Exploratory Learning Process:** The construction of explanatory accounts entails the application of three fundamental inferential processes: deduction, induction, and abduction. Deduction is employed

when a model is simulated to derive the consequences of its underlying assumptions. At the outset, multiple competing explanations may be represented by distinct models, which, when considered collectively as an ensemble, constitute a set of plausible accounts. These models are subjected to simulated experiments in order to examine their behavior. The exploratory phase involves the deliberate selection of experimental conditions and systematic variation of the model, thereby enhancing the capacity of model users to discriminate among alternatives on the basis of their explanatory adequacy. Subsequently, data generated through simulation are generalized via inductive reasoning to establish learning rules that differentiate successful outcomes from those that fail to achieve the intended objectives.

5.6 Implications for Simulation Environments

Explanatory modeling entails an exploratory learning process designed to facilitate the generation and evaluation of explanatory accounts. Effective exploration necessitates systematic variation in model structures and representational forms. The capacity to construct multiple models with diverse architectures, spanning different levels of resolution and perspectives, must be managed in a transparent and methodologically rigorous manner. This management should further extend to computational experimentation, enabling alternative scenarios and model variants to be adaptively integrated within the context of an evolving analytical framework.

The generation of explanatory models requires that the outcomes of exploratory processes be generalized and articulated through visual, causal, and conceptual representations that are intelligible and meaningful to model users. Furthermore, the set of candidate models must be explicitly represented in a manner that enables systematic querying, sampling, and analysis, thereby facilitating incremental and adaptive modeling practices. Models endowed with self-adaptive capabilities introduce additional constraints on the representation and selection of model features, necessitating more rigorous methodological considerations.

5.6.1 Variation of Causal Assumptions in an Evolving Analysis

Models are constructed on the basis of specific underlying assumptions. The rules governing agent behavior typically integrate both theoretical principles and pragmatic considerations. Pragmatic elements often involve simplifying assumptions that are not directly derived from the foundational theory of the system. For instance, in ABMs of ecological dynamics, the movement of prey and predator agents may be specified as either random or intentional in order to approximate collective flocking behavior. To enable flexible variation in model structures and representations, both implicit design strategies and explicit modeling constructs may be employed. Implicit strategies frequently draw upon established design patterns, such as the Strategy and Factory Method, which facilitate the systematic management of behavioral variability in cases where specific agent actions cannot be anticipated in advance.

Alternatively, explicit modeling constructs and paradigms may be employed to design models with the deliberate aim of introducing variability. For example, feature-oriented and aspect-oriented modeling paradigms emphasize relatively simple features and crosscutting aspects that can be composed into aggregate models subject to defined composition constraints. Within such modeling environments, these constraints and feature models can be explicitly represented through declarative specifications. Declarative approaches enable the application of advanced tools capable of manipulating and adapting models in a transparent manner, thereby supporting manipulative abductive reasoning [21]. To accommodate variation and customization, a modeling environment must incorporate an explicit variability model. Such a model should provide mechanisms for specifying the behavior of individual features, defining syntactically and semantically valid feature combinations, and facilitating the iterative evolution of these specifications through simulation-based experimentation.

5.6.2 Controlling for Confounding Factors

A model development environment should be designed to support both interventions and counterfactual reasoning, thereby enabling explanations that extend beyond simple associative relationships. Experimental inquiry serves as a critical tool in this process, as it permits the manipulation of simulation conditions to observe their causal effects. Such interventions allow researchers to identify the consequences of deliberately altered causes. While interventions primarily emphasize prospective reasoning, counterfactual experimentation requires revisiting the foundational assumptions of the model. This involves posing questions such as: "What is the probability of observing the outcome $Y = y$ had the causal mechanism `M1` been employed, given that the observed outcome was $Y = y'$ under the mechanism `M2`?" Intervention-based exploration thus necessitates systematic searches within the experimental space, whereas counterfactual analysis demands parallel exploration across both the model design space and the experimental space.

Conducting experiments on a model entails subjecting it to the oversight of an experimental management framework designed to systematically refine and revise the set of candidate causal mechanisms. The central objective of experimentation is the evaluation of alternative causal explanations, which requires the design of targeted experiments to elucidate the role and significance of specific components within hypothesized mechanisms. One experimental strategy involves assessing whether a particular entity, activity, property, or organizational feature possesses causal relevance to another. This approach assists model developers in identifying the contextual conditions that trigger a phenomenon or in determining whether a given element is sufficient to initiate subsequent stages of the mechanism. A complementary strategy focuses on evaluating whether an entity, activity, or organizational characteristic contributes to the overall behavior produced by the mechanism as a whole. Such experimental practices enable the identification of causal relationships across multiple levels of hierarchically structured models. Once sufficient confidence in these relationships has been established, further inquiries can be pursued by intervening in the model and systematically manipulating experimental conditions.

Mechanism-aware Targeted Experiments: Targeted experiments are designed with explicit awareness of underlying mechanisms, thereby enabling direct forms of interference, stimulation, and activation of specific components within a model. Interference experiments function by inhibiting, attenuating, or disabling elements of the causal structure, whereas stimulation experiments involve interventions that excite or amplify system behavior in order to observe the resulting effects. Both interference and stimulation approaches are classified as bottom-up strategies, as they operate directly on lower-level mechanistic components. In contrast, activation experiments constitute top-down strategies, wherein experimental conditions are manipulated to elicit higher-level behaviors, thereby allowing for the assessment of their influence on subordinate mechanistic processes.

Whereas intervention experiments are designed to elucidate the role of relatively stable mechanisms composed of specific components, exploratory analyses commence by formulating targeted mechanistic questions intended to reveal and characterize these components. A range of experimental strategies of this kind has been identified and examined in detail by Darden [26].

- **By-what-activity Experiments:** The primary objective of by-what-activity experiments is to identify the specific activities that mediate the relationship between causes and effects. Such experiments require systematic reasoning about the alignment of invariants with the preconditions necessary for activities to occur. This process involves determining the preconditions of plausible activities and subsequently comparing contexts in which these preconditions are satisfied with those in which they are not. These comparisons serve to differentiate among competing hypotheses regarding activity features.
- **By-what-component Experiments:** Activities are generated through the coordinated functioning of components that collectively produce observable behavior. In by-what-component experiments, artificial conditions are introduced to selectively disable components in a controlled manner, thereby

assessing whether the intervention prevents the occurrence of the phenomenon. Alternatively, different combinations and organizational configurations of components may be employed to reproduce the desired behavior. The integration of activity-based and component-oriented experimental strategies enables systematic pruning of the space of possible explanatory causal mechanisms.

5.6.3 Stepwise Refinement with Explanatory Modeling

As experiments are performed to intervene in the mechanisms of the target system, the resulting evidence is expected to progressively converge toward increasingly credible causal accounts capable of generating and sustaining the desired behavior. Consequently, confidence in the model's performance develops as it incorporates and explains a broader accumulation of empirical findings. Models that demonstrate success are subsequently refined to provide more detailed and nuanced characterizations of the higher-level mechanisms involved. This iterative learning process should be supported by appropriate methodological tools capable of systematically evaluating experimental outcomes and offering critical feedback to further guide and optimize the search for robust explanatory mechanisms.

The assessment of components and activities necessitates the explicit separation of a learning element from the model. This separation enables the learning element to receive feedback from a critic and to determine how the model should be modified in order to achieve the objectives of the experiment. The critic provides evaluative input in the form of rewards or penalties, contingent upon the extent to which the model's performance aligns with the targeted standard. In addition, the learning process requires a generative component that supports the systematic production of models consistent with the abductive reasoning strategies previously outlined. Absent such a generative mechanism, the model risks converging prematurely on a representation that performs adequately under narrowly defined scenarios but lacks robustness when applied to a broader and more diverse range of conditions.

5.7 Conclusion

The application domains of successful models broaden as successive refinements increase the level of resolution, thereby enhancing accuracy and fidelity throughout the learning process. Nevertheless, the explanation of causal relationships within such models presents a significant challenge, particularly given the autonomous and decentralized decision-making processes of agents and the complexity of their interactions. Strategies for explanation draw upon insights from philosophy and the cognitive science of scientific reasoning. To ensure that explanatory accounts are aligned with the cognitive requirements of the intended audience, a coherence-based cognitive strategy is employed.

The coherence-driven strategy demonstrates the manner in which explanatory models evolve through the refinement of simulation models derived from experimental outcomes. Grounded in the theory of explanatory coherence and its associated framework of reflective equilibrium, this approach illustrates how inquiries in population dynamics can progressively broaden and deepen, leading to the systematic revision of beliefs across multiple levels and scales regarding the causal premises underlying expected behavioral regularities.

Building upon the theory of explanatory coherence, the principles of explanatory modeling are articulated to characterize the dynamic processes underlying model formation and development. The inferential mechanisms that support this process are situated within a framework that integrates both exploratory inquiry and adaptive learning. On the basis of these principles and associated evaluative criteria, specific guidelines are proposed for the design of modeling and simulation environments that effectively facilitate and sustain explanatory modeling.

References

1 Larissa Chazette, Jil Klünder, Merve Balci, and Kurt Schneider. How can we develop explainable systems? Insights from a literature review and an interview study. In *Proceedings of the International Conference on Software and System Processes and International Conference on Global Software Engineering*, pages 1–12, 2022.

2 Kathleen A. Creel. Transparency in complex computational systems. *Philosophy of Science*, 87(4):568–589, 2020.

3 Alison R. Panisson, Débora C. Engelmann, and Rafael H. Bordini. Engineering explainable agents: An argumentation-based approach. In *International Workshop on Engineering Multi-Agent Systems*, pages 273–291. Springer, 2021.

4 Paul K. Davis, Angela O'Mahony, Timothy R. Gulden, Osonde A. Osoba, and Katharine Sieck. *Priority Challenges for Social and Behavioral Research and Its Modeling*. RAND Corporation, Santa Monica, CA, 2018.

5 Jan H. Kwakkel. The exploratory modeling workbench: An open source toolkit for exploratory modeling, scenario discovery, and (multi-objective) robust decision making. *Environmental Modelling & Software*, 96:239–250, 2017.

6 Danny Weyns, Radu Calinescu, Raffaela Mirandola, Kenji Tei, Maribel Acosta, Amel Bennaceur, Nicolas Boltz, Tomas Bures, Javier Camara, Ada Diaconescu, Gregor Engels Simos Gerasimou, Ilias Gerostathopoulos, Sinem Getir Yaman, Vincenzo Grassi, Sebastian Hahner, Paola Inverardi, Dimitri Van Landuyt, Rogerio de Lemos, Emmanuel Letier, Marin Litoiu, Lina Marsso, Angelika Musil,

Juergen Musil, Genaina Nunes Rodrigues, Diego Perez-Palacin, Federico Quin, Patrizia Scandurra, Antonio Vallecillo and, Andrea Zisman. Towards a research agenda for understanding and managing uncertainty in self-adaptive systems. *ACM SIGSOFT Software Engineering Notes*, 48(4):20–36, 2023.

7 David Klahr and Herbert A. Simon. Studies of scientific discovery: Complementary approaches and convergent findings. *Psychological Bulletin*, 125(5):524–543, 1999. https://doi.org/10.1037/0033-2909.125.5.524.

8 Finale Doshi-Velez and Been Kim. Towards a rigorous science of interpretable machine learning. *arXiv preprint arXiv:1702.08608*, 2017.

9 Rudresh Dwivedi, Devam Dave, Het Naik, Smiti Singhal, Rana Omer, Pankesh Patel, Bin Qian, Zhenyu Wen, Tejal Shah, Graham Morgan, and Rajiv Ranjan. Explainable AI (XAI): Core ideas, techniques, and solutions. *ACM Computing Surveys*, 55(9):1–33, 2023.

10 David Gunning. Explainable artificial intelligence (XAI). *Defense Advanced Research Projects Agency (DARPA)*, 2, 2017.

11 Zheng Zhang, Levent Yilmaz, and Bo Liu. A critical review of inductive logic programming techniques for explainable AI. *IEEE Transactions on Neural Networks and Learning Systems*, 35:1–17, 2023. https://doi.org/10.1109/TNNLS.2023.3246980.

12 Levent Yilmaz. Reproducibility in M&S research: Issues, strategies and implications for model development environments. *Journal of Experimental & Theoretical Artificial Intelligence*, 24(4):457–474, 2012.

13 Alejandro Teran-Somohano, Orçun Dayıbas, Levent Yilmaz, and Alice Smith. Toward a model-driven engineering framework for reproducible simulation experiment lifecycle management. In *Proceedings of the Winter Simulation Conference 2014*, pages 2726–2737. IEEE, 2014.

14 Stephan Onggo, Levent Yilmaz, Franziska Klugl, Takao Terana, and Charles M. Macal. Credible agent-based simulation – An illusion or only a step away? In *Proceedings of the Winter Simulation Conference*, pages in–press. ACM, 2019.

15 Levent Yilmaz and Bo Liu. Model credibility revisited: Concepts and considerations for appropriate trust. *Journal of Simulation*, 16:1–14, 2020.

16 Carl G. Hempel and Paul Oppenheim. Studies in the logic of explanation. *Philosophy of Science*, 15(2):135–175, 1948.

17 Wesley C. Salmon. *Statistical Explanation and Statistical Relevance*, Vol. 69. University of Pittsburgh Press, 1971.

18 Nancy Cartwright and Ernan McMullin. How the laws of physics lie. *American Journal of Physics*, 52:474–476, 1984.

19 Paul Thagard. Explanatory coherence. *Behavioral and Brain Sciences*, 12(3):435–467, 1989.

20 Charles Sanders Peirce. *The Essential Peirce: Selected Philosophical Writings*, Vol 2. Indiana University Press, 1992.

21 Lorenzo Magnani. *Abductive Cognition: The Epistemological and Eco-Cognitive Dimensions of Hypothetical Reasoning*, Vol. 3. Springer Science & Business Media, 2009.

22 Paul Thagard. *The Cognitive Science of Science: Explanation, Discovery, and Conceptual Change*. MIT Press, 2012.

23 Michael Friedman. Explanation and scientific understanding. *The Journal of Philosophy*, 71(1):5–19, 1974.

24 Wesley C. Salmon. *Scientific Explanation and the Causal Structure of the World*. Princeton University Press, 1984.

25 Carl F. Craver and Lindley Darden. *In Search of Mechanisms: Discoveries Across the Life Sciences*. University of Chicago Press, 2013.

26 Lindley Darden. Strategies for discovering mechanisms: Schema instantiation, modular subassembly, forward/backward chaining. *Philosophy of Science*, 69(S3):S354–S365, 2002.

27 James Woodward. *Making Things Happen: A Theory of Causal Explanation*. Oxford University Press, 2005.

28 Joseph Goguen and Răzvan Diaconescu. An oxford survey of order sorted algebra. *Mathematical Structures in Computer Science*, 4(3):363–392, 1994

29 Klaus Pohl, Günter Böckle, and Frank Van Der Linden. *Software Product Line Engineering: Foundations, Principles, and Techniques*, Vol. 1. Springer, 2005.

30 Norman Daniels. *Justice and Justification: Reflective Equilibrium in Theory and Practice*, Vol. 22. Cambridge University Press, 1996.

31 Paul Thagard. *Coherence in Thought and Action*. MIT Press, 2002.

32 Michael Weisberg. *Simulation and Similarity: Using Models to Understand the World*. Oxford University Press, 2012.

6

Multi-agent Reinforcement Learning for Decision-making in Contested Environments

*Ishan Honhaga,[1] Anh Viet Do,[1] Eyoel Teffera,[2] Ant Perry,[2] Rob Baker,[2] and Claudia Szabo[1]**

[1] *School of Computer and Mathematical Sciences, The University of Adelaide, Adelaide, Australia*
[2] *Defence Science and Technology Group, Adelaide, Australia*

*Corresponding author: claudia.szabo@adelaide.edu.au

6.1 Introduction

The current landscape of military operations has undergone a profound transformation, moving decisively toward uncrewed aerial systems (UAS)-centric operations. This shift not only showcases a leap in technological advancement but also a fundamental reconceptualization of battlefield dynamics, strategic planning, and the nature of engagement [1]. UAS are increasingly indispensable, with their utility expanding from surveillance and logistical support to direct combat engagement. This trend is not only established but is projected to intensify significantly in the future [1].

The rapid technological advancements of the past few decades have expanded the usage of UAS from niche capabilities to central components of global military arsenals. Today, UAS can be used for various roles such as intelligence gathering, surveillance, and reconnaissance (ISR), precision strike operations, logistical support, electronic warfare, and search and rescue missions [1–3]. Their utility is demonstrated across various conflict zones, from counter-terrorism operations to large-scale conventional warfare [1]. Conversely, as UAS systems can and have been used in various offensives, research into counter-UAS systems is picking up significant interest [3]. A critical challenge in determining the effectiveness and fault tolerance of existing solutions is the high cost of their testing and optimization of pre-deployment configuration in real-life scenarios [3].

The advancements in UAS technology are rapidly evolving, as highlighted by recent works such as [2, 4–6]. The expansion in UAS utility has increased research investment into two primary areas: developing UASs with sophisticated capabilities [4, 6] and building advanced counter-UAS systems [5, 7]. While these works demonstrate significant potential in both domains, a key gap remains: a unified system enabling both UAS and countering systems to learn and adapt by directly playing against each other, thereby pushing the boundaries of autonomous adversarial development. This is crucial for advancing the state of the art in autonomous combat and developing robust solutions against evolving aerial threats.

In this chapter, we describe the ARES environment, developed as a training ground for counter-UAS reinforcement learning (RL) algorithms. Due to the versatility of UAS and the advancement in technology, UAS are getting better and smarter at an accelerated rate [1, 2, 4–6]. To tackle this rapid advancement, a training ground for RL algorithms needs to be highly customizable, capturing diverse scenarios and observations, with an extended ability to evaluate numerous solutions. ARES is a simulation environment based on the OpenAI Gym toolkit [8], designed around the counter-UAS problem with a range of customization options available to allow users to tailor the scenario to their needs. To the best of our knowledge, no simulator of this type of scenario that evaluates and integrates RL algorithms exists.

6.2 Challenges for Decision-making in Contested Environments

Decision-making in contested and dynamic environments is inherently shaped by the properties of the systems that operate in them, which are resilient complex adaptive systems (RCAS). Such environments involve multiple interacting agents—human, technological, and organizational—whose behaviors evolve in response to both adversarial actions and environmental pressures. This interplay generates nonlinear dynamics, emergent outcomes, and systemic fragility, all of which complicate the ability of decision-makers to anticipate, plan, and act effectively.

A key challenge stems from the nonlinearity of interactions. Small perturbations, such as minor disruptions in communication or adversary deception tactics, can propagate and amplify, leading to disproportionate operational consequences. Predictive models that assume linear cause-effect relationships are therefore insufficient; decision-makers must instead operate with an acceptance of uncertainty and prepare for cascading effects across the system.

The emergent nature of adversarial behavior further complicates decision-making. Adversaries, as adaptive agents within the CAS, continuously adjust tactics in response to observed strategies. This co-adaptation creates a moving target problem, where previously effective decisions rapidly lose relevance. Decision-making must therefore incorporate mechanisms for ongoing monitoring, feedback, and rapid adjustment, rather than reliance on static plans.

Distributed cognition and coordination present additional challenges. In contested environments, decision-making authority often needs to be decentralized to the tactical edge to maintain agility under degraded communications. However, such distributed decision-making risks fragmentation if local adaptations diverge from overarching strategic intent. Maintaining coherence requires resilient feedback mechanisms that can reconcile local autonomy with global system objectives, despite adversarial efforts to exploit seams in coordination.

Finally, the dynamic and evolutionary nature of these environments demands resilience-oriented decision frameworks. This involves not only robustness to immediate shocks but also the capacity to adapt and evolve under persistent stressors. Decision-makers must cultivate an ability to recognize weak signals of systemic change, support organizational learning, and adjust doctrine and tactics as the environment co-evolves with technological innovation and adversary adaptation.

6.3 ARES Environment and RL Training

In an ARES scenario, the basic setup positions a Blue Team against one or more Red Teams. The Blue Team is composed of stationary ground agents (tanks), each armed with countermeasures (CM) to intercept incoming Red agents. The specific CM for each Blue unit can be chosen from those available in the ARES framework, with the option to introduce new ones as needed. Scoring is handled through reward structures, which may be predefined or user-defined. Red agents represent mobile aerial units (drones) that attempt to destroy Blue units by colliding with them. Their behavior, speed, and targeting can be configured, while new capabilities can also be developed and added. The competition ends when one side eliminates the other, with the surviving team declared the winner of the episode.

RL relies on the formal structure of a Markov decision process (MDP), which dictates how an agent perceives its environment, takes actions, and receives feedback to learn optimal strategies [9]. ARES provides the MDP framework, allowing RL agents to explore state–action spaces and optimize decision-making. This foundational MDP structure, coupled with ARES's customizable features—configurable environment parameters, reward functions, and agent behaviors—significantly enhances the ability of RL algorithms to adapt and learn robust strategies. The following sections discuss the MDP framework and various customizations available in ARES.

6.3.1 Markov Decision Process Framework

To facilitate the learning of effective strategies by RL algorithms, the problem of countering uncrewed systems is modeled as an MDP in ARES and characterized by five key elements: State (S), Action (A), Transition Probability (T), Reward (R), and Horizon (H) [9].

6.3.1.1 State

A state S represents all relevant observable information an agent perceives within its environment at a given moment. An observation O defines what the agent actually perceives and can be a subset of the true state depending on observability. A Blue agent's observation space includes its position, hitpoints, and specifications of its available CMs (maximum range, hit probability, current ammunition, CM type, area of impact). It can also perceive these features for other Blue agents. Furthermore, the agent has visibility of each Red agent on screen, including their position, velocity, type, hitpoints, and relative distance. A Red agent observes its own location, velocity, type, hitpoints, and its distance from all Blue agents, along with the same features for other Red agents.

6.3.1.2 Action

An action A is the decision applied at a particular state, causing a transition to a new state. ARES uses a joint action space:

1. *Select target and CM*: choose a target moving toward Blue and a valid CM.
2. *No operation*: always available and the default if the agent is destroyed.

6.3.1.3 Transition Probability

The transition function $T = P(s' \mid s, a)$ quantifies the probability of moving to the next state s', given state s, and action a. Modeling T is intractable in large/continuous spaces like ARES; during training we adopt a model-free approach and use observed next states directly [9].

6.3.1.4 Rewards

Executing a in s leading to s' yields reward R, which guides learning of a policy that maximizes long-term return. Blue Team rewards consider: Red UAS destroyed, cost of firing a CM, penalty for Blue elimination, and a penalty for selecting an invalid CM (target out of range or CM out of ammunition). ARES exposes hooks to customize reward shaping.

6.3.1.5 Horizon

H is the episode length (from initial to terminal state or max steps).

6.3.2 Simulator Structure

The Python-based simulator is structured around several classes and wrappers. The *Client* feeds specifications into the *Playground*, which instantiates entities and passes RL requirements to the RL model factory. The environment maintains state, steps the world in response to actions, and computes rewards (see Figure 6.1 for the operational flow).

During each timestep, Blue agents select and apply CMs (subject to rules), Red agent positions update via their speed functions, and Pygame renders the environment for visual runs.

6.3.3 Environment Configurations

6.3.3.1 Map

ARES allows two-dimensional/three-dimensional (2D/3D) maps, configurable physical size and scale (pixels to distance), adjustable visualization frame rate, and *invisibility zones* that hide Red agents (simulating line-of-sight occlusions).

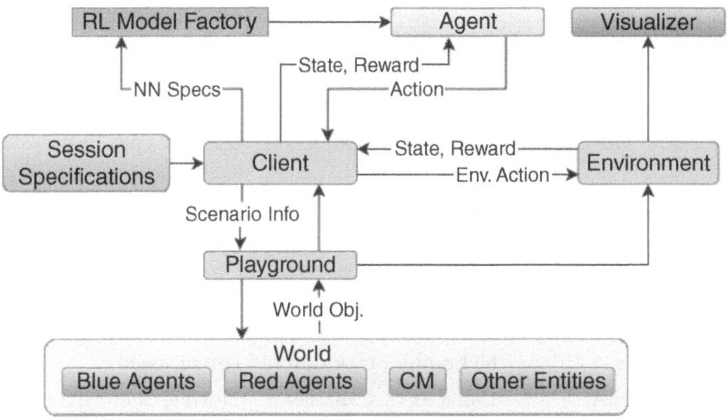

Figure 6.1 ARES operational flow.

6.3.3.2 Scenarios

Scenarios define the positioning and behavior of both teams, enabling policies to learn nuanced behaviors. Table 6.1 lists CM examples and Table 6.2 shows two scenario specifications.

6.3.3.3 Swarm Behavior

Red agents may exhibit evasive swarm maneuvers (orbiting waypoints), complicating targeting and depleting Blue ammunition. This enables learning of anti-swarm tactics. ARES provides predefined CMs that can be adapted (Table 6.1).

Table 6.1 Available countermeasure configurations.

Property	Gun	Jammer	Frag	Laser	Net
Ammunition	100	80	10	150	10
Maximum range	200	60	100	200	100
Hit probability	0.6	1	0.85	0.6	0.85
AOI	N.A.	N.A.	10	N.A.	N.A.

Table 6.2 Scenario 1 (S1) and Scenario 2 (S2) specifications. N.A. indicates unavailable attributes.

2*Attribute	Blue Team		Red Team		
(lr)2–5	S1	S2	S1	S2	2*Description
Max agents	3	3	5	5	Team sizes
Agents on screen	N.A.	N.A.	5	5	Max simultaneous Red agents
Max hit-points	5	5	1	1	Hits an agent can take
CM/agent	Gun	Gun, laser	N.A.	N.A.	CM types per Blue agent
Arrangement	Equidistant	Equidistant	N.A.	N.A.	Single column
Drone type	N.A.	N.A.	Phantom, parrot	Phantom, parrot	Agent types
Size	N.A.	N.A.	30	30	Drone size
Speed	N.A.	N.A.	60 kmph	60 kmph	Drone speed
Speed function	N.A.	N.A.	Callable/null	Callable/null	Flexible or constant
Spawn window	N.A.	N.A.	0.8	0.8	Map breadth percentage
Spawn location	N.A.	N.A.	2	2	1:Top, 2:Middle, 3:Bottom

6.3.4 Episode Rollout

As an MDP as a discrete-time process, an episode is a sequence of states. Each episode starts with initial states where

1. Red UAS are spawned from the map edge opposite to the Blue Team;
2. Each agent is supplied with the maximum hitpoint;
3. Each blue agent is supplied with a preconfigured CM set; and
4. Each CM is supplied with the maximum ammunition.

At each step, each agent commits an action within their individual action space. For blue agents, the action space is described in Section 6.3.1. For red agents, each agent moves across the map via a displacement vector. The episode ends when either side is destroyed, which is considered a win for the other team, or a time limit is reached. An active episode is shown in Figure 6.2.

6.3.4.1 Blue Team

At each step, if the no-op action is selected, nothing is fired. Otherwise, engagement applies only if: (1) a target–CM pair is selected; (2) the target is within the CM range; (3) the CM has ammunition; and (4) the target is visible and alive. The outcome of each engagement is stochastic, as detailed in Table 6.1.

6.3.4.2 Red Team

On spawn, a Red drone targets an operational Blue agent and flies toward it, following the scenario's trajectory/speed. The policy reassigns a target only if the

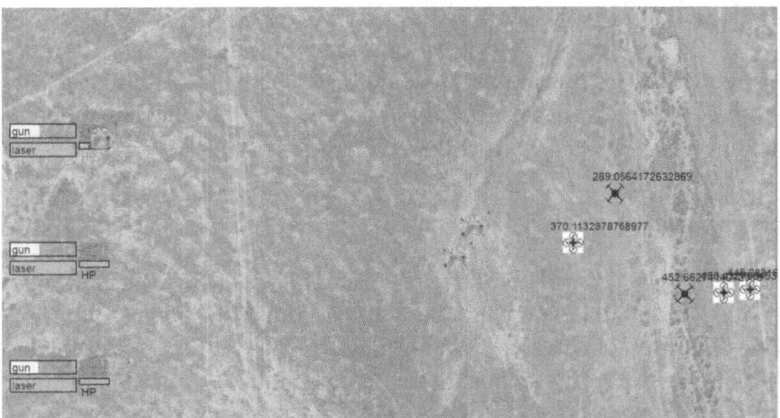

Figure 6.2 ARES simulation of an active episode. Blue has three agents (tanks), each with a gun and a laser, engaging five Red agents (two UAS types).

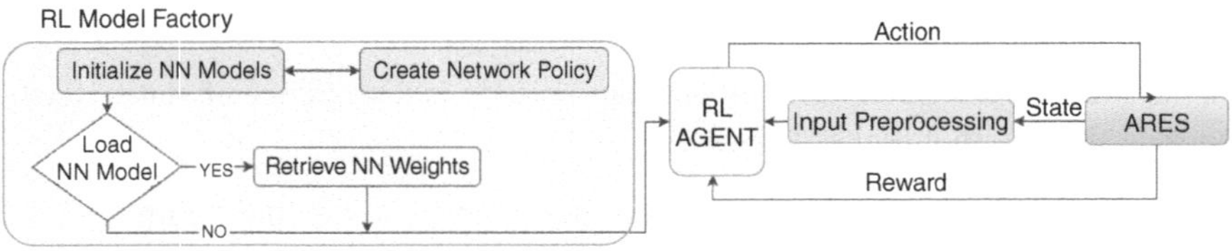

Figure 6.3 ARES interfacing with the RL algorithm.

current target is destroyed. A Red drone is eliminated if it collides with its target (explosion) or is intercepted by an effective CM.

6.3.5 Training Blue and Red Policies

To facilitate integration with RL algorithms, each simulation step produces, for each agent, a transition which is a tuple consisting of the current state, the action taken, the resultant reward, and next state (Figure 6.3). This MDP-based configuration is used in many RL paradigms (see Section 6.3.1). Experiences (transitions from past simulations) are stored in a replay buffer and sampled for policy updates. ARES also supports training Red with RL, enabling adaptive adversaries and richer stress tests.

It is important to note that when terminating conditions are met, a new episode is immediately started, and a terminating signal is recorded in the corresponding experiences for all agents, as this signal is also relevant to training the policies.

6.4 Evaluating Decision-making

To demonstrate ARES' RL integration capabilities, we conducted a training experiment with several RL algorithms. In this section, we present the experimental setting and findings.

6.4.1 RL Algorithms

We employed Deep Q-Networks (DQN) and Soft Actor-Critic (SAC) to showcase the utility of ARES simulation environment and its ability to support RL procedures. These algorithms are used as building blocks in many other RL algorithms, so ARES's ability to work with them indicates its compatibility with a broad range of state of the art.

DQN [10] is a value-based RL algorithm that extends traditional Q-learning by employing a deep neural network to approximate the optimal action-value function, $Q(s, a)$. DQN overcomes the instability issues inherent in combining

Q-learning with nonlinear function approximators through the use of Experience Replay (ER), which stores and samples past transitions to break correlations in the training data, and a separate Target Network, which provides stable targets for the Q-value updates. DQN is highly effective in environments with discrete action spaces.

SAC [11] is a state-of-the-art off-policy actor-critic algorithm known for its stability, high sample efficiency, and effectiveness in continuous control tasks. SAC operates within the maximum entropy RL framework, where the agent is encouraged to maximize a dual objective: the expected return from rewards and the policy's entropy (randomness). This entropy maximization encourages broader exploration, leading to a more robust and resilient policy capable of handling complex decision-making, which is particularly beneficial in multi-agent and competitive environments.

6.4.2 Results

We present results from two scenarios (Table 6.2) to validate ARES for RL research. Blue's reward emphasizes destroying Red while minimizing Blue losses. For comparative analysis across both scenarios, the Blue agent was implemented using DQN and SAC. These RL algorithms are used within a centralized-training-decentralized-execution paradigm: each training run uses experiences from all blue agents to train a singular policy, which outputs actions for all blue agents from their corresponding observations in each rollout step. Red follows a random policy (selects a Blue target and moves toward it).

6.4.2.1 Rewards

Across both scenarios, the Blue agent trained with SAC robustly outperformed the DQN. This showcases that SAC consistently achieves significantly higher cumulative rewards, thereby indicating a superior and more effective learned policy under the tested conditions.

6.4.2.2 Survivability

Episode length served as a proxy for agent survivability. Across both scenarios, SAC agent exhibited longer average episode lengths than DQN, demonstrating better survivability. In Scenario 1, SAC converged and settled on the optimal policy faster than DQN, and in Scenario 2, SAC outperformed DQN.

6.4.2.3 Selection of CM

An invalid CM selection is defined as the action of selecting a CM to engage an incoming drone when that CM is unable to execute the engagement. The conditions of engagement are described in Section 6.3.4. In both scenarios,

SAC demonstrated a consistent and significant reduction in invalid CM selections throughout the training process, indicating successful policy refinement toward the selection of valid CM. Conversely, DQN exhibited an increase in invalid CM selections across both scenarios. This failure to converge on an effective policy is corroborated by the lower successful hit count and higher total ammunition expenditure observed for DQN, which collectively represent suboptimal learning.

6.4.2.4 Hit Counts

The successful hit count, defined as the number of drones successfully neutralized by the selected CM, serves as the primary performance metric. As illustrated in Figures 6.4 and 6.5, the hit count for both algorithms is higher in Scenario 1 than in Scenario 2. This highlights the increased complexity in decision-making introduced by the availability of two CM (Scenario 2) as opposed to just one CM (Scenario 1). In both scenarios, SAC consistently achieved a superior hit count compared to DQN, further validating its more effective learning policy.

6.4.2.5 Red Wins

The Red Team Win Count is an inverse performance indicator, quantifying the number of episodes in which the Red Team was unable to neutralize the Blue Team or the Blue Team was able to survive until the predefined maximum episode length. As demonstrated in Figures 6.4 and 6.5, a decrease in this metric over the training period is observed. This trend signals that the Blue Team's defensive policy is successfully learning to avoid failure states and increasing its survival time. Consistent with the other metrics, SAC maintains a significantly lower Red Team Win Count than DQN across both scenarios.

6.4.2.6 Used ammunition

The Used Ammunition metric quantifies the total number of CM discharged, irrespective of successful hits, i.e., hit count. SAC consistently exhibited superior resource management compared to DQN, minimizing unnecessary expenditure. This effective utilization of limited resources translates directly to increased survivability as depicted by longer episode length.

6.5 Conclusion

ARES is a robust and highly adaptable simulation environment engineered to develop and evaluate RL algorithms for countering UAS. Its core strength is customizability—default CM and configurable environment parameters enable

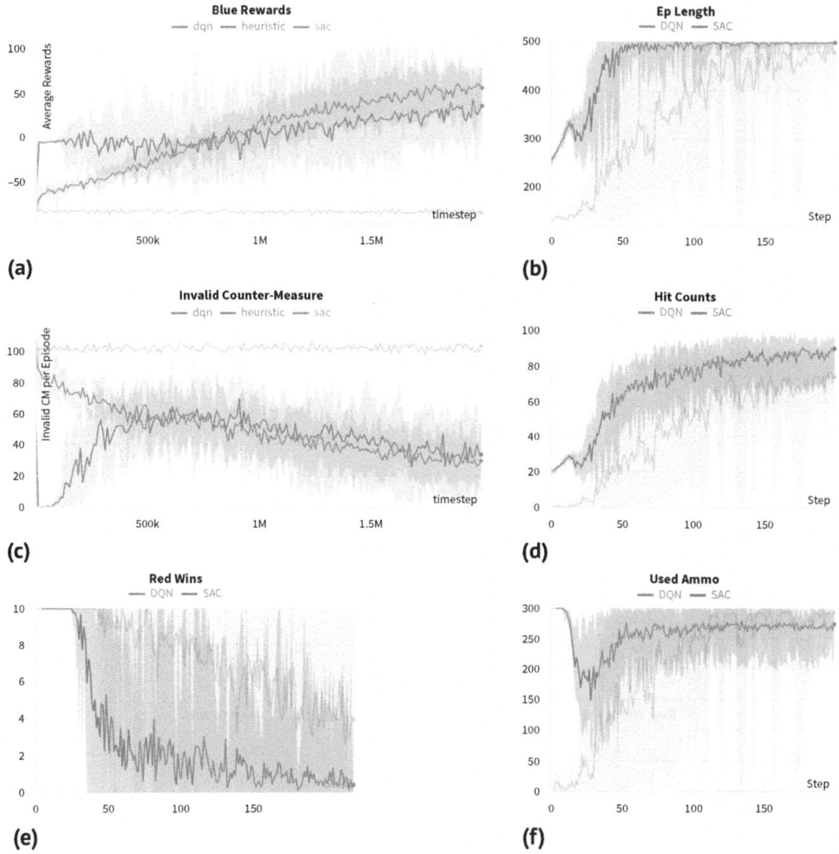

Figure 6.4 Scenario 1: blue with DQN (orange), SAC (purple); red agents approach randomly picked surviving blue agents. (a) Blue rewards, (b) episode length, (c) invalid CM, (d) hit counts, (e) red win counts, and (f) used ammo. For color references, please refer to the online version.

diverse training scenarios that capture real-world complexity. Custom reward structures focus learning on operational outcomes such as survivability and resource conservation. Future work will explore complex swarming behaviors, integrate advanced sensor models, and investigate decentralized control for robust, adaptive counter-UAS. Blue/Red will be extended with ground units and heterogeneous behaviors, positioning ARES as a test bed for the next generation of autonomous defence strategies.

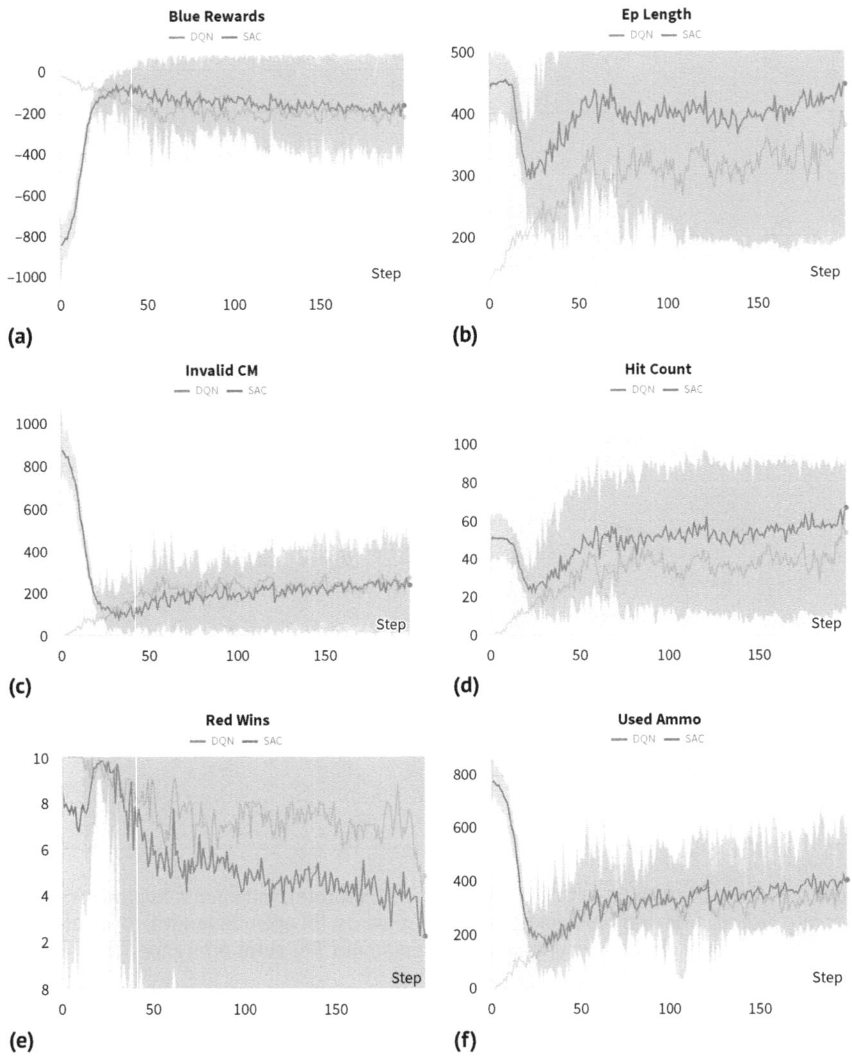

Figure 6.5 Scenario 2: blue with DQN (orange), SAC (purple); red agents approach randomly picked surviving blue agents. (a) Blue rewards, (b) episode length, (c) invalid CM, (d) hit counts, (e) red win counts, and (f) used ammo.

References

1 O. Molloy. Drones in modern warfare: Lessons learnt from the war in Ukraine. *Australian Army Occasional Paper No. 29,* 2024.

2 V. U. Castrillo, A. Manco, D. Pascarella, and G. Gigante. A review of counter-UAS technologies for cooperative defensive teams of drones. *Drones*, 6(3):65, 2022.

3 B. Pong. The art of drone warfare. *Journal of War & Culture Studies*, 15(4):377–387, 2022. https://doi.org/10.1080/17526272.2022.2121257.

4 Y. Tian, F. Lin, Y. Li, T. Zhang, Q. Zhang, X. Fu, J. Huang, X. Dai, Y. Wang, and C. Tian. UAVs meet LLMs: Overviews and perspectives towards agentic low-altitude mobility. *Information Fusion*, 122:103158, 2025.

5 R. K. Nichols, H. C. Mumm, W. D. Lonstein, J. C. H. Ryan, C. Carter, and J. P. Hood. *Unmanned Aircraft Systems in the Cyber Domain*. New Prairie Press, 2019.

6 W. Y. H. Adoni, J. S. Fareedh, S. Lorenz, R. Gloaguen, U. Madriz, A. Singh, and T. D. Kühne. Intelligent swarm: Concept, design and validation of self-organized UAVS based on leader–followers paradigm for autonomous mission planning. *Drones*, 8(10):575, 2024.

7 Samaras, S. Diamantidou, E. Ataloglou, D. Sakellariou, N. Vafeiadis, A. V. Magoulianitis, A. Lalas, A. Dimou, D. Zarpalas, K. Votis, P. Daras, and D. Tzovaras. Deep learning on multi sensor data for counter UAV applications—a systematic review. *Sensors*, 19(22):4837, 2019.

8 G. Brockman, V. Cheung, L. Pettersson, J. Schneider, J. Schulman, J. Tang, and W. Zaremba. OpenAI Gym. 2016. https://arxiv.org/abs/1606.01540.

9 R. S. Sutton and A. G. Barto. *Reinforcement Learning: An Introduction*, Second edition. The MIT Press, 2018.

10 V. Mnih, K. Kavukcuoglu, D. Silver, A. Graves, I. Antonoglou, D. Wierstra, and M. Riedmiller. Playing atari with deep reinforcement learning, 2013. https://doi.org/10.48550/arXiv.1312.5602.

11 T. Haarnoja, A. Zhou, P. Abbeel, and S. Levine. Soft actor-critic: Off-policy maximum entropy deep reinforcement learning with a stochastic actor—arxiv.org. In *Proceedings of Machine Learning Research*, 2018.

7

Intersectoral Labor Mobility in Argentina: Evolution and Resilience

Viktoriya Semeshenko[1,2] and Sergio A. De Raco[2]*

[1] *Instituto Interdisciplinario de Economía Política, CONICET, Ciudad Autónoma de Buenos Aires, Argentina*
[2] *Facultad de Cs. Económicas, Universidad de Buenos Aires, Ciudad Autónoma de Buenos Aires, Argentina*

*Corresponding author: vika.semeshenko@gmail.com

7.1 Introduction

Labor mobility is a complex and captivating economic phenomenon that plays a central role in driving economic adaptability and growth. Studying labor mobility is essential for understanding how economies adjust and evolve in response to structural shifts, changing conditions such as technological advancements, shifting market demands, responses to external shocks, and the reallocation of resources to sustain equilibrium and productivity [1]. It offers valuable insights into addressing skill mismatches [2], improving the distribution of the workforce, and designing evidence-based policies to reinforce economic growth and resilience.

Typically, economists explore labor transitions at aggregated levels, such as one-digit classifications (broad sectors such as agriculture, manufacturing, or services) or two-digit categories (more specific subcategories like machinery manufacturing or transportation equipment). They usually focus on large-scale structural shifts, e.g., transitions from agriculture to services [3], and track overarching trends such as urbanization or industrialization. Although this analysis is useful for identifying broad patterns, it often oversimplifies the complexity of labor markets, because it tends to overlook critical details related to skills, industry-specific dynamics, and the subtle impacts of innovation and technological advancements.

At higher levels of granularity, it becomes possible to identify distinct transitions between niche industries and analyze specialized skill ecosystems, understand the specific contributions of narrowly defined sectors to economic growth, and identify new patterns and phenomena that emerge [4]. It allows researchers and policymakers to better understand the intricate relationships within the labor market [5–7], identify opportunities for targeted workforce training and skill development, and design policies that account for the complexities of modern economies.

Here, we are interested in studying the formal private sector labor mobility dynamics in Argentina at four-digit level of disaggregation details for economic activities, for pre-pandemic, pandemic, and post-pandemic periods, and analyze these new dimensions that emerge in the labor dynamics. The COVID-19 pandemic introduced an unprecedented global shock that disrupted labor markets, reshaped sectoral interdependencies, and accelerated shifts toward remote work and digitalization, and the prevalence of essential economic activities [8]. We extract and analyze information from highly disaggregated data on intersectoral labor flows, exploring a level of detail that has been largely unexamined before. In particular, we explore the mesostructures (communities) within these labor flow networks and their persistence, and identify key sectors that define the main groupings and those that shift between communities during these periods. Additionally, we seek to uncover insights into the dynamics and interactions of these communities and sectors before, during, and after the pandemic, shedding light on how labor markets adapt to and recover from such unprecedented disruptions.

Our analysis reveals a modular structure of communities of economic sectors, emerging from the backbone of labor flow networks, which distinctly identifies groups of sectors with mutual interactions. This structure captures the stress experienced during the pandemic years, which degraded and fragmented the system's cohesion, as well as the gradual recovery in the post-pandemic period, albeit with a shift in the internal structure of labor flow interactions. This recovery can be understood as a form of systemic resilience in a broad, evolutionary context [9]. We identified a complementary structure within subgroups (or clusters) of economic sectors, which serves as a tool for identifying and interpreting activities that maintained persistent interactions across different subperiods within these communities. These findings demonstrate that valuable, detailed insights can be derived from labor flow data in administrative records, enhancing our understanding of labor markets as complex adaptive systems and offer promising applications for employment and industrial policy design by monitoring intersectoral linkages and flow changes.

This chapter is structured as follows. In Section 7.2, we conceptualize the labor flow network within the framework of complex adaptive systems. Section 7.3

details the data used, its disaggregation, preprocessing, and the methods used to filter and prepare the original labor flow data for analysis. Section 7.4 presents the results, and we conclude in Section 7.5.

7.2 Labor Mobility as CAS

Labor mobility is particularly intriguing due to its nature as a complex adaptive system (CAS) [10] and can be conceptualized as a complex network, wherein the nodes represent industries, sectors, or specific economic activities,[1] and the links signify the flow of workers transitioning between these industries, capturing interdependencies and the dynamics of the labor market. This perspective emphasizes the interactions among individual workers, firms, and industries, which collectively produce emergent behaviors that cannot be easily predicted by examining individual components in isolation. Industries are deeply interconnected, meaning shifts in one sector often trigger ripple effects that influence others, creating widespread economic impacts. Workers, on the other hand, are highly heterogeneous, with diverse skills, preferences, and barriers shaping their mobility decisions. Together, these elements form a dynamic and adaptive network of interactions, which continuously evolve in response to economic, technological, and policy changes, making labor mobility a rich, multifaceted, and challenging area of study [5, 6].

From the CAS perspective, we study the properties of the dynamics network, its evolution, topology, and the emergent patterns. Analyzing changes over time helps to uncover how labor markets respond to shocks and adapt to structural shifts in the economy [7, 11]. The interconnectedness among sectors acts in a way that individual disruptions can propagate throughout the entire network and potentially lead to widespread the economic instability of the entire system [12, 13]. Examining the structure of the network in terms of centrality and density is important to identify key industries that serve as pivotal hubs for worker transitions and for disruptions. Understanding the emergence and persistence of patterns, such as the clustering of sectors with high interindustry mobility or the formation of hubs where transitions concentrate, is important for market performance, because these insights help identify recurring mobility flows and emergent clusters of interconnected sectors, revealing communities that may indicate specialization or shared labor dynamics. Such patterns shed light on the structural organization of labor markets and the underlying forces driving worker transitions. Thus, treating labor mobility as a CAS permits us to explore how economies

---

1 Hereafter, we will use indistinctly the terms industry, sector, and economic activity.

adapt to shocks and transitions, and therefore, derive actionable insights to guide economic policy and enhance resilience.

7.3 Data and Methodology

7.3.1 Data

We use administrative data from the Argentine Integrated Pension System provided by the Secretary of Employment of Argentina. It offers thorough information classified at the four-digit level International Standard Industrial Classification (ISIC) of all economic activities, providing an in-depth view of industrial activities across various sectors. The dataset consists of private sector employment records from employer affidavits submitted by employers to the Revenue Collection and Customs Control Agency. These administrative records capture formal employer–employee relationships, including detailed economic and social characteristics.

The analysis focuses on interindustrial flows, tracking interannual worker transitions between different sectors. These sector-to-sector transitions provide insights into structural labor market changes, economic reallocation, and industry-specific employment trends. The data is preprocessed to construct interannual transition matrices, tracking individual employment switches across various economic activities on a year-to-year interannual panel of individuals. These matrices offer detailed insights into workforce dynamics, including occupational mobility and industrial transitions.

The analysis covers 2017–2023, encompassing transitions in three significant periods: pre-pandemic (2017–2018), the COVID-19 pandemic (2019–2020), and post-pandemic recovery (2021–2022). This time frame enables a detailed examination of long-term trends, disruptions caused by the pandemic, and labor market transformations. So far, we are left with the set of productive activities which includes 412 sectors or branches of economic activity at four-digit ISIC Rev. 4 organized in 21 sections (i.e., letters). These industries account for almost total employment in Argentina as the industries left out of the analysis are those with less than 20 employees.

It is important to note that working at this level of granularity allows distinguishing essential from nonessential activities, pertinent for understanding the labor market's response to COVID-19. At higher levels of aggregation, such as one-digit or two-digit classifications, these distinctions turn unpracticable, as essential services are often grouped with nonessential ones. This emphasizes the need for disaggregation to assess labor market resilience and adaptability during the pandemic.

7.3.2 Labor Flow Networks

We use transition matrices to construct labor flow networks[2] for each year. These networks are characterized by high and persistent connectivity, and a single connected component with short paths, a hallmark of "small world" properties. Unlike many social and financial networks, which are typically sparse, labor flow networks are notably dense, with core-periphery sectoral organizational structures [14].

A dense labor flow network with a core-periphery structure reflects the distinct roles that sectors and workers take in the broader economy. Core sectors are highly interconnected, serving as hubs of activity and central nodes within the network. They typically have robust worker inflows and outflows, connecting with many other sectors and facilitating extensive labor mobility. These sectors are often foundational to the economy, such as manufacturing, healthcare, or logistics, where significant skills, resources, and employment opportunities are concentrated. Their dense interconnections indicate their centrality in sustaining economic stability. Workers transitioning into or out of these sectors experience higher mobility opportunities and skill transfer across industries. Disruptions in core sectors can cascade throughout the network, magnifying their impact on peripheral sectors.

Peripheral sectors have fewer connections and are less integrated into the network. They act as "spokes" to the core sectors but do not exhibit dense interlinkages with one another. These sectors often represent specialized industries. While they may employ fewer workers or have limited influence on other industries, they often rely heavily on the core for their supply chains, workforce, or market demand. Workers in peripheral sectors may face more barriers to mobility or fewer opportunities for skill transfer compared to core sectors. Peripheral sectors can be more vulnerable to economic shocks. Given that labor networks are dense, and are also directed and weighted, it complicates to identify the most relevant connections. This highlights the complexity of the analysis and emphasizes the need for appropriate methodologies to uncover meaningful patterns and insights.

7.3.3 Methods

There exist different network reducing techniques, each with its pros and cons [15]. Here, we adopted a two-step approach to find their underlying structures and

2 The system is preserved in terms of size. The number of nodes is given from a predefined classification system—a taxonomy of economic activities, the emergence of "new nodes" (new economic activities) is inherently conditioned by the classification system and granularity level chosen. However, the system is open in terms of flows. This highlights the importance of interactions and their configurations, which defines the system's resilience, one of the properties that we anticipate observing in the data.

explore their temporal dynamics. First, we use the Disparity Filter (DF) method [15], which in the context of labor flow networks is particularly valuable for reducing the complexity of the original dense graph, by eliminating weak or incidental connections, allowing for a clearer and more accurate analysis of the system's underlying structure and dynamics. The DF method identifies and retains the most statistically significant edges in the network, effectively reducing noise while preserving the critical connections that represent meaningful labor flows between sectors, emphasizing economically meaningful flows that influence workforce transitions, skill transfer, and sectoral interdependencies. The resulting backbone contains labor mobility patterns, highlighting sectors that function as hubs of activity, those that depend heavily on flows from others, and potential bottlenecks or vulnerabilities in the system. Next, we use the Leiden algorithm [16], a widely used method for community detection that optimizes modularity to identify groups of sectors with dense intracommunity connections. By applying this algorithm to the backbone networks, we identify cohesive clusters or communities that represent sectoral interdependencies and mobility patterns.

We conduct this two-step analysis for each of the three distinct periods to examine how the structure and composition of identified sectoral communities changes over time, explore sectoral relationships, detect resilience or vulnerabilities, revealing how labor flow dynamics adapted to varying economic conditions and external shocks such as the COVID-19 pandemic, and trying to understand how transitions shaped the economic recovery across different stages of the underlying analysis.

7.4 Results

7.4.1 Community Structure

The results are analyzed using Sankey diagrams, visualizing the structure and dynamics of the obtained communities.[3] This allows to track the evolution of sectors and their patterns of interaction, identifying stable or shifting community structures, and uncovering patterns of resilience or adaptation within the labor market. In Figure 7.1a, a Sankey diagram depicts the community structure extracted from the backbones corresponding to the six transitions under analysis. It shows sectors assigned to each community and tracks their transits between years, allowing for the observation of sector reallocations to different communities whenever such changes occur. In addition, Figure 7.1b illustrates the aggregated sectoral employment shares for each community, while Figure 7.1c provides

3 To ensure robust and reliable results, we perform 1,000 iterations of Leiden algorithm for each period, employing tuning parameters specifically designed to generate larger and more representative communities.

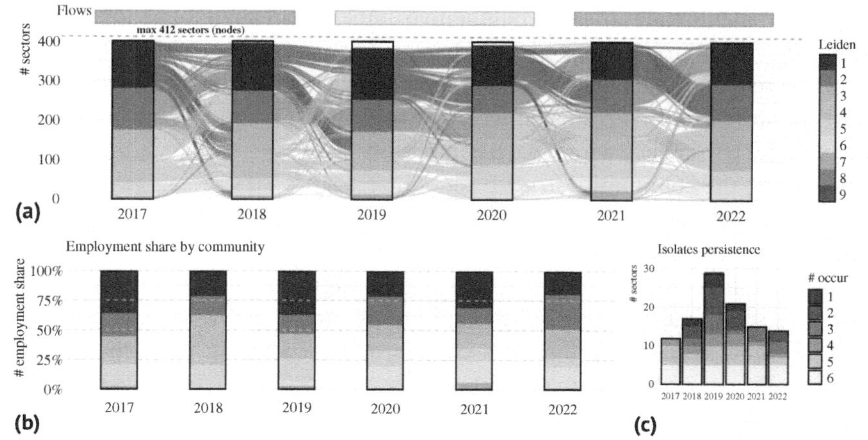

Figure 7.1 Communities. (a) Sankey diagram of Leiden communities extracted from backbones. Communities at the top (bottom) represent more (less) connected sectors, with significant interactions within in terms of modularity. Obtained by progressively refining node assignment in a partition while optimizing the modularity metric, the sequential labeling (C1, C2, and so on) represent groups with decreasing order of modularity and cohesion. Horizontal color bars indicate the pre-pandemic, pandemic, and post-pandemic subperiods for subsequent analysis. (b) Aggregate private employment share per community. (c) Histogram of isolated sectors per period, with identified persistence in color scale. For color references, please refer to the online version.

a summary of the count and persistence of isolated sectors[4] which are inherently excluded from the community detection process. The diagram represents groups of economic sectors with different connectivity and cohesion patterns, derived from their backbones through the whole available period. Each bar represents the number of sectors assigned in each community, while the transits between show the reallocation of sectors to the same or a different community in the subsequent time period, eventually changing partners and patterns of interactions.

In Figure 7.1a, the first three communities dominate the *community structure*, with C1, C2, and C3, encompassing 60–75% of the sectors and 55–70% of total employment. Along with C4–C6, these communities together capture nearly the entire structure of the system through the analyzed period. Communities C7–C9, representing groups of sectors with lower connectivity levels, emerge predominantly during periods of heightened stress, such as the transitions in 2019 and 2021, which correspond to the onset of the pandemic and the post-pandemic subperiod. This indicates a less cohesive and more fragmented structure of connectivity following the pandemic, characterized by an increased number of connectivity "layers."

4 These sectors represent less than 0.1% of total employment.

Figure 7.1b illustrates that the cumulative employment share of sectors within C1–C3 declined during the pandemic years, only recovering in 2022. Notably, C1 members retained just half of their 2017 employment share. This suggests that the most connected sectors in 2022 differ from those in 2017, and leads us to inquire into sectoral persistence within each community.

Regarding the community *structure evolution*, Figure 7.1a reveals two distinct types of transitions: (1) "mass migration," where groups of sectors shift predominantly between adjacent communities, and (2) "lone moves," individual sectors transition to communities with higher (upward) or lower (downward) cohesion, sometimes making substantial jumps. Additionally, these transitions highlight how certain communities change by restructuring through sectoral changes—some sectors exit in one period only to reappear in later periods, either as large-scale movements (*en masse*) or through the gradual reintegration of a few sectors. These dynamics stem from fluctuations in connectivity and the intensity of intersectoral interactions, as reflected in the directionality and magnitude of weighted connections.[5] To gain deeper insights into the dynamics driving these movements and to enhance the interpretability of the identified communities, the next section examines which sectors tend to transition together most frequently.

7.4.2 Sectoral Persistence

Next, we investigate co-occurrences of sectors in the identified communities to assess how stable their composition is over time. Which (groups of) sectors are found most frequently interacting in this system? Persistent communities highlight sectors with strong, long-term interdependencies, while shifting communities suggest adaptation or disruption in response to, for example, the pandemic shock. In particular, these dynamic responses may be influenced by whether a sector has been defined as an "essential activity" during the pandemic or not.

We assess sector persistence within communities by tracking the frequency with which each sector is assigned to a given community over time. Figure 7.2a presents

5 The pre-pandemic years, 2017 and 2018, were already marked by economic instability in Argentina, with declining economic performance, rising political uncertainty, and a tense labor market ahead of the 2019 elections. This is reflected in Figure 7.1a, where the core communities (C1–C3) underwent internal reorganization, with some sectors shifting toward less connected groups. During the pandemic, transitions in 2019 and 2020 reinforced less connected communities (C5–C7), while the most integrated ones (C1 and C2) shrank, increasing fragmentation and weakening overall network connectivity. Post-pandemic, the structure began to reassemble, with 2022 showing a configuration similar to 2017. The largest communities (C1–C3) regained nearly 75% of sectors and 70% of employment, indicating a partial recovery, with more (and less connected) communities. This indirectly evidences that the system managed to cope with the stressful events, but still struggles to recover previous connectivity levels.

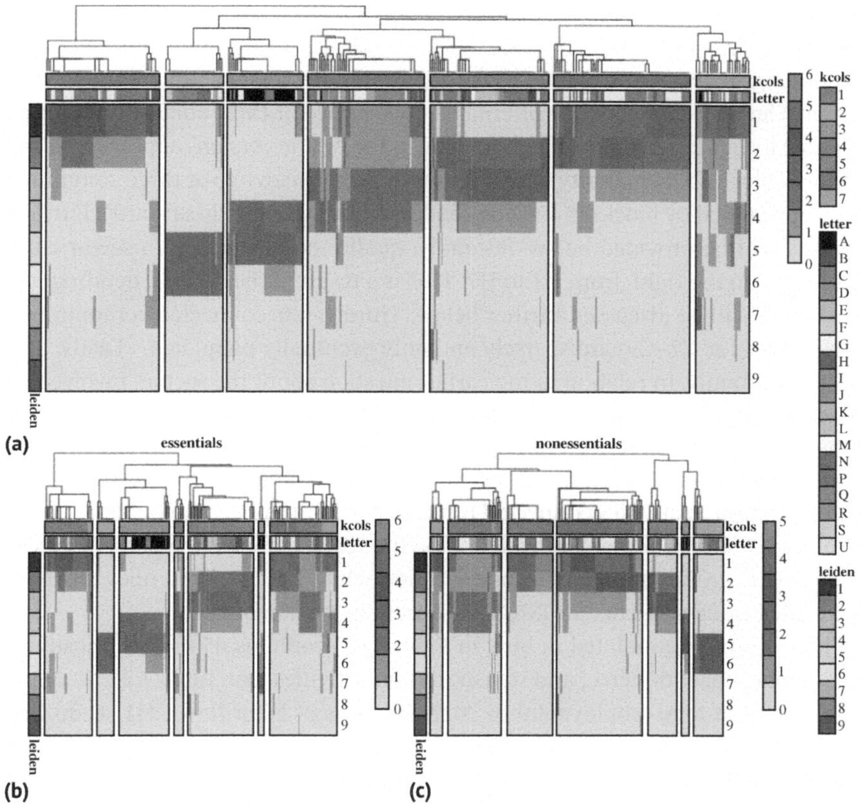

Figure 7.2 Persistence. Leiden labeling (rows) counts per sector (columns) over the period of analysis. (a) Heatmap of sectors arranged by hierarchical clustering. Colors indicate the frequency of each sector's membership in the communities throughout the period ($n \in \{0, 6\}$). Letter is ISIC Rev. 4 sectoral division. $kcols$ is best $k = 7$ PAM identified groups of maximum average silhouette width. (b) and (c) Heatmaps showing the persistence of sector partition based on the classification of essential activities, as explained in the text, according to regulations in Argentina during the pandemic. For color references, please refer to the online version.

a heatmap illustrating this distribution, where columns represent sectors that were part of the largest connected component used in community detection each year (comprising 384 sectors), while rows display the labeled communities (numbered from 1 to 9).

Additionally, in Figure 7.2b,c, we present the same persistence matrix for all periods, partitioned into essential and nonessential activities, to provide insights into the significance of this distinction for the desired mesostructural analysis.

Firstly, we observe that only a small fraction of sectors remains in the same community for most of the period (indicated by the darker, higher-intensity cells, corresponding to the first group H1 (red in the online version), in the upper-left corner. The majority of sectors alternate between two or three communities (represented by lighter gradient (shades of blue in the online version) appearing in the diagonal blocks. The majority of sectors alternate between two or three communities (represented by blocks in varying shades of blue, diagonally upward). Further analysis will be provided below. Secondly, quality of persistence in sector clusters lowers to the right, from H1 to H7. This is a result of the column dendrogram cut, which will be discussed further below. Thirdly, less connected communities (lower rows for C7–C9) are sparsely and only eventually populated.[6] Lastly, and most importantly, in relation to the earlier question about the sectors involved in the evolution of the community structure and the types of changes observed in transitions, many of these groups appear in some communities but not in others. Next, we characterize the sectoral persistence groups identified, labeled as **H1–H7**, which represent activities connected through partial supply-chain linkages.

Tables 7.1 and 7.2 present information on the sector composition of hierarchical clustering H-groups and Leiden communities, while Table 7.3 describes the most aggregated category (letter, or ISIC Rev. 4 section) for activities.

The **H1** group (annotated in blue in Figure 7.2) contains 65 sectors, mostly of manufacturing, commerce, and transportation activities (see Table 7.2), accumulating 14.9% of total employment in 2022. In terms of labor flows, H1 seems not much affected by the pandemic showing flows of steady growth, from around 15% of total flows in 2017 to 20% in 2022. From these flows, between 6 and 8% of total flows occurred within sectors of H1, which means that the interactions with sectors of other groups take on greater relative importance. This group includes manufactures and wholesale activities of food and beverages (F&B), wood, paper, pharmaceuticals, chemicals, soap and detergents, consumer electronics, domestic appliances and furniture, manufacture and sales of motor vehicles, parts and accessories; air, maritime and coastal freight transportation and related services; and warehousing, storage and related services. This group participates persistently in the C1 community, having the most cohesion, as can be seen in Figure 7.2.

Cluster **H2** contains 33 sectors, including all the letters (see description of the top-level categories used in 7.1) of activities for education, health, and social assistance, accumulating 18.4% of the total 2022 employment. Also, in this group, we find activities of real estate on fees or contracts, social and natural science R&D, and other social organizations activities, but no manufacturing sectors. Labor flows for H2 accelerated in the pandemic subperiod (from 9 to 10% of

6 This occurs due to the use of a modularity-based optimization algorithm, which forms new communities when the algorithm discards the assignment of nodes to existing communities.

Table 7.1 HC7 groups. Count of activities per section (letter).

HC7	A	B	C	D	E	F	G	H	I	J	K	L	M	N	P	Q	R	S	U	Total
H1			37		2		15	8						3						65
H2											1	1	2		8	12	3	5	1	33
H3	28		10				4							1						43
H4	1	2	11	1			2	2		23	9	1	7	5				1		65
H5		9	27	3	3	13	3	2					2	3			1	2		68
H6			26				29	2	4	1			1	7			3	5		78
H7	3		14		1		4	3			2		1	1			1			30
Total	32	11	125	4	6	13	57	17	4	24	12	2	13	20	8	12	8	13	1	382

Table 7.2 Communities (2022). Count of activities per section (letter).

Leiden	A	B	C	D	E	F	G	H	I	J	K	L	M	N	P	Q	R	S	U	Total
C1		4	67		3		10	10		1				2						97
C2			17				39	6	4	1		1	2	9			5	5		89
C3	1	7	34	4	3	13	3	1					2	3						71
C4			1				2			22	10		7	4				3		49
C5	28		5				3							1						37
C6											2	1	2	1	8	12	3	5	1	35
C7	3		1																	4
Total	32	11	125	4	6	13	57	17	4	24	12	2	13	20	8	12	8	13	1	382

Table 7.3 Section aggregation of sectors (letter).

Code	Description	Code	Description
A	Agriculture, forestry, and fishing	L	Real estate
B	Mining and quarrying	M	Professional, scientific, and technical
C	Manufacturing	N	Administration and support
D	Electricity, gas, steam, and a/c supply	O	Public administration and defence
E	Water supply; sewage, and waste management	P	Education
F	Construction	Q	Human health and social work
G	Wholesale and retail trade; motor repairs	R	Arts, entertainment, and recreation
H	Transportation and storage	S	Other service activities
I	Accommodation and food service	T	Households
J	Information and communication	U	Extraterritorial organizations and bodies
K	Financial & insurance		

Source: Adapted from United Nations.

total flows to 12 to 13%) while later returned to the previous levels. Flows occurring within sectors of H2 grew from 7% of total flows to 10% in pandemic years and returned later to 7%. This evidences strong interactions of sectors within H2, which also increased at the beginning of the pandemic subperiod—as education, health and social assistance where predominantly declared as essential activities—while interactions with other groups took on less relative importance. This appears as a group of stronger interactions within activities of H2 relative to those between H2 and the other groups. We will refer to this group as the *Education, Health, Social assistance and R&D services* cluster. H2 participates alternatively in the C5 and C6 communities.

The **H3** group consists of 43 sectors that comprise almost all the activities for the letter A (except for fishing, in H7), the "primary sector," with the addition of derived and related ancillary manufacturing and services to those activities, including pesticides and agri-chemicals, wholesale of cereals and seeds, and packaging. This group accumulates 8.4% of total employment (2022). Labor flows for H3, as the case of H2 accelerated growth during the pandemics, from around 7 to 8% of total flows in 2017 to 12 to 13% in 2020 and later returned to the previous levels. Flows within sectors of H3 fluctuated along the whole period. In 2017 represented about 5% of total flows, then scaled up to 8% in 2019 and ended around 4% in 2022. Possibly, this oscillating dynamics may have ground in the usual annual fluctuations of primary production. With respect to the interactions with sectors of other groups, H3 took on less relative importance, behaving like a center (albeit weaker than H2) in the sense that the flows within the group are relatively equivalent to the flows between this and the rest of the economy. We will refer to this group as the *primary activities* subcore. H3 participates alternatively in the C4 and C5 communities.

Cluster **H4** is composed of 65 sectors, including almost all the activities in letters J (Information and communication), K (Financial and insurance activities) and M (Professional, scientific, and technical activities). Among the rest of the sectors included, we can mention gas extraction, manufacturing and distribution; real estate (own or leased); passenger air transport; and tourism. This group represents 10.7% of total employment. Flows involving H4 sectors account for 15% of total flows in 2017 and fluctuate upwards to achieve 18% in 2020, decreasing to 12% in 2022. We refer to H4 as the *specialized services* subcore. This group participates more persistently in the C3 and C4 communities, but almost all its members participated at least one period in C1 and C2.

The **H5** group consists of 68 sectors, including oil and mining, construction and related activities and services, electricity generation, transmission and distribution, machinery and equipment, creative arts and entertainment, as the most relevant. This group accounts for 16.1% of total employment (2022). Labor flows with sectors of H5 decelerated from the pre-pandemic levels (28–29% of total flows in 2017 to 19–21% in 2020), and in 2022 are still recovering (22–24% in 2022). We refer to H5 as the *industrial core*, a highly dynamic group of economic activities, the most vibrant in terms of interactions within the group which represented 22% of total flows (pre-pandemic) and more than 15% on average throughout the whole period. H5 participates predominantly in C2 and C3, with many sectors participating one period in C4.

Cluster **H6** comprises 78 sectors and accounts for 22.5% of total employment (2022), making it the biggest cluster in both dimensions, thus having great relevance for the system functioning. This group includes all activities belonging to letter I (Accommodation and food service, which accounts for 4.2% of total employment), almost half of the sectors in G (Wholesale and retail trade; repair of motor vehicles and motorcycles) and the broad textiles sector, including manufacturing, wholesale and retail of textiles, clothing, and footwear. Among others included, a very particular and relevant subgroup are private security and building cleaning (4.3% of total employment), and also relevant retail of F&B, games and toys, cosmetics, pharmaceutical goods, gambling, betting and recreational services. In terms of flows, H6 showed high growth during the whole period, similar to H1, participating with 16–18% of total flows in 2017, increasing to 21–23% in 2019, and to 21–24% in 2022. Of these flows, the interactions within sectors of H6 presented some fluctuations, decreasing at the beginning of the pandemics, but with a later burst of growth, accounting for 7.5% of total flows in 2019, 6.5% in 2019 and 11.3% in 2022. This acceleration in the inside dynamics of the group pushes to relabel it from a *proximate periphery* to a *proximate subcore*, in the relative terms used before. We will refer to H6 as a *complementary activities* periphery. This group participates most of the time in C1 and C2, with many sectors participating one period in C3.

Lastly, the cluster **H7** contains 30 activities, including fishing and supporting services, processing and preserving of fish and meat, manufactures of chocolate, sugar, macaroni, noodles, and other food products, industrial services and wholesale for F&B, and others. In terms of flows, H7 showed a mild expansion in the pandemic period (from 6 to 7% of total flows in 2017 to 7 to 8% in 2020), and in 2022 returned to the previous levels. This group accounts for 7.9% of total employment and, regarding labor flows, behaves like a distant periphery, in the sense that has low within interactions and relatively higher interactions between the rest of the groups. We will refer to H7 as a *fishing, confectionary and baking* distant periphery. H7 participates half of the time in C1, while the other half varies between almost all the rest of the communities.

During the pandemic, the Argentine government defined essential activities, gradually expanding the list over time. This special status was assigned to highly specific activities, which can only be identified using high-granularity data like that employed in this study. Ultimately, nearly half of the four-digit activity codes were classified as essential. How does this distinction impact the previous analysis? Essential activities are not confined to any single community, but are distributed across all of them. Within the defined persistence groups, the main variation lies in their share of essential activities. In decreasing order of participation, the primary activities cluster H1 was almost entirely classified as essential, followed by H3 and H7, where nearly two-thirds of their members were deemed essential. Next, H1, H4, and H2 had shares ranging from 50 to 45%, while H6 had nearly 30% of its members classified as essential.

Persistence analysis for the complete time period available is useful to detect affinity between sectors based on labor flows. But, as noted earlier, the whole period comprises three qualitatively different subperiods regarding the pandemics. It is useful then to dive into the persistence analysis within each defined subperiod. We repeat the same persistence analysis for each subperiod and also to an additional between-ends to contrast the first and last transition periods available (e.g., 2017 vs. 2022). This temporal partitioning allows for the observation of large-scale *en masse* changes in the persistence of sector groups across different communities. We highlight the most relevant patterns, starting with cluster H1, which consistently exhibits high persistence in C1 across all subperiods, except in the post-pandemic period, where some members shift to C2. Group H2 (health and social assistance cluster) spends half the time in C5 during the pre-pandemic years but transitions to C6 during the pandemic period, remaining there in the post-pandemic years. Similarly, group H3 (primary activities) is associated with community C4 in the pre-pandemic period but transitions to community C5 during the pandemic and remains there in the post-pandemic years. Additionally, part of this group fragments and integrates into community C1. Finally, group H5 (industrial core) is primarily associated with communities C2 and C3 during both the pre-pandemic and pandemic years. However, in the post-pandemic period, it transits to community C4. The between-ends transition analysis reveals that most members of this group remain in C3 before and after the pandemic, indicating a resilient behavior among the sectors involved. These observations altogether provide major contributions to explain the most relevant community reallocations observed as transits between years in the Figure 7.1 diagram. Particularly, they provide functional explanations and economic interpretations to the *relabeling* occurring between two instances of community detection.

Returning to the analysis of labor flows, Figure 7.3 shows the observed interannual flows across the extracted mesoscales. Figure 7.3a displays the share of total flows between Leiden communities, revealing a declining pattern of accumulated flows, as expected, with communities C1–C3 accounting for an average of 60% of the flows. In contrast, Figure 7.3b presents a different picture for flows between HC groups, depicting a core-periphery structure. Here, H5 emerges as the most dynamic core, interacting with the growing peripheries H1 and H6, as well as the subcore H4. It is important to note that both mesostructures complement each other. The community structure captures modules with varying levels of interactions, but overlooks more specific functional roles. In contrast, the HC groups

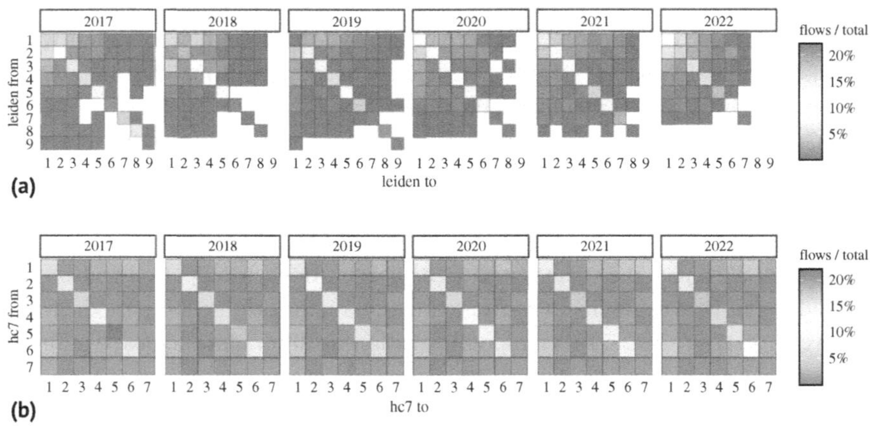

Figure 7.3 Flows between groups. (a) Leiden communities. Modular structure showing more interactions within each community than between communities. (b) Persistence clusters. Core-periphery structure with H5 as the most dynamic core interacting with growing peripheries H1 and H6, and with H4 subcore.

identify common associations of sectors based on functional characteristics shared among their members.

7.5 Discussion

We explore labor mobility in Argentina's formal private sector at the four-digit level across pre-pandemic, pandemic, and post-pandemic periods. Using intersectoral labor flow data, we examine how the labor market adapts to pandemic, evolves, and recovers. The analysis focuses on mesostructures (communities and clusters) within labor flow networks, tracking their persistence and changes over time, while identifying key sectors and those transitioning between communities. The aim is to understand how labor markets respond to and recover from major disruptions like the pandemic.

Our findings reveal a modular structure of sector communities within labor flow networks, highlighting mutual interactions and capturing the pandemic's impact on system cohesion. The post-pandemic recovery, though gradual and marked by shifts in labor flow patterns, reflects systemic resilience. Additionally, sector subgroups help track persistent interactions over time, offering insights into labor market dynamics and production structure and evolution. These results underscore the value of high-granularity labor flow administrative data for understanding labor markets as complex adaptive systems (CAS), and provide a foundation for designing employment and industrial policies by monitoring subtle changes in intersectoral linkages and flows. Empirical networks often display various mesoscale structures, such as community and core-periphery patterns [17]. Our findings align with the literature on core-periphery structures, which can be richly explored in terms of statistical significance and quality and will be our next step to assess their further robustness [18].

The analysis reveals that the intersectoral labor mobility system in Argentina exhibits systemic resilience, categorizing it as a resilient complex adaptive system. This is evidenced by its demonstration of three fundamental capacities: absorption, adaptation, and recovery/evolution. The system's absorption capacity was notably apparent in how it maintained its structural integrity during the initial shock. The main central groups consistently kept the core of the system stable despite significant instability. A specific industrial core was the most dynamic part of the system, exhibiting a high level of internal activity and demonstrating resilience that avoided total stagnation while absorbing a significant portion of the pandemic's impact.

The labor flow network remained as a single interconnected unit. This indicates that despite higher fragmentation (caused by the emergence of less connected communities), a complete network collapse was successfully avoided. This strong structure provided the foundation for the subsequent adaptive capacity. This adaptation was visible in the dynamic reorganization of sectors, characterized by both "mass migration" (large-scale shifts) and "lone moves" (individual shifts). These movements illustrated how sectors were actively changing their interaction patterns in a functional reorganization process. A specific core in education, health, and social assistance was declared an essential activity. It strengthened its internal interactions and moved between communities to meet sectoral needs, reflecting an effective adaptive response. The system's capacity to rearrange itself through these interactions (with essential sectors distributed across all main groups) reflects a localized form of adaptation that fosters resilient systemic behavior.

Ultimately, the system demonstrated its ability to recover and evolve following the pandemic, as the structural configuration began to reorganize and the main central groups regained their leading roles. While this resurgence indicated a recovery, it did not signify a complete return to the previous state. Instead, it highlighted systemic resilience within an evolutionary framework characterized by ongoing, albeit shifted, labor flow patterns. The sustained presence of key clusters, such as the Industrial Core in one of the central groups, both prior to and following the pandemic, provides strong evidence of this persistent and resilient behavior at the functional core level over time.

Additionally, we intend to link how these changes can be connected to employment policy design and economic recovery efforts for Argentina. Resilience has traditionally been examined through an equilibrium-based lens, emphasizing a system's capacity to return to a pre-crisis state or transition to a new equilibrium. More recently, an evolutionary approach to regional resilience has gained traction, framing resilience as a region's ability to adapt to adverse shocks [9]. This perspective shifts the focus from recovery to evolution and adaptation, underscoring the importance of dynamic responses over a return to the status quo [19]. Through this analysis, we seek to deepen our understanding of the labor market's resilience and transformation in response to unprecedented challenges, providing valuable insights for future policy and economic planning.

References

1 D. H. Autor. Work of the past, work of the future. *AER*, 109(5):1–32, 2019.

2 P. Esposito and S. Scicchitano. Educational mismatch and labour market transitions in Italy: Is there an unemployment trap? *Structural Change and Economic Dynamics*, 61:138–155, 2022.

3 W. Nowack, T. R. Popp, J. C. Schmid, and H. Grethe. Does agricultural structural change lead to a weakening of the sector's social functions? A case study from north-west Germany. *Journal of Rural Studies*, 100:103034, 2023.

4 D. Acemoglu and P. Restrepo. Artificial intelligence, automation, and work. *NBER Working Paper 24196*, pages 1–43, 2018.

5 R. Boschma and K. Frenken. Technological relatedness and regional branching. *Papers in Evolutionary Economic Geography (PEEG) 0907*, Utrecht Uni, Dep of H Geog and Spat Plann, Group Econ Geo, 2009.

6 F. Neffke and M. Henning. Skill relatedness and firm diversification. *Strategic Management Journal*, 34(3):297–316, 2013.

7 S. A. De Raco and V. Semeshenko. Identificaci6n de diferencias y similitudes estructurales de las redes interindustriales del empleo de Argentina. *Electronic Journal of SADIO (EJS)*, 23(2):200–220, 2024.

8 R. M. del Rio-Chanona, P. Mealy, A. Pichler, F. Lafond, and J. D. Farmer. Supply and demand shocks in the COVID-19 pandemic: An industry and occupation perspective. *Oxford Review of Economic Policy*, 36(Supp. 1):S94–S137, 2020.

9 R. Boschma. Towards an evolutionary perspective on regional resilience. *Regional Studies*, 49(5):733-751, 2015.

10 J. H. Holland. Studying complex adaptive systems. *Journal of Systems Science and Complexity*, 19(1):1–8, 2006.

11 C. Macchiarelli, V. Monastiriotis, and N. Lampropoulou. *Transitions in the EU labour market: Structure, Crisis and Recovery*. Pub. Off. of the EU, 2018.

12 M. D. König, A. Levchenko, T. Rogers, and F. Zilibotti. Aggregate fluctuations in adaptive production networks. *PNAS*, 119(38):e2203730119, 2022.

13 E. Atalay. How important are sectoral shocks? *American Economic Journal: Macroeconomics*, 9(4):254–280, 2017.

14 S. A. De Raco and V. Semeshenko. Labor mobility and industrial space in argentina. *Journal of Dynamics & Games*, 6(2):107, 2019.

15 M. Serrano, M. Boguná, and A. Vespignani. Extracting the multiscale backbone of complex weighted networks. *PNAS*, 106(16):6483–6488, 2009.

16 V. Traag, L. Waltman, and N. J. Van Eck. From Louvain to Leiden: guaranteeing well-connected communities. *Scientific Reports*, 9(1):1–12, 2019.

17 R. J. Gallagher, J. G. Young, and B. F. Welles. A clarified typology of core-periphery structure in networks. *Science Advances*, 17(7), 2021.

18 A. Elliott, A. Chiu, M. Bazzi, G. Reinert, and M. Cucuringu. Core–periphery structure in directed networks. *Proceedings of the Royal Society A*, 476(2241):20190783, 2020.

19 D. Straulino, D. Diodato, and N. O'Clery. Economic crisis, urban structural change and inter-sectoral labour mobility. *Structural Change Economic Dynamics*, 71:135–144, 2024.

8

A Novel Integrated Agent-based Computational Economics and Multi-agent Reinforcement Learning Framework: Application to Macrocrisis Phenomena

Rodrigo Castro,[1] Viktoriya Semeshenko,[2,3] and Pablo Mira[3]*

[1] *Facultad de Cs. Exactas y Naturales and Instituto de Ciencias de la Computación, Universidad de Buenos Aires and CONICET, Ciudad Autónoma de Buenos Aires, Argentina*
[2] *Instituto Interdisciplinario de Economía Política, CONICET, Ciudad Autónoma de Buenos Aires, Argentina*
[3] *Facultad de Cs. Económicas, Universidad de Buenos Aires, Ciudad Autónoma de Buenos Aires, Argentina*

*Corresponding author: rcastro@dc.uba.ar

8.1 Introduction

A macrocrisis is a system-wide breakdown in the performance of a macroeconomy. To paraphrase an old quote, although there is no precise definition of a crisis, society knows it when it happens. Public life convulses, output collapses, employment plunges, credit dries up, and economic, social, and political institutions may remain under stress for a long time.

These events interrupt extended periods of normal growth and have occurred repeatedly across centuries, countries, and regions [1, 2]. According to [3], between 1970 and 2016 the world experienced 329 recessions defined as a cumulative GDP loss above 4%. Of the 166 countries analyzed, 135—around 80%—suffered at least one, and some experienced up to eight such episodes in less than 50 years.

A useful way to approach macrocrises is to classify them by the nature of their initial shock. First, *exogenous shocks* arise outside the economic domain, such as wars or pandemics. In these cases, the features of the economic system do not appear to be the root cause of the disturbance.

However, the scale and duration of the damage still depend on how the system responds to the original shock. Economic agents may or may not act appropriately, domestic financial buffers may be sufficient or inadequate, and feedback mechanisms may or may not amplify the initial disturbance. Even when an initial shock is external, the crisis may worsen through internal mechanisms such as fire sales, bank runs, or rapid shifts in expectations.

Another example of an exogenous shock is that originated from outside the affected economy but still reflecting a malfunction somewhere else in the system as a whole. For example, a collapse in the terms of trade or a sudden stop in global liquidity are shocks that still reflect some economic malfunctioning, but they can be external to the economy analyzed. This is a typical situation experienced by emerging markets, when bad news arrives from abroad, but it can also be seen as a pecuniary externality, reflecting flaws or excessive risks in the global economic structure.

Second, *endogenous crises* originate within the system itself. Credit booms, asset price bubbles, maturity or currency mismatches, unsustainable fiscal behavior, and policy regimes that encourage excessive leverage tend to develop gradually through several mechanisms, and then unravel abruptly. Typically, endogenous shocks originate from the interplay of agents' decisions, which in a given institutional arrangement, may lead to perverse market dynamics. They are often the product of cumulative processes such as leverage build-up, speculative bubbles, or persistent current account or fiscal imbalances that create latent vulnerabilities. When certain thresholds are crossed, these vulnerabilities can trigger abrupt adjustments without any obvious external catalyst.

Endogenous shocks are characterized by the presence of multiple self-reinforcing feedback mechanisms, including asset price declines, which typically lead to balance sheet deterioration, which in turn depresses credit and investment. A key property of these events is uncertainty; their probability of occurrence is shaped by the system's own history and path dependence, and cannot be established *a priori* with exactitude. Unlike exogenous shocks, endogenous ones challenge the notion of shocks as purely random disturbances, instead reflecting structural vulnerabilities and the potential for instability that builds up in the system's evolution.

We can think of macroeconomic crises (in the following, macrocrises), at their core, as *coordination failures across time and space*. In this view, disturbances cannot be fully understood using timeless general equilibrium models. Instead, they require frameworks that describe how heterogeneous agents form, revise, and attempt to reconcile intertemporal plans under uncertainty.

As we explore in this chapter, agent-based computational economics (ACE) and multi-agent reinforcement learning (MARL) can help explain crises and their properties. In particular, economic disturbances and their dynamics align neatly with the framework of resilient complex adaptive systems (RCAS).

Many characteristics from this modeling approach appear as useful to interpret macrocrises. First, interconnected agents operate with incomplete information and revise their strategies in response to changing circumstances. Second, feedback loops, like those seen in credit expansions and contractions, can amplify or stabilize shocks, depending on the structure of the system. In some contexts, agents may anticipate critical thresholds, such as debt sustainability limits, and develop buffers to mitigate the risk of tipping points and cascading failures. Third, the system evolves through learning, both in the private sector and through public policy. As several authors observed long ago, markets penalize inefficiency through selection, while crises can also prompt reforms that reflect institutional learning. Finally, macrocrises rarely return automatically to the pre-shock state. They often trigger processes of transformation—such as creative destruction in key sectors—that illustrate the adaptive, co-evolutionary nature of resilience in complex economic systems.

The remainder of this chapter is structured as follows. Section 8.2 introduces socioeconomic systems through the looking glass of RCAS, motivating the need for advanced modeling approaches and making the case for a new, integrated framework capable of studying macrocrises phenomena. Section 8.3 presents the ACE framework as an alternative to traditional equilibrium models. Section 8.4 details the MARL framework and its relevance to economic adaptation and crisis dynamics. Building on this, Section 8.5 examines macrocrises through the lens of Leijonhufvud's corridor hypothesis, and discusses their resilience mechanisms. We propose the core argument on how to study macrocrises phenomena by leveraging the combined strengths of ACE and MARL. Section 8.6 presents the integrated ACE+MARL framework for macrocrises, including its schematic structure and dynamic layers. Finally, Section 8.7 concludes with reflections on the framework's implications, applications, and future research directions on economic resilience.

8.2 Socioeconomic Complex Adaptive Systems

Socioeconomic systems are complex and dynamic by nature [4–6]. Their behavior is influenced by often intricate interdependent interactions among various heterogeneous agents, (i.e., individuals, firms, governments, and institutions). These agents interact through networks that comprise social, economic, and institutional connections, and are characterized by featuring feedback loops and different degrees of mutual influence. Local interactions among heterogeneous agents contribute additional layers of complexity that can result in substantial consequences at the systemic level. Consequently, nonlinear dynamics and emergent behaviors can arise, wherein patterns at the macrolevel emerge from interactions at the microlevel.

The interconnected nature of socioeconomic systems can make them highly adaptive, yet also delicate and fragile. Consequently, these systems enable innovation, growth, and resilience while simultaneously exposing them to vulnerabilities. On the one hand, we observe emergent trends such as the ongoing growth of wealth inequality [7], the wide effects of technology transitions [8, 9], or the transformative forces of urbanization dynamics [10, 11].

The vulnerabilities of these systems appear also in stressful events such as large economic macrocrises, widespread unemployment, or financial disruptions [12–14], which can spread through connected networks with serious effects. This combination of adaptability and fragility underscores the need for robust methodologies to analyze how these systems change, respond to stress, and recover from disturbances. It seems therefore adequate to adopt methodologies capable of capturing nonlinear dynamics, multi-agent interactions, and intricate interplays between local choices and global emergent events.

To examine such features effectively, it is essential to employ modeling and simulation instruments that go beyond rigid economic relationships, and cannot be solved with simplified mathematical models. These tools must take into account the adaptive and strategic behavior of agents acting in these systems, reflecting how individual choices, learning, and interactions influence larger socioeconomic results. This dual emphasis on structure-level patterns and agent-level adaptability is imperative for elucidating the mechanisms that drive the system into tipping points, precipitating abrupt system-wide changes. This emphasis may also prove advantageous in the design of interventions that are both effective and resilient in dealing with complexity and unpredictability.

We consider the development of a unified modeling and simulation reference framework for the study of socioeconomic systems a goal of considerable importance. Such an approach should facilitate the exploration of emergent socioeconomic phenomena and the complex processes underlying disruptive events. It should also provide a deeper understanding of how microlevel behaviors and adaptations combine to form macrolevel patterns. This will offer new insights into the causes, progression, and potential solutions of undesirable disruptive events.

In the case of macrocrises, this framework can provide a novel approach to comprehending the phenomenon by providing insights into its causes and development, enabling the exploration of novel emergent patterns, feedback effects, and critical points that characterize systemic disruptions. Ultimately, the goal of applying this framework is to facilitate the design and evaluation of policies that can foster resilience to the functioning of macroeconomies, and mitigate the effects of future crises.

8.3 Agent-based Computational Economics vs. General Equilibrium Economics

In Fagiolo and Roventini [15], the authors make the case for the parallel emergence of a novel, radically different approach as a reaction to the limitations of general equilibrium strategies: the ACE. As the authors state, given the acknowledged theoretical and empirical problems of micro-founded, general-equilibrium-based neoclassical models, a group of scholars has begun to devise ACE as an entirely new paradigm. The philosophical underpinnings of ACE largely overlap with those of similar, complementary approaches known in the literature as "Post Walrasian Macroeconomics" [16] and "Evolutionary Economics" [17, 18].

ACE aims at building models based on more realistic assumptions for agent behaviors and their interactions both among themselves and with the environment they are imbued into, rooted in empirical and experimental microeconomic evidence [6]. The assumptions of perfect rationality and foresight are replaced with those of bounded rationality and adaptive behavior. The term ACE can be traced back to 1996 with Leigh Tesfatsion coining the idea in the context of a congress about computing in economics and finance [19].

ACE sees the model of an economy as

> ... an open-ended system of interacting agents, each characterized by its own state (data, attributes, and/or methods) [...] These agents can represent any entity capable of affecting the trajectory of system outcomes: e.g., individual life-forms, social groupings, institutions, and/or physical phenomena. The interactions of these agents induce all dynamics (state changes) for the modeled system, starting from initial agent states configured and set by the modeler [19].

The main linkages with the broader concept of agent-based modeling (ABM, a previously existing discipline applied since the 1970s to various fields in the social, natural, and exact sciences) can be consulted in Tesfatsion and Judd [20].

The ACE paradigm removes or significantly relaxes most of the controversial core assumptions of the general equilibrium framework. ACE leaves the field open for out-of-equilibrium dynamics to develop and, of particular interest in this chapter, for emergent behavior to take place, that may in turn give rise to a macrocrisis. Also, the flexibility and capabilities afforded to agents in ACE indicate that they can draw on learning capabilities, including group learning dynamics, by assuming a role within the environment. This is despite the fact that the environment itself is affected by the actions of other agents, which in turn adapt their learning in response.

The mainstream research program on general equilibrium starts from the premise that the economy is always at (or quickly returns to) a single, well-defined equilibrium in which rational agents make optimal intertemporal plans. In the original real business cycle (RBC) paradigm [21], the typical shock of interest is technological, and its properties and fluctuations are treated as efficient responses to these exogenous disturbances. Therefore, recessions are interpreted as nothing more than temporary productivity slowdowns to which rational agents respond efficiently. Needless to say, this is not the best environment to discuss macrocrises.

The dynamic stochastic general equilibrium (DSGE) and its New Keynesian variant [22] soften some of RBC's assumptions by adding ad hoc features such as sticky prices, monetary rules, and even a "financial accelerator" to help explain extreme endogenous shocks. Yet, these tools cling to the similar architecture: a representative household–firm, log-linearized equations around a unique steady state, and forming model-consistent expectations. As a result, crises appear only when an exogenous shock of sufficient size is injected or when ad hoc wedges (credit frictions, occasionally binding constraints) are bolted on.

The predictive record of these approaches underscores the critique. Benchmark DSGE specifications failed to trigger basic warnings of the 2008–2009 collapse and, even after asset markets had already seized up, projected a swift recovery that never materialized. Joseph Stiglitz argues that this strategy "hid much that is of first-order macroeconomic significance" such as information asymmetries, defaults, balance-sheet interactions. So the models were "particularly poorly designed to analyze the origins and evolution" of the 2008 financial crisis [23].

The main reason seems to be methodological. The requirements to solve these models often involve linearization and certainty-equivalence in financial decisions, which wipe out the nonlinear feedbacks that turn small shocks into cascades. Meanwhile, the representative agent construction rules out the very heterogeneity, leverage, and coordination failures that characterize real world crises. Even the more elaborate post-crisis "financial DSGE" models retain a rational expectations backbone, a concept that has been criticized in both the theoretical and the practical arenas.

Also, the granular and reductionist logic of standard macroeconomic models aligns with what [24] might describe as the economist's "under-socialized" approach, emphasizing the bottom-up aggregation of atomistic individual decisions. This view pays little attention to the dense web of social networks, shared norms, and institutional frameworks that shape, enable, and constrain economic behavior, and underestimates how these social structures influence both the feasibility and the aggregate outcomes of individual choices.

In short, the equilibrium and optimization framework makes genuine macro-crises look like statistical outliers, not endogenous properties of a complex adaptive system. This motivates the search for alternative frameworks.

8.4 The MARL Framework

Reinforcement Learning (RL) algorithms enable an agent to learn the best course of future actions (an *optimal policy*) in a *sequential decision-making process*. The learning takes place by means of maximizing cumulative rewards over a finite set of *repeated exploratory experiences*, which in practice are numerical simulations.

This iterative process of trial-and-error agent-environment interactions allows individuals to incrementally refine their long-term strategy, ultimately leading to improved decision-making performance in complex environments. Throughout the learning process, the agent's actions influence its own rewards and internal states, while also altering the state of its environment, in turn influencing its own future choices.

This dynamic interplay between actions, rewards, and states is what distinguishes RL from other, more popularized, machine learning paradigms, as RL emphasizes the importance of **exploration** (of novel strategies) versus **exploitation** (of known, high-reward strategies) in achieving optimal overall outcomes.

Formally, RL relies on the dynamic framework of Markov decision processes (MDPs), and distinguishes itself from supervised and unsupervised machine learning through its reliance on interactive, reward-driven adaptation, rather than on static training data. In *supervised learning*, models learn from labeled input–output pairs, explicitly trained to map inputs to correct answers (e.g., image classification). *Unsupervised learning* identifies hidden patterns in unlabeled data (e.g., clustering). In contrast, RL agents learn by receiving scalar reward signals, rather than predefined labels, to guide behavior toward long-term goals [25]. As an example, Q-learning uses Temporal Difference methods to estimate Value Functions that predict long-term rewards, enabling the agent to learn without prior knowledge of the environment [26].

Several issues studied in economics can be articulated within the RL paradigm, opening up the door to address intricate behavioral dynamics that cannot be captured with traditional aggregated equilibrium models. By framing economic decisions as a series of *state-action-reward* interactions, it can be better captured how economic agents learn from their environment and adapt to changing circumstances over time.

In this chapter, we suggest the possibility of characterizing macroeconomic crises. As said, a crisis is a sudden shock to the system emerging from uncoordinated, but mutually influenced, actions of many agents. Thus, it becomes evident that RL can provide insights into how these agents adjust their strategies in response to new (yet imperfect/incomplete) information and altered incentives.

This application shifts the focus of analysis to a *system of collective interacting learning agents*, which is better known as MARL. In this framework, each agent not only learns from its own experiences but also can take into account the actions, strategies, and results of others, dealing also with a shared environment that acts as an indirect information channel. Such a collective dynamic type of adaptation may manifest as sudden disruptions or avalanches, highlighting how spatiotemporal localized behaviors can lead to widespread systemic effects, thereby serving as a suitable framework for the analysis of economic crises.

The integration of MARL with ACE represents a significant advancement in modeling complex economic systems. While MARL enables agents to learn optimal strategies through agent-environment dynamics, ACE provides a conceptual framework to deal with heterogeneous economic agents.

Our perspective is part of a recent and prolific trend of efforts that apply the ACE framework to macroeconomics [27] propose a "Rational Macro ABM" (R-MABM) framework incorporating capital and credit along with banks, households, and capital- and consumption-goods firms, with the latter substituted by RL agents that can be either "bounded rational" or "fully rational". In Evans and Ganesh [28], agents with varying levels of strategic skills go beyond strict utility maximization, in an effort to overcome the limitations of traditional ABMs, which rely on manually defined behavioral rules. The model introduces parameterized shared policy learning, allowing agents to grasp from a common policy while still displaying a range of bounded rational behavior.

Dwarakanath et al. [29] present a general purpose MARL simulator for economic systems meant to be versatile, so that users can specify the number and types of economic agents along with their heterogeneity parameters. The capability of each economic agent to learn objective maximizing strategies from interaction with others allows for the simulation of various counterfactual scenarios. Some scenarios, for example, include the interplay between household skill and their preference to work at different firms, and the response of firms to positive exogenous production shocks.

Zheng et al. [30] develop a two-level deep MARL framework, termed the AI Economist, meant for economic policy design. It highlights how economic agents (producers and consumers) and a social planner co-adapt utilizing structured curriculum learning to address taxation policy design. The AI Economist is shown to recover optimal tax policies from economic theory in one-step economies, and to significantly improve social welfare and the balance between equality and productivity in spatiotemporal economies.

8.5 Macrocrisis, the Corridor, and Resilience Mechanisms

8.5.1 Macrocrises

In his quest to understand macrocrises, Axel Leijonhufvud proposed many years ago that the economy should be seen as a challenge related to the *coordination of interdependent plans* made at different points in time [31]. In an idealized frictionless Walrasian (general equilibrium) world, says Leijonhufvud, those plans are consistent by construction: household saving decisions align with firm investment plans, prices instantly validate expectations, and the economy follows a full employment growth path. In the real world, though, (real) time is key. Agents form expectations under uncertainty, act on those expectations today, and only later discover whether their plans were compatible. Coordination is therefore an intertemporal phenomenon. What matters is not whether the identity between savings and investment holds **after** the fact, but whether savings and investment **intentions** are aligned beforehand in a way that ensures sufficient effective demand. If not, explains Leijonhufvud, we then have an effective demand failure.

Of course, the odds that every plan in a whole economy will be consistent with each other tends to zero. But when the system functions properly, this is not a significant issue, because feedback mechanisms exist that may accommodate differences. For example, when planned saving exceeds planned investment, interest rates may vary in order to match plans. This is the "macro equivalent" of prices bringing equilibrium to specific markets.

However, adjustments sometimes occur not only in prices, but also in quantities. Now, when planned saving exceeds planned investment, inventories accumulate, firms reduce output, and income falls until saving intentions adjust downward. Leijonhufvud's main point is that during this state of disequilibrium, prices may not function properly, or even in the expected direction, bringing more confusion to the workings of the stabilizing mechanisms in the economy. When a significant shock hits the economy, this kind of maladjustment can persist for a long time, resulting in a crisis.

This dynamic, central to macroeconomic analysis since Keynes, becomes even more complex in an open economy. Here, the coordination problem among economic agents extends the gap between domestic spending and domestic output, reflected in a current account imbalance. Again, this imbalance may adjust without trauma if the price system (in this case, the exchange rate) works, bringing the right decisions to the agents involved. But sometimes this is not the case, and an unsustainable current account deficit adjusts with a sharp crisis.

Daniel Heymann analyses these cases in detail. He emphasizes that the same current account deficit may reflect very different stories [32] about the future of the economy. It might signal foreign optimism about tomorrow's productivity, or it could reveal an unsustainable consumption boom built on fragile expectations. As long as international lenders believe the optimistic narrative, voluntary financing keeps the economy working as if nothing bad could happen, and builds dangerous debt dynamics. Once expectations change, capital inflows vanish and the burden of adjustment falls on domestic output and employment. The open economy version of an effective demand failure reveals that saving and investment plans were never truly aligned.

The Leijonhufvud's "corridor" hypothesis [33] offers a middle-ground between perpetual full-equilibrium optimism and ever-present instability.[1] The economy,

1 Although this idea has struggled to find its way into mainstream macroeconomics, many authors developed models incorporating corridor stability properties. See for example Refs. [34–36]. For an examination of possible precursors to the corridor concept, see Assous [37]

he argues, is held together by a web of compensatory mechanisms—price changes, inventory adjustments, credit lines, informal norms—that counteract local shocks and steer behavior back toward a path consistent with full employment. So long as disturbances remain within a tolerable quantitative range, these feedbacks absorb them and the system stays inside what Leijonhufvud calls the *corridor of normal macroeconomic functioning*.

Outside that corridor, those feedbacks lose traction or even turn destabilizing. Inventories that normally buffer demand swings become unsaleable stocks (they are too cheap to be sold!); credit and economic policy may become strongly procyclical, so that smooth expenditure morphs into a deleveraging spiral; price signals that guide allocation degenerate into downward wage-price dynamics provoking deflationary expectations and freezing production. Once the economy has traveled far enough from its notional equilibrium, effective demand failures arise and self-correcting forces are overwhelmed, setting the stage for a macrocrisis.

The implication for modeling is straightforward. Equilibrium-based frameworks like RBC, DSGE, and New Keynesian variants may give sensible guidance to understand some processes inside the corridor, where deviations are small and compensatory buffers dominate. The key mechanism in these frameworks is the price system. In these circumstances, prices function as sufficient signals to summarize the information required for agents to make consistent intertemporal plans, thereby ensuring coordination across decentralized decisions.

But these frameworks are ill-suited to track the economy as it exits the corridor, diagnose how far it has strayed, or predict the nonlinear dynamics that may eventually follow. In those regions, the assumption of unique, globally stable equilibria with rational expectations becomes a limited artifact if left to its original bricks.

In particular, during crises, prices may be poor signals to act rationally, because everything is changing in the economy. As Leijonhufvud [31] explains, transactions are often made in a disequilibrium state, since agents do not postpone their economic decisions until the system has reached equilibrium, the typical assumption in DSGE models. Even if prices are meaningful during crises, they may be only one among several signals available to agents operating under bounded rationality and incomplete information. During most crises, prices can not guarantee equilibrium consistency, because they become noisy signals, confusing local feedback that agents interpret through adaptive learning processes. Coordination, therefore, is not enforced by prices alone but emerges (often imperfectly) through dynamic interaction, experimentation, and institutional mediation.

In our view, to be helpful during crisis times, a model needs to include heterogeneous agents, balance sheet interactions and state-dependent feedbacks, key variables driving socially unwanted outcomes. As will be shown soon, a modern MARL perspective may be a vehicle to understand how the economy behaves during these events.

8.5.2 Resilience Mechanisms

Buffers are central to the functioning of the economy. Some of them are the result of the rational actions of households. For example, when faced with a negative shock (e.g., the price of the country's main export commodity drops sharply), national income contracts and firms lay off workers. Yet, the damage may be mitigated if households hold liquid precautionary savings or have access to credit to keep consumption near pre-shock levels. Firms and banks, in turn, may enter the shock with some inventories, or high capital buffers and stable retail-deposit funding. These financial and private-saving cushions let consumption fall only modestly. If developed enough, financial system may help cash-flow constrained exporters obtain short-term credit instead of fire selling assets. In this setting, where the private sector behaves with rational perceptions about the future, aggregate demand remains inside Leijonhufvud's corridor.

However, households cannot always be so rational. Sometimes families don't know how or where to invest, while access to credit is limited to a few financially sound actors. Also, the private sector has to learn when and how to deploy savings. Still, over successive cycles, families may internalize simple heuristics to act properly, learning from their environment. When export prices tumble, some households may automatically cut discretionary spending and dip into liquid assets rather than defaulting, cushioning the macroshock.

Institutional developments behind macroeconomic prudential policies are central to help bounded rational agents. Because perfect foresight is unrealistic, institutions embed resilience into routine choices. So, in the example above, we may find that automatic stabilizers, as is the case of progressive taxes or unemployment insurance, inject without legislative delay purchasing power exactly when private income falters. Also, targeted credit lines facilitated by national and development banks may preapprove contingent credit for exporters whose collateral values swing with world prices, preventing a credit crunch.

These institutional rules translate complex optimization problems into simple heuristics that bounded rational agents can follow ("draw on the credit line only if export revenue drops more than a certain threshold", "maintain a three-month liquidity buffer at all times"). Together, buffers, learned heuristics, and rule-based institutions interact to absorb the shock, keep expectations anchored, and prevent the economy from tipping outside the corridor into a full-blown macrocrisis.

In emerging economies (such as Argentina) more problems may arise. Output, credit and the exchange-rate regime can move erratically through recurrent "macro-regime switches" (high inflation vs. stabilization, fixed vs. floating, capital-inflow booms vs. sudden stops). Each regime shift re-draws the safe operating zone, so the distance from normal conditions to the corridor's edge is shorter, and the edge can shift overnight.

Two structural facts contribute to a thinner and unstable corridor. First, in these countries, buffers are typically thinner. Private savings are low, sovereigns carry high interest debt, and the financial system is shallow, so even moderate external shocks translate into large domestic swings. Second, domestic amplification mechanisms may be powerful if labor markets are too rigid, currency mismatches exist, and fiscal macro-policy turns procyclical. In these cases, disturbances reinforce each other, shrinking the corridor further.

Because cushions are scarce under uncertainty, agents must rely more heavily on learning. Yet informational frictions are larger and forecasting errors more frequent, so individual trial and error alone cannot provide enough resilience. Here institutions may become the key layer of defense by establishing, for example, macro-prudential leverage caps, hard budget constraints for sub-national governments, automatic fiscal stabilizers that work even when politics grid-lock. Well-designed rules stimulate prudent behavior into simple heuristics that bounded rational households and firms can follow. This is the kind of scaffold Leijonhufvud had in mind when he referred to compensating feedbacks.

8.5.3 Expectations and the Corridor

Expectations play a central role in the functioning of complex economic systems, as they condition how individual agents perceive shocks, adjust their behavior, and coordinate with others over time. In a nonlinear, interdependent environment, expectations can either amplify disturbances—if pessimism spreads and feedback loops reinforce contractionary dynamics—or help stabilize the system when agents anticipate compensatory mechanisms at work. Thus, expectations act as a coordinating device, shaping the boundaries within which adaptive behavior sustains systemic coherence.

Leijonhufvud's corridor hypothesis captures this idea by suggesting that economies possess self-correcting mechanisms that keep fluctuations within a tolerable range, provided agents expect them to be effective. Once shocks push expectations outside of that corridor—whether through loss of confidence in credit, prices, or institutional support—the very mechanisms that previously ensured resilience may collapse. The corridor is therefore not only a structural property of the economy but also an *expectations-driven domain of stability*, where beliefs about feedback mechanisms are as decisive as the mechanisms themselves. So a key property in macroeconomic systems is the modeling of agents' expectations about economic variables.

In mainstream models (as DSGEs), the main assumption is that of "rational expectations". The hypothesis implies that forecasts made by agents in the model align with the predictable (systematic) results of the (real) functioning of the economy. But as Heymann and Pascuini [38] explain, the typical interpretation of rational expectations rises ambiguities with the model-consistency notion, especially in macroeconomics. These issues affect the logic and significance of analytical procedures for treating expectations, and seem particularly pertinent when studying crises.

The core idea behind rational expectations is the assumption that individuals will not leave information "on the table" because this contradicts (intertemporal) optimization decisions. Then, the argument goes, there are incentives to try to understand the economic environment using, in the best possible way, readily available information.

However, merely incentivizing agents to gather data does not necessarily dictate how they will process them in practice. Then, a more natural way to address expectations formation is assuming a learning process. Agents update their expectations based on their perceptions, compare them against appropriate criteria, and try to identify the trends of external variables. Under certain conditions, the rational expectations hypothesis may be seen as a plausible limit of a learning process [39, 40]. But this potential asymptotic property does not apply when agents must recalculate their expectations in novel situations, for instance, during a "regime change" situation. From the agents' perspective, the problem of making forecasts becomes complicated when the economy faces a novel and permanent shock, limiting the usefulness of prior experience. And of course, it also becomes complex when this change is such that determining its consequences would require a highly complex model.

At the same time, under conditions of fundamental uncertainty, such as those associated with a regime change, the assumption of fully rational expectations is excessively demanding, as it requires levels of information processing and foresight that exceed the plausible cognitive capacities of human agents. Yet, portraying decision makers as merely inertial or unreflective is equally unsatisfactory. In crisis situations, individuals, and especially firms, often engage in deeper forms of reflection out of necessity. It is therefore important not to disregard potential adaptive responses and learning dynamics. MARL offers an intermediate framework, as it enables agents to adjust their behavior through iterative interaction and feedback, thereby capturing bounded but evolving forms of rationality.

8.6 An Integrated Framework for Modeling and Simulation of Macrocrises

In this section, we present our main contribution, outlined in two subsections. First, we show how ACE and MARL can be combined to locate and analyze the corridor in crisis scenarios. Then, we present a schematic structure that formalizes the multilayered interactions among agents and illustrates how shocks, learning processes, and feedbacks shape macroeconomic resilience.

8.6.1 ACE+MARL for Macrocrises

Let's summarize the main points discussed so far. Macrocrises can be understood as failures of intertemporal coordination. Leijonhufvud's corridor provides a useful metaphor: within it, buffers and feedbacks preserve coherence, but outside it, amplification mechanisms dominate and crises erupt. And expectations are one of the critical elements that define whether the system remains resilient or falls into instability.

Note that, in the general equilibrium settings outlined above, the question of whether a system may or may not be resilient to crises cannot even be asked. Instead, Leijonhufvud's corridor provides a conceptual bridge: it explains why conventional general-equilibrium tools assume resilience most of the time, yet look highly vulnerable during crises.

Modeling approaches that treat the corridor's boundaries—and the mechanisms that can push the system beyond them—as endogenous and learnable may help to understand, monitor signs of occurrence, and eventually help mitigate real macrocrises. In particular, these tools may shed light on our guiding question: which are the key mechanisms that can strengthen or smooth resilience to macrocrises?

In our view, Leijonhufvud's corridor can be reinterpreted in the language of resilience engineering. Inside the corridor, the economy displays absorptive resilience: routine compensating mechanisms—price changes, inventory smoothing, lender of first resort facilities dampen shocks and restore pre-crisis trajectories. Outside the corridor, those same mechanisms become ineffective or even self-reinforcing, so the system must rely on adaptive resilience (agents revise strategies, institutions redesign rules) or transformational resilience (policy regimes, financial structures, and even sectoral compositions change).

A unified conceptual framework that merges consistently the ACE and MARL approaches can be a useful way to study that progression. Because agents are heterogeneous learners, the model can endogenously reproduce the compensatory feedbacks that keep the system within the corridor, while still allowing for coordination failures that drive it beyond the boundary. Specifically, in simulations it would be possible to:

- **Map the Corridor:** Run many stochastic scenarios and identify state-space regions where learned policies reliably stabilize output and credit; those regions trace the corridor's "safe-operating" set.
- **Detect Tipping Points:** Observe when local learning rules (e.g., leverage targets, pricing heuristics) flip from stabilizing to destabilizing as aggregate indicators cross critical thresholds, exiting Leijonhufvud's corridor.
- **Design Resilience Policies:** Embed macroprudential or fiscal agents whose reward functions include both performance and distance-to-boundary.

The MARL algorithm then discovers interventions (capital buffers, countercyclical taxes, emergency liquidity) that enlarge the corridor or shorten excursions outside it.

Thus, the corridor hypothesis offers the conceptual lens, while ACE+MARL provides the computational laboratory to identify that corridor, analyze how and why economies slip beyond it, and test adaptive policies that rebuild resilience and guide the system back inside.

Moreover, an ACE+MARL model can capture some emerging market realities by (1) coding thinner initial buffers and higher shock volatility, (2) letting the corridor's boundaries depend on the prevailing policy regime, and (3) allowing institutional agents—central bank, treasury, bank supervisor—to learn policies that stretch the corridor over time (e.g., building reserve cushions in booms, tightening foreign exchange mismatch limits as leverage rises). Simulations then may show how, relative to advanced-economy calibrations, the system exits the corridor more often, but also how smart institutional design can widen the safe zone and shorten excursions outside it, pointing to a practical way to enhance resilience even when private buffers and rationality are constrained.

As stated in the introduction, macroeconomic crises can emerge as a result of external shocks or, alternatively, from an endogenous mechanism. Often, these events lead to significant disruptions in the functioning of the economy, affecting decisions of the government, firms, and households. Every agent in the economic system must review former decisions and take new directions influenced by their own set of beliefs and expectations about future economic conditions.

In the government arena, disturbances force a reevaluation of existing policies in order to adapt to the novelties in the system. This policy response is often the result of "lessons" that policymakers obtain from the crisis. Firms must review production decisions, investment schedules, and financing strategies. Finally, households must reassess their labor, consumption, and saving plans.

But this is hardly all that needs to be reviewed. Each different economic agent (and its groups of interests) can be conceptualized as interacting *decision makers* operating under distinct policy frameworks. Each agent type may pursue its own objectives, hold unique perceptions, and face specific constraints both in accessing economic information and in enacting policy changes.

Interdependent sequences of decisions, perceptions, belief revisions, expectation adaptations, and consequent responses by agents are then manifested across micro, meso, and macrolevel hierarchies. This complex network of relations and events interacts and can determine whether the economy becomes susceptible or not to amplify volatility and fluctuations that may ultimately extend the crisis and its effects.

At this point, it is clear that the ability of economic agents to learn based on interpretations of limited information about the past is, at the very least, a considerably constrained process that may well lead to miscalculations and unintended consequences, such as exacerbating instability within the economic system. This is a crucial point at which MARL can introduce a pivotal change in how agents learn and plan: to learn also from information about "possible futures."

Integrating scenario planning within a MARL framework augments the decision-making dimension by providing an additional informational layer that can enhance the agents' capacity to deal with inherent uncertainties and mitigate systemic risks. In this context, *learning from possible futures* refers to the systematic exploration of an extensive ensemble of simulated outcomes representing feasible future developments, where *all pertinent agents interact* in pursuit of their objectives. Consequently, the prevention of crises can be explicitly incorporated as an objective within the agents' goal sets. Given its inherently exploratory simulation-based architecture, MARL is particularly well-suited to implement this approach.

Nevertheless, substantial challenges persist. The robustness of any outcomes is inherently contingent upon the soundness of the underlying model structure and its parameter estimates. The integration of heterogeneity and multilayered interactions further introduces complexities. Consequently, these issues necessitate the development of a framework to systematically organize and integrate the diverse concepts discussed.

8.6.2 Schematic Framework

In general terms, an economic system can be conceptualized as a multilayered structure composed of interacting levels, reflecting an organization that spans from microlevel dynamics to macrolevel phenomena and back through feedback processes. This national economy is populated by three primary layers of agents, as shown in Figure 8.1.

The **Households** layer represents the microlevel of the economy, consisting of individual agents interconnected through a particular network structure (N_H). Interactions at this level can be social and informational. The **Firms** layer, representing the mesolevel, consists of productive units with their own internal network (N_F). The interactions here are primarily economic and market-based. Finally, the **Government** layer, at the macrolevel, acts as a top-down regulator and policymaker within its own network ($_G$). Interactions at this level are typically political and institutional.

Figure 8.1 reflects the multi-directional causality and feedback mechanisms that define this complex system. Solid line arrows denote bottom-up interactions (micro-to-meso and micro-to-macro), illustrating how the aggregate actions and

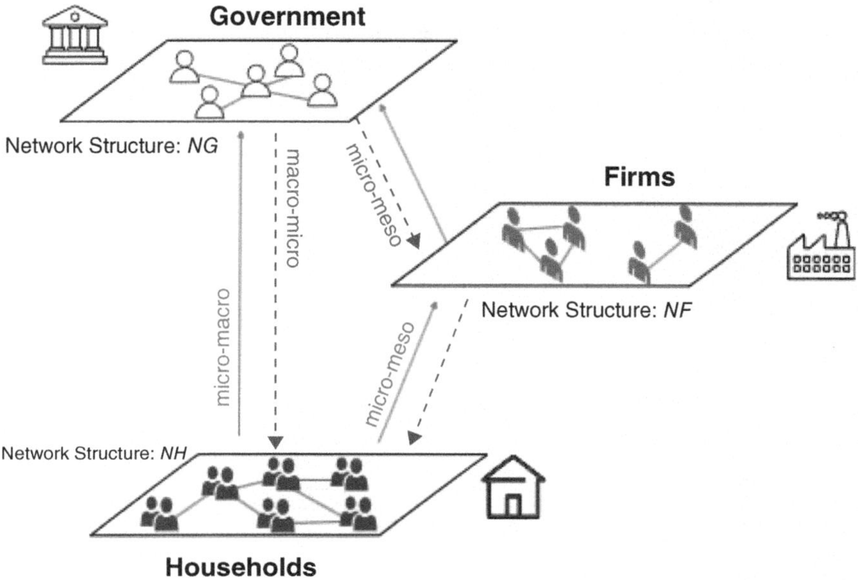

Figure 8.1 Global multilayered structure of an economic system. Multilayered framework of interacting agents, representing the micro (households), meso (firms), and macro (government) levels of a national economy, each defined by its unique network structure (N_H, N_F, and N_G respectively). Arrows indicate the feedback loops and interactions that connect these levels (see text).

emergent behaviors of households influence firms and the overall national economy. In contrast, dashed arrows denote top-down interactions (macro-to-meso and macro-to-micro), representing the impact of government policies and regulations on firms and, subsequently, on households. This visualization captures the interplay between individual actions and system-wide outcomes, which is key to modeling the complex, nonlinear dynamics of macroeconomic crises.

Moreover, these structures can evolve in time, where links between agents are constantly created and removed, either through slow transformations or through sudden disruptions. Layers are meant to capture the distinct spatiotemporal scales and degrees of granularity within the whole system dynamics.

To understand the mechanisms allowed by our framework, we define two main stages. First, we establish the structure, where agents are organized into distinct layers and information channels are established between them. Second, we present the temporal dynamics of each agent within this proposed structure.

Figure 8.2 provides a closer look at each layer and its temporal dynamics. Each parallelogram includes the decision structure of each agent. Agents have (different) goals. For example, government may want to expand GDP or reduce inflation, firms to make profits, and households to maximize utility (of consumption). Goals are the criteria for choosing among a set of possible plans.

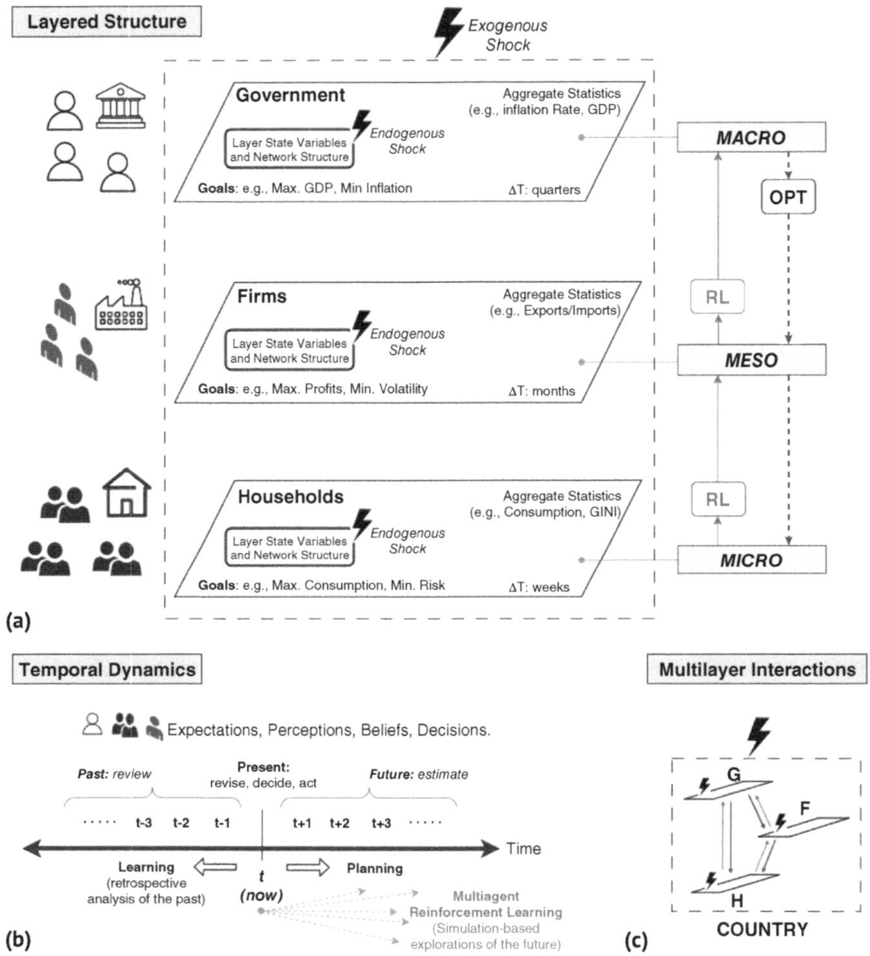

Figure 8.2 Multilayered agent-based framework for modeling macroeconomic social systems. (a) A three-tier agent hierarchy consisting of: *Government* agents operating at a macrolevel with quarterly decision cycles focused on optimizing GDP and inflation; *Firms* operating at a meso level and making monthly decisions to maximize profit while monitoring trade flows; and *Households* operating at a microlevel and making weekly choices regarding consumption and risk. Each layer contains heterogeneous agents with distinct state variables and network structures, which are subject to both endogenous and exogenous shocks. (b) Temporal cognitive dynamics architecture in which agents employ retrospective learning from historical experiences and forward-looking planning based on expectations and beliefs to create path-dependent adaptive dynamics. The MARL mechanism enables strategies to evolve based on feedback about outcomes in a shared environment. (c) A synthesis of the multilayer network structure highlighting intra-layer and cross-scale interactions.

Agents have access to two distinct types of information: Local and Global. Local Information is represented in the graph as the **Layer State Variables**. This information comes mostly from individual experience, originating in an immediate environment, and can be only accessible within each layer. Global Information, in contrast, is information accessible to all layers and is depicted, as a simplification, as Aggregate Statistics (the interest for each layer may not be the same, which is why we exemplify with different variables) and trends observable across the entire system.

Each layer unfolds with characteristic temporal scales. In this case, they are defined as the smallest discrete step times ΔT for each layer: weeks, months, and quarters respectively. Combined, the set of layers allows capturing the effects of agent interactions, the emergence of shocks, and their cumulative impact on the system over time.

In terms of structure (as presented in Figure 8.1) each layer of the system features a unique **Network Structure** where different agents interact. This structure is a local property, meaning it is specific to that layer. The system's overall dynamics emerge from how agents interact within their distinct, yet interconnected, networked layers.

The Government layer, operating at the macrolevel, serves as a top-down regulator and policymaker. Its network structure, N_G, is defined by political and institutional interactions, where nodes represent government bodies and policymakers. The links signify official channels and hierarchical structures. This layer acts as a crucial feedback mechanism, with its interventions designed to influence lower-level agents and stabilize the economic system.

The Firms layer, representing the mesolevel, is a network N_F where productive units are the nodes. The connections within this network are a direct reflection of economic and market-based interactions. These links are not just conceptual; they can be formalized to model a variety of critical relationships.

The Households layer, operating at the microlevel, is a network N_H comprised of individual households as its nodes. The links within this network signify social and informational interactions, reflecting how information, behaviors, and opinions are exchanged and propagated through personal relationships. This structure is best modeled as a small-world or scale-free network, which accurately captures the decentralized and often non-transitive nature of human connections.

Finally, we consider two types of shocks: exogenous and endogenous. While exogenous shocks impact over the whole economy, endogenous shocks may originate at any specific layer and can then spread to others, creating spillover effects.

At the right corner of Figure 8.2a, we portray the interactions between layers. Agents can employ RL mechanisms to explore possible future paths through simulated trial and error mechanisms. This helps inform their present planning and decision-making. This distinction helps us understand how different kinds of disturbances interact with the system's structure and agent behaviors, leading to more complex, multilayered responses.

Figure 8.2b illustrates the temporal dynamics followed by agents. We assume that, at the macrolevel, the Government solves a classical optimization problem. Then, the MARL type of strategy is only followed by agents at the micro and mesolayers.

Each agent makes decisions at present time t based on information taken from three time ranges: past, present, and future. Regarding the past, the agent has access to the history of (both local and global) variables and to its own history of expectations regarding its economic performance. About the present, the agent observes its current situation, and evaluates the risks and margins of action for different decisions that could be taken. To consider the future—and this is of special interest to the MARL approach—the agent evaluates a future horizon of possible sequences of actions, and then learns to choose the best path of future actions that will be regarded as *optimal* (in some sense). This task yields a new possible *plan*, a candidate to be enacted during a finite time horizon. Having access to all this information, the agent makes a decision on what is the best course of action to be followed at the present time, which will in turn act as the new expectation formation to be evaluated in the next cycle.

In Figure 8.2c, we represent a synthetic systemic view to stress the fact that agents at any layer have access to some aggregated information on other layers (e.g., inflation rate, balance of payments, household debt ratio, etc.). Thus, agents at each layer don't make decisions in isolation, and decisions made by agents at a layer can affect (as an aggregate) the decisions made at other layers. Furthermore, the framework is flexible and can accommodate other intermediate layers (e.g., unions of workers, sectoral confederations, trade associations, etc.) leveraging the techniques of multilayered networks [41].

Let's now examine the processes that could take place in this framework when applying it to macrocrises. As we have explained, the modern macroeconomy can be conceptualized as a complex, multilayered system where agents with different goals, time horizons, and information sets interact.

Let us assume that our sequence of events begins with an exogenous shock originating from outside the model. For example, this could be a sharp fall in terms of trade, which alters cost structures and expectations across all three layers of the domestic economy. A first wave of chained events can be the following:

- **Government Layer:** Officials in the government layer observe the shock through its key state variables: exporters start to complain, and signs of slowing GDP growth and drain of reserves start to unravel. If agents in the government layer interpret the fall in the terms of trade as a major threat, and tries to avoid losing reserves, the administration may decide—for example—to let the exchange rate depreciate. This can be viewed as an "endogenous shock" developed at the macrolevel, but taken by other layers as an exogenous one. This decision may or may not consider other Aggregate Statistics from other layers, particularly the relative impact of international prices and internal devaluation in firms' costs (mesolayer) and consumer spending (microlayer).

- **Firms Layer:** Companies operate in our example on a monthly basis cycle and receive the information of government's endogenous shock (the devaluation) through their local state variables: their cost of capital rises, and their business may suffer from a lower demand. To protect profits in the face of the new situation, firms may pass through their increased costs to consumers by raising output prices, and/or delaying or canceling planned investments. This can be seen as a second "endogenous shock" at this layer level, fearing lower future consumer demand due to higher interest rates and their own price increases. Firms base these decisions on its own Aggregate Statistics, considering the new relative prices (between external and internal) and the behavior of other firms, watching the development of total exports and competitiveness of the economy.
- **Households Layer:** Finally, households make decisions on a weekly basis, and are the last to feel this chain of events. But their response is crucial to complete feedback loop. Their state variables change dramatically: they see higher prices for tradable goods, and their expectations on the exchange rate may very well be that the situation has just began to unravel. In fact, their decision to discretionary (not survival) spending may confirm the expectations of firms, and some household members may decide to seek for second jobs to maintain income levels.

The chain of events is now complete and begins to feed back on itself, in a number of ways. A traditional way of seeing this is the "keynesian multiplier". Firms observe a contraction in consumer spending from the Households layer, confirming their fears, which naturally leads to expect further cuts in production and investment, potentially leading to layoffs. Consumers then get a new round of reductions in income and jobs, reinforcing the reduction in there spending.

Now, as explained in the corridor subsection, a macrocrisis may happen or not. If compensatory mechanisms are strong enough—for example, if precautionary savings are deployed, automatic stabilizers inject income, or firms can smooth cash flows with inventories or short-term credit—the system may remain inside the corridor, and the multiplier effect is contained. Learning dynamics at the household and firm levels, together with stabilizing rules at the institutional level, prevent expectations from turning self-fulfillingly pessimistic. The shock is absorbed, the system proved resilient, and the economy eventually reverts to a sustainable trajectory within the corridor.

If, however, buffers are thin, learning is myopic, or institutional credibility is weak, the same sequence of shocks and reactions can push the system outside the corridor. In that case, the stabilizing feedbacks flip sign, becoming amplifiers of instability: firms retrench further, households cut consumption more aggressively, financial constraints bind harder, and government policy may turn procyclical. What began as a localized disturbance evolves into a macrocrisis.

The schematic framework proposed here shows how ACE+MARL can be used to simulate both paths. By embedding heterogeneous, bounded rational agents into a multilayered learning environment, the model can generate the endogenous emergence of either resilience or fragility. It can also test how alternative institutional designs or policy interventions alter the width of the corridor and shorten the length of excursions outside it. In this way, the corridor hypothesis provides the conceptual lens, while the ACE+MARL framework offers the computational laboratory for identifying early-warning signals, evaluating alternative responses, and ultimately designing policies that strengthen resilience to macrocrises.

8.7 Conclusions and Outlook

We proposed a novel conceptual framework that integrates ACE with MARL to model, simulate, and analyze the complex dynamics of macroeconomic crises. This framework is specifically designed to capture both the intricate complexity of agent interactions and the adaptive, path-dependent learning processes that characterize real-world economic systems.

This mapping process involves two interconnected steps. The first step focuses on translating crisis features into an ACE Model. This involves the identification and formalization of the structural and behavioral characteristics of a specific crisis to an ACE model, identifying and modeling key components of the system, where heterogeneous agents (e.g., firms, households, regulators) interact, each with unique behavioral rules, across micro, meso, and macrolevels and with external factors, assuming that they have limited information and bounded rationality, which is a key driver of emergent behavior of crisis phenomena. By formalizing these elements, the ACE model provides a simulated environment where the fundamental components and their interdependencies are clearly represented.

The second, more advanced step, involves embedding adaptive learning using MARL. This simulates how agents learn and adapt their strategies over time based on past outcomes and future expectations. Through repeated interactions, agents collectively generate emergent behaviors like bubbles, panic, and recovery paths, which are common features of real-world crises. This process allows the framework to move beyond static behavioral rules and capture the dynamic, path-dependent nature of economic systems.

By combining historical crisis patterns with dynamic learning and networked agent interactions, this framework can be used to study key characteristics of macroeconomic crises, explore "what-if" scenarios, and evaluate the resilience of economies under different policies.

Within the ACE-MARL framework, the "expectation of the outcome" for an agent refers to the anticipated utility that an agent predicts from taking a specific action in a given state. It is a core component of the agent's learning process. Unlike a perfect forecast, this expectation is a dynamic, learned belief that is continuously updated as the agent gains more experience and as the simulated environment evolves. The agent's goal is to select actions that maximize this long-term expected reward. This process allows the model to simulate an aspect of "real-world" economic behavior: agents making decisions under uncertainty, guided by learned, but imperfect, mental models of the world. This rule can be related to the outcome of trial and error learning—captured in an ACE+MARL model as a policy that maximizes long-horizon utility under uncertainty.

This framework presents a novel coherent approach to policy evaluation by providing a computational tool for analyzing economic resilience. Rather than seeking to inherently stabilize the system, the framework serves as a tool for probing its dynamic responses to various interventions.

We suggest that this framework be employed to conduct ex ante evaluations of policies, such as specific regulatory, monetary, or fiscal interventions, before their real-world implementation. By simulating "what-if" scenarios, this computational tool can be used to identify critical vulnerabilities, analyze potential failure points, and explore diverse recovery paths under different policy regimes.

Ultimately, the framework's value lies in its capacity to serve as a laboratory for institutional learning, providing a structured environment for policymakers to gain insights into the complex, nonlinear behaviors of complex economic phenomena.

As an imperative future step, the framework must be empirically validated using practical tools and real-world data. Its agnosticism to specific simulation platforms or programming languages is a key strength, as it ensures broad applicability and theoretical flexibility. However, this very characteristic necessitates the adoption of a sufficiently expressive simulation environment—one capable of implementing the framework's core features, including multilayer architecture, emergent behavior, adaptive learning, agent-oriented design, and multiscale temporal dynamics. Selecting such a platform will be essential to translating the conceptual model into a functionally rigorous and scientifically reproducible computational instrument. We postulate that the DEVS formal framework for modeling and simulation, and in particular its recent extension for emergent behavior (EB-DEVS, [42]) and its associated simulation toolkit are strong candidates that fulfill the core requirements.

Finally, beyond its economic applications, the proposed framework can be applied to study other types of social systems where emergent behavior is a relevant property of interest, albeit often of a different nature than those in economic systems (for instance, the classic Schelling model of residential segregation). The original Schelling model is driven by a simple, non-learning rule where agents move to avoid being a minority in their neighborhood. How this model would change if agents were allowed to learn, plan, and interact (bottom-up and top-down) with a market and the government at different spatiotemporal dynamics? This would enable the exploration of interesting new scenarios, and explore how adaptive learning processes contribute to the emergence of different social patterns. Such explorations remain as part of several future works stemming from the present proposal.

References

1 Carmen M. Reinhart and Kenneth S. Rogoff. This time is different: A panoramic view of eight centuries of financial crises. *NBER Working Paper 13882*, National Bureau of Economic Research, 2008. https://www.nber.org/papers/w13882.

2 Luc Laeven and Fabian Valencia. Systemic banking crises revisited. *IMF Working Paper 18/206*, International Monetary Fund, 2018. https://www.imf.org/en/Publications/WP/Issues/2018/09/14/Systemic-Banking-Crises-Revisited-46232.

3 Daniel Heymann, Juan Pablo Brichetti, Pedro Juarros, and Gustavo Montero. Expectations, coordination failures and macro crises. *Journal of Globalization and Development*, 11(1):1–25, 2020. https://doi.org/10.1515/jgd-2020-0001.

4 P. W. Anderson, K. Arrow, and D. Pines. *The Economy as an Evolving Complex System*. Santa Fe Institute Series. Avalon Publishing, 1988.

5 B. W. Arthur. Chapter 2: Complexity and the economy. In *Handbook of Research on Complexity*. Edward Elgar Publishing, 2009. https://doi.org/10.4337/9781781952665.00007.

6 A. P. Kirman. The economy as an interactive system. In *The Economy as an Evolving Complex System II (First published 1997)*, pages 1–41. CRC Press, 2017.

7 V. M. Yakovenko and J. B. Rosser. Colloquium: Statistical mechanics of money, wealth, and income. *Reviews of Modern Physics*, 81:1703–1725, 2009. https://doi.org/10.1103/RevModPhys.81.1703.

8 B. W. Arthur. *The Nature of Technology: What It Is and How It Evolves*. Free Press, New York, 2009.

9 J. D. Farmer and D. Foley. The economy needs agent-based modelling. *Nature*, 460:685–686, 2009.

10 M. Batty. *The New Science of Cities*. The MIT Press, 2013.

11 L. M. A. Bettencourt, J. Lobo, D. Helbing, C. Kühnert, and G. B. West. Growth, innovation, scaling, and the pace of life in cities. *Proceedings of the National Academy of Sciences*, 104(17):7301–7306, 2007. https://doi.org/10.1073/pnas.0610172104.

12 Frank Schweitzer, Giorgio Fagiolo, Didier Sornette, Fernando Vega-Redondo, Alessandro Vespignani, and Douglas R. White. Economic networks: The new challenges. *Science*, 325(5939):422–425, 2009. https://doi.org/10.1126/science.1173644.

13 Prasanna Gai, Andrew Haldane, and Sujit Kapadia. Complexity, concentration and contagion. *Journal of Monetary Economics*, 58(5):453–470, 2011. https://doi.org/10.1016/j.jmoneco.2011.05.005.

14 Stefano Battiston, Domenico Delli Gatti, Mauro Gallegati, Bruce Greenwald, and Joseph E. Stiglitz. Liaisons dangereuses: Increasing connectivity, risk sharing, and systemic risk. *Journal of Economic Dynamics and Control*, 36(8):1121–1141, 2012. https://doi.org/10.1016/j.jedc.2012.04.001.

15 Giorgio Fagiolo and Andrea Roventini. Macroeconomic policy in DSG and agent-based models redux: New developments and challenges ahead. 2016. (available at SSRN 2763735).

16 David Colander. *Post Walrasian Macroeconomics: Some Historic Links*. Cambridge University Press, 2006.

17 Richard R. Nelson and Sidney G. Winter. *An Evolutionary Theory of Economic Change*. Harvard University Press, 1985.

18 David J. Teece, Richard Rumelt, Giovanni Dosi, and Sidney Winter. Understanding corporate coherence: Theory and evidence. *Journal of Economic Behavior & Organization*, 23(1):1–30, 1994.

19 Leigh Tesfatsion. Agent-based computational economics: Overview and brief history. *Artificial Intelligence, Learning and Computation in Economics and Finance*, pages 41–58, 2023.

20 Leigh Tesfatsion and Kenneth L. Judd. Handbook of computational economics, Vol. 2: Agent-based computational economics. *Technical Report*, Lowa State University, Department of Economics, 2006.

21 Finn E. Kydland and Edward C. Prescott. Time to build and aggregate fluctuations. *Econometrica*, 50(6):1345–1370, 1982. https://doi.org/10.2307/1913386.

22 Frank Smets and Rafael Wouters. An estimated dynamic stochastic general equilibrium model of the euro area. *Journal of the European Economic Association*, 1(5):1123–1175, 2003. https://doi.org/10.1162/154247603770383415.

23 Joseph E. Stiglitz. Where modern macroeconomics went wrong. *Oxford Review of Economic Policy*, 34(1–2):70–106, 2018. https://doi.org/10.1093/oxrep/grx057.

24 Mark Granovetter. Economic action and social structure: The problem of embeddedness. *American Journal of Sociology*, 91(3):481–510, 1985.

25 Richard S. Sutton and Andrew G. Barto. *Reinforcement learning: An introduction*, 2nd Edition. The MIT Press, 2018. Adaptive computation and machine learning.

26 Christopher J. C. H. Watkins and Peter Dayan. Q-learning. *Machine Learning*, 8(3):279–292, 1992.

27 Simone Brusatin, Tommaso Padoan, Andrea Coletta, Domenico Delli Gatti, and Aldo Glielmo. Simulating the economic impact of rationality through reinforcement learning and agent-based modelling. In *Proceedings of the 5th ACM International Conference on AI in Finance*, pages 159–167, 2024.

28 Benjamin Patrick Evans and Sumitra Ganesh. Learning and calibrating heterogeneous bounded rational market behaviour with multi-agent reinforcement learning. *arXiv preprint arXiv:2402.00787*, 2024.

29 Kshama Dwarakanath, Svitlana Vyetrenko, Peyman Tavallali, and Tucker Balch. Abides-economist: Agent-based simulation of economic systems with learning agents. *arXiv preprint arXiv:2402.09563*, 2024.

30 Stephan Zheng, Alexander Trott, Sunil Srinivasa, David C. Parkes, and Richard Socher. The AI economist: Taxation policy design via two-level deep multiagent reinforcement learning. *Science Advances*, 8(18):eabk2607, 2022.

31 Axel Leijonhufvud. Keynes and the keynesians: A suggested interpretation. *American Economic Review*, 57(2):401–410, 1967. https://www.jstor.org/stable/1821641.

32 Daniel Heymann. Sobre la interpretación de la cuenta corriente. *Desarrollo Económico*, 34(135):323–342, 1994.

33 Axel Leijonhufvud. Effective demand failures. In *Macroeconomic Instability and Coordination: Selected Essays of Axel Leijonhufvud*, pages 59–78. Edward Elgar, Cheltenham, 2000.

34 James Tobin. Keynesian models of recession and depression. *The American Economic Review*, 65(2):195–202, 1975. http://www.jstor.org/stable/1818852.

35 Peter Howitt. The limits to stability of a full-employment equilibrium. *The Scandinavian Journal of Economics*, 80(3):265–282, 1978. https://doi.org/10.2307/3439581.

36 J. Bradford De Long and Lawrence H. Summers. Is increased price flexibility stabilizing? *The American Economic Review*, 76(5):1031–1044, 1986.

37 Michaël Assous. "Corridor stability" in early history of macroeconomics. *Preprint*, 2024. https://hal.science/ffensl-04604537f. HAL Id: fensl-04604537f.

38 Daniel Heymann and Paulo Pascuini. On the (In) consistency of re modeling, *Industrial and Corporate Change*, 30(2):347–356, 2021.

39 Albert Marcet and Thomas J. Sargent. Convergence of least squares learning mechanisms to rational expectations equilibria. *Macroeconomic Dynamics*, 1(1):1–23, 1989.

40 George W. Evans and Seppo Honkapohja. *Learning and Expectations in Macroeconomics*. Princeton University Press, Princeton, NJ, 2001.

41 Mikko Kivelä, Alex Arenas, Marc Barthelemy, James P. Gleeson, Yamir Moreno, and Mason A. Porter. Multilayer networks. *Journal of complex networks*, 2(3):203–271, 2014.

42 Daniel Foguelman, Philipp Henning, Adelinde Uhrmacher, and Rodrigo Castro. EB-DEVS: A formal framework for modeling and simulation of emergent behavior in dynamic complex systems. *Journal of Computational Science*, 53:101387, 2021. https://doi.org/10.1016/j.jocs.2021.101387.

9

DEVS as a Method to Design Resilient Complex Adaptive Systems

Hazel Griffith, Cristina Ruiz-Martin, and Gabriel Wainer*

Department of Systems and Computer Engineering, Carleton University, Ottawa, ON, Canada

*Corresponding author: gwainer@sce.carleton.ca

9.1 Introduction

A complex system is a real phenomenon with many interacting components whose combined behavior is more than the sum of its parts. For example, an airplane in flight is a complex system that consists of many parts, such as the engine, the wings, and the landing gear. When these parts are combined correctly, the plane can successfully fly, but when disconnected, these parts cannot achieve flight. To study or design such complex systems, researchers would create models they can use to run simulations with the objective of better understanding the system.

Creating such models comes with many challenges. The first challenge is ensuring that every part of a complex system works on its own and when connected to other parts. This is accomplished through rigorous verification and validation. Though this process can be different depending upon the method used to model the system, it consists of ensuring that the model's component functions without errors and behaves in a way that is consistent with the real system. The second challenge lies in how easily the model can be changed to conduct different experiments. For example, if researchers wanted to study how well an airplane would fly with a new engine that is being developed, an important feature of the model would be to allow us to configure the airplane's engine model. Another challenge is to consider the reusability of the model's components for a different context. For example, if instead of studying an airplane in flight, we wanted to study the interactions between flight staff and passengers, would we be able to use parts of our existing airplane model in tandem with models of the people on board the airplane? Being able to reuse components would reduce the number of new models that would need to be developed and tested. With these challenges in mind, we need to select a modeling and simulation formalism that can overcome these obstacles and run efficient simulations of our models. Discrete event system specification (DEVS) modeling and simulation formalism would overcome the challenges explained above.

DEVS is a formal method of modeling and simulating discrete-event systems. A real system modeled with DEVS is described as a composite of submodels, each of them being behavioral (atomic) or structural (coupled) [1]. Each behavioral model consists of a time base, inputs, states, outputs, and functions to compute the next states and the outputs. Each model is seen as having input and output ports to communicate with other models. The input and output events will determine the values to appear in those ports. The input external events are received in an input port, and the model specification should define the behavior under such inputs. The internal events produce state changes, whose results are spread through the output ports. The ports' influences will determine if these values should be sent to other models. As the formalism is closed under coupling, both behavioral and structural models can be integrated into a model hierarchy to build more complex models. A full description of the formalism can be found in [1]. Diagrams depicting a DEVS atomic and coupled model can be found in Figure 9.1.

The benefits of modeling a system with DEVS or any of its extensions include reduced simulation execution times, highly modular and reusable parameterized models, and easier model testing for verification and validation [2]. These three features make DEVS ideal for modeling complex systems.

Complex systems contain many components that work in tandem, so trying to represent such a system with a monolithic model is cumbersome. Since DEVS is both hierarchical and modular, it allows modeling the system in a much more natural way, focusing on the different components individually (submodels) and connecting them to create the model of the complex system. These submodels can be disconnected and reconnected in any valid configuration. This means that if a particular component of the model should be changed, it is simply a matter

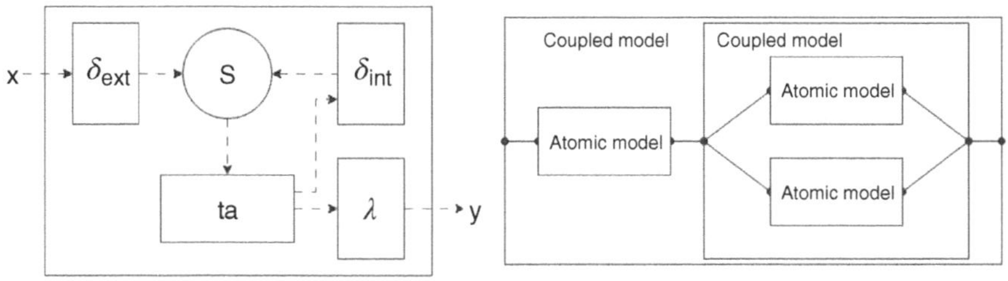

Figure 9.1 (Left) DEVS atomic model. (Right) DEVS coupled model.

of swapping out the specific submodel. This separation of concerns between components simplifies modeling complex systems and allows for easy changes and testing.

DEVS models can be ordered in a hierarchy thanks to the property of closure under coupling that allows coupled models to be treated as atomic or behavioral models. This makes modeling complex systems easier, as we can organize components of the model by their scale in the system. For example, when modeling an airplane, the entire airplane could be at the top of our hierarchy. Below that, we may have the different sections of the airplane, from the cockpit to the fuselage, tail, and wings. We can continue breaking down the components until we reach the appropriate level of detail. Therefore, as the complexity of the system increases, we can continue to add an unlimited number of layers to the hierarchy of models, ensuring that a model of any complexity can be modeled with the DEVS formalism.

Another benefit of having a hierarchy of models is that it allows for the easy parallel execution of models within a simulation. This is because the singular program that is normally responsible for determining which model may execute next and passing messages between models can now be broken down into smaller programs that only need to coordinate the models on their level of the hierarchy. This distributes the computations of the simulation being executed across multiple programs, therefore allowing for easy parallel execution of each model [3].

Another benefit of the modularity and hierarchy of DEVS models is that it eases testing by using unit testing and integration testing. Unit testing is the process of testing an individual model to ensure that it behaves correctly given any input. This process is made easier with DEVS models, as the formalism's transition functions and state variables outline every event and state transition that can occur. To evaluate a model, a researcher can trace the execution of a model when it is supplied with any input it can receive. And with the DEVS property of closure under coupling, any coupled model can be treated as a black-box atomic model, so any coupled model can be unit tested for correct behavior in the same way.

Integration testing focuses on whether coupling two or more models elicits correct behavior. This is a straightforward process with DEVS, as with the external output/input couplings and internal couplings provided by the coupled model definition, we can ensure that messages are being passed from and towards the correct models. By using the output functions of each model, we can verify that messages with the correct information are being passed to the correct ports. Like with unit testing, the executions of a coupled model can be traced when messages are passed along different couplings to ensure the model behaves correctly.

After we have rigorously tested the DEVS models, we will possess highly modular verified DEVS models. This benefits researchers in many ways, as we can now exploit the modularity of DEVS to adapt the models to fit new experiments and systems.

With an adaptable and easy to test model of a complex system, we can perform experiments of rare or dangerous events that could occur in the real system. We can use what we learn from these experiments to prepare the real system for these challenging events and make it more resilient. This is accomplished by judging the effectiveness of different policies for controlling the rare or dangerous situations in the experiments we perform on the model. Then, when a policy's effectiveness meets our standards with an acceptable cost, we can implement it in the real system. Thereby protecting the system from events that would be difficult to account for with other methods.

Some complex systems (floods, forest fires, traffic) can be better studied using a spatial notion of the systems under study. In those cases, a DEVS model might not be the best possible tool for analysis. For instance, cellular automata (CA) [4] might lend itself more naturally to depicting these kinds of systems. But we still want to retain the benefits obtained by using DEVS. To do this, we can use extensions of DEVS that include these aspects and make modeling these systems easier. Cell-DEVS [5] uses the general concept of CA while handling some of the shortcomings of classic CA, which include reduced performance and precision and time management.

The tools used to implement and execute the DEVS and Cell-DEVS models shown are known as CD++ [6] and cadmium [7]. CD++ is a standalone executable toolkit, while cadmium is a C++ header library. In CD++ and cadmium you can define DEVS atomic models in C++ by extending an atomic model class with some small differences. In both, you overload functions that are analogous to DEVS model functions like the state transition functions and the output function. To define coupled models in CD++, you would use a configuration file written in a coupled model specification language. This allows you to connect atomic models by their input and output ports to create DEVS coupled models. In cadmium you write coupled model classes in a similar way to atomic model classes, where you create instances of the atomic model classes you defined and couple them by their input and output ports. Both the CD++ toolkit and cadmium header library offer extensions of the atomic and coupled models to create cell-DEVS atomic and coupled models. In CD++, you define cell-DEVS models with a specification language in a configuration file. In this definition you indicate the cell type, delay function used, neighborhood, whether the border is wrapped or not, and the dimensions of the grid. The local transition function is also specified as a list of rules. Each rule is of the form VALUE DELAY {CONDITION}. This means that when the local transition function activates, it goes down the list of rules testing the Boolean conditions. If it is false, then it goes to the next in the list; otherwise, it sets the new values of the state variables indicated and sets the amount of time to delay transmitting this cell's new state to its neighbors. In cadmium you define cell-DEVS atomic models in C++ as a class that extends the cell-DEVS atomic model class. Within it you

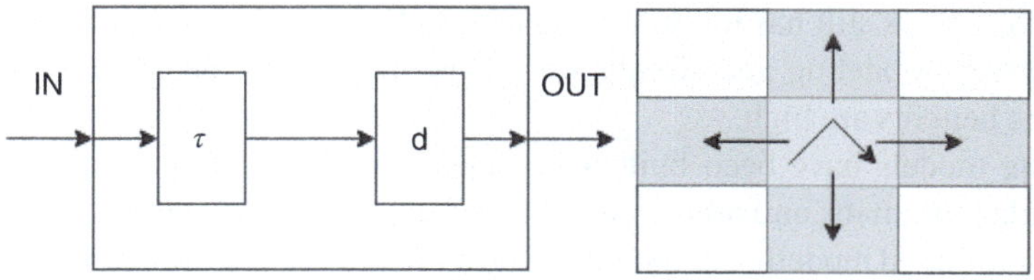

Figure 9.2 (Left) Cell-DEVS atomic model. (Right) A cell transmitting its state to four neighbors.

overload the local computation and output delay functions to define the behavior of the cell. In CD++, a cell-DEVS coupled model is defined in the same configuration file as the cell, where the dimensions parameter indicates the number of cells to be coupled together. Whereas in cadmium you define the coupled model as different scenarios in JSON configuration files. Here you can define the initial values of each cell, the type of delay to be used, and the neighborhood of each cell type, along with the grid dimensions and whether the borders are wrapped.

Using DEVS to build cellular models is accomplished by modeling the cells of a cell space using DEVS. A single cell receives inputs from a configurable neighborhood or other DEVS atomic models. These inputs are processed by a function that determines the state of this cell, but unlike CA, with DEVS we can set a timer to delay when the cell broadcasts its state to its neighboring cells. Various kinds of delay functions can be used. In Figure 9.2, you can see a depiction of an atomic model in Cell-DEVS and an example of a cell neighborhood.

The remaining sections of this chapter illustrate how to use DEVS and Cell-DEVS to design resilient complex systems in an efficient way. Different tools have been used to build and execute these models, including CD++ [6] and cadmium [7].

9.2 City Planning

Policymakers of municipal governments need to study why people live in various places with different goals that go from planning up to improving the resiliency of their municipality under different scenarios. For example, a city needs to establish zoning laws around what areas are for building residential homes, what areas are for commercial properties, and what locations are for industrial facilities.

Traditional experimentation for city planning suffers from various problems, including accuracy, costs, and ethics. For example, conducting highly accurate experiments by changing a law to see if it impacts the city's resiliency could cause harm, negatively impact people, and disrupt the city's economy. Gathering information via surveys still has unknown impacts and the intrinsic cost of conducting the survey. Modeling and simulation can be difficult to model human choices, but the benefits are high.

City-planning models have been built using agent-based modeling, scenario planning, cellular automata, and others. Agent-based modeling is a common technique. With agent-based modeling, a system can be represented by the interactions between decision-making agents. These agents are entities whose behaviors are commonly defined by systems of rules, though they could be designed in many ways. The entities can vary in scale from individuals to communities to cities themselves [8]. Scenario planning uses flexible strategies by examining how different scenarios may unfold [9]. Cellular automata are frequently used to model how urban centers grow over time. In combination with multi-agent systems, they can realistically model the changes in a city over time [10].

Our research uses an alternative to these methods based on DEVS and Cell-DEVS. The next subsections show how to apply these methods to distinct aspects of city planning.

9.2.1 Urban Growth

In recent years, there have been numerous efforts on the application of social simulation for the prediction of urban growth. Individuals' decisions on where to live can create people's concentration in big cities as well as in unpopulated areas. There are diverging policies in the management and control of the environment, land use, scale, and arrangement of settlements that lead to specific urban growth. We will show how to build a Cell-DEVS model of urban growth, focusing on a case study for the Changjiang delta region in China. This area covers 75900 km^2 of land with 10200 km^2 of water bodies. There are 16 regional-level cities, 28 county-level cities, and 1700 towns, as shown in Figure 9.3.

Several factors for city growth in the subsections of this area are considered, in particular: development corridors, corridors with big city growth, ecological system concerns, and disaster prevention with development corridors. All these factors impact the development of the area [11].

Let us see how to model the impact of big cities on urban growth (called "corridors with big cities" above). We build a Cell-DEVS model in which each cell in the grid calculates city growth based on the following formula:

$$G_{\text{total}}^{t+1} = \sum \Big(G_{\text{spontaneous}(i,j)}^{t+1} + G_{\text{spread}(i,j)}^{t+1} + G_{\text{edge}(i,j)}^{t+1} + G_{\text{road}(i,j)}^{t+1} + G_{\text{corridor}(i,j)}^{t+1} + G_{\text{city}(i,j)}^{t+1} \Big), \tag{9.1}$$

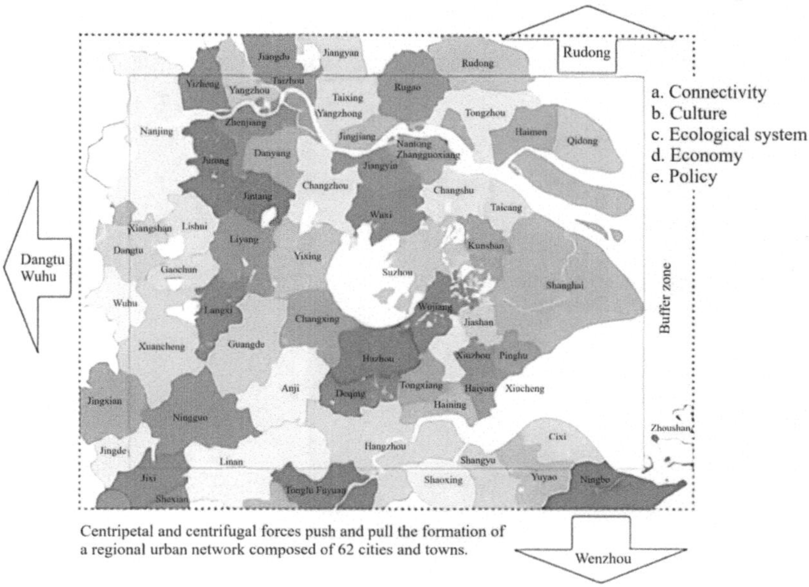

Figure 9.3 Study area of the Changjiang delta region.

where G_{total}^{t+1} is the total urban growth prediction, at year t+1; $G_{\text{spontaneous}(i,j)}^{t+1}$ is the spontaneous growth, the occurrence of random urbanization of land, at year t+1; $G_{\text{spread}(i,j)}^{t+1}$ is the urban spreading of a newly urbanized land cell, at year t+1; $G_{\text{edge}(i,j)}^{t+1}$ edge growth, the further expansion of a newly spread cell, at year t+1; $G_{\text{road}(i,j)}^{t+1}$ is the road influence at year t+1; $G_{\text{corridor}(i,j)}^{t+1}$ is the development corridor growth, at year t+1; and $G_{\text{city}(i,j)}^{t+1}$ is the big cities growth, at year t+1.

A cell space of 900 cells (30 × 30) was used to experiment with this model; each cell has its own information about its growing potential defined in the model execution. We conducted an experiment with real values from the region, as shown in Figure 9.4. Blue cells represent basic settlements, turquoise cells are roads, orange cells are development of the corridors, and the pink cells are cities.

Figure 9.4 shows how the model evolves in time. To compare our model's results with the real map of the region, we over impose the simulation results and the area map in Figure 9.5. There is a similarity in the big urban centers' locations as well as in the unpopulated areas. Although the overlapping figures can be difficult to understand at first glance.

With the experiments conducted on this urban growth model, a policymaker of the Changjiang delta region in China could assess how their urban centers will grow over time and implement policies that meet the needs of the expanding population.

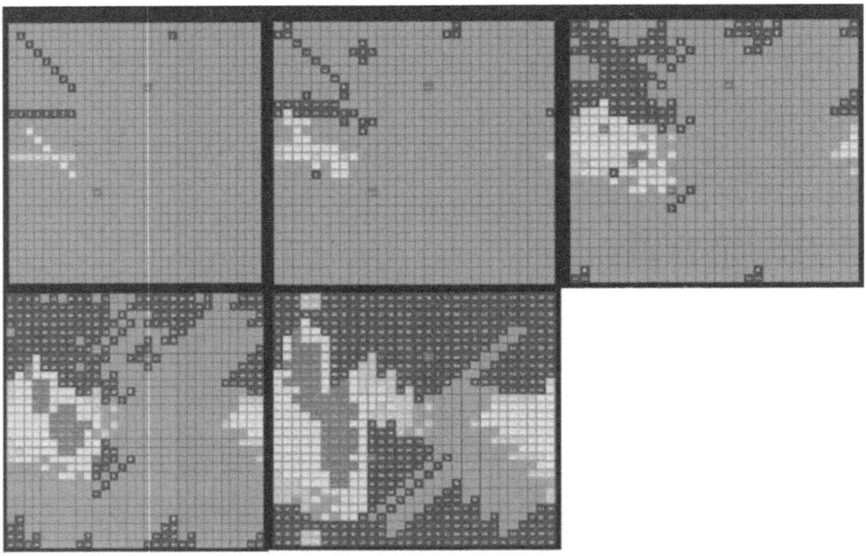

Figure 9.4 Model evolution. For color references, please refer to the online version.

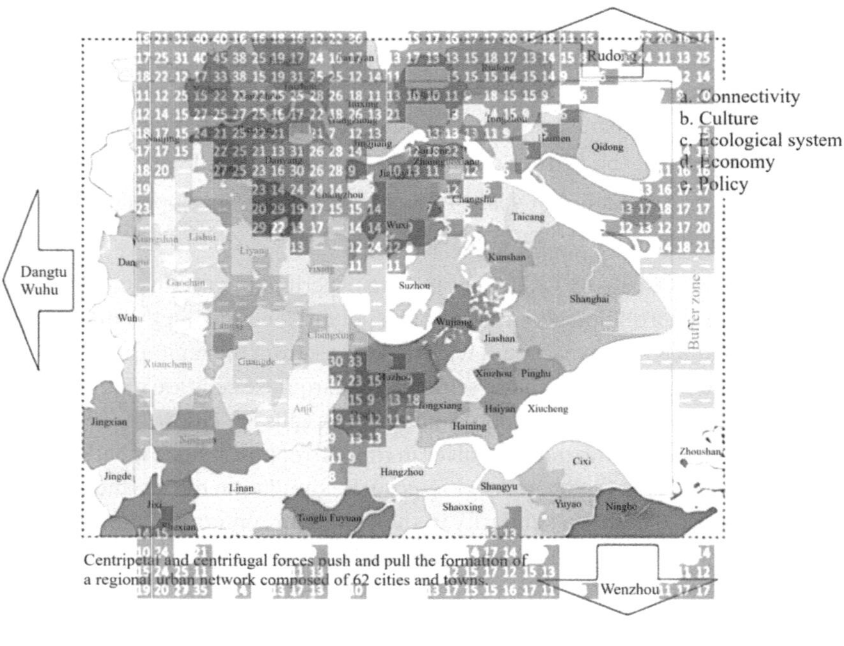

Figure 9.5 Comparison of our model's result with region.

9.2.2 Segregation

Complex social patterns often emerge based on hard-to-define parameters and factors. One example of these patterns being generated in day-to-day observable phenomena can be observed in residential segregation. Cities and neighborhoods often end up divided based on racial, cultural, or socioeconomic divides. These patterns can often be attributed to urban planning or intentional, codified discrimination but can emerge because of simple, decentralized decisions made by individuals based on their unique sets of personal preferences. Segregation modeling explores these ideas, where small-scale interactions generate large-scale geographic structures and patterns. Modeling segregation can give better insights on how localized interactions among individuals give rise to broader social patterns, enabling the exploration of various scenarios and uncovering emergent patterns that may otherwise remain hidden. We can better understand the underlying mechanisms of segregation, inform urban planning and policymaking, and potentially devise strategies to promote more integrated, resilient, and inclusive communities.

The first model demonstrating how individual preferences, even those of tolerant people, would lead to unintended and widespread segregation came about in 1970 by economist Thomas Schelling [12]. This model of segregation is based on the internal behavior of each individual agent, which can be translated into cell behavior and extended with other factors. Schelling's model demonstrated that even mild preferences would lead to a highly segregated society.

There are many individual preferences that factor into the decision to stay, which might increase or counteract the desire for a tolerant or intolerant person to segregate themselves from a specific group. The model uses two opposing groups, and we look at how they move in space based on different factors, such as their proclivity to live near others like or unlike them, the effect of obstacles like a highway or other terrain that separates space, and economic factors like the quality of the residence and its impact on the individual's decision-making. The model includes:

- Configuration of cells to have variable ratios of tolerance (for instance, in one experiment, the groups might want to be surrounded by at least 50% similar cells, while in another, this ratio might drop to 10%). This would, according to Schelling's segregation model, still result in segregation, but the asymmetric model will allow for observation on the strength of that model.
- A resident cannot be expected to move instantly if they relocate to an undesirable neighborhood. Adding a minimum stay period for a cell adds the dynamic that a resident will slowly see its neighborhood transform into a desirable or undesirable one, thus affecting their decision to relocate.

- Since humans need social interactions, if a resident is surrounded by empty cells, it will move as soon as possible. This prevents patterns of individual segregation from appearing.
- Another common occurrence of segregation in cities is the stark segregation along landmarks, natural geographic features like rivers, or even human-caused separation with walls and highways.
- Lastly, the quality of the location is often a high indicator of a resident's desire to stay and will offset some amount of intolerance to a different group. Adding states for specific cells to be desirable or not also plays into the desire of a resident to move. Such examples could be that the building itself is nice, the schooling system in the location is great, or a combination of many factors.

These behaviors are defined in each cell's rules. First, a cell must wait for five time units before it can decide to move. Then, if a cell is surrounded by empty cells, it will decide to move. If a cell is empty, it will do nothing; otherwise, if it is in group A with a value of 1 or in group B with a value of −1, it will decide whether it should move or stay where it is by calculating the ratio of percentage of neighbors that belong to the same group, A or B. Every cell has a Moore neighborhood with a radius of 1 cell. If this percentage is less than the threshold set for the experiment, known as the tolerance ratio, the cell will move; otherwise, it will stay where it is. To represent the cell's willingness to stay on a valuable location, a value of 1 is added to the sum of neighboring cells that belong to the same group as it before dividing by the total number of nonempty neighbors to get the percentage. If the cell is not on a valuable location, then the weight added is equal to 0. A cell's states can be summarized in Table 9.1.

In the following case studies, we show the segregation running in a 10×10 square grid that represents the community. As this represents a top-down view of a community, borders are not wrapped.

Our first case study is used to show the execution of a basic configuration, where a single cell belonging to group positive 1 (1) is around 3 group negative 1 (−1)

Table 9.1 Segregation model cell states.

Value	Meaning
0	Empty location
0v	Valuable location, empty
1	Group A
1v	Group A, valuable location
−1	Group B
−1v	Group B, valuable location

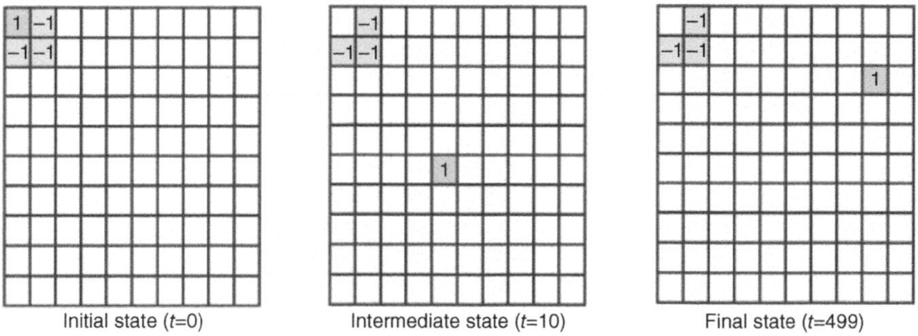

Figure 9.6 Experiment 1—single-cell infinite movement.

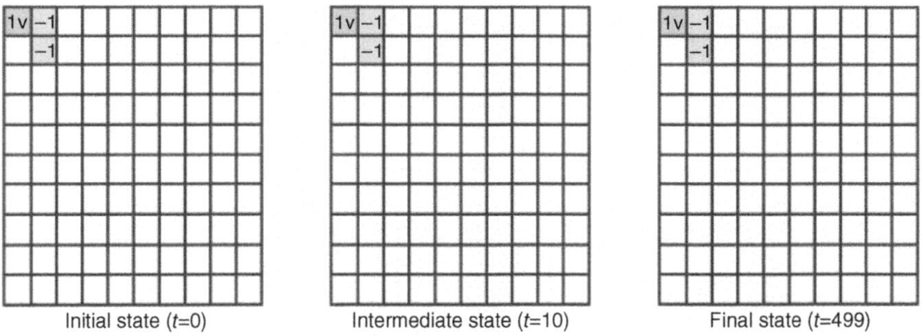

Figure 9.7 Experiment 3—remaining in valuable location.

cells with a 50% ratio (Figure 9.6). Based on the basic rules, the positive 1 cell must move, as it is a strong minority. After five time units, the cell moves to a random location. Since most of the grid is empty, the cell will have to move forever, as there are no suitable locations for it to stay long term.

Now, if we introduce one valuable location on the map (i.e., we have one positive 1 cell in a valuable location at a ratio of 50%, and we add weight to its decision to move), the positive 1 should want to move, as it is underrepresented. Because of the added value of the location, however, positive 1 will stay there until the spot below it is changed to negative 1. As can be seen in Figure 9.7, the negative 1 cells also do not move, as they are both in higher proportions in their respective neighborhoods.

Thomas Schelling's findings highlight that even in scenarios where groups are tolerant of one another, segregation can occur. That is, smaller tolerance ratios also lead to strong segregation within the residential space. The case study in Figure 9.8 shows different ratios of tolerance from the individual cells to observe if and how they end up segregating within the map. At a 50% tolerance ratio, the map is segregated.

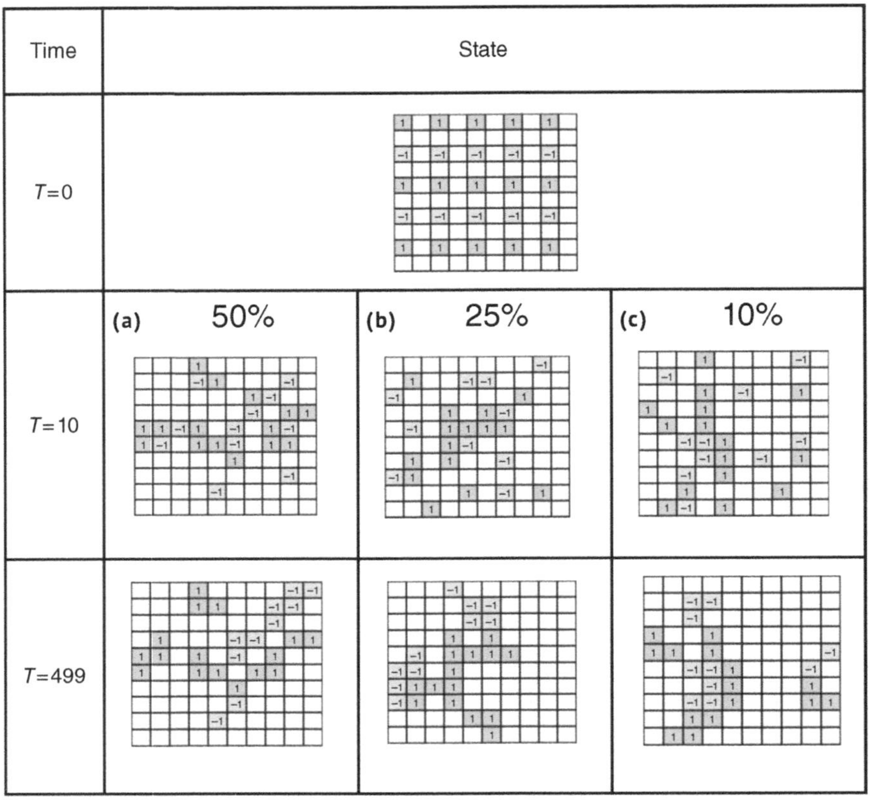

Figure 9.8 Segregation behavior. Initial state, (left) 50% tolerance ratio, (middle) 25% tolerance ratio, and (right) 10% tolerance ratio.

At a 25% ratio, there is no observable difference to the scenario with a 50% ratio. The final state in Figure 9.8b) looks even more segregated than the 50% ratio map, due to the randomness of the moves. Lastly, the 10% ratio seems to differ slightly from the other two. While cells are still segregated, there is much more distance between the groups that they end up forming, as can be seen in the final state of Figure 9.8c).

This residential segregation model shows how simple individual behaviors, when repeated across a population, can give rise to strong and persistent segregation patterns, showing that even mild preferences for similar neighbors can lead to large-scale spatial separation, even when initial conditions seem tolerant, as shown by Schelling's research. These experimental findings may help a policymaker decide which policies help to discourage segregation, building an adaptive model for this complex system. If it is true that even tolerant residents will contribute to segregating communities, a policymaker may make it a government policy to incentivize programs that provide opportunities for people with backgrounds that are in the minority. This could ensure that members of any existing segregated communities have equal opportunities.

9.3 Social Network

A social network model is a representation of social interactions among a group of people and how these interactions influence the people involved. There are many benefits that come with studying how and why people socially interact with each other. The benefits include but are not limited to understanding how information is spread throughout a group of people, how to discourage negative behavior (NB) in a formal setting, and even how advertisers should publicize their brand.

We can study these social systems by modeling them as social networks where we can run experiments on. Assuming the scenario being modeled is not dangerous due to other risks, a real experiment with people would be inaccurate, time-consuming, and costly. Real people would need to be compensated for every experiment, which would take a long time for preparation, execution, and data collection. Also, as we would study their social interactions, it would be difficult to know if the participants are behaving as they would normally do, or if they would be affected by the experiment. Although there are also drawbacks to using modeling and simulation techniques when modeling social interactions (e.g., it is difficult to account for unexpected human behaviors and reactions), this can be partially overcome by using traditional experimentation methods alongside modern modeling and simulation techniques.

Social systems are usually studied using agent-based or mathematical models. With agent-based modeling, the social system is represented by the interactions between agents, whose systems of rules could be designed in many ways [13]. Some examples of agent-based models for social systems include, for instance, Cohen et al. [14], which studies the role of social structure in maintaining social regimes. To be specific, they had adaptive agents play pairwise prisoner's dilemma and classified sustained cooperative behavior that occurs despite this behavior being costly in the short run as a cooperative regime. To study the evolutionary trait of cooperation without reciprocity, Riolo et al. [15] used agent-based modeling to show how such cooperation can occur by using a tag system. Agents were marked as having similar characteristics to each other with tags and were tasked with making costly donations to one another. This gave rise to cooperative behavior without needing any memory of previous donations.

Mathematical models for social systems use equations to describe the change of model's variables. Mathematical modeling for social systems gave rise to the discipline of sociodynamics, the method of modeling social phenomena. Weidlich [16] provides a systematic approach to mathematically modeling social phenomena in a quantifiable manner. Sorenson [17] gives an overview of well-established uses of mathematical modeling to study social systems.

With such complicated representations of social systems, it is important for them to be easily verifiable and reusable. Policymakers who oversee social settings like workplaces or social events put their trust in these models and do not want models that quickly become obsolete. The use of DEVS and Cell-DEVS helps to overcome these problems, as discussed in this section.

9.3.1 Employee Behavior

Employees' behavior can have a significant impact on the success of any organization, and studying the employees' behavior is important to lead them from a NB to a positive one [18]. It is not easy to study and see the employees' behavior from general mathematical models. However, cellular models have been used to simulate the employees' behavior, because each employee affects and is affected by its neighbors, and this process is self-reproducing. As such, a policymaker may use a cellular model to study the effectiveness of policies pertaining to the promotion or discouragement of behavior. The model may show whether the policies elicit positive or negative changes in the employees.

Based on a model proposed by Jiao et al. [19], our Cell-DEVS model considers three options: positive behavior (PB), the behavior that the organization wants to encourage; NB, the behavior the organization wants to avoid; and zero behavior (ZB), a behavior not encouraged or forbidden. Employees in the organization are organized in a cell space, where each employee is a cell. Distance between cells is not the physical distance but the distance in psychology and behavior. Each cell is influenced by its neighbors and at the same time influences others, which causes the evolution of employee behavior. The degree of influence an employee has on another is related to the distance between their cells.

In the cell space, every cell at the i, j position can be represented using three state values representing behavior at time t, namely $S^t_{i,j} = \{-1, 0, 1\}$, where 1 is the PB, 0 is the ZB, and -1 is the NB. Similarly, two characteristics related to the employees' behavior are modeled: influence and insistence. Influence represents the extent to which the employee affects their neighbors. Insistence is the degree at which the employee refuses to change their own behavior. High-insistence employees are rarely affected by their neighbors. In this way, each cell has two characteristics: influence $\text{INF}_{i,j} = \{1, 2, 3\}$ and insistence $\text{INS}_{i,j} = \{1, 2, 3\}$, and each characteristic has three degrees.

Different neighbor behaviors have a different influence on the cell. The cumulative influences of PB, NB, and ZB neighbors on one given cell are separately called positive, negative, and zero environmental disturbances (ped, ned, and zed) degree, formulated by

$$\text{ped}_{i,j}^{t} = \sum_{i'=i-2}^{i+2} \sum_{j'=j-2}^{j+2} \frac{\text{INF}_{i',j'}}{\sqrt{(i'-i)^2 + (j'-j)^2}}, \quad S_{i',j'}^{t} = 1, \tag{9.2}$$

$$\text{ned}_{i,j}^{t} = \sum_{i'=i-2}^{i+2} \sum_{j'=j-2}^{j+2} \frac{\text{INF}_{i',j'}}{\sqrt{(i'-i)^2 + (j'-j)^2}}, \quad S_{i',j'}^{t} = -1, \tag{9.3}$$

$$\text{zed}_{i,j}^{t} = \sum_{i'=i-2}^{i+2} \sum_{j'=j-2}^{j+2} \frac{\text{INF}_{i',j'}}{\sqrt{(i'-i)^2 + (j'-j)^2}}, \quad S_{i',j'}^{t} = 0, \tag{9.4}$$

where *i'* and *j'* are indices for the neighborhood of each cell, *i* and *j* represent the cell of interest. Similarly, the following local rules are defined:

(a) When $S_{i,j}^{t} = 1$, if $\text{ped} + \text{INS}_{i,j} = \max\{\text{ped} + \text{INS}_{i,j}, \text{ned}, \text{zed}\}$, then $S_{i,j}^{t+1} = 1$; else, if $\text{ned} > \text{zed}$, then $S_{i,j}^{t+1} = -1$; if $\text{zed} > \text{ned}$, then $S_{i,j}^{t+1} = 0$; if $\text{zed} = \text{ned}$, then $P\{S_{i,j}^{t+1} = 0\} = 0.5$, $P\{S_{i,j}^{t+1} = -1\} = 0.5$.

(b) When $S_{i,j}^{t} = -1$, if $\text{ned} + \text{INS}_{i,j} = \max\{\text{ned} + \text{INS}_{i,j}, \text{ped}, \text{zed}\}$, then $S_{i,j}^{t+1} = -1$; else, if $\text{ped} > \text{zed}$, then $S_{i,j}^{t+1} = 1$; if $\text{zed} > \text{ped}$, then $S_{i,j}^{t+1} = 0$; if $\text{ped} = \text{zed}$, then $P\{S_{i,j}^{t+1} = 0\} = 0.5$, $P\{S_{i,j}^{t+1} = -1\} = 0.5$.

(c) When $S_{i,j}^{t} = 0$, if $\text{zed} + \text{INS}_{i,j} = \max\{\text{zed} + \text{INS}_{i,j}, \text{ped}, \text{ned}\}$, then $S_{i,j}^{t+1} = 0$; else, if $\text{ped} > \text{ned}$, then $S_{i,j}^{t+1} = 1$; if $\text{ned} > \text{ped}$, then $S_{i,j}^{t+1} = -1$; if $\text{ped} = \text{ned}$, then $P\{S_{i,j}^{t+1} = 1\} = 0.5$, $P\{S_{i,j}^{t+1} = -1\} = 0.5$.

Any organization wants to encourage good behavior and to forbid NB. In order to do so, the following rules were proposed. To encourage the PB of employees, Formula 9.5 is used instead of Formula 9.2, and to punish the NB of employees, Formula 9.6 is used instead of Formula 9.3.

$$\text{ped}_{i,j}^{t} = \sum_{i'=i-2}^{i+2} \sum_{j'=j-2}^{j+2} \frac{\alpha\text{INF}_{i',j'}}{\sqrt{(i'-i)^2 + (j'-j)^2}}, \quad S_{i',j'}^{t} = 1, \tag{9.5}$$

$$\text{ned}_{i,j}^{t} = \sum_{i'=i-2}^{i+2} \sum_{j'=j-2}^{j+2} \frac{\beta\text{INF}_{i',j'}}{\sqrt{(i'-i)^2 + (j'-j)^2}}, \quad S_{i',j'}^{t} = -1, \tag{9.6}$$

where alpha is the PB factor and beta is the NB factor: $\alpha \in R$, $\alpha > 1$, and $\beta \in R$, $0 < \beta < 1$. These rules can be formally modeled as Cell-DEVS equations using a 3D model in which each state variable is associated with a different plane in the model, as seen in Figure 9.9. Plane 0 defines the employee behavior, plane 1 the influence characteristic, and plane 2 the insistence characteristic.

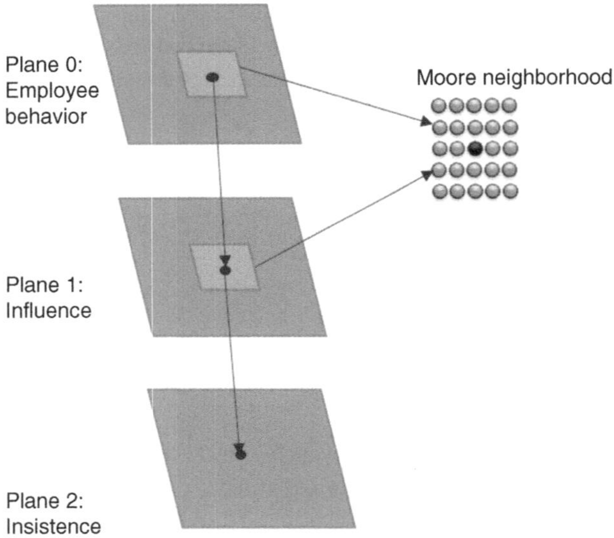

Figure 9.9 Neighborhood of cells in plane 0.

Each cell in plane 0 represents an employee that can have three states: −1, 0, and +1, corresponding to PB, ZB, and NB. Each cell in plane 0 uses an extended Moore's neighborhood with 25 cells on the same plane, plus a Moore's neighborhood with 25 cells in plane 1, and its corresponding cell in plane 2, as shown in Figure 9.9. Each cell in the influence plane (plane 1) can have values 1, 2, or 3, representing the degree of influence the employee has on its neighbors. A value of 1 means that the employee has little influence, 2 means it has medium influence, and 3 means it has major influence. Each cell in the insistence plane (plane 2) can have values 1, 2, or 3, and this value represents the degree of insistence the employee has (the capacity of being affected by other employees' behavior). A value of 1 means the employee is little affected by others, a value of 2 means they are moderately affected by others, and a value of 3 means they are highly affected by others.

The rules of the employee behavior model are defined such that only the cells in plane 0 can change state, as it represents the employee's behavior. The influence and insistence values for each employee (planes 1 and 2, respectively) are always the same. However, note that the influence and insistence of a cell are used to determine the employee behavior of that cell (in plane 0).

A cell space of 225 cells (15 × 15) was used to experiment with this model. As this number of cells corresponds to a small- to medium enterprise (SME). The initial values of the cells in Plane 0 were specified such that although the state of the cells is set randomly, there is a 1:1:1 proportion of PB, NB, and ZB states.

The insistence and influence planes were defined with the cell values 1, 2, and 3 distributed uniformly.

The diagram is shown in Figure 9.10. In plane 0 on the left, the 1 cells are set to PB, the 0 cells are set to ZB, and the −1 cells are set to NB. In the 2 remaining planes, red cells have the lowest value of 1, 2 cells have the next highest value of 2, and the 3 cells have the highest value of 3.

We show the results of three different experiments that demonstrate how the model executes under different assumptions. In the first one, no encouragement policy factors or punishment policy factors were considered. The state of the cell space at the end of the simulation period is shown in Figure 9.11. Note that the Influence and Insistence planes remain constant, but the employee behavior plane changes to mostly PB 1 cells, some ZB 0 cells, and no NB −1 cells.

The second experiment was simulated using equation *ped* as defined in Eq. (9.5) with $\alpha = 1.1$, while *ned* and *zed* are as defined in Eqs. (9.3) and (9.4), respectively. The resulting cell state space is shown in Figure 9.12. The circled area shows the new PB cells relative to the previous experiment.

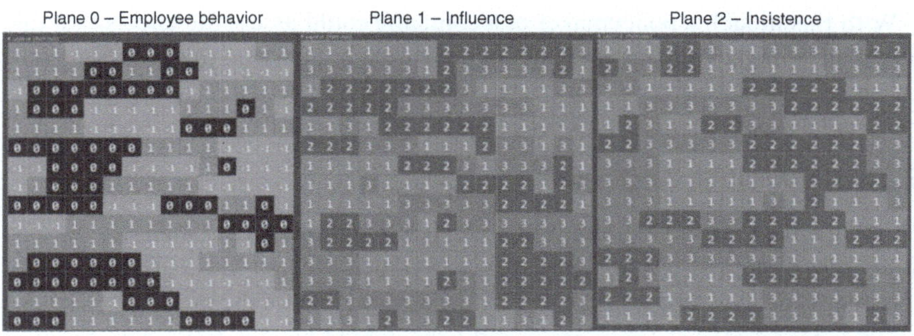

Figure 9.10 Initial values of the cell space. For color references, please refer to the online version.

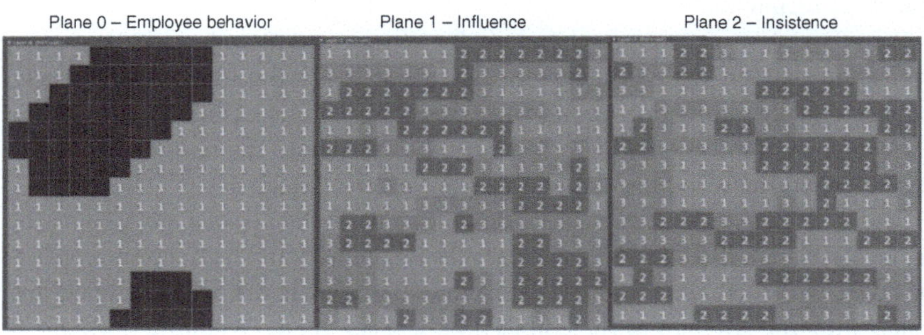

Figure 9.11 Final cell values after experiment 1.

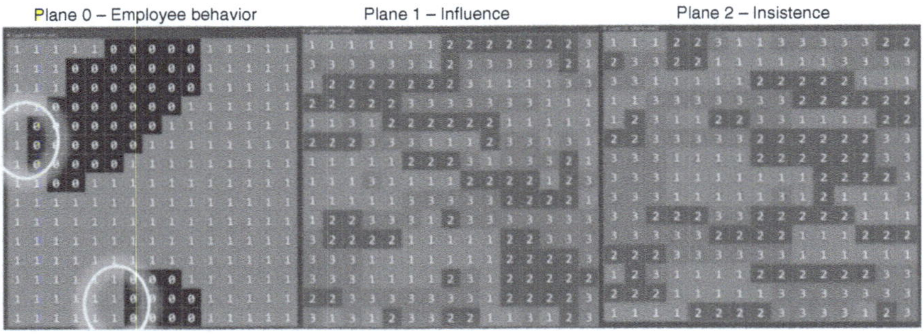

Figure 9.12 Final cell values after experiment 2.

Finally, an experiment was executed using *ped* as defined in Eq. (9.5) with $\alpha = 1.1$ and *ned* as defined in Eq. (9.6) with $\beta = 0.9$, while *zed* is used as defined in Eq. (9.4). In this test, we checked the effect of discouraging NB and encouraging PB simultaneously. For this case, there was not an improvement in the PB cells if we compare it with experiment number 2.

With this model, a policymaker of a company could assess the effects that different encouragement and punishment policies might have on the behavior of the company's employees.

9.3.2 Information Propagation

Human behavior can be difficult to understand and predict; as such, we need to study how people behave when they receive a message in a social group. Do they ignore the message or pass it along? If they do disseminate the message, to whom do they pass it along? These are useful behaviors to be able to predict, as some emergency situations could require the rapid propagation of information across a community. Or a malevolent agent may be spreading false information that could harm the recipients.

We adapted an existing model made by Wainer and Giambiasi [20] of virus spreading defined with Cell-DEVS [21]. There is a basic analogy (Table 9.2) between the dissemination of information among a group of individuals and the transmission of infectious disease between the individuals themselves. In the two cases, something is communicated through a given contact.

This DEVS model in Figure 9.13 is describing the influence of a message on the behavior of an individual and potentially its dissemination. The first state consists of being in contact with another agent in its social network neighborhood and calculating the strength of the connection between them. When the message is

Table 9.2 Analytical model parameters in epidemiology versus diffusion.

Symbol	Epidemiology	Information diffusion
N	Number of individuals	Number of individuals
$S(t)$	Number of susceptible individuals at time t	Number of susceptible individuals at time t
$I(t)$	Number of infectious individuals at time t	Number of individuals at time t who carry the information
X	Contacts per unit of time and per individual	Contacts per unit of time and per individual
β	The probability of transmission in a contact between an infective and a susceptible	The probability of transmission in a contact between an information holder and a susceptible
$a = \beta x / N$	Infection rate	Infection rate
$\alpha = a^{*}\mathrm{i}(t)$	Force of infection: The probability per unit of time for a susceptible to become infective	Transmission Strength: The probability per unit of time for a node to receive the information

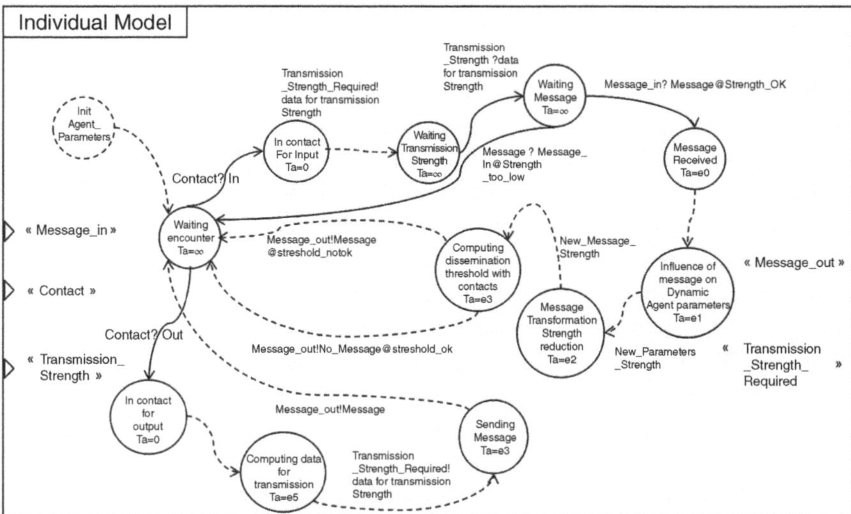

State Variables :

- Agent_Parameters [**Static agent parameters**{ethnic group, Language, Religion...}, **Dynamic agent parameters** {Maslow's needs, vulnerability, interest}]
- Social Network [Family, Ethnic Group, Religion and Conjunctural network]
- Geographical position [tribe, Village, City and remote place]

Figure 9.13 DEVS individual model.

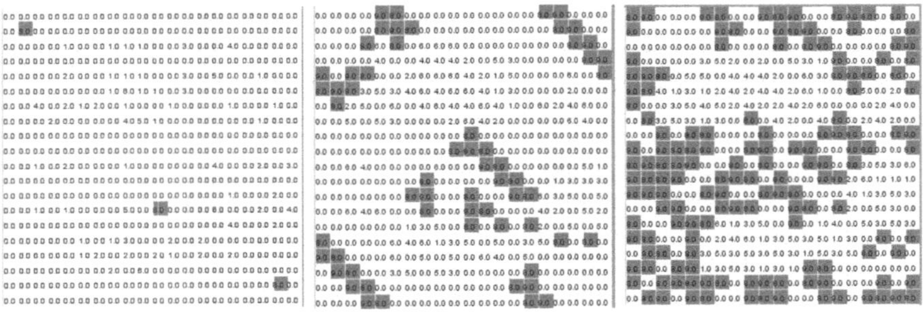

Figure 9.14 Execution result at time 00:00:00:000; 00:00:00:500; 00:00:01:000.

received, it creates an impact on the individual, which changes its behavior eventually depending on the strength of the message [22]. Then, if the message's strength is strong enough, the receiver will prepare on its turn to transmit the message to its network neighbors, considered as target info.

In the experiment we conducted, there is a 20×20 mesh of cells. Each individual, residing in (i,j) cell, is equipped with a number ranged from 0 to 9 indicating different parameters of the individual. Each cell is assigned one of two categories: one is a person who did not receive the information, represented by white cells; the other is the person who received the message, represented by shaded cells. The individual can be opposed to the information and/or the person that is transmitting the message and/or favorable to the emitter and message content. The simulation results show in the right picture of Figure 9.14 that, depending on the opinion of individuals and the configuration of the social networks, some individuals can be reached by the information whereas others are not. This test is very simple; it is based on the abstract geographical situation of individuals, and it considers only one dimension of a social network. We have not introduced yet the graph representation for defining the multilevel social network links between individuals or groups. Nevertheless, the approach already validates the possibility to reuse epidemic algorithms for information spreading. Figure 9.14 shows the model executions.

At the individual level, the DEVS model proposed is simply keeping raw Maslow parameters as state variables and being simply influenced by arriving messages. In its turn, the individual model can transmit the message after having potentially transformed its strength according to several criteria. The Cell-DEVS structure then uses these models of extremely simplified humans and starts building on top of its connected modules to form a Cell-DEVS network. The model shows the impact of human behavior regarding information perception and treatment. The information-spreading model reusing the epidemic approach seems accurate after some adaptations to propagate the information over the network.

9.4 Infectious Diseases

Infectious diseases can affect large swathes of the population. In those cases, we need to build adaptive models for this complex phenomenon to answer different questions, for instance: what is the best way to avoid being infected? What locations or events are at a higher risk of facilitating infections? How can businesses prevent outbreaks? How do we prevent it from spreading?

To answer these questions, epidemiology models can be used to predict how a disease will spread. Scale can vary from transmission inside a household, to over communities, and even entire countries. These models can be used to determine what policies can be implemented to mitigate the spread of disease while having minimal impact on the lifestyles of the population.

Different diseases spread in different ways, so the policies used for one may not be effective for another. For example, the avian flu is transmitted by people who have contact with living infected poultry, so we can limit contact by enforcing the use of personal protective equipment (PPE) and ensuring farms regularly test their livestock for disease. Whereas with a disease like cholera, where transmission is from consuming contaminated food or water, the best policy for reducing infections is to ensure the population has access to clean water and has good hygiene practices. The severity and strictness of the policy implemented depend on the dangers of the disease being mitigated. As such, it is necessary not only to study how the diseases spread but also their impact on the communities.

Traditional experiments to study the spread of the disease would either be too inaccurate to be worthwhile or too dangerous to ethically conduct. We must also consider how people behave differently, so it becomes much more difficult to tell if a policy would be effective from the results of traditional experiments, where people are biased by the experiment, generating inaccurate results. Although using modeling and simulation is difficult to fully represent the range of possible human behaviors derived from stress in potentially dangerous situations, experiments can be conducted on a model with any number of simulations, and different scenarios can be tested with greater speed and far fewer resources than if we used traditional experimentation, obtaining statistically significant results much faster and at a smaller cost than if we used traditional experimentation techniques.

Consequently, many different methods exist for modeling the spread of diseases. Some of the most popular methods are extensions of the foundational susceptible-infected-recovered (SIR) type of models [23], as well as Monte Carlo models [24] and the Wells-Riley model [25, 26].

In SIR models, people are considered to change between three states to represent how the population is infected. This uses an aggregated manner based on parameters that determine how values of each SIR compartment change over time, using parameters like transmission and recovery rates, which reflect disease-specific traits such as contagiousness and recovery time. These models operate at the population level and do not track individuals.

Over time, they have evolved to include additional compartments and spatial dynamics. SIR-type models have evolved, and now they include other states (e.g., exposed, deceased). Hethcote [27] introduced MSEIR and SEIR models, adding maternal immunity (M) and exposure (E) states. Barman et al. [28] applied a SEIR model with a social contact matrix to simulate COVID-19 spread in India, capturing age-specific interactions. Lin et al. [29] used SQIARD and SIARD models to forecast COVID-19 in several countries, incorporating quarantined (Q), asymptomatic (A), and deceased (D) populations. Sen et al. [30] employed a SIRD model to analyze time-series data across five countries, enabling simultaneous temporal analysis. Davidson et al. [31] explored spatial disease transmission using a geography-based Cell-DEVS model, accounting for movement restrictions and compliance behaviors.

Monte Carlo models simulate disease dynamics by feeding deterministic systems with randomly generated inputs based on known probability distributions. By comparing outcomes across many input sets, these models estimate the likelihood of specific events. Xie [32] applied this approach to forecast peak daily COVID-19 cases in Australia and the UK, producing results that aligned well with observed data and suggesting broader applicability. Similar methods have been used to study measles and influenza outbreaks [33], showing consistency with other modeling techniques. Foss and Carter [34] employed Monte Carlo simulations to assess the accuracy of lifetime risk estimates for Krabbe disease, aiming to refine future epidemiological studies.

The Wells-Riley model estimates the probability of airborne infection in well-mixed indoor environments, based on the emission rate of infectious particles. Aganovic et al. [35] adapted the model to assess how relative humidity and ventilation affect the removal of COVID-19 droplets, finding ventilation to be significantly more effective. Li et al. [36] extended the model using computational fluid dynamics to simulate pathogen movement indoors, demonstrating that effective ventilation combined with respiratory protection is essential for controlling the spread of COVID-19 variants.

In the rest of this section, we present three different epidemiology models defined using DEVS and Cell-DEVS and how policymakers can use them.

9.4.1 Avian Influenza

H5N1 is a highly contagious viral disease that spreads among poultry and may attack other species of animals as well. The virus first emerged in 1996 in Eastern Asia and has been spreading among birds' populations since then. Outbreaks caused by the virus resulted in animal health and economic crises worldwide.

The virus also attacks humans through direct contact with living infected poultry or animals, and in unusual cases, through human-to-human interaction [37]. When it attacks humans, the mortality rate becomes 60% [38]. Preemptive actions in terms of control and immunization can help to avoid the grave consequences of the virus spread. H5N1 has received attention from scholars as a potential pandemic threat. Due to the threats, consequences of the virus spread, and the unique nature of the virus, different attempts have been made to model the spread of the virus [39, 40]. The Cell-DEVS Influenza model we introduce is based on [41]. We model the transmission of the virus among birds living in the same area. We limit the discussion to birds for simplicity; however, including humans and other animals in the model is possible with modifications.

We model a population of 1225 agents (i.e., individual birds) represented as a 35 x 35 grid, where each cell of the grid represents a bird. We assume that the birds are residing in the same area, and we consider direct contact with other birds in neighboring cells. The specific states available for each cell depend on the scenario under study, being the basic ones susceptible and infected. The state of a given cell changes based on the state of the cell's neighborhood and a disease diffusion rate.

We study four different scenarios: (1) a basic case where we have only two states for the cells (i.e., infected and susceptible) and each cell is affected by a neighborhood of nine near neighbors; (2) the introduction of boundary cells to control the spread of the disease; (3) an expanded neighborhood consisting of 24 neighbors classified in two different proximities from the core cell; and (4) the expanded neighborhood along with three more states: questionable, immune, and dead.

In the basic scenario, we consider one bird per cell, and each bird can be in two states: susceptible (represented as 0) and infected (represented as 2). A susceptible cell has not been infected, but there is a probabilistic potential that it can be infected in the future, while an infected cell has already caught the virus and can transmit it. The neighborhood for the core cell is the Moore neighborhood, which includes the cell's nine near neighbors.

The change of states in Cell-DEVS is modeled in the transition function that can be written as a set of rules using CD++ notation as follows:

```
rule: 2 100 {(0,0) = 0 AND stateCount(2) > 0 AND random
             < diffusion_rate}
rule: {(0,0)} 100 {t}
```

Here, the first rule indicates that if the core cell is not infected ((0,0) = 0), and there is one or more neighboring cell infected (stateCount (2)), then the core cell will change based on the diffusion rate (random < "diffusion_rate") to infected (2) or will remain susceptible (0) after a given time (in this case, 100).

The second rule means that in all other cases the cell will keep its current value.

Figure 9.15 shows the progress of the simulation results with a diffusion rate of 0.3, where the susceptible cells are shown in light shade and the infected cells are dark. The simulation starts at time 0, where there are two cells initially infected (Figure 9.15a) and ends at time 4400 (Figure 9.15d), when there are no more events to execute. Figure 9.15b and c show the spread of the virus at times 100 and 1000, respectively.

A scenario with a control barrier shows the spread of the virus when the bird population is divided using barriers to control the spread. The cells representing the barriers have no birds, and they are always in the state barrier (represented by 5). Figure 9.16 shows how the virus spreads with a 0.5 diffusion rate at various times, ending at time 2300 (Figure 9.16d) when there are no more events to execute. The simulation results show how introducing a boundary area can stop the disease from spreading to the rest of the population.

We then considered an extended neighborhood with 24 neighbors in total, divided into two types as shown in Figure 9.17: 8 "adjacent neighbors," which are

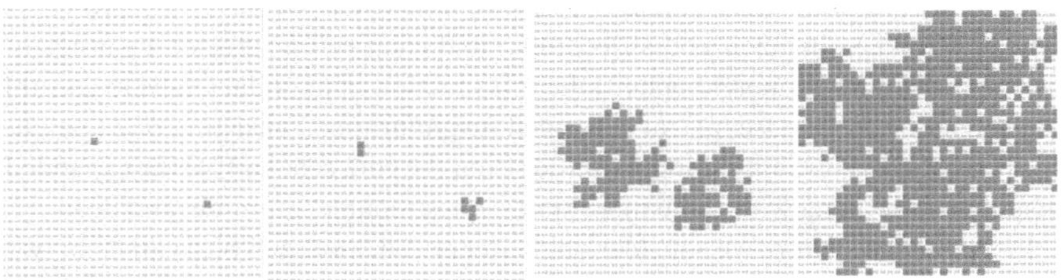

(a) (b) (c) (d)

Figure 9.15 Avian flu basic scenario: snapshots of the simulation at different times with diffusion rate 0.3. (a) time = 0 ms, (b) time = 100 ms, (c) time = 1000 ms, and (d) time = 4400 ms.

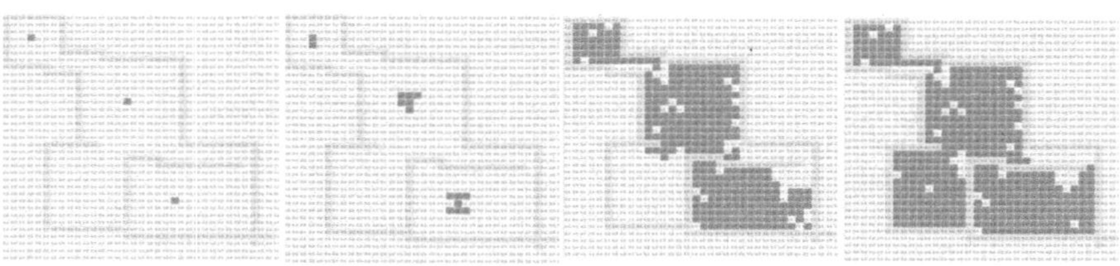

(a) (b) (c) (d)

Figure 9.16 Avian flu boundary scenario: snapshots of the simulation at different times with diffusion rate 0.5. (a) time = 0 ms, (b) time = 100 ms, (c) time = 1000 ms, and (d) time = 2300 ms.

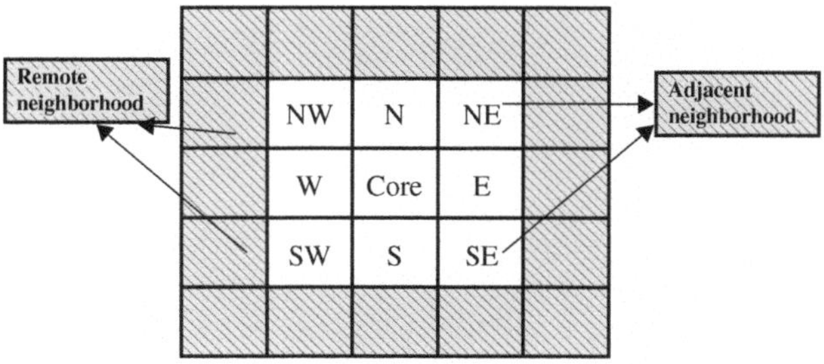

Figure 9.17 Avian flu: 5 × 5 neighborhood.

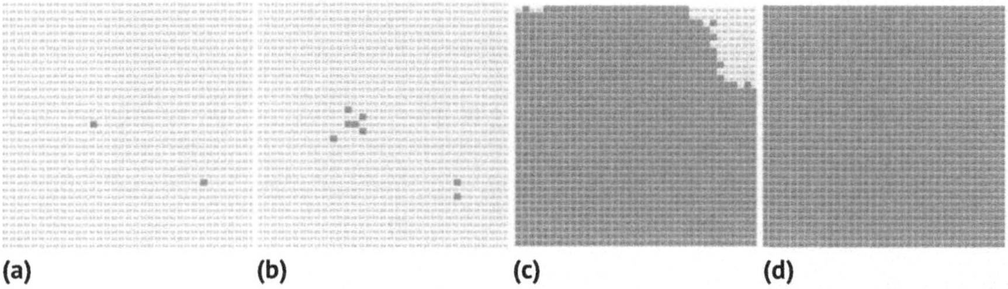

Figure 9.18 Advanced avian flu: diffusion rate calculated based on the outer and inner neighbors. (a) time = 0 ms, (b) time = 100 ms, (c) time = 400 ms, and (d) time = 1400 ms.

the cells next to the core cell, and 16 "remote neighbors," which are one cell apart (adjacent neighbors have more chances of infection than the remote ones).

Figure 9.18 shows the results when the new neighborhood is introduced. As shown in the figure, when introducing remote neighbors, the disease spreads faster compared to the basic model of Figure 9.15, where only adjacent neighbors influence infection.

We then consider a scenario with an extended neighborhood and additional states: questionable cell (represented as 1), immune cell (represented as −1), and dead cell (represented as 3). Questionable cells represent birds who may be showing symptoms of the flu. Immune cells are those who have received treatment and thus, cannot become infected. Lastly, dead are the birds who did not receive treatment, as we assume they will die due to the high mortality rate of H5N1. We assume for simplicity that the dead birds will not transmit the disease. The possibility of the virus being transmitted is calculated based on the number of infected cells in the neighborhood and the diffusion rate, where a higher probability of transmission is used when the infected cells are closer to the core cell. Birds may receive treatment after flu symptoms have been noticed, and the state of the cell becomes questionable. The probability that a cell's state transitions to questionable is determined by an equation that assumes an invigilation policy implemented as

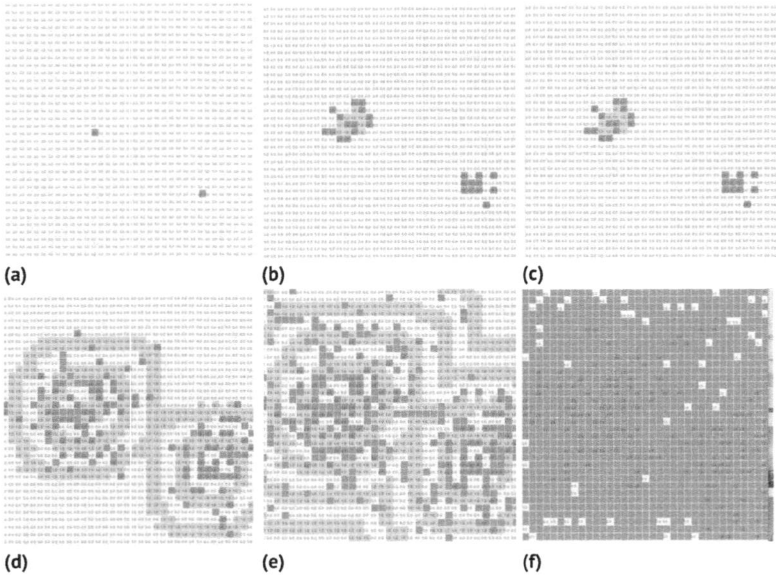

Figure 9.19 Avian flu: 24 cells neighborhood: immunization and vigilance of the inhabitants is considered after noticing symptoms. (a) time = 0 ms, (b) time = 100 ms, (c) time = 200 ms, (d) time = 600 ms, (e) time = 1000 ms, and (f) time = 12100 ms.

a certain probability of considering checking for symptoms in other birds in each of the cell's 24 neighbors' states and considering the status of neighboring cells.

Figure 9.19 shows the simulation results of the scenario assuming there are no deaths. In Figure 9.19a, the simulation starts with two cells initially infected (2.0). Then, in Figure 9.19b, the virus spreads, and some cells become questionable (1.0). In Figure 9.19d, immunization (−1.0) starts to appear. This limits the spread of the disease, as we can observe by comparing Figure 9.19f and Figure 9.18d. While in Figure 9.18d all the population is infected, Figure 9.19f shows a few infected individuals surrounded by the immune population.

Figure 9.20 shows the simulation results when considering death. The simulation starts with two infected cells (Figure 9.20a). Then, the disease starts to spread among the neighboring cells (Figure 9.20b). The dead cells (3.0) start to appear in Figure 9.20c at the center of the infected population. The dead cells are surrounded by questionable cells at this point in the simulation. Then in Figure 9.20d, some immune cells start to appear, but also the number of dead individuals increases. Then, gradually the number of immune cells increases at a faster rate (Figure 9.20e). At the end of the simulation (Figure 9.20f), no more questionable cells are present.

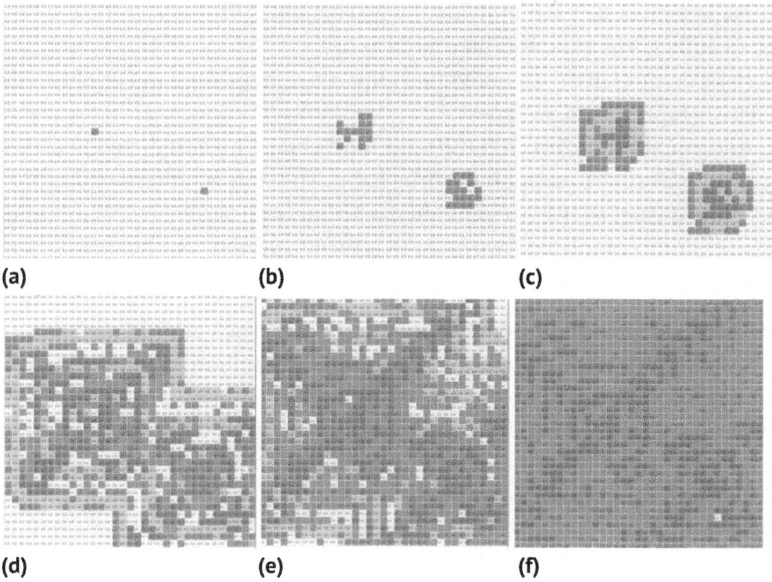

Figure 9.20 Avian flu: neighborhood of 24 cells: inhabitants who are not immunized die due to the dangerous flu. (a) time = 0 ms, (b) time = 100 ms, (c) time = 200 ms, (d) time = 600 ms, (e) time = 1000 ms, and (f) time = 4200 ms.

9.4.2 Effects of Addiction in Criminality

Within a community that is at an elevated risk of hard drug use, such as heroin or cocaine, addiction can spread similarly to any other disease. In the case of drug use, the spread occurs through social interactions, and much like a virus, through repeated exposure, it leads to addiction. Drug addiction has been identified as a major cause of different criminal acts committed for the sake of covering the cost of the expensive drugs that they are dependent on [42]. In this section we model the spread of addiction to hard drugs in a high-risk community. We include the probability that addiction can lead to criminal acts and consider how treatment can mitigate the spread of addiction.

We create increasingly complex Cell-DEVS models based on a previous model proposed by Dabbaghian et al. [43]. In later iterations of this model, we consider the incapacitation of drug users (i.e., treatment of drug users in hospitals), which can reduce drug usage but also criminal activities. Each cell contains an individual, and the cell value is the state of that individual. Each individual can be in one of the following five states depending on the scenario:

- Susceptible Person (SP): An individual who currently does not consume drugs but is vulnerable or likely to start taking drugs. The state is represented as "0."

- Low-risk Person (LRP): An individual who consumes drugs occasionally but is not an addict. The state is represented as "1."
- High-risk Person (HRP): An individual who consumes drugs and is addicted. The state is represented as "2."
- Drug Addict Criminal Person (DCP): An individual who is addicted to drugs and commits a crime to support the drug habit. The state is represented as "3."
- Incapacitated Person (IP): An individual who has been treated by doctors and has successfully completed treatment to return to society as a healthy member. The state is represented as "4."

The Cell-DEVS models we introduce analyze the influence of drug pattern/influence, peer associations, drug usage patterns, and criminality dependency. We will discuss three different cases: (1) a basic scenario where someone starts consuming drugs; (2) an addiction scenario where the drug consumer becomes addicted; and (3) a criminal scenario when an addict commits crimes to support their addiction and can receive treatment.

First, we consider two states: SP and LRP. When individuals start consuming drugs, they transition from SP to LRP based on the states of their neighbors. We use the Von Neumann neighborhood (i.e., we only consider the North (N), East (E), West (W), and South (S) neighbors) as we assume that a person does not get affected by all individuals surrounding them. Instead, everyone is influenced by those with whom they are in social contact [43].

The cell's behavior is dictated by the following three rules, expressed in CD++ notation:

```
rule: 1 200 {(0,0)=0 AND (0,-1)=1 AND (-1,0)=1 AND
            (0,1)=1 AND (1,0)=1}
rule: 1 200 {((0,0)=0 OR(0,0)=1) AND normal(0.4,0.3)
            > 0.7 }
rule: { (0,0) } 200 {t}
```

The first rule represents a cell becoming LRP (i.e., 1) if the four neighbors with whom the individual is in social contact consume drugs. In the second rule, the individual can randomly start consuming drugs based on a normal distribution randomization function. This second rule models a high-risk community where individuals could start consuming drugs even if not all their social peers are drug users. The third rule represents the default case, where the state of the individual remains unchanged.

Figure 9.21 depicts a simulation of the first model where 0.0 cells are SP and the 1.0 cells are LRP. The simulation starts with every cell as an SP individual, but as it is a high-risk community, there is a chance that any SP member may transition to an LRP member despite not having neighbors who are LRP. As more cells transition to LRP and surround SP cells, those SP cells transition to LRP cells.

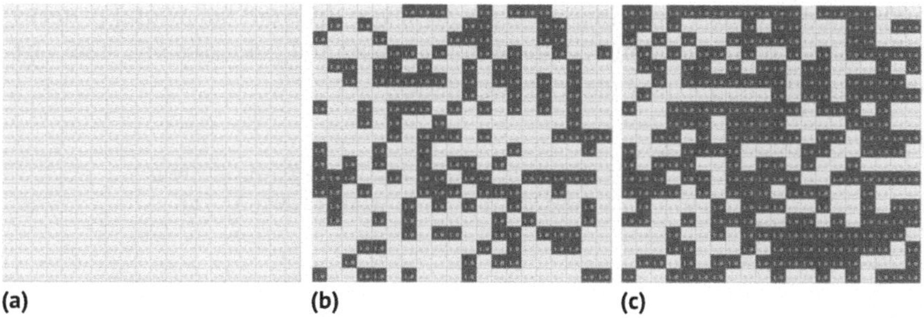

Figure 9.21 Drug addiction basic model: transitioning from healthy individual to drug consumers. (a) Initial placement, (b) mid-point, and (c) end of simulation.

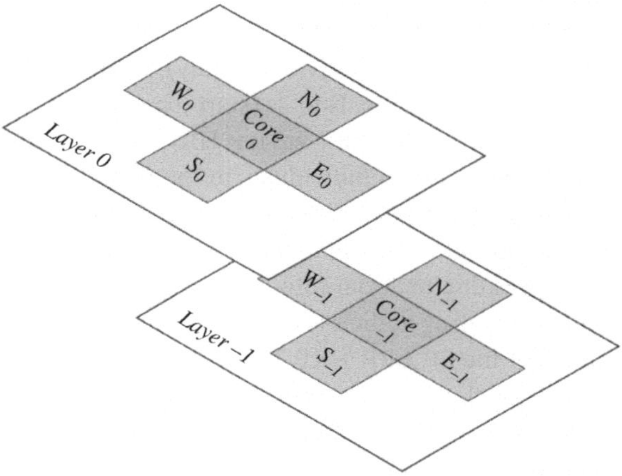

Figure 9.22 Drug addiction 3D model: two Layers neighborhood.

The results of the simulation therefore show how the SP members of a high-risk community become LRP as their social neighbors start consuming hard drugs and become LRP.

We then consider a scenario where a healthy person (SP) starts consuming drugs and changes state to LRP, and it can potentially become a drug addict (HRP). We introduce a three-dimensional Cell-DEVS model with two layers (Figure 9.22). The first layer is the same as in the previous scenario. However, with the second layer, we consider five neighbors. This means that the neighborhood of the core cell in the new layer has four neighbors from its layer, and the core cell of the layer immediately below it is a fifth neighbor. The cell in the first layer still only considers the original four cells as its neighbors, as there is no layer below it. In this way, a cell must be in the previous state for it to advance to the next state.

The model defines two different zones with two different behavioral patterns. Each zone is represented in one plane in the model and possesses one state variable. The rules for the new plane are as follows:

```
[DrugUser-transition]
rule: 2 150 {(0,0,0)=0 AND (0,0,-1)=1 AND (0,-1,-1)=1 AND
             (-1,0,-1)=1 AND (0,1,-1)=1 AND (1,0,-1)=1}
rule: 2 150 {(0,0,0)=2}
rule: 0 150 {t}
```

The new group of rules (DrugUser-transition) represents the behavior of a cell (individual) whose neighbors (social influencers) are drug users. If the neighbors in the other layer belong to the LRP group, this individual's state becomes state 2 (HRP). In other words, this individual becomes a drug addict.

Figure 9.23 shows the snapshots of a simulation's results. Figure 9.23a shows the first layer of the model (SP in 0.0 and LRP in 1.0 individuals only). Figure 9.23b shows the point in time when a few HRP individuals in 2.0 start to appear in the second plane. The yellow cells here indicate that they are not HRP, as this is the second layer. Figure 9.23c shows the state of the individuals in the first plane of the mode at the end of the simulation, where most of the population is in the LRP state while a small percentage of the population is still in the SP state. Figure 9.23d shows how this situation can progress due to social interaction with more drug consumers, and individuals can become drug addicts (HRPs).

We extended the model to consider the SP, LRP, HRP, DCP, and IP states, represented as a four-layer Cell-DEVS model. We model individuals who are initially in an SP state and can potentially transition to the four other states. Each of them is modeled in a different plane, and there are four transition functions for calculating the state of each variable in the cell using a 3D Von Neumann neighborhood with four layers. The variables of LRP, HRP, and IP stages are supplied to the

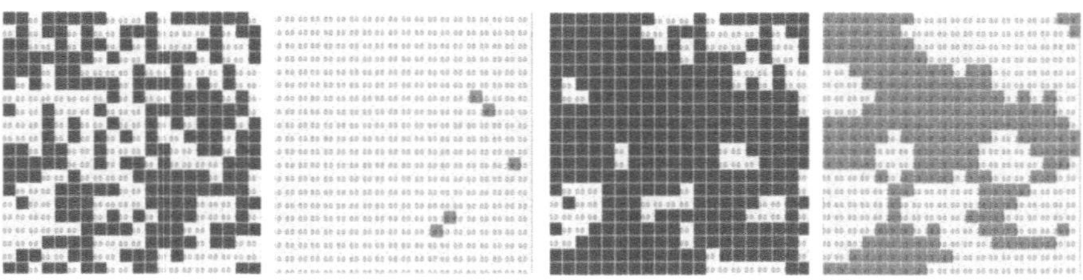

(a) (b) (c) (d)

Figure 9.23 Drug addiction model: transitioning from healthy individual to drug consumer, then to a drug addict. (a) Plane 0 at t = 450 ms, (b) plane −1 at t = 450 ms, (c) plane 0 at t = 1950 ms, and (d) plane −1 at t = 1950 ms.

model to simulate complicated drug patterns and crime rates in an area. To generate random drug behavior, in the beginning, the variables in different cells are changed from 0 to 1 following a normal probability distribution. Then cells, which are surrounded by the persons who consume drugs, are susceptible and will start to consume drugs.

The first new set of rules for criminal activities is responsible for switching the state of the cell to a DCP state if the core cell in layer −1 (relative to the neighborhood) is 2 (HRP), and at least one of its four direct neighbors in the same plane (N_{-1}, S_{-1}, E_{-1}, W_{-1}) is in the HRP state.

```
[crimedrug-transition]
rule: 3 100{(0,0,0)=0 AND normal(.4,.1) > .6 }
rule: 3 100 {(0,0,0)=0 AND (0,0,-1)=2 AND ((0,-1,-1)=2 OR
             (-1,0,-1)=2 OR (0,1,-1)=2 OR (1,0,-1)=2)}
rule: 3 100 {(0,0,0)=3}
rule: 0 100 {t}
```

The second group of new rules handles transitioning the cell to the IP state. This happens when the value of the core cell in layer −1 is HRP and the value of the core cell in layer −2 is IP using a random normal distribution generator.

```
[Incapacitation-transition]
rule: 4 100 {(0,0,0)=0 AND (0,0,-1)=3 AND (0,0,-2)=2
            AND normal(0.4,0.3)>0.1}
rule: 4 100 {(0,0,0)=4}
rule: 0 100 {t}
```

Figure 9.24 shows a snapshot of the simulation at the middle of the execution of the model for the four different planes representing the four transition functions: Figure 9.24a: transition from SP to LRP as explained in the basic scenario, Figure 9.24b: drugUser transition as explained in the second scenario, Figure 9.24c: crime drug transition as explained in this section, where black represents DCP, and Figure 9.24d Incapacitation transition as explained in this section, where IP is represented in white. However, at this time of the simulation, the last plane does not have any IPs yet.

Figure 9.25 shows the end of the simulation, where more individuals are in the LR state (low-risk addicts). In addition, many individuals transition into the HRP state (high risk). Finally, comparing the fourth plane of the simulation at a midpoint (Figure 9.25d) and at the end of the simulation (Figure 9.25d) shows the transition of more than half of the population to the IP state (individuals who have been incapacitated).

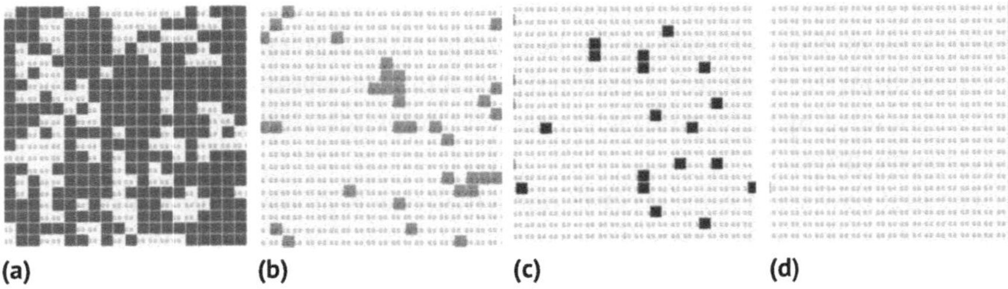

Figure 9.24 Crime drug model: midpoint of simulation. (a) Plane 0 at t = 400 ms, (b) plane 1 at t = 400 ms, (c) plane 2 at t = 400 ms, and (d) plane 3 at t = 400 ms.

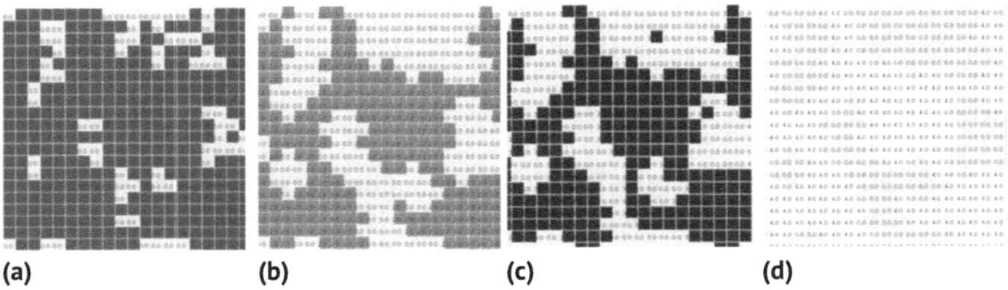

Figure 9.25 Crime drug model: end of simulation. (a) Plane 0 at t = 1300 ms, (b) plane −1 at t = 1300 ms, (c) plane −2 at t = 1300 ms, and (d) plane −3 at t = 1300 ms.

9.4.3 Airborne Disease Spread in Workplaces

Policymakers of a workplace are responsible for setting policies that can benefit the workplace, including disease mitigation policies such as wearing masks or setting a required minimum ventilation rate. However, implementing any policy comes with a cost. Thus, policymakers need to prove that these policies are worth the cost of implementing them. These policies can be evaluated by comparing the simulation results of models that depict the spread of airborne diseases with the policy in place and without. Unfortunately, to have such a model of their workplace, they would need to devote considerable resources to developing and evaluating the models.

To improve this process, we developed a new method. We built two parameterized abstract DEVS models that represent the fundamental parts of any model depicting the spread of an airborne disease in a workplace. These include the people spreading the disease and the rooms of the workplace they use. The models were designed so that a policymaker could use their own data to instantiate a model of their workplace without needing to develop any new DEVS models.

To develop these abstract models, we used an architecture that determines how the models would interact, adapting the one presented in [44] to study complex diffusion processes like this one. In our case, the diffusion process represents the disease being spread through a population of people. The original architecture includes many abstract components that can be used to help design complex diffusion processes. We adapted the architecture by instantiating the components we needed for our system of interest. First, we used the node component, which represents the medium that carries the element, to create our model of a person, as it is people that carry the disease. In the architecture, there are direct and indirect links between nodes that describe how the element may spread between nodes. An indirect link represents a connection that is only possible because of some intermediate object or device, whereas a direct link occurs without any assistance. In the context of airborne diseases, a direct link is represented by two people sharing an airspace, like in an enclosed room, whereas an indirect link might represent a person sneezing on their hand, touching an object, and then another person later touching that same object. We chose to focus on airborne diseases, where the chance of indirectly spreading the disease is negligible. Therefore, we instantiated the direct link component as our model of a room. Also, we used the node updater component, which can change properties of the nodes during a simulation, to depict the daily changes in the weather, as this can influence a person's decisions as they move about a large workplace with outdoor paths.

In any workplace, there may be various interconnected rooms and people traveling between them. As such, we designed our person and room models to be instantiated as agents with unique parameters that dictate their behaviors and characteristics that influence the spread of the disease. To design the person agents, we determined what behaviors each person might perform to influence the spread of the disease. We defined a variety of behaviors within the model that people can exhibit, including wearing masks, social distancing, entering a room at maximum occupancy, and traveling on a route that is more or less likely to expose them to the virus. The first three behaviors are determined by each person's relationship status with every other person in the model (as we assume that individuals will act differently around a close friend than around a stranger). Every person's behavior is defined as a DecisionMakerBehaviour object using an XML file. This model also includes the person's connections. The model can be easily extended to account for any additional behaviors a policymaker would want to include. They would only need to make minor changes to the models.

For the room model, we also defined each room in the workplace's model with a list of unique characteristics that could influence the chances of a person becoming exposed to the virus. These are characteristics like the ventilation rating, room dimensions, and the room's maximum occupancy.

With the person and room models connected, we can model the spread of an airborne disease if no external factors affect the behavior. We take it a step further and consider the weather as a possible external factor on each person's behavior.

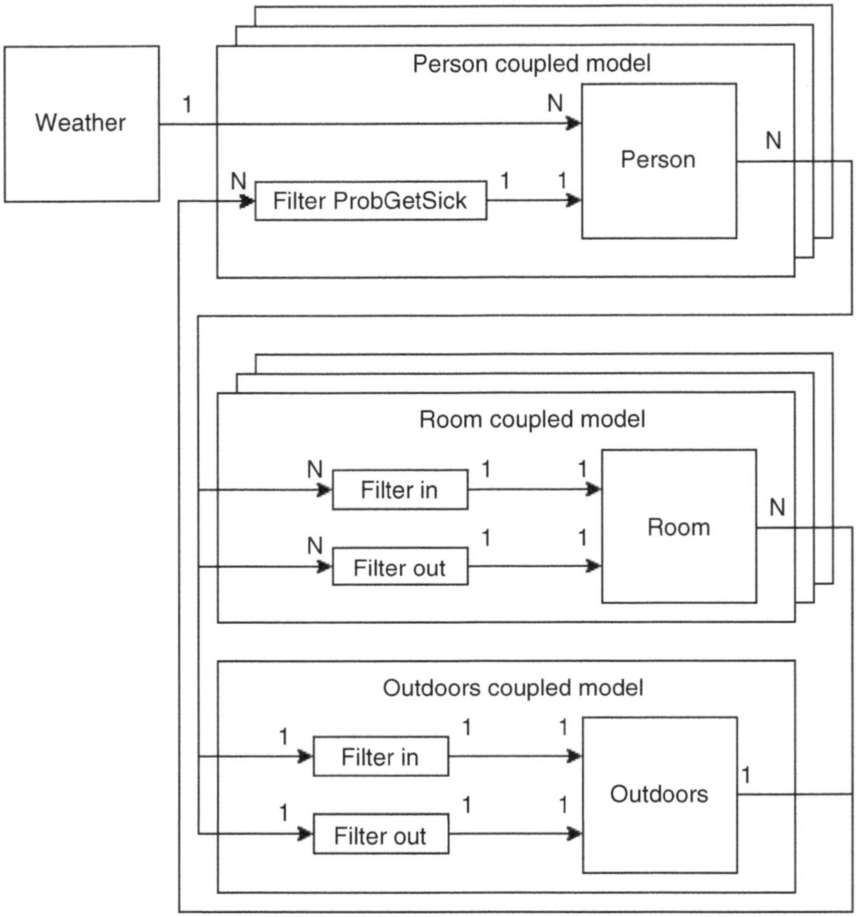

Figure 9.26 DEVS top model.

To be specific, the weather influences each person's decision to take a safer route by moving outdoors (where they are much less likely to be exposed to the virus but are exposed to the weather), or to take cover from the weather and risk being exposed to the virus while moving to their destination. The interconnected model can be seen in Figure 9.26, where each component is represented as a DEVS model.

The outdoors model is an instance of the room model where the probability of becoming exposed to the virus is always zero, as we consider the probability of becoming exposed to the virus while traveling outdoors to be negligible. The filter models do not contribute to the behavior of any models; they just facilitate message passing between many instances of each model. These DEVS models were implemented in C++ with the cadmium header library. Each model was unit tested, and then integration testing was performed on the top model. With these models, we have created a tool set for policymakers to study the spread of an airborne disease in their workplace. It simplifies the process, as policymakers who do not know how to program only need to compile the data necessary to instantiate the models at their workplace. If a policymaker has a basic understanding of C++ and the formalism, they could easily add new features such as additional behaviors to the person model.

Here, we are not interested in creating an accurate model for these experiments; instead, we wanted to show how a policymaker would use this method to create a model of their workplace and evaluate different policies. Thus, we made arbitrary choices with the goal of getting reasonable results where we did not have the necessary data to accurately instantiate the model. With that in mind, we constructed a model of COVID-19 spreading through the student population of Carleton University's campus. We obtained information for 30 rooms on Carleton's campus and randomly generated student schedules based on our personal knowledge of an undergraduate engineering student's daily schedule. Carleton has a unique feature in that it has an underground tunnel network that connects every building on Carleton's campus, such that a person does not need to travel outdoors to walk across campus. We represented the tunnel system with an arbitrarily large room with an average ventilation rating. We then conducted experiments on two different policies: (1) high ventilation rating for every room in the workplace; and (2) not allowing sick people to attend university campus.

We first study the probability of exposure to the virus if a policy requires every room to have high-quality ventilation. To analyze the policy, we ran two simulations, one where we set every room to have a ventilation rating of just 1 air change per hour (ACH), and another one where we set every room to have a ventilation rating of 8 ACH. Figure 9.27 shows the results of these experiments as the average probability of exposure for each room in comparison to that room's ventilation rate in L/min. The ventilation rate is calculated as the ventilation rating in ACH converted to air changes per minute and multiplied by the room's volume in liters.

It can be seen how in the graph on the right, where the ventilation rate of each room is much lower, there is around a 5% increase in the average probability of exposure in comparison to the graph on the left, where the ventilation rate in each room is much higher. The dot in the bottom right corner of both graphs is the tunnels, which remained constant.

We also studied how requiring students who are sick with COVID-19 to stay away from Carleton's campus affects the spread of the disease. We ran simulations where students would return to campus if they became sick and other simulations

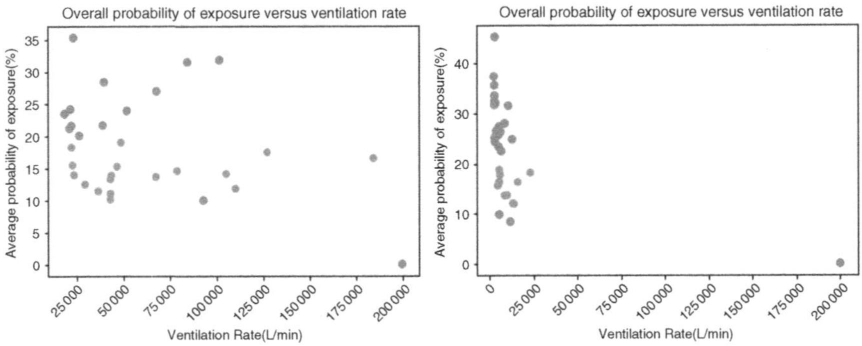

Figure 9.27 Average probability of exposure generated in each room versus ventilation rate per room. (Left) ACH value = 8. (Right) ACH value = 1.

where students would not arrive on campus if they became sick. Unsurprisingly, the results show a decrease in the probability of exposure to the virus by about 4%.

With these results, a policymaker could weigh the benefits of implementing each policy against its costs. If the costs seem too high for the benefits received, then a policymaker could change values for the Person and Room models that pertain to the policy being evaluated and run new experiments to see if these changes result in a more favorable result. This process can be repeated any number of times until the policymaker finds a disease mitigation policy that is effective for use in their workplace.

9.5 Conclusion

In this chapter, we showed how the DEVS and Cell-DEVS formalisms can be used to create well-tested and easily adaptable models for designing resilient and adaptable complex systems. To summarize our discussion, we showed how DEVS can ease the modeling process with its modular and hierarchical properties. These properties make it easier to perform unit testing and integration testing of DEVS models. And they, in turn, make it easier to run efficient simulations of new systems at a reduced cost. We illustrate these benefits by explaining how researchers have built DEVS models for city planning, epidemiology, and social systems.

With such models, policymakers could run experiments that help them determine which policy would be better suited for handling different situations. For instance, in the case of the spread of diseases, policymakers are interested in finding mitigation policies that limit its spread over a community. With the method we presented, a policymaker could execute such a model without the need to develop any new models with limited experience in programming or modeling.

DEVS allows for developing well-tested and adaptable models of complex systems. We showed multiple DEVS models of complex systems, such as epidemiology models that are used to study the spread of disease, social system models that allow researchers to study complicated human interactions, and city-planning models that are used to study how a city will change over time.

These properties make the DEVS formalism a good choice for building resilient, complex, and adaptive systems. DEVS is great for building such systems that are highly modular and reusable, while still remaining easy to verify and validate.

References

1 B. P. Zeigler, A. Muzy, and E. Kofman. *Theory of Modeling and Simulation.* 3rd edition, Academic Press, 2019.

2 A. C. Chow and B. P. Zeigler. Parallel DEVS: A parallel, hierarchical, modular modeling formalism. In *Proceedings of the Winter Simulation Conference*, pages 716–722. IEEE, 1994.

3 G. A. Wainer, Q. Liu, and S. Jafer. Chapter 2: Introduction to the DEVS modeling and simulation formalism. In Pieter Mosterman, editor, *Discrete-Event Modeling and Simulation*. CRC Press, 2011.

4 K. Bhattacharjee, N. Naskar, S. Roy, and S. Das. A survey of cellular automata: Types, dynamics, non-uniformity and applications. *Natural Computing*, 19:433–461, 2018.

5 Gabriel A. Wainer. *Discrete-Event Modeling and Simulation: A Practitioner's Approach.* CRC Press, 2009.

6 A. Lopez and G. Wainer. Improved cell-DEVS model definition in CD++. *ACRI*, pages 803–812, 2004.

7 L. Belloli, D. Vicino, C. Ruiz-Martin, and G. Wainer. Building DEVS models with the cadmium tool. In *Proceedings of the Winter Simulation Conference*, pages 45–59, 2019.

8 Pascal Perez, Arnaud Banos, and Chiristopher J. Pettit. Agent-based modelling for urban planning current limitations and future trends. *Lecture Notes in Computer Science*, Springer, 2017.

9 G. A. Jaoude, O. Mumm, and V. M. Carlow. An overview of scenario approaches: A guide for urban design and planning. *Journal of Planning Literature*, 37:467–487, 2022.

10 P. M. Torrens. Cellular automata and multi-agent systems as planning support tools. In S. Geertman and J. Stillwell editors. *Planning Support Systems in Practice*, pages 205–255. Springer, 2003.

11 ChengHe Guan and Peter G. Rowe. Should big cities grow? Scenario-based cellular automata urban growth modelling and policy application. *Frontiers of Architectural Research*, 5(3):248–262, 2016.
12 T. C. Schelling. Dynamic models of segregation. *The Journal of Mathematical Sociology*, 1(2):143–186, 1971.
13 F. Bianchi and F. Squazzoni. Agent-based models in sociology. *Wires Computational Statistics*, 7(4):284–306, 2015.
14 M. D. Cohen, R.L. Riolo, and R. Axelrod. The role of social structure in the maintenance of cooperative regimes. *Rationality and Society*, 13(1):5–32, 2001.
15 Rick L. Riolo, Michael D. Cohen, and Robert Axelrod. Evolution of cooperation without reciprocity. *Nature*, 414:441–443, 2001.
16 Wolfgang Weidlich. *Sociodynamics: A Systemic Approach to Mathematical Modelling in the Social Sciences*. Dover Publications, 2006.
17 A. B. Sorenson. Mathematical models in sociology. *Annual Review of Sociology*, 4:345–371, 1978.
18 Mark J. Somers. Ethical codes of conduct and organizational context. *Journal of Business Ethics*, 30(2):185–195, 2001.
19 Y. Jiao, S. Sun, and X. Sun. Simulation of employee behavior based on cellular automata model. In *Computational Science – ICCS*, pages 134–137, 2007.
20 Gabriel A. Wainer and Norbert Giambiasi. Cell-DEVS/GDEVS for complex continuous systems. *SIMULATION*, 81(2):137–151, 2005.
21 Gabriel Wainer. Applying cell-DEVS in modeling of complex systems. In *Discrete-Event Modeling and Simulation: Theory and Applications*. CRC Press, 2014.
22 C. Faucher, G. Zacharewicz, A. Hamri, and C. Frydman. PSYOPS and CIMIC operations: from concepts to G-DEVS models. In *Proceedings of the 2012 Symposium on Theory of Modeling and Simulation - DEVS Integrative M&S Symposium*, 42, pages 1–8, 2012. https://www.researchgate.net/publication/235423926_PSYOPS_and_CIMIC_operations_from_concepts_to_G-DEVS_models
23 W. O. Kermack and A. G. McKendrick. A contribution to the mathematical theory of epidemics. In *Proceedings of the Royal Society A*, 115(772):700–721, 1927. https://doi.org/10.1098/rspa.1927.0118.
24 Nicholas Metropolis and Stanislaw Ulam. The Monte Carlo method. *Journal of the American Statistical Association*, 44(247):335–341, 1949. https://doi.org/10.1080/01621459.1949.10483310.
25 W. F. Wells. *Airborne Contagion and Air Hygiene*. Harvard University Press, Cambridge, MA, 1955.
26 E. C. Riley, G. Murphy, and R. L. Riley. Airborne spread of measles in a suburban elementary school. *American Journal of Epidemiology*, 107(5):421–432, 1978.
27 Herbert W. Hethcote. The mathematics of infectious diseases. *SIAM Review*, 42(4):599–653, 2000. http://doi.org/10.1137/S0036144500371907.

28 M. Barman, S. Nayak, M. K. Yadav, S. Raha, and N. Mishra. Modeling control, lockdown & exit strategies for COVID-19 pandemic in India. *medRxiv*. 2020. https://doi.org/10.1101/2020.07.25.20161992.

29 B. Lin, Y. Chen, Y. Hung, C. Chen, H. Wang, and J. Chern. The data forecast in COVID-19 model with applications to US, South Korea, Brazil, India, Russia, and Italy. *arXiv*, 2021. https://doi.org/10.48550/arXiv.2011.04738.

30 Pei Lijun and Mengyu Zhang. Long-term predictions of COVID-19 in some countries by the SIRD model. *Complexity*, 2021:18, 2021.

31 G. Davidson, A. Fahlman, E. Mereu, C. R. Martin, G. Wainer, P. Dobias, and M. Rempel. A methodological approach for modeling the spread of disease using geographical discrete-event spatial models, *Simulation*, 100:39–70, 2023.

32 Xie. G. A novel Monte Carlo simulation procedure for modelling COVID-19 spread over time. *Scientific Reports*, 10:13120, 2020.

33 P. D. O'Neill, D. J. Balding, N. G. Becker, M. Eerola, and D. Mollison. Analyses of infectious disease data from household outbreaks by Markov chain Monte Carlo methods. *Journal of the Royal Statistical Society Series C: Applied Statistics*, 49(4):517–542, 2000.

34 Duffner Foss and Carter. Monte Carlo methods to estimate lifetime risk at birth of Krabbe disease. *Genetics in Medicine*, 15(12):997–1005, 2013.

35 A. Aganovic, Y. Bi, G. Cao, F. Drangsholt, J. Kurnitski, and P. Wargocki. Estimating the impact of indoor relative humidity on SARS-CoV-2 airborne transmission risk using a new modification of the Wells-Riley model. *Building and Environment*, 205: 108278, 2021. https://doi.org/10.1016/j.buildenv.2021.108278

36 X. Li, D. Lester, G. Rosengarten, C. Aboltins, M. Patel, and I. Cole. A spatiotemporally resolved infection risk model for airborne transmission of COVID-19 variants in indoor spaces, *Science of the Total Environment*, 812:152592, 2022.

37 M. C. Paul, T. Vergne, P. Mulatti, T. Tiensin, and I. Iglesias. Editorial: Epidemiology of avian influenza viruses. *Frontiers in Veterinary Science*, 6:1–2, 2019.

38 WHO. Avian influenza A(H5N1) virus, 2025. https://www.who.int/teams/global-influenza-programme/avian-influenza/avian-a-h5n1-virus (Accessed May 9, 2025).

39 Ansgar Kirchner and Andreas Schadschneider. Simulation of evacuation processes using a bionics-inspired cellular automaton model for pedestrian dynamics. *Physica A*, 312(1–2):260–276, 2002.

40 A. W. Roddam. Mathematical epidemiology of infectious diseases: Model building, analysis and interpretation. *International Journal of Epidemiology*, 30(1):186–186, 2001.

41 Hokky Situngkir. Avian flu: A cellular automata approach. *Working Paper, Bandung Fe Institute*, 2004.

42 K. Soothill, B. Francis, and R. Fligelstone. Patterns of offending behaviour: A new approach, 2002. https://www.ojp.gov/ncjrs/virtual-library/abstracts/patterns-offending-behaviour-new-approach.

43 Vahid Dabbaghian, Alexander Rutherford, Valerie Spicer, Suraj K. Singh, PeterBorwein, and Patricia Brantingham. The social impact in a high-risk community: A cellular automata model. *Journal of Computer Science*, 2:238–246, 2011. https://doi.org/10.1016/j.jocs.2011.05.008.

44 C. Ruiz-Martin, G. Wainer, and A. Lopez-Paredes. Discrete-event modeling and simulation of diffusion processes in multiplex networks, *ACM Transactions on Modeling and Computer Simulation*, 31:1–32, 2018.

10

A Framework for Characterizing, Understanding, and Predicting Behavior in Complex Edge Computing Systems

Kevin S. Chan[1]* *and Frank T. Johnsen*[2]

[1] *U.S. Army Combat Capabilities Development Command Army Research Laboratory, DEVCOM ARL, Adelphi, MD, USA*
[2] *Norwegian Defence Research Establishment (FFI), Kjeller, Akershus, Norway*

*Corresponding author: kevin.s.chan.civ@army.mil

10.1 Introduction

As of today, there is no agreed definition of "edge computing," as different sources describe and define this term in different ways, depending on their perspective. The main purpose of edge computing is to improve time-critical and/or data-intensive applications, such as industrial automation, real-time traffic analysis, smart homes, and the internet of things (IoT) [1, 2]. This means that edge computing is about performing computing at or near some edge, where the edge is often understood as the interconnection between two networks.

Seen from a communication perspective, the term edge computing is often used to refer to multi-access edge computing (MEC) [3] or fog computing [4]. MEC is a realization of edge computing, which is linked to one specific communication technology, namely mobile technology. Fifth-generation mobile technology (5G) is the first version of mobile technology to support edge computing [5].

In this work, we consider edge computing in two distinct but related variations: the edge computing network (edge network) and the cloud-supported edge computing network (cloud-edge network), with the obvious distinction that edge networks must function without cloud support. Edge networks are groups of computing devices of varying capacities and are tasked with processing data at the edge, and they are isolated from any other external resources. Edge network computing platforms can function independently or coordinate with other edge computing platforms to collaboratively execute the tasks. The systems are connected via some peer-to-peer or ad hoc network capability that creates a network independent of any backhaul or external management entity. Cloud-edge networks are edge networks with access to cloud resources to augment data processing capabilities and for resource coordination/management. We assume that cloud resources are a vast set of resources physically separated from edge networks. If resources are available and job requirements allow, then edge nodes might leverage cloud resources for data processing. Figure 10.1 illustrates the distinction between the edge networks and cloud-edge network types.

This work takes an edge network perspective, so therefore, we assume that cloud-edge networks comprise both the cloud and fog networks. Edge computing and cloud services are two important computing concepts that complement each other, rather than replace each other. The interaction between the two concepts appears relevant to various civilian applications and military operations. Edge computing makes it possible to process information quickly and continuously where the users are located, even during periods when the cloud is not available. Further processing of data can be done by sending it to the cloud, which is better suited for heavy data analysis and long-term storage. Understanding edge computing in this way, independent of a specific communication technology but seen from the need to use specific software services, can be used to describe edge

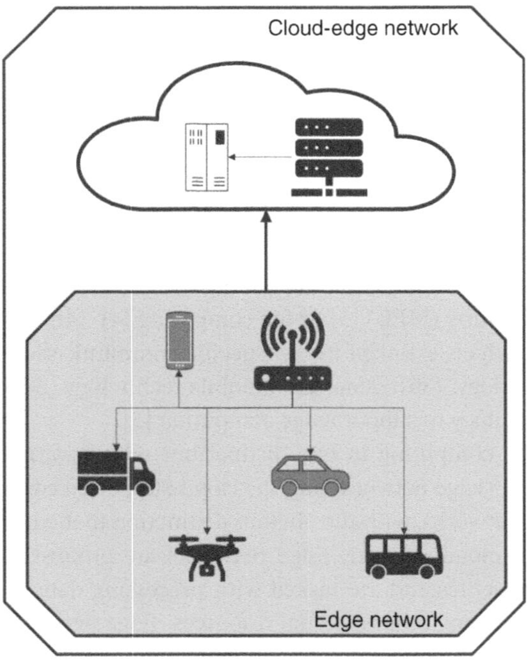

Figure 10.1 Illustration of cloud-edge and edge networks.

computing as a concept for delivering services: "Edge computing is the extension of cloud computing where the computing services are brought closer to the end users at the edge of the network" [2].

The edge implies that resources are not necessarily abundant as they might be when considering cloud computing or core/enterprise networks. Performance may also be hindered by limitations in bandwidth, data sharing issues or policies, and dynamics of the systems. The main challenge of edge networks is focused on providing the best possible performance—maximizing performance and reliability despite challenges caused by complexity and dynamics of the environment, and spatially distributed deployment of the involved systems and networks.

10.2 Motivation

Edge networks, as resilient complex adaptive systems (RCASs), can be applied to a wide range of applications and scenarios, including smart homes, military networks, and disaster response. These systems should exhibit resilience and adaptability, key characteristics of RCAS, enabling them to function effectively in diverse and dynamic environments. Refs. [1, 2] cite varied application settings, including smart cities, industrial operations, environmental monitoring, and smart utilities. Some common examples of edge computing are with mobile computing with smartphones, where smartphones offer substantial computational capabilities and can perform a wide range of applications on their own. Another example is smart homes, where home automation senses activities or events in homes and makes control decisions (e.g., regulate thermostat, unlock doors, turn on floodlights). These applications can be run locally if sufficient computing is available. A more dynamic and critical situation for edge computing is disaster response operations, where local infrastructure may be partially or fully destroyed as a result of a natural disaster or another catastrophic event. First responders need to set up local computing and communications capabilities for temporary command and control (C2) and information processing for quick, local responsiveness and situation awareness.

Today, even relatively small devices can offer significant computing power. This computing power can be used internally in edge nodes to make it possible to use advanced algorithms, such as artificial intelligence (AI), close to where the users are located, so-called *Edge AI* (see, e.g., Refs. [6, 7]). A key limiting factor is resource constraints of the computing platforms and bandwidth-limited nature of the networking available for reach back to the cloud or other locations with more available and reliable computing resources. Smartphones are unable to run large AI models like ChatGPT4, as the model is over 500 GB, beyond the current memory capacity of common mobile devices [8].

However, we can also describe edge computing as a distributed system where the edge nodes cooperate to utilize the overall processing, storage, and communication capacity in and between the edge nodes in such a way that it is possible to perform larger calculation tasks or store larger amounts of data close to the users. Edge computing can thus help give users the opportunity to solve tasks that previously required access to external resources, such as cloud services, to solve them. Other challenges in these situations include some system or information security issues, including data privacy, where sending information outside of where the data is generated is a risk. For example, smart home video feeds may contain sensitive personal information and may become vulnerable if the data is stored in the cloud. Perhaps the most difficult challenge involves complex systems operating in a dynamic environment where time is of critical importance. In situations such as disaster response, critical decisions must be made rapidly using available information and deploying available resources to meet needs as best as possible.

In addition to these challenges, it is important to consider varied configuration and deployment of these edge computing systems such that they are appropriate for the deployment environment. It is not a "one size fits all" situation given the complexities presented by the diversity of compute platforms and deployment scenarios with varying requirements and environment constraints. Moreover, it may be important to also account for the dynamics of the environment. The quality of the deployment will depend on the coupling between the deployment configuration and the environment, particularly if the situation is continuously changing. These dynamics may result in disconnections, drops in service, and faulty applications. To maintain sufficient system performance, it is important to develop agile methods that detect the dynamics that may be causing suboptimal performance, identify reconfiguration, and deploy new configurations.

We must consider how to design and implement systems that can efficiently handle the demands of distributed environments with the necessary scalability, elasticity, and flexibility. These principles are key to ensuring that edge computing platforms can deliver high-performance and responsive services. Here is an elaboration on how software architecture, orchestration, and technologies such as SDX (software-defined everything) can help achieve this.

10.2.1 Microservices in Edge Computing

The traditional monolithic approach to software architecture often does not scale well. The edge requires applications to be modular, distributed, and able to respond dynamically to different loads. This is where microservices come into play [9]. Microservices break down applications into smaller, loosely coupled services that can independently scale based on demand.

10.2.1.1 Scalability and Elasticity

Microservices enable horizontal scaling (scaling out), which is especially important at the edge where workloads can vary significantly. For instance, edge devices like IoT sensors or mobile networks may only need a small amount of computing power during off-peak times, but during periods of high activity, additional computing resources must be brought online. Microservices can be deployed across multiple edge nodes and can scale individually based on traffic, ensuring elasticity—the ability to automatically adjust resource allocation in response to changing workloads.

10.2.1.2 Resilience and Flexibility

With microservices, even if one service or node fails, the others continue functioning. This fault-tolerant behavior is critical at the edge, where network instability or device failures are common. Microservices also allow for greater flexibility in evolving applications since individual services can be updated without affecting the whole system.

In edge computing, orchestration plays an important role in managing the deployment, scaling, and operation of microservices across a distributed environment. Tools like Kubernetes [10] or custom edge orchestrators can manage these microservices, ensuring that they are deployed efficiently across edge nodes based on workload and resource availability.

10.2.2 SDX in Edge Computing

SDX is an advanced concept where all aspects of the network and infrastructure are abstracted and controlled via software. In the context of edge computing, SDX allows for greater flexibility and control over the resources at the edge. It can be particularly beneficial for managing highly dynamic and decentralized edge environments. Through SDX, the network infrastructure can adapt in real-time to changing conditions, improving the scalability and elasticity of the entire edge computing environment. For example, in a communications platform, SDX can enable dynamic adjustments in network slicing, where different types of traffic (e.g., video, IoT data, AI models) are handled in specific slices, each optimized for performance. This helps improve the efficiency and scalability of the network, ensuring that high-demand applications at the edge (like real-time video streaming or machine learning inference) have the resources they need while maintaining optimal performance for other services.

10.2.3 Edge Computing Supporting IoT and AI

A main motivating use case of edge computing is in the combination of IoT and AI [11]. IoT devices generate vast amounts of data at the edge; sensors, cameras, and other devices continuously feed data that needs to be processed quickly to deliver actionable insights. Traditionally, this data would be sent to centralized cloud servers for processing, but with edge computing, much of the processing can occur closer to the devices, significantly reducing latency and bandwidth requirements.

AI models can be deployed directly to edge devices, enabling real-time decision-making. For example, edge devices in a smart factory can use AI models to monitor machinery and predict failures before they occur, allowing for preventive maintenance. The AI models may be trained on historical data, but real-time data is processed at the edge to make predictions immediately.

As IoT devices grow in number and sophistication, the edge infrastructure must be capable of handling massive concurrent data streams. This is where AI models can be applied to filter and analyze data at the edge before it is sent to centralized systems. For instance, an edge device might use AI to detect anomalies in sensor data or to identify patterns without needing to send all raw data to the cloud. This not only reduces the volume of data transmitted but also enhances system responsiveness.

10.3 Edge Network Components

Edge systems comprise many integrated components, so we can consider them as RCAS. Therefore, we need to understand the individual components and their capabilities before assessing how they will operate as a RCAS. For edge systems to sufficiently execute their intended purpose, the system must identify where the data resides, how and at what nodes the data can be processed within application requirements, and then how to execute the application in the system. These systems might involve platforms that are extremely heterogeneous, comprising a diverse set of sensing and compute devices, applications, information management services, networks, and communications. This is illustrated in Figure 10.2.

10.3.1 Sensor Platforms

Edge networks, functioning as RCAS, support processing of data generated at the edge. These systems handle diverse data modalities, formats, and volumes from IoT devices, leveraging decentralized processing and local interactions to enhance resilience and adaptability. Examples of sensors in edge environments include those that measure some aspect of the physical environment, including video

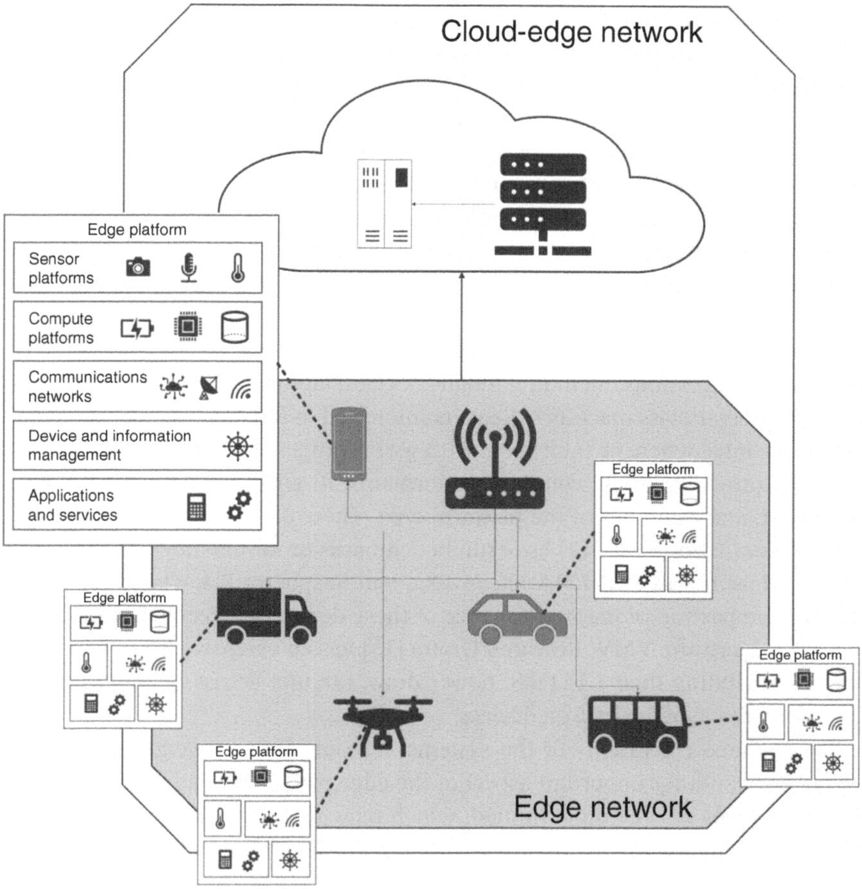

Figure 10.2 Illustration of edge and cloud-edge networks with edge platform components.

cameras (including electro-optical and infrared sensors [EO/IR] and wide-area motion imagery [WAMI]), seismic sensors, microphones, and temperature monitors. Examples of cyber data sources include intrusion detection systems, firewalls, network monitors, information databases, RFID readings, and endpoint sensors.

10.3.2 Compute Platforms

The edge compute platform is the main building block of edge compute systems, where they are deployed to process data available in the edge environment, including data generated by co-located sensor platforms, possibly with assistance from cloud resources. Each platform provides a set of services to a localized region of the environment. For example, compute platforms will include a range of compute, data storage, and available energy/power. These systems might include servers with significant storage, compute, and energy and possibly be substantially powered; laptops with significant compute capability with central processing unit (CPU) but with limited battery; graphics processing unit (GPU) compute devices that have significant compute capability but require significant power to run; and resource-constrained devices such as Raspberry Pis (RPi), which have limited compute and storage but are quite energy efficient. Based on resourcing and configuration, the edge platforms will demonstrate a performance profile that may or may not suffice for the situation.

We report on several classes of edge compute platforms or devices that could be used in edge situations, including mobile devices, laptops, or PC-class machines, and up to server-class machines. Readers interested in technical details may refer to the appendix, where in Tables 10.3–10.5 we list ranges of several aspects of compute platforms, including compute performance in gigaflops (GFLOPS), power draw, cost, and examples of the platform type. These data were drawn from several sources. Süzen et al. [12] have run benchmarks on several devices, including NVIDIA Jetson Nano, NVIDIA Jetson TX2, and Raspberry Pi4, which provides a relative comparison of the performance of these devices to execute a deep learning (DL) algorithm. VMW Research Group [13] has an extensive list of computer devices, including their GFLOPS, power draw, random access memory (RAM), cores for a range of devices, on devices.

The resource constraints of the systems, termed low size weight and power (SWaP), are another important aspect of the edge platforms. The physical size of these platforms may be quite limited, which may be determined by needing the device to be hidden or not disruptive to the deployment environment, or it may need to be very light if it is being deployed from a remote location. The power of the edge platforms may be quite limited if they are dependent on a battery rather than being plugged into a power source; the compute, memory, and data storage capacity will potentially be less than typical compute devices due to size and power constraints. As a result, these constraints affect which information management and services may run on these systems.

10.3.3 Communications Networks

If edge platforms can communicate with other edge nodes or with cloud resources, the compute capability of these systems is greatly increased. As stated previously, edge compute platforms possess a significant amount of compute capability, deployment of a single node may be sufficient in certain circumstances. Further, deployment of independently operating edge nodes may also suffice, or the environment may restrict coordination between devices. One example is remote environmental sensing, where a node may be deployed to track some environmental activity, and data is collected and processed locally but retrieved manually after the job is finished. This section provides an overview of communications and network technologies that facilitate such capabilities in edge networks.

Networking in edge situations enhances system performance by implementing principles of RCAS: Collaborating edge platforms increase area coverage, provide robustness and resilience to environmental dynamics, and reduce manual maintenance requirements. These networks adapt to changing conditions, ensuring continuous and efficient operation. In cloud-edge networks, networking cloud resources with edge resources will also significantly enhance the capabilities of these systems.

In edge computing, communications between edge platforms ensure seamless data synchronization and collaborative processing across a distributed network. These communications involve various edge devices and computing resources that need to interact efficiently, often relying on low-latency, high-bandwidth connections. One key aspect of these communications is inter-edge communication, where edge platforms work together to process, share, or aggregate data. This can be especially important for load balancing or coordinating processing tasks.

Networking technologies like 5G and Wi-Fi 6 provide high-speed, low-latency connections that are fundamental to supporting the dynamic and distributed nature of edge computing. Further, cloud-edge communication plays a role when edge devices need to offload heavier processing tasks or store long-term data in a central cloud or data center. Hybrid architectures, where real-time, mission-critical data is processed at the edge, and nonurgent data is sent to the cloud, are common in many edge computing systems.

To optimize and manage communication between edge devices, a subset of SDX, namely SDN can be employed. SDN allows for dynamic management of network traffic and communication pathways, adapting in real-time to varying network conditions and ensuring that bandwidth is used efficiently while reducing congestion.

The quality of communication depends on factors related to the underlying radio frequency (RF) technologies. For example, the frequency band used for communication can affect the range, capacity, and latency of connections between edge devices. Edge platforms may employ bands like sub-6 GHz for 5G or millimeter-wave for higher speeds, with each having different trade-offs in terms of coverage and speed. Throughput (often denoted "bandwidth") is another important factor, as higher throughput implies faster data transfer for more data and is an important aspect of system performance. Latency, the time delay in communication, is particularly critical in edge computing, where real-time processing is often required. 5G, for example, offers ultra-low latency, with potential delays as low as 1 millisecond, making it ideal for edge scenarios where immediate data processing is needed, such as in AI applications or autonomous systems. Of course, processing time adds to latency, as well as the amount of data that needs to be transferred. Therefore, employing AI-enabled preprocessing near/at the data source reduces communication overhead in central parts of the system and thereby also latency.

Other elements of radio performance that impact communication include signal strength and interference. In distributed edge environments, especially those located in remote or crowded areas, signal degradation can occur, affecting the consistency of data transmission. Techniques like mesh networking, where multiple devices collaborate to relay data, or network slicing, which partitions the network for specific use cases, can help mitigate these issues. Moreover, interference from other wireless devices or environmental obstacles can also degrade communication quality. To mitigate interference, Edge Networks often use advanced technologies like MIMO (multiple-input multiple-output) and beamforming, which optimize signal quality by directing it where it is needed.

Finally, asserting quality of service (QoS) is essential in any network, thereby also in edge computing, where different applications may have varying requirements for bandwidth, latency, and reliability. For example, critical applications like healthcare monitoring or autonomous driving systems need guaranteed low latency and high reliability, even in congested network conditions. QoS mechanisms help prioritize these essential services, ensuring they get the necessary resources in a shared, dynamic environment.

Overall, a combination of high-performance communication protocols, advanced radio technologies, and adaptive networking strategies may enable edge computing platforms to achieve the low-latency, high-bandwidth, and resilient communications that will be required for real-time processing and decision-making applications. The appendix lists several communications technologies for use in edge networks. Given the wide range of edge network deployment environments, these values provide a sense of the ranges and data rates that networks can be used to support. As shown, there is a general trade-off of communications ranges versus data rate.

While communications network technologies are pervasive, some of these technologies could be leveraged to enhance data exchange at the edge. Cloud-edge networks may operate in a similar fashion to commercial networks, where commercial 5G or wireless technologies can be used with minimal reconfiguration necessary. In other situations, edge networks may provide challenges significantly different from commercial networks. Edge networks might not have access to any fixed infrastructure and may not have access to an Internet backhaul. In these cases, setup of these networks might require ad hoc or peer-to-peer communications using in situ methods or predefined processes.

To exploit communication technologies at the edge, networked edge compute platforms can create advanced distributed edge computing capabilities that leverage connected (potentially over multiple hops) edge platforms and cloud resources in the case of cloud-edge networks. However, traditional networking techniques may not adequately suit the edge environment and may require custom solutions that provide flexibility and situation-dependent configurations. Recent advances in softwareization have resulted in the SDX concept [14], which includes SDN [15] and software-defined radios (SDR) [16], that provides flexibility beyond traditional networking implementations. This flexibility potentially facilitates computing in the edge scenarios. One flexibility that SDX techniques offer is a reconfiguration of networking waveforms that might be a better fit for deployment at the edge compared with traditional off-the-shelf protocols.

The deployment environment poses a challenge for the operations of these networks. The terrain may be very complex, including urban situations with many obstructions hindering line-of-site communications and physical access to certain locations. The information environment may be very cluttered with noise and interference of various kinds, including other RF causing communications interference, physical clutter increasing the difficulty of some sensors to perform, and dynamics of the environment, which might require the system to adapt and deal with constant change.

A primary aspect of the deployment environment is that it can be characterized as disconnected, intermittent connectivity, and limited bandwidth (DIL) network environment. It is therefore not possible to use the assumptions cloud or enterprise networks use in terms of the predictability and performance of the systems. Disconnected implies that networked systems lose active connections to other systems in the environment and the connection may not be reliable. Intermittent connectivity is the case where communications links are not stable and reliably exhibiting their configured throughput or latency. Limited bandwidth concerns the difference between edge networks and enterprise networks, where the throughput is considerably lower due to operating over multiple wireless links.

In edge computing, a DIL environment leads to several challenges that must be mitigated through the edge computing platform: In the case of being disconnected, devices lack continuous access to the cloud or central servers, requiring them to process data locally. In the case of intermittent connectivity, devices experience sporadic network connectivity, necessitating local storage and delayed synchronization with the cloud. Finally, limitations on throughput mean that network bandwidth is constrained, requiring efficient local data processing and transmission of only essential data when possible.

10.3.4 Device and Information Management

As RCAS, edge platforms, and networks require coordinated management to function effectively. Hastily deployed systems, such as in disaster response situations, benefit from adaptive management strategies that can dynamically configure networks and services based on real-time conditions and needs. Further, as the deployment environment might be dynamic, the edge systems will need management and orchestration services to initiate the applications, establish the network, and adapt any system configuration as necessary. Current approaches for facilitating the management of services and networks have been in the virtualization and containerization of services. One common platform for container management is Kubernetes [10].

Past approaches have attempted to provide uniformity to the computing environment to enable interoperability across different systems. Through the use of microservices, the concept is that a common application can be deployed across a variety of systems and have some guarantees that the application will function properly.

While Kubernetes is suited to cloud or data center computing infrastructures where compute and networking are plentiful, these orchestration services may require additional compute and memory overhead as well as network overhead for coordination across Kubernetes pods. That said, there exist lightweight versions of Kubernetes that are more amenable to edge situations. Kubernetes variants include microK8s, Minikube, K3s, and KubeEdge. The appendix also describes the recommended memory (for the edge devices) and storage footprints and the differentiating features between the versions. Other versions of Kubernetes provide lightweight and simpler versions, which can facilitate deployment and development at the edge [17].

In the context of edge computing, information management is needed as a facilitator of data processing, trustworthiness, and privacy. At the edge, data is often generated by numerous distributed devices (e.g., IoT sensors, cameras, etc.), and managing this data effectively ensures that the system operates efficiently and securely. However, edge computing, exemplified through service orchestration solutions like Kubernetes, typically manages compute resources rather than storing and handling the data itself. This distinction is important because edge devices are often resource-constrained and may not have the capacity to store vast amounts of data. Instead, they perform real-time data processing, analysis, and decision-making close to where the data is generated, with the actual data either stored at the edge or sent to centralized cloud servers or other storage systems.

Information management in edge computing involves efficiently handling the flow of data between edge devices and servers, ensuring that data is processed, transmitted, and acted upon appropriately. The complexity increases with data provenance—the tracking of the origins, transformations, and movements of data throughout its lifecycle. This traceability allows organizations to verify the accuracy and reliability of the data, especially when dealing with applications such as healthcare or autonomous systems, where the source and quality of data must be validated to ensure trustworthy results.

Furthermore, trustworthiness is key in edge computing. Ensuring that the data hasn't been tampered with or corrupted before being processed is critical. Edge computing platforms need to incorporate mechanisms to validate data at the point of origin, as well as ensure secure communication between devices, ensuring that unauthorized access is prevented. Blockchain and encryption techniques are sometimes used to ensure the trustworthiness of data, and these could also be implemented at the edge to track the provenance of data across its lifecycle. One area of future work in edge computing is advancing information management systems that provide better support for the dynamic and decentralized nature of edge environments. This includes improving data provenance tracking, developing more efficient orchestration frameworks for distributing compute tasks, and enhancing privacy-preserving techniques that ensure secure processing of sensitive data at the edge. Another area of development is model management, where model zoos (collections of pretrained AI models) become increasingly important. Edge platforms could benefit from being able to easily access a model zoo to download and deploy the most suitable machine learning models based on the task at hand. This would allow edge devices to perform complex AI tasks (like object detection or natural language processing) without needing to send all data to the cloud. Instead, they could leverage pretrained models tailored to their specific use cases, ensuring efficient and rapid processing at the edge while respecting privacy concerns by keeping data local.

10.3.5 Development/Deployment

Deployment of edge systems will invariably need updates when the system configuration no longer matches the job/task/mission requirements. On one hand, as previously stated, hasty deployments may prevent preplanning and configuration and initialization before deployment. Another situation may involve a long-term deployment at the edge, where the applications, network, or system may become out of date. For adequate system performance, management or administration of these systems might need to update one or more of these system components. These updates are potentially performed manually and require some on-site maintenance or could be done over the air (OTA). Management of software may employ continuous deployment processes (e.g., DevSecOps) of the deployed software applications [18]. These processes allow for the in situ update of software versions, network configurations, and system configurations. DevSecOps implements a continuous development pipeline that enables coordinated development of software applications, validates their performance and compatibility with legacy software, and deploys the new versions to platforms that need the update. For example, machine learning-based systems may deploy pretrained models, and based on data gathered at the edge, the models can be fine-tuned using a variety of methods, including personalization, continuous learning, and service migration/model deployment [19]. Management and development of the edge systems will also enable elasticity and scalability of the applications as the demands and environment evolve. There may be cases of increased data processing demand, and new data processing applications must be deployed or activated. This may require distribution of additional applications to edge platforms or may require some load rebalancing of network resource allocations.

In resource-constrained environments, this is another layer of the system that must consider the overhead, both compute of nodes involved in the DevOps process, the memory necessary to store the software updates, and the network overhead required to execute the process with all involved systems.

10.3.6 Applications/Services Types

Since the edge platforms and network services have been described, we can now describe the applications or services that these systems can execute. These applications tasked to process the data generated at the edge will consume some portion of the available resources. This includes compute resources (e.g., CPUs, RAM) at edge and cloud platforms and network resources, including network bandwidth, to realize some data throughput. In congested computing and information processing situations, these resources must be coordinated between multiple jobs requesting the same resources. Some "classic" or core services must be deployed to support data processing applications. Some services must exist to support the data processing services: These services include Kubernetes for container management, network control services (e.g., SDN controllers), and network routing protocols. For the user services, the performance of these applications can be predicted when given specified amounts of the resources [20, 21]. Example applications include machine learning-based approaches such as computer vision tasks, event recognition (such as some scene or anomaly detection) [22], structural health monitoring, and physical access control [23, 24]. Other applications include preprocessing of the data, which may occur on the edge device even in situations where the majority of the compute is offloaded. This includes compression of the data, data filtering or sampling, or fusion of data even across data modalities. To facilitate deployment of these applications/services, they can be containerized and managed via Kubernetes. Further, information management that enables rapid deployment and reconfiguration of containerized services provides flexibility and adaptability of these services on any system under the purview of the management service.

10.4 Edge System Properties and Performance Metrics

Edge networks will be deployed with some selected technology for each component, and the overall system configuration will demonstrate some emergent behavior as they operate as a RCAS. In addition to expressing the challenges presented in the need for edge systems, this section describes the design choices in the deployment of these systems and their notional system performance in several scenarios. The design of edge networks will not be a "one-size-fits-all" approach, so when the need and environment are understood (as much as possible), the design and deployment choices can be made to execute with desired performance metrics or to maximize performance despite system constraints.

In terms of the performance of these networks, we consider general system performance metrics such as data processing rates, which is a downstream metric of the more objective network metrics such as data throughput or system uptime, or system resource usage metrics like power usage, job capacity, or system overhead. Systems can also be evaluated based on application-specific metrics related to system performance and resource usage metrics. For imagery or video processing tasks, metrics might include frames per second (fps), detection accuracy, and processing latency.

10.4.1 Our Framework

As edge systems can be considered as a variation of distributed systems, the properties and design choices are very well-studied [7, 25]. However, we can apply relevant distributed systems properties to edge systems. Naik [25] identifies the following properties: resource sharing, openness, concurrency, consistency, idempotency, scalability, availability, reliability, fault tolerance, and transparency. Meuser et al. [7] identify several challenges, including resource limitations, privacy and trust, sustainability and energy efficiency, programmability and interoperability, dependability and resilience, and measurability. Mancini [26] suggests autonomy, flexibility, robustness, and performance from a system security perspective. The cost to deploy networks is also an important aspect, as it will dictate what devices are used and how they are employed. In turn, this will directly affect the performance and priorities of the edge networks.

Given the vast frameworks that have attempted to characterize distributed system properties, we categorize them, then use the categories to establish an edge computing framework for several edge network use cases. Let us define the following categories: *robust and resilient* as performance of the network despite unexpected, uncertain circumstances or operating environment; *adaptive* as performance in a dynamic environment; *secure* as performance in the presence of adversarial or untrusted entities; and *realizable and practical* as the resource requirement necessary to operate the system. Table 10.1 summarizes subproperties of the robust and resilient, adaptive, secure and realizable, and practical categories.

The categories represent complex interactions with the subproperties, but we use this in Section 10.5 to compare examples of edge computing networks.

10.4.2 Aligning the Framework with RCAS

Thus, our framework categorizes edge computing systems into four domains: (1) Robust and resilient; (2) adaptive; (3) secure; and (4) realizable and practical. Next, we align our framework with the 10 properties of RCAS discussed in Chapter 1: (1) Adaptability; (2) redundancy; (3) fault tolerance; (4) scalability; (5) self-healing; (6) diversity; (7) decentralization; (8) robustness; (9) modularity; and (10) monitoring and feedback. Table 10.1 describes edge system properties for each category, and we map the distributed edge categories into the RCAS properties in Table 10.2.

10.4.2.1 Resilient and Robust

Being robust and resilient ensures continuous operation in unpredictable conditions. Key RCAS principles include fault tolerance, redundancy, robustness, and self-healing.

Table 10.1 Distributed edge system property categories.

Category	Subcategory
Robust and resilient	Performance, consistency, fault tolerance, idempotency, concurrency
Adaptive	Flexibility, adaptability, heterogeneity, autonomy
Secure	Safety, privacy, transparency, openness, trustworthiness
Realizable and practical	Platform acquisition, upkeep/maintenance, and deployment costs

Sources: References [7, 25, 26].

Table 10.2 Distributed edge system property categories mapped to RCAS properties.

Category	RCAS property
Resilient and robust	Fault tolerance, redundancy, robustness, and self-healing
Adaptive	Adaptability, scalability, modularity, and monitoring and feedback
Secure	Diversity, decentralization, self-healing, and adaptability
Realizable and practical	Modularity, scalability, redundancy, and monitoring and feedback

Edge systems exhibit fault tolerance through decentralized adaptability rather than predefined error correction. Nodes, often in unstable environments, can dynamically adjust to handle disruptions, maintaining performance despite node or network failures.

Redundant nodes and communication paths ensure continuity by providing backup options when components fail. This helps sustain operations by redistributing computational loads or replicating functions.

Edge systems' decentralized, self-organizing nature allows them to withstand fluctuations in workload, cyber threats, or hardware issues. Emergent coordination ensures performance and stability under stress.

With self-healing, the system autonomously recovers from failures by redistributing tasks, rerouting data, or replicating services, minimizing downtime and maintaining functionality without centralized control.

10.4.2.2 Adaptive

Systems must adjust to dynamic environments, workloads, and resource limitations. Key RCAS principles include adaptability, scalability, modularity, and monitoring and feedback.

Edge systems modify their behavior in real time, using methods like predictive analytics or reinforcement learning to optimize workload distribution, network routing, and service placement based on evolving conditions.

Edge systems seamlessly scale with demand, integrating new nodes or resources without disrupting existing operations, maintaining stability and efficiency as they grow.

By structuring systems into independent subsystems, modularity supports localized updates and heterogeneous deployments, enhancing flexibility and longevity in diverse edge environments.

Continuous performance monitoring of parameters like latency, energy use, and throughput provides feedback loops that enable autonomous optimization to ensure the system copes with changing conditions.

10.4.2.3 Secure

Security is critical in edge computing, where data and computation span multiple nodes. Relevant RCAS principles include diversity, decentralization, self-healing, and adaptability.

The possible heterogeneity of edge nodes—varying in hardware, software, and functionality—makes the security problem more complex, requiring distinct security configurations tailored to the diversity of capabilities.

By decentralizing security functions like authentication and encryption across nodes, we prevent single points of failure. When security breaches occur, self-healing properties of the system can isolate compromised nodes, reroute sensitive data, and replicate services in secure zones, minimizing attack impact and restoring normal operations.

Continuous monitoring enables edge nodes to dynamically adapt security measures in response to emerging threats, ensuring an evolving defense posture.

10.4.2.4 Realizable and Practical

Edge computing systems must not only be robust and adaptive but also feasible to realize and efficient to be practical for real-world deployment. Key RCAS principles include modularity, scalability, redundancy, and monitoring and feedback.

Modular architecture enables incremental deployment, simplified maintenance, and localized upgrades. Independent components can be replaced or enhanced without disrupting overall operation, supporting flexibility and technological evolution.

Scalable designs allow systems to handle variable workloads and expand smoothly as demand grows, integrating new nodes or services while maintaining stable performance and manageable overhead.

Redundancy ensures reliability in mission-critical applications by preserving functionality during component failures or local disruptions.

Continuous monitoring and automated feedback loops support proactive fault detection, resource optimization, and adaptive control, enabling manageability in practical deployments.

10.4.3 From Design Properties to System Requirements

For each edge system design property, we can assume that some combination of *microscopic* system components results in some *macroscopic* emergent system behavior. Then, based on the goals of the edge network deployment, one can configure the system to perform with the desired performance or exhibit the desired properties. We note several methods to jointly consider these properties for multicriteria decision-making for system configuration, including the analytic hierarchy process, skew-symmetric bilinear representation, and large language models [27–29]. We show each of the categories in a radar chart, as shown in an example in Figure 10.3. Each category can be given values from [0, 1] and is independent, where a higher value indicates a higher design requirement for that category. Three examples are shown, including a balanced case and one each for a security and robust application.

In the example, the robust application has a maximum resilient and robust property value, and the security application has a maximum security property value. These two examples would require the system designer to expend resources to implement capabilities that address that one property and spend relatively less attention addressing the other properties. However, in the balanced application,

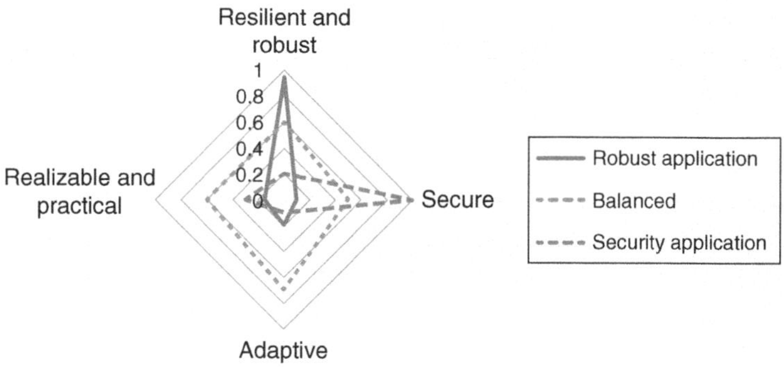

Figure 10.3 Radar chart example of complex adaptive systems properties.

it has requirements for each of the properties, requiring the system designer to implement aspects addressing all system requirements. The complexity of the use case defines the difficulty in designing a system to meet requirements.

10.5 Edge Networks: Examples

Now that we have described the RCAS components and characteristics of edge computing networks, this section provides several examples of these systems and compares their properties and configurations. This illustrates the claims of edge networks operating in varying situations and computing environments. These visualizations exemplify applying the framework as shown throughout the following example scenarios and radar charts.

10.5.1 Edge Computing for Disaster Response

The first few hours of a natural disaster are considered mission critical, because lives are at stake. A quick and organized response is necessary to save lives, and giving the response teams a wider range of tools (like edge computing) may prove beneficial for that cause.

For this example, we consider an actual disaster that occurred in Norway a few years ago. There was a massive landslide that happened just north of Oslo in a small town called Ask. The slide took out a significant part of a residential area and dragged a dozen buildings with it. It started in the early hours of the morning and continued expanding. In circumstances like this, infrastructure is completely torn away. Applying the idea of edge computing and a disconnected operation mode can be beneficial in these types of scenarios where multiple teams of responders must perform a variety of tasks.

At the time, we had an emergency response network based on the TETRA standard utilizing only voice communication. TETRA is a European standard for both mobile and two-way communication [30].

10.5.1.1 Enter 5G

Recently, innovations in the field of 5G have pushed researchers to experiment with replacing the TETRA solution altogether toward the more favorable aspects that 5G would offer. In 5G, the network is separated into multiple slices of communication for compartmentalization, separation of concerns, and allowing priority assignment to certain slices. Working under the presumption that 5G mobile base stations will be available for deployment directly at future operation sites, we could consider 5G as the backbone communications component enabling edge computing. The EU project 5G-VINNI is a perfect example of this, allowing for mobile base stations to be deployed as needed [31].

We have worked on different components and architectures, but, in essence, the goal was to explore the capabilities and opportunities of an ad hoc networking solution for disaster response operations. In conjunction with 5G for communications, there is also a need for a service platform that is scalable, flexible, resilient towards partial failure, and has self-healing properties. Therefore, we started investigating Kubernetes, and especially the K3s implementation of it, as a possible solution.

10.5.1.2 Enter IoT

With IoT, we are experiencing a new dawn of computing with increased automation and ubiquitous computing. When a natural disaster strikes, existing infrastructures are, in most cases, torn away, making communication difficult for the first responders. With the recent developments in fields such as 5G and IoT, we believe that opening more channels of communication instead of simply relying on voice can improve the efforts of those responding. Video and sensory data can offer a new valuable source of information for first responders during a natural disaster.

Applying these principles to create an improvised, ad hoc networking solution that can be deployed as a temporary measure before surrounding infrastructure is restored could provide new opportunities for the people on the ground. IoT has contributed to an increased effort in uncovering capabilities that exist on networks, even adopting concepts such as edge computing, which could power a new alternative computational platform. Instead of relying solely on data centers, which are regionally located and require sustained broadband access for utilization.

Instead, we may attempt to leverage 5G and concepts of embedded computing, as exemplified in the research drive behind IoT development in general.

Edge computing has also posed interesting questions for researchers and companies alike as to how we can best use the existing infrastructures, along with bringing our ideas, such as ours, of using small single-board devices to host an ad hoc network for disaster response. Our work has taken inspiration from earlier research by The North Atlantic Treaty Organization (NATO) regarding the applicability of IoT [32] and edge computing [33].

10.5.1.3 On MQTT and Event-driven Architecture

MQTT is an event-driven paradigm that is based on the publish and subscribe pattern (see [34]). If we consider a system that gathers massive amounts of information per second, then an event-driven architecture helps structure and create logical timelines of that information.

MQTT messages also offer ease of use in terms of parsing the information to JSON or any other formats more readily used within modern computing. MQTT is based on the idea of clients and brokers. A broker is a central entity responsible for coordinating the messages between the participating clients on the network. Further research has also identified the MQTT paradigm to play a key role in interoperability between military C2 systems and IoT infrastructures [35]. We foresee this can also be extended to first responders so that the very same information flowing into the military C2 system may be seen and acted upon by police, medical personnel, etc. Different departments would likely have different profiles in the edge computing system, ensuring that they get the data and computing resources they need. There must also be information management to mitigate privacy issues, a major concern not only in Norway but also in most European countries through the proliferation of GDPR (General Data Protection Regulation). For example, position data is likely needed by all actors, whereas individual medical details are likely only needed by the medical staff.

Edge computing must function as an ad hoc networking solution to be deployed until the surrounding infrastructure becomes accessible again. With this approach, the solution must be able to seamlessly operate with the surrounding infrastructure if it is available. The approach we pursue, combined with the idea of a locally deployed 5G core, would essentially create an edge network. This edge network would be self-contained for communication, storage, and processing power and function disconnected from other infrastructure, thus in essence realizing edge computing. Here, the 5G core would have the role of a MEC, where additional IoT devices would be enrolled as edge devices in a Kubernetes cluster, in which the MEC functions as the cloud controller.

10.5.1.4 On Kubernetes

Kubernetes, as introduced in Section 10.3.4, is an open-source orchestration tool used to manage containerized environments. Containerized environments are simply a way of deploying applications where the applications or services are packaged as containers. It is a popular approach in many Infrastructures as a Service (IaaS) solutions such as Amazon Web Services (AWS).

There are many different Kubernetes distributions. Due to our edge computing and IoT focus, we opted for a lightweight implementation, K3s, that is specifically targeting edge computing applications of both x86 and ARM 64 architectures, such as the RPi.

In our architecture, we opted for the core of our system to be operated with K3s and subsequently host an MQTT broker along with its clients of the cluster. This would allow the devices to coordinate efforts with each other.

10.5.1.5 On RPi

We chose RPi as the platform. It is a ubiquitous small computing device introduced in Section 10.3.2, with the appearance of a circuit board costing around 15–40 dollars. We discovered very early on in development, when all of our nodes were of the Model 3 type, that K3s exhibited intermittent failures. Investigating this, we discovered that the upper memory limit of the master node was being exceeded. It was simply not feasible to run the master node with only 1 GB of RAM. Upgrading to one of the newer versions of the RPi, Models 4 and 5, solves this issue. We chose a Model 4 with 4 GB of RAM, which was fully capable of functioning as a K3s controller. Model 3 is capable of functioning as a worker nodes (edge devices).

This experience must be seen in conjunction with Table 10.5, which lists the worker node requirements for K3s, which indeed are sufficiently accommodated by a lower end device like the RPi Model 3. The controller node, on the other hand, requires more than 1 GB RAM.

10.5.1.6 Experimental Evaluation

The setup was evaluated using several emulated radio links. We included tactical networks like NATO narrowband waveform and combat network radio (CNR) which are typical configurations for military deployments and are significantly weaker compared to 5G, but they could also be relevant, which is why we wanted to include them in our test. The complete results are published and described here [36] and support earlier research with performance evaluation of MQTT [37] (though not in Kubernetes) and edge computing-optimized Kubernetes distributions, including K3s [33]. In general, the trouble experienced in low bandwidth communication boils down to TCP and how it works with retransmissions, e.g., CNR. Naturally, the higher throughput networks, like 5G, work without such communication issues and would be the preferred type of communication provided for edge computing. This would accommodate both the industry standard MQTT and the maintenance traffic of K3s to keep the cluster healthy.

10.5.1.7 Capability Summary

In this use case, the fixed infrastructure is gone. The question becomes how to establish something quickly. The answer to this is edge computing, where we may set up a 5G mobile base station, leverage small form factor devices (e.g., RPi) as edge devices. This disconnected edge can then provide communications and computational resources while conducting the search and rescue mission, without having to wait for the reestablishment of the damaged infrastructure. Eventually, when infrastructure is reestablished, one can foresee switching to that or using it in synergy with the edge computing capability. The accompanying radar chart shows these design properties accordingly in Figure 10.4.

10.5.2 Edge Computing for Smart Cities and Smart Homes

Smart homes and smart cities are two concrete examples of edge computing systems [23, 24]. Building on desires of home automation capabilities and as the internet grew, commercial use cases of home automation grew, including smart electronics (e.g., televisions, lights, doors, toasters) and human assist devices (e.g., voice assistants, thermostats). At the city level, municipalities are working toward smart cities or smart grids to enhance the services provided to their citizens and ways to enhance maintenance and monitoring of infrastructure.

10.5.2.1 Characteristics and Motivation

Applications of edge computing in these situations resemble cloud-edge networks as infrastructure can provide communications links to cloud resources and also provide power to edge platforms so these systems can operate as long as needed.

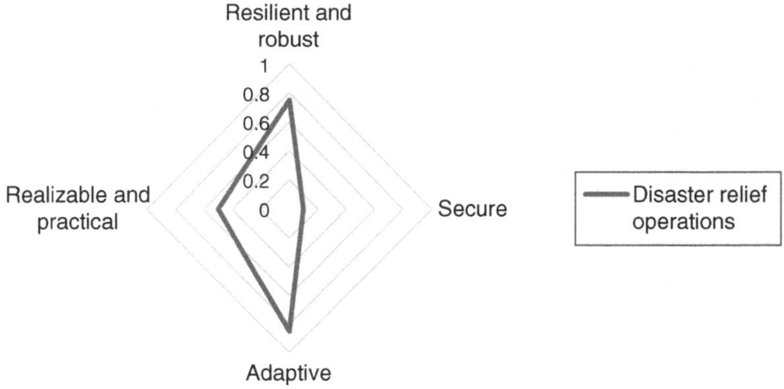

Figure 10.4 Radar chart of edge metric properties: disaster response example (with cost).

However, the computing resources at the edge may be quite limited, as compute-capable devices are more expensive and prohibitive for setup in homes and cities. As a result, it may be preferred to push data up to the cloud to perform complex compute jobs. Furthermore, the systems here can leverage existing residential Wi-Fi networks for reliable access to the cloud. However, privacy may be a hindrance to this, as data collected in homes and in cities may contain personal information and invade the privacy of an individual. Sending this to the cloud may be very controversial. One may consider preprocessing steps to remove or whiten the potentially sensitive information before transmitting it outside the edge platform that collected it. Powering the devices is likely less of a concern, as they can connect to infrastructure, as the devices are in a less dynamic deployment situation. Additionally, services and management are likely preinstalled by the manufacturers, which might result in interoperability or reconfiguration issues between devices from different manufacturers.

Home residents see smart homes as means to increase security with cameras, doorbell cameras, and floodlights for surveillance and antitheft. This situation highlights the need for information privacy for individuals captured on video. Configuration of these systems should include sufficient security and privacy and the ability to perform the intended functionality. Edge applications also are used to save energy and lower operating costs by using remote-controllable and automated smart thermostats and connected appliances. These cases highlight the need for cloud compute offload, as some of these appliances and hardware likely will not possess significant compute, so these situations must consider some trade-off of edge versus cloud compute.

City planners desire to provide useful services to the population, and they also can exploit edge computing as a means to facilitate municipal functions that otherwise require significant manual intervention. This might include utility monitoring (through smart meters), saving operating costs with automated lights and other utilities, enhancing transportation, optimizing traffic patterns and pedestrian mobility. However, there is some caution about attempting to pursue too many smart city initiatives [38]. In these situations, privacy is critical, but reliability is also important given the public's reliance on these services.

10.5.2.2 Example Use Cases

Home automation for convenience, efficiency, and security are prototypical examples of the smart home. Some applications include using cameras, motion or infrared sensors, or microphones to detect if anyone is in a room to control temperature, lighting, or speaker volume. In these cases, the sensors are potentially always on, so substantial data is generated to control the home. Performance and reaction are not necessarily critical, but residents likely desire guarantees that video or voice recordings are protected. For home security systems, applications will use the same sensors outside and inside the house to alert residents and security services or police of a break-in. In these cases, reliability is utmost, while security of the systems is also important such that sophisticated criminals are unable to tamper with the sensors or any other devices. In cases of agility or reactivity, continuous updates of the system are not likely critical, as the application is somewhat narrow in its functional scope. However, firmware and software updates can be performed by users or the manufacturers using reliable network infrastructure. For smart homes, [39] identifies several edge applications, including proactive maintenance. For example, sensors can be deployed on buildings at potential stress points or in high-usage areas to monitor structural integrity or any malfunctioning equipment to proactively address issues before they become major issues. In these cases, the reliability is paramount, while the other issues are less of an issue. These monitoring services may require very low information exchange between nodes and the cloud.

10.5.2.3 Capability Summary

Smart homes and smart cities serve to benefit significantly from edge computing capabilities, mainly by enhancing automation and security capabilities that leverage recent advances in AI and cloud-based technologies. These applications generally desire high security and reliability system properties but relatively low adaptability. The scenarios described here represent situations with lower dynamics and changes in applications. These properties are reflected in Figure 10.5, where either the robustness and resilience or security of the system is the critical performance category.

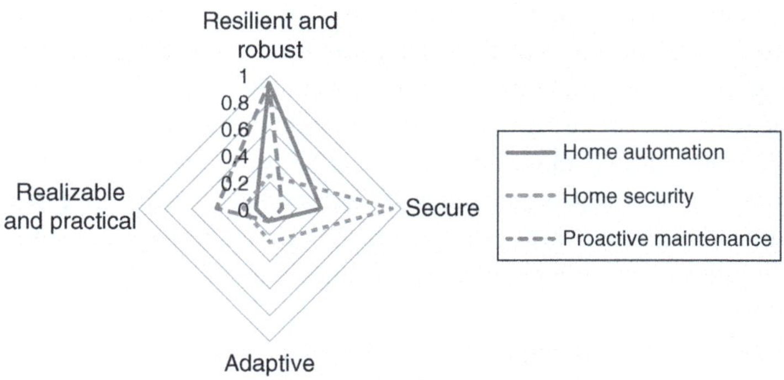

Figure 10.5 Radar chart of edge metric properties: smart automation examples.

10.5.3 Edge Computing for Environmental Sensing

Edge networks can be useful in monitoring phenomena in the environment and nature providing extended insight into natural phenomena such as tracking animal migration patterns, detecting abnormal volcanic activity, and monitoring dangerous gas concentration densities. Monitoring the environment will be difficult to physically monitor directly, such as in the case of high temperatures or dangerous conditions, or may be prohibitive to monitor larger numbers of animals that move great distances. Edge computing augments the potential performance of these applications through use of offload of compute to the edge and local compute when communications are difficult.

10.5.3.1 Characteristics and Motivation

These systems will have to deal with extreme conditions, so the platforms and networks must be robust for the application. In animal tracking situations, the platforms must be appropriately sized to enable tracking but also limit the disruption to the wildlife and bias the sensing application. Additionally, the applications might have to be tracked for long periods of time or great distances. However, tracking these behaviors can provide great empirical understanding through direct sampling as opposed to remote indirect sensing.

10.5.3.2 Example Use Cases

Yu et al. [40] provide an overview of the benefits of edge computing in ecological sensing. They also underscore the challenges of using resource-constrained devices at the edge. The paper describes several examples, including smaller biologging devices to summarize acceleration, body posture, prey capture events, and other activities of birds. Edge computing is envisioned to provide real-time event detection in these settings; however, it suggests a trade-off between edge processing and loss of raw data for post hoc processing. This example suggests that size, system lifetime, and system reliability are paramount given the extremely lightweight footprint and long flight times. Another example includes edge computing in aquatic wildlife monitoring to track migration patterns and other activities, where underwater radio communications are difficult. In an edge network setting, data from the edge sensors attached to the animals can be collected through communications with nearby autonomous unmanned underwater vehicles or sensors, but they must gather sufficient data given limited throughput and intermittent connectivity.

Roostaei et al. [41] is a good example of environmental sensing that compares the performance of edge networks and cloud-edge networks (called IoT sensor networks with edge computing (IoTEC) in this paper) for a volatile organic compound (VOC) monitoring application. The paper shows the performance of the

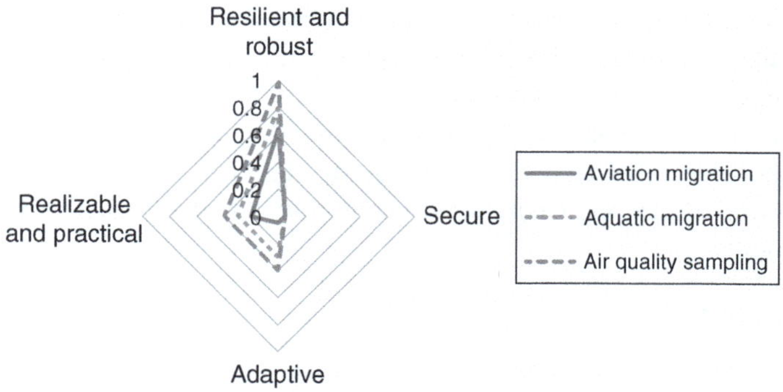

Figure 10.6 Radar chart of edge metric properties: environment sensing examples.

system when edge networks are used and when cloud-edge networks assist in computing only when VOCs surpass a threshold. The paper compares data latency, energy consumption, and economic cost between the IoTEC approach and a conventional sensor monitoring approach. The system consists of Raspberry Pi and ARM devices communicating over Wi-Fi.

10.5.3.3 Capability Summary

In these situations, the primary concern is coping with the extreme operational environments, including having limited data storage, compute, and networking resources at the edge, and needing long system lifetimes. The remote deployment of these devices on wildlife, while unattended for prolonged periods of time, is not subject to adversarial influences, but rather must deal with austere network and compute devices deployed in unstable, potentially uncooperative organic platforms. These environmental and wildlife example cases are represented in Figure 10.6, where system robustness and resilience are the predominant design properties.

10.6 Future Vision: Distributed Edge Computing

The future vision of edge computing aligns with the principles of RCAS, where distributed resources are leveraged in a cloud-like manner near the user. Adaptive services, central to RCAS, dynamically adjust their behavior in response to changes in network conditions, resource availability, and user demands, ensuring optimal performance and resilience. Adaptive services will play a central role in the continued evolution of distributed edge computing by ensuring optimal performance and user experience, even in environments with fluctuating resources or unpredictable conditions.

As edge computing environments evolve, the demand for systems that can intelligently manage and adapt resources across diverse applications will only grow. The integration of adaptive services into this distributed ecosystem is key to addressing the challenges posed by limited resources, volatile network conditions, and real-time application requirements. By enabling systems to dynamically adapt to changing conditions, we foresee edge computing providing greater flexibility, efficiency, and responsiveness. This vision is particularly significant for edge applications in military and healthcare, where service continuity and real-time performance are paramount.

10.6.1 Key Characteristics

We emphasize four key characteristics of adaptive services that will drive their importance in the future of distributed edge computing:

1. Dynamic Adjustment: Adaptive services will need to modify their behavior in real-time based on feedback from the system, user, or environment. For instance, network congestion or fluctuations in resource availability may require the system to adjust QoS settings, communication protocols, or application parameters to maintain optimal performance. This dynamic adjustment will be especially important in environments with unpredictable conditions, such as tactical edge networks, as discussed in the work on cross-layer optimization [42]. Services will need to respond to changes in both the hardware and software environment to ensure continuous, efficient operation.
2. Resource Optimization: In distributed edge environments, resources such as bandwidth, processing power, and storage are often limited. Adaptive services will play a key role in dynamically allocating these resources to ensure they are used efficiently. This will involve adjusting resource usage based on demand and the current state of the system. Middleware solutions will be vital to manage this adaptation, as discussed in the context of adaptive networking [43], where services dynamically adjust data transmission rates based on network conditions. Such optimization is essential in scenarios where network and computational resources are shared by multiple applications with differing requirements.
3. Context-awareness: As edge environments become more diverse, adaptive services will need to be aware of the broader context in which they are operating. This includes not only network conditions but also hardware capabilities, power constraints, device mobility, and application-specific requirements. Future adaptive services will need to react to changing contexts, ensuring that applications are able to continue operating optimally despite environmental challenges. This will be particularly critical in mobile edge computing environments, where devices may move between networks with varying characteristics, or in environments with fluctuating power levels, such as remote healthcare services.

4. Enhanced User Experience: Ensuring a consistent and reliable user experience will remain a driving force behind adaptive services. These services will adjust to the operating environment, dynamically modifying aspects like video resolution or data compression to maintain acceptable performance levels, even when bandwidth or device capabilities are constrained. As shown in prior studies on QoS adaptation in dynamic environments [44], adaptive services that can react to changes in network conditions are essential for applications like real-time video streaming or remote consultations. This flexibility ensures that users continue to receive high-quality services, even in challenging environments.

10.6.2 Toward Adaptive Services in Distributed Edge Computing

In the context of distributed edge computing, adaptive services will not operate in isolation but as part of a broader, more integrated system. The future of edge computing will require sophisticated orchestration and management frameworks that can dynamically allocate resources across a wide variety of applications. These frameworks will ensure that applications with competing resource needs—such as network bandwidth, computational power, and storage—are allocated resources in an efficient and prioritized manner. Middleware solutions that expose real-time network information to higher layers, as described in the research on cross-layer optimization [42], will be integral to enabling this flexibility.

For instance, in tactical edge environments, where multiple applications may need to share a limited network or processing capacity, adaptive services will allow for the dynamic allocation of resources based on task priority. A middleware layer will enable communication between the application and network layers, ensuring that services adjust to network conditions or resource availability in real-time. This will allow higher-priority applications, such as mission-critical communications, to be given preferential treatment, while lower-priority tasks may be deferred or run with reduced resource usage.

As described in the context of adaptive networking [43], the future vision of distributed edge computing will require the orchestration of compute and network resources across heterogeneous systems. In coalition operations, where different organizations may be using varying standards and protocols, adaptive services must adjust their behavior to ensure interoperability across systems, as highlighted in the work on NATO coalition operations [45]. By enabling cross-layer adaptation, these services will ensure that communication protocols, data formats, and security measures are compatible, facilitating effective information exchange across different edge devices and systems.

10.6.3 Supporting Experiments, Ongoing, and Future Work

Recent experimental work has demonstrated the potential for adaptive services to enhance edge computing performance in real-world environments. For instance, test beds for 5G networks [44], open radio access networks (ORAN) [46], and tactical networks [21] have shown that adaptive resource management can be successfully implemented to optimize network performance, particularly in environments with dynamic conditions and limited resources. These experiments have demonstrated the feasibility of applying cross-layer optimization and adaptive resource management techniques to improve QoS, reduce latency, and ensure service continuity.

However, further research is needed to refine these capabilities and fully integrate adaptive services into distributed edge computing environments. Future work should focus on enhancing the flexibility of edge devices and networks, ensuring that they can respond to changing conditions without sacrificing performance. In particular, advances in middleware and orchestration techniques will be essential to managing complex interactions between the network, resources, and applications. As the demand for edge computing services grows, the integration of adaptive services will be vital for enabling the next generation of intelligent, responsive, and efficient distributed computing systems. These systems will be able to support a wide range of applications, ensuring reliable, high-performance service delivery in challenging environments.

10.7 Conclusion

This chapter provided an overview of adaptive edge computing networks, discussing such systems as a prime example of RCAS. These networks have strong potential to enhance computing in complex, remote, and extreme situations by leveraging resilience, adaptability, and decentralized control. We described the current state of the components of the complex edge computing system, which highlighted the complexity of the system but also suggested the potential of these systems, if configured appropriately, to substantially enhance computing at the edge.

Our framework positions edge computing systems within four interrelated categories that collectively define the attributes of a sustainable and high-performing distributed infrastructure: (1) *Robust and resilient* systems emphasize fault tolerance, redundancy, and self-healing to sustain operation under disruption. (2) *Adaptive* systems leverage scalability, modularity, and feedback-driven learning to adjust dynamically to environmental changes. (3) *Secure* systems employ diversity, decentralization, and adaptability to protect data and functionality against evolving threats. (4) *Realizable and practical* systems integrate modularity, scalability, and redundancy with continuous feedback to ensure feasibility and efficiency in deployment.

By assessing the robustness and resilience, adaptability, security, practicality, and realizability of the systems, it serves as a first step to understand the applicability of such deployments into a range of edge environments. We applied this approach to several use cases, including disaster response, home automation and security, and environmental sensing. Lastly, we cast the vision of the future of edge computing, specifically asserting the importance of dynamically adaptive and distributed capabilities.

10.8 Appendix: Edge Network Component Tables

Table 10.3 Description of edge compute platforms with approximate ranges of performance characteristics.

Device class	GFLOPS	Avg. Power (W)	RAM (GB)	Cost ($)
IoT	0.07–31	1–8.6	0.5–	≤ 100
CPU	3–2000	25–200	1–80	200–3000
GPU	1000–20 000	100–500	4–32	400–5000

Source: References [12, 13, 47, 48].

Table 10.4 Description of communication capabilities and approximate range of capabilities.

Communications	Data rate	Communication range	Communication frequency	Transmit power
LoRA, LoRaWan	≤250 kbps	2–6 km	100–900 MHz	70 mW
TETRA1/2	7.2–16/250 kbps	50/83 km	160/400 MHz	≤2 W
Combat net radio	9.6 kbps	10 km	30–108 MHz	0.25, 20 W
Tactical broadband	1–2 Mbps	100 m	225–400 MHz	100 mW–5 W
WiFi	1–9608 Mbps	100 m	2.4, 6 GHz	≤1 W
SATCOM	0.01–133 Mbps	1000 km	100 Mhz–75 GHz	5–100 W
5G	100–700 Mbps	0.5–3 km	0.45 kHz–52 GHz	260 mW–120 W
mmwave	≤1 Gbps	1 km	30–300 GHz	0.5–1.5 W
Narrowband IoT	26–159 kbps	1–10 km	180–200 kHz	0.03–0.2 W

Source: References [30, 49–56].

Table 10.5 Comparison of Kubernetes variations and key properties.

Variant	Binary size	RAM recommendation	Compatibility with edge/cloud
K8s	300 MB	2 GB	Scalable and lightweight
K3s	65 MB	≤512 MB	Both cloud and edge environments
KOs	165 MB	0.5 GB	Distributed, high fault tolerance
microK8s	192 MB	512 MB–1GB	Centralized control plane
KubeEdge	76 MB	256 MB	Server-agent paradigm
Minikube	15 MB	2 GB	Only single-node supported

Source: References [10, 17, 57].

References

1 V. K. Sarker, J. Peña Queralta, T. N. Gia, H. Tenhunen, and T. Westerlund. A survey on LoRa for IoT: Integrating edge computing. In *2019 Fourth International Conference on Fog and Mobile Edge Computing (FMEC)*, pages 295–300, 2019. https://doi.org/10.1109/FMEC.2019.8795313.

2 Wazir Zada Khan, Ejaz Ahmed, Saqib Hakak, Ibrar Yaqoob, and Arif Ahmed. Edge computing: A survey. *Future Generation Computer Systems*, 97:219–235, 2019. https://doi.org/10.1016/j.future.2019.02.050.

3 Quoc-Viet Pham, Fang Fang, Vu Nguyen Ha, Md. Jalil Piran, Mai Le, Long Bao Le, Won-Joo Hwang, and Zhiguo Ding. A survey of multi-access edge computing in 5G and beyond: Fundamentals, technology integration, and state-of-the-art. *IEEE Access*, 8:116974–117017, 2020. https://doi.org/10.1109/ACCESS.2020.3001277.

4 Mung Chiang and Tao Zhang. Fog and IoT: An overview of research opportunities. *IEEE Internet of Things Journal*, 3(6):854–864, 2016. https://doi.org/10.1109/JIOT.2016.2584538.

5 Sami Kekki, Walter Featherstone, Yonggang Fang, Pekka Kuure, Alice Li, Anurag Ranjan, Debashish Purkayastha, Feng Jiangping, Danny Frydman, Gianluca Verin, Kuo-Wei Wen, Kwihoon Kim, Rohit Arora, Andy Odgers, Luis M. Contreras, and Salvatore Scarpina. MEC in 5G networks. ETSI White Paper No. 28, 2018.

6 Javier Mendez, Kay Bierzynski, M. P. Cuéllar, and Diego P. Morales. Edge intelligence: Concepts, architectures, applications, and future directions. *ACM Transactions on Embedded Computing System*, 21(5):1–41, 2022. https://doi.org/10.1145/3486674.

7 Tobias Meuser, Lauri Loven, Monowar Bhuyan, Shishir G. Patil, Schahram Dustdar, Atakan Aral, Suzan Bayhan, Christian Becker, Eyal de Lara, Aaron Yi Ding, Janick Edinger, James Gross, Nitinder Mohan, Andy D. Pimentel, Etienne

Riviere, Henning Schulzrinne, Pieter Simoens, Gurkan Solmaz, and Michael Welzl. Revisiting edge AI: opportunities and challenges. *IEEE Internet Computing*, 28(04):49–59, 2024. https://doi.ieeecomputersociety.org/10.1109/MIC.2024.3383758.

8 Xin Jin, Jonathan Larson, Weiwei Yang, and Zhiqiang Lin. Binary code summarization: Benchmarking chatGPT/GPT-4 and other large language models, 2023. https://arxiv.org/abs/2312.09601.

9 Md Delowar Hossain, Tangina Sultana, Sharmen Akhter, Md Imtiaz Hossain, Ngo Thien Thu, Luan N. T. Huynh, Ga Won Lee, and Eui Nam Huh. The role of microservice approach in edge computing: Opportunities, challenges, and research directions. *ICT Express*, 9(6):1162–1182, 2023. https://doi.org/10.1016/j.icte.2023.06.006.

10 Kubernetes. Kubernetes documentation. https://kubernetes.io/docs/home (accessed January 08, 2025).

11 Haochen Hua, Yutong Li, Tonghe Wang, Nanqing Dong, Wei Li, and Junwei Cao. Edge computing with artificial intelligence: A machine learning perspective. *ACM Computing Survey*, 55(9), 1–35, 2023. https://doi.org/10.1145/3555802.

12 Ahmet Ali Süzen, Burhan Duman, and Şen Betül. Benchmark analysis of Jetson TX2, Jetson Nano and Raspberry PI using deep-CNN. In *2020 International Congress on Human-Computer Interaction, Optimization and Robotic Applications (HORA)*, pages 1–5, 2020.

13 VMW Research Group. The GFLOPS/W of the various machines in the VMW research group, 2024.

14 Arpit Gupta, Laurent Vanbever, Muhammad Shahbaz, Sean P. Donovan, Brandon Schlinker, Nick Feamster, Jennifer Rexford, Scott Shenker, Russ Clark, and Ethan Katz-Bassett. SDX: A software defined internet exchange. *SIGCOMM Computer Communication Review*, 44(4):551–562, 2014. https://doi.org/10.1145/2740070.2626300.

15 Hend Abdelgader Eissa, Kenz A. Bozed, and Hadil Younis. Software defined networking. In *2019 19th International Conference on Sciences and Techniques of Automatic Control and Computer Engineering (STA)*, pages 620–625, 2019. https://doi.org/10.1109/STA.2019.8717234.

16 Tejas Subramanya, Leonardo Goratti, Shah Nawaz Khan, Emmanouil Kafetzakis, Ioannis Giannoulakis, and Roberto Riggio. SDEC: A platform for software defined mobile edge computing research and experimentation. In *2017 IEEE Conference on Network Function Virtualization and Software Defined Networks (NFV-SDN)*, pages 1–2, 2017. https://doi.org/10.1109/NFV-SDN.2017.8169849.

17 Rafael Vaño, Ignacio Lacalle, Piotr Sowiński, Raúl S-Julián, and Carlos E. Palau. Cloud-native workload orchestration at the edge: A deployment review and future directions. *Sensors*, 23(4):2215, 2023. https://doi.org/10.3390/s23042215.

18 Jan Bauwens, Peter Ruckebusch, Spilios Giannoulis, Ingrid Moerman, and Eli De Poorter. Over-the-air software updates in the internet of things: An overview of key principles. *IEEE Communications Magazine*, 58(2):35–41, 2020. https://doi.org/10.1109/MCOM.001.1900125.

19 Anna Ferrari, Daniela Micucci, Marco Mobilio, and Paolo Napoletano. On the personalization of classification models for human activity recognition. *IEEE Access*, 8:32066–32079, 2020. https://doi.org/10.1109/ACCESS.2020.2973425.

20 Scott Brown, David Harman, Cleon Anderson, and Matthew Dwyer. Measuring data transmissions from the edge for distributed inferencing with gRPC. In *2023 IEEE International Conference on Big Data (BigData)*, 2023.

21 Jake Perazzone, Matthew Dwyer, Kevin Chan, Cleon Anderson, and Scott Brown. Enabling machine learning on resource-constrained tactical networks. In *MILCOM 2022 - 2022 IEEE Military Communications Conference (MILCOM)*, pages 932–937, 2022.

22 Muhammad Babar, Mian Ahmad Jan, Xiangjian He, Muhammad Usman Tariq, Spyridon Mastorakis, and Ryan Alturki. An optimized iot-enabled big data analytics architecture for edge–cloud computing. *IEEE Internet of Things Journal*, 10(5):3995–4005, 2023. https://doi.org/10.1109/JIOT.2022.3157552.

23 Austin Harris, Jose Stovall, and Mina Sartipi. MLK smart corridor: An urban testbed for smart city applications. In *2019 IEEE International Conference on Big Data (Big Data)*, pages 3506–3511, 2019. https://doi.org/10.1109/BigData47090.2019.9006382.

24 Rahmadi Trimananda, Ali Younis, Bojun Wang, Bin Xu, Brian Demsky, and Guoqing Xu. Vigilia: Securing smart home edge computing. In *2018 IEEE/ACM Symposium on Edge Computing (SEC)*, pages 74–89, IEEE Computer Society, Los Alamitos, CA, USA, 2018. https://doi.org/10.1109/SEC.2018.00013.

25 Nitin Naik. Demystifying properties of distributed systems. In *2021 IEEE International Symposium on Systems Engineering (ISSE)*, pages 1–8, 2021.

26 F. Mancini. A security perspective on the military application of edge computing. *IST-208-RSY*, Lillehammer, Norway 15–16 October 2024, 2024.

27 Silvia Carpitella, Václav Kratochvíl, and Miroslav Pištěk. Multi-criteria decision making beyond consistency: An alternative to AHP for real-world industrial problems. *Computers & Industrial Engineering*, 198:110661, 2024. https://doi.org/10.1016/j.cie.2024.110661.

28 Kevin Chan, Kelvin Marcus, Lisa Scott, and Rommie Hardy. Quality of information approach to improving source selection in tactical networks. In *2015 18th International Conference on Information Fusion (Fusion)*, pages 566–573, 2015.

29 Igor Svoboda and Dmytro Lande. Enhancing multi-criteria decision analysis with AI: Integrating analytic hierarchy process and GPT-4 for automated decision support, 2024. https://arxiv.org/abs/2402.07404.

30 Mehmet Ulema. *Terrestrial Trunked Radio (Tetra)*, pages 87–106. John Wiley & Sons, 2019. https://doi.org/10.1002/9781119369554.ch6.

31 5G PPP. 5G-VINNI, https://www.5g-vinni.eu (accessed January 30, 2021).

32 Manas Pradhan. Interoperability for disaster relief operations in smart city environments. PhD thesis, Department of technology systems, University of Oslo, Norway, 2021.

33 Thomas Kudla, Mattia Fogli, Sean Webb, Geert Pingen, Niranjan Suri, and Harrie Bastiaansen. Quantifying the performance of cloud-oriented container orchestrators on emulated tactical networks. *IEEE Communications Magazine*, 60:74–80, 2022. https://doi.org/10.1109/MCOM.003.00975.

34 MQTT.org. MQTT is an OASIS standard, specifications, 2025. https://mqtt.org/mqtt-specification.

35 Marco Manso, Fernando Freire, Janusz Furtak, Bárbara Guerra, James Michaelis, Daniel Ota, Reinhard Claus, Niranjan Suri, Roberto Fronteddu, Edoardo Di Caro, Konrad Wrona, Emil P. Andersen, and Frank T. Johnsen. IoT in coalition federated operations: Multi-national C2 integration and technical interoperability experiments. In *MILCOM 2023 - 2023 IEEE Military Communications Conference (MILCOM)*, pages 219–224, 2023. https://doi.org/10.1109/MILCOM58377.2023.10356285.

36 Nikolai Vatne. Edge computing for disaster relief operations. Master's thesis, Department of Informatics, The Faculty of Mathematics and Natural Sciences, University of Oslo, Norway, 2022.

37 Norman Jansen, Marco Manso, Andrew Toth, Kevin S. Chan, Trude H. Bloebaum, and Frank T. Johnsen. NATO core services profiling for hybrid tactical networks — results and recommendations. In *2021 International Conference on Military Communication and Information Systems (ICMCIS)*, pages 1–8, 2021. https://doi.org/10.1109/ICMCIS52405.2021.9486415.

38 Karrie Jacobs. Toronto wants to kill the smart city forever. https://www.technologyreview.com/2022/06/29/1054005/toronto-kill-the-smart-city/amp (accessed January 30, 2022).

39 Nikhil Gupta. Smart homes are getting smarter. https://transmitter.ieee.org/smart-homes-are-getting-smarter (accessed January30, 2024).

40 Hui Yu, Guillermo J. Amador, Antoine Cribellier, Marcel Klaassen, Henrik J. de Knegt, Marc Naguib, Reindert Nijland, Lukasz Nowak, Herbert H. T. Prins, Lysanne Snijders, Chris Tyson, and Florian T. Muijres. Edge computing in wildlife behavior and ecology. *Trends in Ecology & Evolution*, 39(2):128–130, 2024. https://doi.org/10.1016/j.tree.2023.11.014.

41 Javad Roostaei, Yongli Z. Wager, Weisong Shi, Timothy Dittrich, Carol Miller, and Kishore Gopalakrishnan. IoT-based edge computing (IOTEC) for improved environmental monitoring. *Sustainable Computing: Informatics and Systems*, 38:100870, 2023. https://doi.org/10.1016/j.suscom.2023.100870.

42 Emil P. Andersen, Mariann Hauge, and Frank T. Johnsen. Cross-layer optimization: middleware-enabled adaptive services at the tactical edge. In *IST-205-RSY – ICMCIS*, Koblenz, Germany, 2024.

43 Frank T. Johnsen, Mariann Hauge, Lars Landmark, Emil P. Andersen, Terje M. Mjelde, Jostein Sander, and Martin Rytir. Adaptive networking at the tactical edge. In *2023 International Conference on Military Communications and Information Systems (ICMCIS)*, pages 1–10, 2023. https://doi.org/10.1109/ICMCIS59922.2023.10253558.

44 Angel Martin, Jon Egaña, Julián Flórez, Jon Montalbán, Igor G. Olaizola, Marco Quartulli, Roberto Viola, and Mikel Zorrilla. Network resource allocation system for QoE-aware delivery of media services in 5G networks. *IEEE Transactions on Broadcasting*, 64(2):561–574, 2018. https://doi.org/10.1109/TBC.2018.2828608.

45 Frank T. Johnsen and Mariann Hauge. Interoperable, adaptable, information exchange in NATO coalition operations. *Journal of Military Studies*, 11(1):49–62, 2022. https://doi.org/10.2478/jms-2022-0005.

46 Leonardo Bonati, Pedram Johari, Michele Polese, Salvatore D'Oro, Subhramoy Mohanti, Miead Tehrani-Moayyed, Davide Villa, Shweta Shrivastava, Chinenye Tassie, Kurt Yoder, Ajeet Bagga, Paresh Patel, Ventz Petkov, Michael Seltser, Francesco Restuccia, Abhimanyu Gosain, Kaushik R. Chowdhury, Stefano Basagni, and Tommaso Melodia. Colosseum: Large-scale wireless experimentation through hardware-in-the-loop network emulation. In *2021 IEEE International Symposium on Dynamic Spectrum Access Networks (DySPAN)*, pages 105–113, 2021.

47 The Raspberry Pi Foundation. Raspberry Pi, 2024. https://www.raspberrypi.com (accessed December 04, 2024).

48 Maurizio Capra, Beatrice Bussolino, Alberto Marchisio, Guido Masera, Maurizio Martina, and Muhammad Shafique. Hardware and software optimizations for accelerating deep neural networks: Survey of current trends, challenges, and the road ahead. *IEEE Access*, 8:225134–225180, 2020. https://doi.org/10.1109/ACCESS.2020.3039858.

49 A. J. Vigil, John Kirk, Patrick Rayermann, Jessica L. Gregris, Dahesh A. Khalil, and Carl O. Morris. DOD SATCOM growth measurement and planning. In *MILCOM 2022 - 2022 IEEE Military Communications Conference (MILCOM)*, pages 841–848, 2022. https://doi.org10.1109/MILCOM55135.2022.10017764.

50 John Gieske. 5G mmWave vs. Sub-6: A comparative study of two 5G frequency bands, https://nybsys.com/5g-mmwave-vs-sub-6 (accessed August18, 2024).

51 Rui Ma and Zhuxi Zhang. A case study of starlink and 5G in future space based internet services. In *2022 4th International Symposium on Smart and Healthy Cities (ISHC)*, pages 64–67, 2022. https://doi.org/10.1109/ISHC56805.2022.00022.

52 AVSystem. IoT communication protocols with measurements for NB-IoT - expert guide. https://avsystem.com/blog/iot/iot-communication-protocols-with-calculat ions (accessed February 29, 2024).

53 Alexandru-Gabriel Romaniuc, Vlad-Cosmin Vasile, and Monica-Elena Borda. Experimental determination of the area coverage of a tetra base station in a dense urban area. In *2022 International Symposium on Electronics and Telecommunications (ISETC)*, pages 1–4, 2022. https://doi.org/10.1109/ISETC56213.2022.10009957.

54 Elbit Systems. Tadiran CNR-9000, https://elbitsystems.com/media/CNR-9000-1.pdf (accessed January 08, 2025).

55 Motorola. MTP6650 TETRA portable radio data sheet. https://www.motorolasolutions.com/content/dam/msi/Products/dimetra-tetra/tetra-terminals/hand-portables/mtp6000/mtp6650_data_sheet_en.pdf (accessed January 08, 2020).

56 OFCOM. Tetra factsheet. Switzerland, https://www.bakom.admin.ch/dam/bakom/en/dokumente/faktenblatt_tetra.pdf.download.pdf/factsheet.pdf (accessed January30, 2015).

57 Heiko Koziolek and Nafise Eskandani. Lightweight Kubernetes distributions: A performance comparison of microk8s, k3s, k0s, and microshift. In *Proceedings of the 2023 ACM/SPEC International Conference on Performance Engineering, ICPE '23*, pages 17–29, Association for Computing Machinery. New York, NY, USA, 2023. https://doi.org/10.1145/3578244.3583737.

Part III

RCAS Engineering

11

Resilience-aware Development Tools: Translating Resilience Concepts into Engineering Practice

Kelly J. Neville,[1] Tracy Sanders,[1] Emily Barrett,[1] Theresa Fersch,[1] Glenn Lematta,[1] Xuan Chau,[1] Andrew Langone,[1] and Emily S. Patterson[2]*

[1] *The MITRE Corporation, McLean, VA, USA*
[2] *The Ohio State University, Columbus, OH, USA*

*Corresponding author: kneville@mitre.org

11.1 Introduction

Resilience has been defined as the ability of a system to adaptively adjust its functioning prior to, during, and following events, changes, disturbances, and opportunities, and thereby sustain operations required to achieve the goals and mission of the system [1], and it is an inherent capacity of complex adaptive systems [1–6]. In this chapter, we focus on high-consequence sociotechnical work systems, a type of CAS responsible for critical operations such as air traffic management, health care, military, and utility operations. In the operation of such systems, failures to respond to challenges quickly and adaptively can have grave consequences, including threats to human life, system survival, and the inability to achieve a critical mission.

In CASs, humans are typically the system elements that enable adaptivity and hence system resilience [7, 8]. As technology becomes more capable, it is increasingly allocated to performing functions that had previously been the purview of humans, potentially impairing a system's resilience capacity. A goal of technology development and modernization efforts should be to preserve or improve the resilience capacity required to thrive and continue achieving the system's critical mission or purpose. Often, the focus of system improvements, however, is limited to efficiency gains, cutting expenses, or strengthening certain system capabilities while not explicitly attending to resilience requirements or opportunities to strengthen it. Possibly, engineering teams assume the system will retain its resilience capacity while they make changes to other system aspects. Or possibly they lack methods and tools to recognize and address resilience in an adequately rigorous way.

Kauffman [9] in 1980 wrote, "everything is connected to everything else....in addition to the immediate effects of an action, there will always be other consequences of it which ripple through the system" (p. 38). That is, new technologies or other changes may undermine the coordinated resourcefulness and adaptivity necessary among system agents and units to safely and successfully achieve critical work operation goals across conditions, i.e., with resilience. We maintain that changes made without attention to resilience can easily result in resilience losses [10, 11] and missed opportunities for resilience gains.

11.2 Challenges

Because high-consequence work systems require a capacity for resilience, systems engineering activities aimed toward modernizing and improving them should include techniques for attending to their resilience capacity. This is especially true when technologies are given human roles or roles on which humans in the system depend. As the elements of a CAS that are most capable of adapting, replacing or introducing constraints on humans can have unintended consequences for CAS resilience. Negative impacts on system resilience may be an acceptable trade-off for the capabilities gained, but such a trade-off should be deliberate and subject to assessment, particularly for a CAS that is a high-consequence work system. There should be methodological and tool support for considering resilience impacts during the development of technology for high-consequence work systems. These work systems need to perform successfully across a range of conditions, including volatile, unpredictable, complex, and ambiguous conditions [12]. This means the work system elements must be capable of performing in those conditions while in synchrony with the work system and aligned with its goals. "When organized into a complex, integral whole, parts become correlated as a function of context-dependent constraints imposed on them by the newly organized system in which they are now embedded" [13]. New technologies must be able to operate as a part of an integral, whole CAS as it responds with resilience to diverse, including challenging, demands and conditions.

Systems engineering activities that facilitate the development of technologies supportive of resilience in CAS operations include iterative evaluation and development cycles (e.g., in agile development endeavors). An underdevelopment technology's exposure to work system operations within those iterative cycles provides a feedback loop that supports co-evolution of a technology and a complex adaptive work system. Although efforts to improve the situation are underway (e.g., [14]), current systems engineering practice lacks effective means for processing and prioritizing feedback of work system personnel, which can be substantive, during development (e.g., [15, 16, 14, 17, 18]).

Efforts to support operational resilience during technology development include Sherwood et al. [19]. They used a risk management approach to assess and address user feedback describing potential resilience-related risks to operations that emerged across iterations of a technology's conceptual design (e.g., [19]). Minai et al. [20] offer an evolution-based engineering paradigm called *Complex Engineered Systems* for developing and modernizing socio-technical work CASs capable of resiliently adapting and evolving. They propose designing systems with the variety and flexibility to self-adapt and self-evolve as conditions and demands change. Variety and flexibility enable adaptive agents—operations personnel in our case—to use certain technologies, functions, and features over others and to use and interact with them and each other in different ways in response to operational contexts. Although operations personnel are largely responsible for driving system adaptation and evolution in this paradigm, technologies in the system must also be capable of participating. The result is a capacity for mutual responsiveness and co-evolution.

As just noted, appropriately designed iterative engineering approaches can support the co-evolution that enables a new technology to integrate into a work system's resilience mechanisms. An important distinction for iteration that supports resilience is that iterations focus on using feedback to holistically improve the technology's fit into the operational CAS and not be used to develop different technology features and functions in turn. Co-evolution depends on opportunities for all system elements and the operational environment to interact, influence, and shape one another. Engineering complex systems is challenging, and there is a tendency toward making it more manageable by falling back on linear, sequential, piece-by-piece engineering approaches [21, 22]. Attempts to gain adoption of holistic iterative development, such as the spiral development model [23], have experienced only limited success [24]. Agile development is another development approach with the potential to support co-evolution. However, it typically unfolds as the development or enhancement of a series of features and functions assigned to sequences of sprints [25] and prioritizes feedback from customers and "business people" over users [26].

The abstract and emergent nature of resilience creates challenges for incorporating it into the systems engineering of CASs and their technical elements. In work systems, resilience is the emergent product of many system agents and units coordinating over sensemaking and behavior [27]. We tend to know resilience when we see it, but the factors that contribute to it have been difficult to pin down. This means resilience is incompatible with conventional approaches that center on meeting testable requirements and stated needs of customers [20, 28].

Being difficult to test should not prevent attention to resilience. Engineering teams who do not find ways to consider the impacts of their technologies on work system resilience risk unknowingly compromising that resilience. They set the stage for surprise technology malfunctions and coordination breakdowns when performance conditions exceed the routine conditions anticipated by requirements, a time when effective performance matters most (Woods [29] referred to this as "dark debt").

11.3 Operationalizing Resilience

Tools exist for assessing resilience in socio-technical work systems (for a review, see [30]). The majority of these tools were designed to assess resilience enablers and the capacity of an organization (e.g., [27, 31, 32, 54, 55, 56]). For the most part, they were not developed for use in technology development and unsurprisingly are not typically used for that purpose [58]. Exceptions include methods for evaluating how well a new technology interacts with humans (e.g., [59, 60, 61]). The quality of this interaction likely contributes to resilience at the work system level. However, we assert that what matters most is whether the work system is able to accomplish its mission despite adversity. This requires resilience at the work system level, across all interacting system agents and elements. One tool that can be used to evaluate a technology concept or prototype in terms of work system resilience impacts is the functional resonance analysis method (FRAM, [33]). FRAM can be used to model a work system in terms of relationships among elements during work operations (i.e., during "work as done") such that the direct and indirect effects of a new technology on those operations can be seen and evaluated [21]. Although FRAM is time consuming to use and does not explicitly facilitate the identification of resilience enablers nor impacts on them, it does support the representation and assessment of certain key resilience enablers, such as a variety of means to accomplish work and response flexibility.

We seek to build on these efforts to improve the ability of engineers, customers, and other stakeholders to demonstrate due diligence and attend to work system resilience during the development of new technologies and other system modernization activities. To this end, we developed a set of tools designed to help address work system resilience requirements of our high-consequence work systems during technology development. These tools, called Resilience-Aware Development (RAD) tools, are designed to support the translation of abstract resilience concepts into technology development practice. The resilience concepts and thus the RAD tools are likely applicable to CASs in general, although future work is required to evaluate this hypothesis.

Note that our aim is not to support the development of resilient technologies but rather the development of technologies that preserve or improve the resilience capacity of the work system they are to become embedded within it. As observed by Van der Merwe et al. [62], "Different infrastructure systems are dependent on and connected to one another resulting in critical infrastructure interdependency," as are the different individual technologies within those systems. We suggest that by supporting attention to work system resilience, we can reduce the threat of dark debt and retain or improve the ability of the work system to handle both routine and severe nonroutine challenges, thereby earning operator trust and improving adoption likelihood.

11.4 Tool Development Approach

The primary goal of RAD tool development was to facilitate the explicit consideration of resilience during technology development and system modernization. A secondary goal was that the tools be compatible with technology development and systems engineering practices, with a particular focus on agile development. More specifically, tools should be lightweight and produce results or outputs that can be readily used as inputs to various systems engineering artifacts and processes (e.g., as user stories or requirements).

The RAD tools were built to operationalize system resilience enablers, i.e., system characteristics and capabilities that contribute to resilience capacity, specified in the Transform with Resilience during Upgrades to Sociotechnical Systems (TRUSTS) Framework ([63], [64], https://trusts.mep.mitre.org [65]). The TRUSTS Framework was developed to make abstract resilience concepts more relatable to technology development and more tractable for technology developers. It synthesizes into five resilience enabler categories research, theory, and case studies from across multiple bodies of work, including complex system science, resilience engineering, high reliability organizations (HROs), naturalistic decision-making, cognitive systems engineering, and joint cognitive systems [34]. These are the framework's five interrelated, Tier 1 CAS resilience-enabling capabilities ("the big five"), and they consist of:

- **Shared Demand and Deviation Awareness**: All work units and agents, including personnel and software agents, recognize, process, and make others aware of new demands and deviations from expectations and "business as usual."
- **Response Coordination**: Coordination and cooperation among work units avoids coordination breakdowns and brings to bear the full set of resources and capabilities of the work system to achieve its mission across challenging events and conditions.

- **Progressive Responding**: In dynamic work systems and environments, the best response is one that is quickly initiated and then continuously adapted to and used to influence the unfolding situation.
- **Maneuver Capacity**: The work system and its work units have the ability to be responsive to variety in the conditions and challenges they encounter.
- **Guided Local Control**: Frontline work units and personnel are able to respond directly and rapidly to the situations they face. Guided local control is benefited by a nested-layer control scheme where each layer takes primary responsibility for a different granularity of operations. The layers operate in complementary ways to enable local and polycentric control, both contributing to system resilience.

The TRUSTS Framework has three tiers of resilience factors. Each of the five Tier 1 resilience enablers decomposes into two to five contributing resilience enablers. Each Tier 2 source in turn decomposes into two to eight contributing Tier 3 enablers, which include some enablers specific to high-consequence work systems and not just CASs in general. Figure 11.1 shows a partial decomposition of the Tier 1 Shared Demand and Deviation Awareness resilience enabler.

A premise of RAD tool design is that direct and indirect resilience effects of a new technology become recognizable as the technology is developed and can be considered or evaluated in terms of the operational context. Many effects are impossible

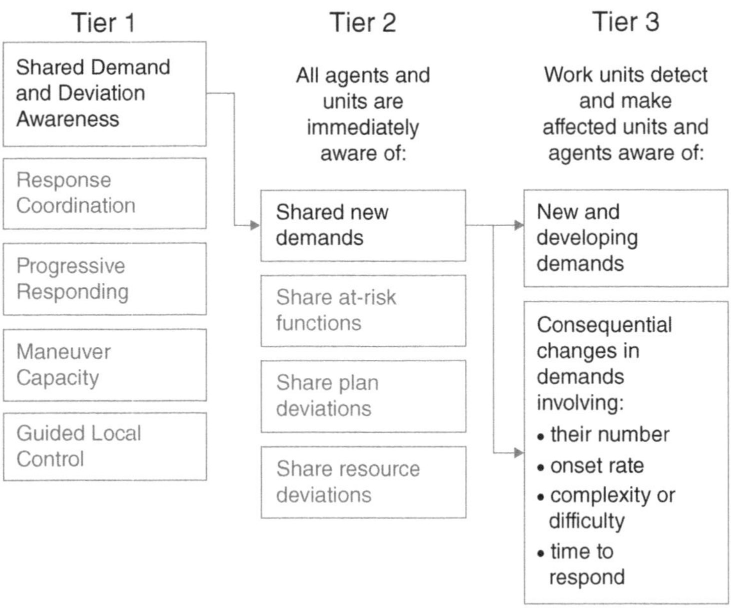

Figure 11.1 Partial decomposition of the Tier 1 TRUSTS framework source shared demand and deviation awareness.

to predict ahead during early phases of development, such as during conceptualization and requirements development [35]. Once a technology has been developed and, according to the traditional systems engineering timeline (e.g., [66]), is ready for operational test-and-evaluation activities, it is typically too costly to make changes (e.g., [16]). So, the best time to start evaluating the effects is when the technology is in an early mockup or prototype form and then iteratively across its development. A related premise is that co-evolution involving interactions of system elements and the operational environment enables the development of technology that contributes to, and will not impair, work system resilience [36].

Based on the goals and premise above, RAD tools were proposed over the course of a two-year project focused on improving the rate of new technology adoption. Each proposed tool was drafted and iteratively refined by a team of two to four of this chapter's authors. Each tool was tried out in at least one technology development or evaluation effort with two exceptions: the development reference guide and the last tool to be developed, the RAD-Simulation Probes tool. Four tools were critiqued by experts in systems engineering, high-consequence work systems, or both. Feedback from all sources was used for tool refinements. Some tool uses resulted in adapted versions available for others to use. Current tool versions and select adaptations are available on the website https://trusts.mep.mitre.org/toolkit/all-tools [37]. Tool overviews are presented below, along with information about their outputs and development. Development activities are included with each overview rather than presented in this *Tool Development Approach* section, so they can be considered in the context of the tool description and purpose.

11.5 RAD Tool Descriptions

In this section, we present the seven RAD tools we developed to provide technology developers with multiple options for thinking about the resilience needs of work system operations. We expect tool users to adapt the tools based on factors such as available resources, specifics of the development effort, and the development team's ability to iteratively adapt the technology design.

The RAD tools are not software, but are more aptly described as protocols, guides, and surveys. Some are both web- and document-based, while others are document-based only (all are available at the website above). We view technology development teams as the primary users; however, all tools are most effective if work system operations personnel, including end users of the technology, participate in or initiate their use. We recommend involving work operations personnel and domain experts to adequately represent operational demands and resilience capabilities. Although we focus on technology development, the tools are also applicable to new policies, concepts of operations, and other changes to be designed and incorporated into a high-consequence work system.

The seven RAD tools consist of, in order of conceptualization:

- **RAD-Survey (RAD-S)**: A survey with 30 questions asking if and how a technology design, in its current state of development, affects core aspects of the five major enablers of work system resilience represented by the TRUSTS Framework.
- **RAD-Exercise Protocol (RAD-XP)**: A protocol for lightweight (1- to 2-hour) tabletop exercises (TTXs) conducted between agile development cycles to produce feedback and user stories for integration into upcoming development sprints.
- **RAD-Explore (RAD-E)**: A lightweight 60- to 90-minute group broadening exercise development teams use to build their awareness of the variety of enablers of work system resilience and ways they are or could be supported by technology.
- **RAD-Observe (RAD-O)**: A guide and worksheet for the targeted collection and synthesis of resilience-related exercise observation data.
- **RAD-Levels (RAD-L)**: A tool for guiding and benchmarking technology development and acquisition projects to ensure operational resilience is being given adequate attention.
- **RAD-Critical Risk (RAD-CR)**: A 20-question survey focused on resilience enablers judged to be critical, show-stopper types of enablers based on the potential to work around their absence and their role in past mishaps; when used across a technology development effort, it could reveal likely technology design impacts on work system mechanisms critical to its resilience capacity.
- **RAD-Simulation Probes (RAD-SP)**: A guide for introducing challenges (probes) that test operational resilience into simulation exercise scenarios; can be used in the design of RAD-Xs to focus on work system resilience.

11.5.1 RAD-Survey

Overview: The RAD-Survey (RAD-S) [37] was designed to help obtain and prioritize resilience-related feedback from work system personnel (including users) about a technology under development, compare competing technologies, and evaluate new technology to ensure it is not inadvertently interfering with operational resilience. It consists of 30 statements, 15 positive and 15 negative, that map directly to Tier 3 TRUSTS Framework resilience enablers systematically sampled across the framework's Tier 2 enablers. Responses are made on a 5-point Likert Scale ranging from *strongly disagree* to *strongly agree*. Statements describe work system capabilities and are to be answered for future envisioned operations that include the new technology. The RAD-S can be completed by observers or participants who have experience working in the target work system immediately following their participation in or observation of an activity simulating or critiquing a proposed or under-development technology in the context of challenging work system operating conditions.

Example Survey Item: *The new technology interferes with the detection of time-critical situations requiring attention.*

Time: The RAD-S can be completed by a participant in 10–30 minutes, and survey data can be processed and summarized in less than an hour. If using the web-based version, results are provided immediately for each individual participant. Completing the survey will take longer if participants document the reasons for their ratings or use the survey as an analysis worksheet.

Outputs: The outcome of the analysis includes the identification of work system resilience enablers that are positively affected and those that may be negatively affected by a technology design or prototype. The web-based version produces two scores accompanied by lists of supported resilience enablers (positive impacts), at-risk resilience enablers (negative impacts), and at-risk items considered to be critical high-importance enablers (showstoppers), organized by Tier 1 resilience source. One of the two scores is the *Combined Resilience Impact Score*. This score ranges from 1 to 6 and its values are found in a lookup table (see Table 11.1) featuring percentages of items assessed positively (*Resilience Benefits*) and negatively (*Resilience Risk*) and the number of showstoppers identified as a risk. The *Adjusted Coordinated Adaptivity Score* ranges from 1 to 8 and is derived using a series of lookup tables featuring counts of positively assessed items (that become zero if one

Table 11.1 Lookup table used to calculate the combined resilience impact score.

Resilience risk:	0–0.25 low risk			0.26–0.50 med-low risk			0.51–0.75 med-high risk			0.76–1 high risk		
Showstoppers:	0	1–2	3–5	0	1–2	3–5	0	1–2	3–5	0	1–2	3–5
	4	1	1	3	1	1	2	1	1	1	1	1
RB of 0.26–0.5	5	2	1	4	2	1	3	1	1	2	1	1
RB of 0.51–0.75	6	3	1	5	2	1	4	2	1	3	1	1
RB of 0.76–1 (High)	6	4	1	6	3	1	5	2	1	4	2	1

Note: Low scores indicate high risk. A score of "1" indicates a very high risk that should be investigated and mitigated. RB, resilience benefits.

of the five showstopper items is assessed as a possible risk). The first lookup table in the series combines Progressive Responding and Maneuver Capacity counts to produce a *Potential for Adaptivity* score. The second lookup table combines Shared Demand and Deviation Awareness and Response Coordination counts to produce a *Potential for Coordinated Awareness and Action* score. These two scores are combined in a third lookup table to produce a *Potential for Coordinated Adaptivity* score, and this score is increased by the number of positively assessed Guided Local Control survey items. Table 11.2 shows the two final scores mapped to risk assessments and associated recommendations. The scoring schemes are intended to produce conservative estimates of risk and bring attention to aspects of a technology that should be evaluated further.

Table 11.1 features percentages of items assessed positively (Resilience Benefits) and negatively (Resilience Risk) and the number of showstoppers identified as a risk. The Adjusted Coordinated Adaptivity Score ranges from 1 to 8 and is derived using a series of lookup tables featuring counts of positively assessed items (that become zero if one of the five showstopper items is assessed as a possible risk). The first lookup table in the series combines Progressive Responding and Maneuver Capacity counts to produce a Potential for Adaptivity score. The second lookup table combines Shared Demand and Deviation Awareness and Response Coordination counts to produce a Potential for Coordinated Awareness and Action score. These two scores are combined in a third lookup table to produce a Potential for

Table 11.2 Scores produced by the two scoring schemes mapped to risk-based recommendations.

Combined resilience impact score	Adjusted coordinated adaptivity score	Risk-based recommendations
1	1–2	High risk: investigate possible showstoppers and additional ways to either improve benefits or reduce risks to resilience
2–3	3–4	Medium risk: investigate possible showstoppers and take steps to improve the balance between resilience benefits and risks
4	5–7	Low risk: investigate possible showstoppers; no other action needed as resilience benefits are otherwise adequate for mitigating new risks
5–6	7–8	Resilience gains assessed as outweighing any new risks. No action needed.

Note: Any showstopper survey items identified as a problem by three or more evaluators must be investigated.

Coordinated Adaptivity score, and this score is increased by the number of positively assessed Guided Local Control survey items. Table 11.2 shows the two final scores mapped to risk assessments and associated recommendations. The scoring schemes are intended to produce conservative estimates of risk and bring attention to aspects of a technology that should be evaluated further.

Development: Over the course of its development, the RAD-S has been deployed as a survey of varying lengths. Earlier versions have varied from 10 to 75 survey items. The longest version contained items for each Tier 3 resilience source. This was used by three of the authors as a framework for analyzing observational and interview data to assess decision support technology for a military time-critical targeting operation, but was judged by two experts in high-consequence military operations to be too cumbersome for their use as a survey. The 10 items of the shortest version were judged to be too few to adequately represent the resilience enablers. The current 30-item version was used by a team of a systems engineer, human factors engineer, and aviation operations professional to evaluate 737-MAX aircraft accidents. Their feedback resulted in the wordsmithing of a small number of questions. An 18-question version was used to compare two competing user interface design capabilities for a military drone-management work system. This latter version is available on the tools website to serve as an example of how the survey can be adapted. The scoring schemes have been peer reviewed by a colleague with a PhD in mathematics and 7 years of experience as a data analyst, and it is a goal to evaluate their value in future work.

11.5.2 RAD-Exercise Protocol

Overview: TTXs are often conducted over the timespan of a day or week; however, the RAD-Exercise Protocol (RAD-XP) [57] specifies a method for planning and conducting a one- to two-hour TTX to generate user stories that positively impact work system resilience. The TTX, which we call a RAD exercise (RAD-X), is kept short to fit into the accelerated pace and lightweight philosophy of agile engineering practices. RAD-Xs are conducted iteratively after agile development sprints or deliveries to assess progressive versions of a technology in terms of its readiness for operational demands and to produce user stories, requirements, and concepts of operations.

RAD-Xs are planned and conducted by members of the technology development teams with work system operations personnel, including users, serving as participants. During planning, RAD-X scenarios are designed to expose participants to a proposed, prototype, or maturing technology in an operational work context. Scenarios include challenges (see example below), and RAD-X discussions center around how those challenges would be handled currently and how the new technology might affect their handling. Once a RAD-X has been planned, little effort is required to conduct subsequent RAD-Xs, allowing developers to regularly engage with users in follow-on iterations of their technology. This repeatability, combined with the TTX format and provision of results in the form of user stories, enables development teams to incorporate RAD-Xs into agile development cycles. Figure 11.2 presents the main elements of RAD-XP: TTX development, TTX conduct, and analysis of TTX outputs.

Example Scenario Challenge: *Present erroneous data that do not fit the rest of the picture.*

Time: Three days for planning; one to two hours to conduct, including introduction, icebreaker, RAD-X, and hot wash (i.e., after action review) phases; and one day for processing and analyzing notes to produce requirements and "desirements" in the form of user stories (Less planning time will be required after the first in a series of RAD-Xs.)

Outputs: The primary outputs of a RAD-X are requirements or user stories that are traceable to technology use in relevant use cases or scenarios. Additionally, RAD-XP supports stakeholder analysis, journey-mapping, and the design of implementation strategies and concepts of operation.

Development: The RAD-XP was developed by working with two TTX experts with a combined 30 years of experience designing and conducting TTXs. The RAD-XP developers asked the experts to walk through their TTX planning and development process and identify critical facets of TTX development. Second, the experts were asked to design a game together while the developers observed, codified, and documented their workflows and verbalized considerations. The results

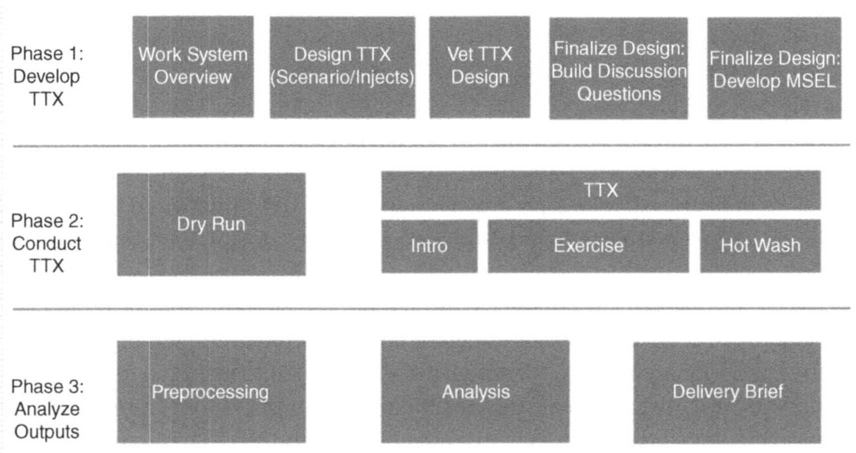

Figure 11.2 A high-level view of RAD-X planning, conduct, and analysis activities described in the RAD-XP.

of these activities were used to determine the structure and content of a 90-minute RAD-X and the process for planning and developing it. The RAD-XP was used to help a team design telemedicine decision support capabilities to identify new user stories, new stakeholders, and new concepts of operation to increase the value of the tool. A version of the RAD-XP was also used to identify a resilience-supportive design concept for a surgery schedule adaptation tool. For a complete description of RAD-XP development, see Dorton et al. [38].

11.5.3 RAD-Explore

Overview: RAD-Explore (RAD-E) [39] is a semi-structured brainstorming and broadening exercise during which a group of technology development team members and, optionally, end users comprehensively consider operations resilience enablers and ways a given technology could impact them. Unlike most other RAD tools, it does not require the involvement of operations personnel or users (although they could contribute resilience assessments that are better grounded in the realities of their operations). At minimum, the RAD-E requires a description of a technology concept. This means that RAD-E can be used at the earliest stages of technology maturity, potentially before initial features and designs are determined. RAD-E may also be used later in development and still have an impact on a technology's design, or implementation, or concept of operation.

The RAD-E exercise uses a shared whiteboard in a virtual or physical collaboration space. The whiteboard is divided into five segments that map to the five Tier 1 resilience enablers of the TRUSTS Framework. An example designed for use in an online collaboration space is shown in Figure 11.3. The exercise is led by a facilitator using the RAD-E protocol available at the tools website previously provided. The facilitator introduces each resilience source by defining it and giving examples of it in the context of restaurant operations and baseball, then invites participants to spend 6 minutes putting their ideas on the associated whiteboard section. Ideas include ways a given technology could benefit or detract from a work system's resilience capacity. As the whiteboard section for each factor has been completed, participants vote on the most important technology impacts from those identified.

Protocol Restaurant Example: *The restaurant staff can keep up with demands on a busy night because they sought out relevant information by confirming upcoming reservations and using it to update the strategic assignments of guests and wait staff to tables throughout the night.* (Example of the Progressive Responding resilience source)

Time: 60–90 minutes, depending on the pace set by the facilitator.

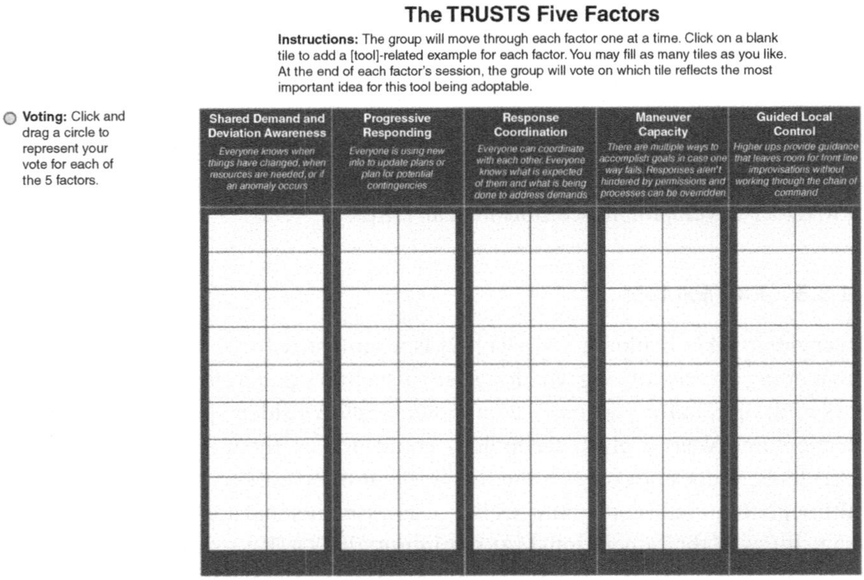

Figure 11.3 An example RAD-E exercise collaborative brainstorming template.

Outputs: Outcomes of a RAD-E exercise include (1) a development team that is familiarized with and better prepared to address work system resilience requirements and (2) a set of prioritized positive and negative resilience impacts that may be used to derive technology requirements or user stories.

Development: The RAD-E exercise was inspired by the lotus blossom brainstorming technique. The petals of the classical lotus blossom template were replaced with areas for recording ideas related to TRUSTS Framework resilience enablers. A RAD-E exercise protocol featuring the adapted lotus blossom template was developed and then refined based on its earliest uses. After each of its first two uses, both with development teams, the protocol was revised to better align with TRUSTS Framework definitions of resilience enablers, examples were refined, and details were added so that all information needed to use the tool was fully encapsulated within the protocol document. A third RAD-E exercise was conducted with a development team working on technology capabilities to improve the handling of surgery schedule changes. A fourth RAD-E exercise was conducted in person with 22 participants to generate design requirements for semi-automated air taxis. Participants were divided into five groups, each with its own whiteboard. As an addition to the protocol, each group shared its main insights during the last part of the exercise.

11.5.4 RAD-Observe

Overview: RAD-Observe (RAD-O) [40] is a data collection guide for an observer or analyst to use to collect resilience-related observational data during an exercise or to filter and organize collected data after the exercise. Relevant exercises include computer agent-based, human-in-the-loop, field, in situ, and tabletop exercises in which a new technology functions in an operational scenario. These evaluation activities can (and should) be conducted periodically across a development effort as a technology matures, using RAD-O among other methods.

An exercise observer uses the RAD-O data collection guide to document observed work behavior and situations that demonstrate the presence or absence of each of the five Trusts Framework Tier 1 resilience enablers in the work system. A second part of the RAD-O guide can be used to document observed behavior and situations indicating the presence or absence of Tier 2 resilience enablers. Alternatively, the second part can be used to sort and analyze collected observation data after an exercise.

Example Tier 2 Resilience Source: *Maneuver Capacity: The work unit can override built-in limits when necessary.*

Time: Use of the RAD-O does not add time to an exercise or other event in which observation data are collected.

Output: Examples of resilience in work system operations, with a focus on the effects of a new technology.

Development: The RAD-O guide was developed to prepare for a resilience-focused data collection and evaluation opportunity that involved the scenario-based user testing of a military logistics decision support tool. Data collected and assessed using the RAD-O guide revealed opportunities to improve the logistics tool design to strengthen the resilience of logistics planning and operations. In addition, the guide was presented to a healthcare research panel that included experienced simulation exercise practitioners, where feedback was collected [40]. Following the panel, additional front matter was added to the tool to provide clarification regarding purpose and usage.

11.5.5 RAD-Levels

Overview: RAD-Levels (RAD-Ls) benchmark the resilience maturity of a technology [41]. The levels are intended to guide development of and categorize technology with respect to its likelihood of contributing to operational resilience. Resilience maturity is represented using six RAD Levels. The levels are listed in Table 11.3, along with RAD tools that can optionally be used to meet the requirements of each. RAD-Ls provide flexible scaffolding to allow the use of methods,

Table 11.3 Recommended RAD tools based on RAD level.

RAD level	Recommended RAD tools
Level 1: Resilience Support Conceptualized	RAD-E, RAD-XP, RAD-CR
Level 2: Resilience Requirements Established	RAD-XP, RAD-S, RAD-CR
Level 3: Iterative Planning and Assessment Established	RAD-XP, RAD-S, RAD-O, RAD-SP
Level 4: Additional Requirements Generated	RAD-XP, RAD-S, RAD-O, RAD-SP
Level 5: Resilience Support Verified	RAD-SP, RAD-CR
Level 6: Post-deployment Resilience Sustained	RAD-O, RAD-S, RAD-CR

Note: Other methods and tools may also be used to meet the requirements of each RAD Level.

tools, and project planning that make sense for the specific cost, scheduling, and performance requirements of a project.

Each RAD level defines a set of required activities that must be accomplished to qualify for the associated resilience maturity level. Additionally, each level specifies a set of necessary inputs and expected outputs of the required activities. Required activities prepare a technology to handle the demanding realities of operations and contribute to operational resilience during challenging situations. The levels are cumulative in that the required activities of each include the required activities of the lower RAD levels. Levels, their descriptions, goals, process guidance, inputs, outputs, and requirements are available at the RAD-Ls website, https://trusts.mep.mitre.org/rrlevels [37].

A project manager may use RAD-Ls as a checklist to ensure accomplishment of the work necessary for achieving their technology resilience maturity goal. Alternatively, an independent reviewer may assess the maturity of a project and provide guidance for using resilience-aware development practices and tools to achieve the next level. A sponsor or customer interested in supporting operational resilience may work with technology developers to ensure development effectively considers technology impacts on resilience. The RAD-L of a given technology may inform the confidence and likelihood of use and adoption by work domain professionals. A contracts specialist can use them to specify a required RAD level and required resources (e.g., access to operations personnel), project activities, and data products. For example, meeting RAD-Level 3 requires operations personnel to assess and give feedback on designs based on realistic challenging scenarios. Stakeholders may agree to document the test scenarios and test cadence in a Test and Evaluation Master Plan and provide results of the assessment as a formal data product, providing evidence of their attention to operational resilience.

Time: The required activities for each RAD Level may increase time required for technology development depending on ease of access to operations personnel involvement and project risk management practices.

Example RAD Level Goal: *Ensure requirements are implemented in ways that support resilience.* (RAD Level 3)

Development: RAD-Ls were developed collaboratively and iteratively by a chapter author and project management and technology acquisition subject-matter expert (SME). Development was informed by existing technology maturity frameworks such as Technology Readiness Levels [42] and Human Readiness Levels [67], the experiences of the authors in systems development, and feedback from acquisitions SMEs. Early versions were reviewed by four project management and acquisition SMEs and a technical editor.

11.5.6 RAD-Critical Risks

Overview: RAD-Critical Risks (RAD-CR) [43] is an assessment worksheet for evaluating essential (critical) enablers of resilience in high-consequence system operations. The worksheet lists 20 critical risks framed as *resilience requirements.* Analysts are asked to rate the likelihood that work system units can meet each requirement on a 5-point Likert scale ranging from *strongly disagree* to *strongly agree.* Loss or compromise of these critical enablers, indicated by requirements rated as *disagree* or *strongly disagree*, could interfere with currently effective resilience strategies and infrastructure to such an extent that mission goals cannot be met. At the same time, critical resilience requirements rated as *agree* or *strongly agree* with respect to contributions of a new technology provide a measure of the technology's resilience capacity contribution.

The RAD-CR worksheet is to be used by people with operations experience in the target work system, analysts with data about work system operations (e.g., data obtained from mishap investigation reports, exercise observation, or operations expert interviews), or a combination of the two. Ideally, a technology is evaluated across the development effort as it matures to detect changes in its potential to negatively impact a critical risk. Other uses are also possible, including periodic assessments of high-consequence work operations to ensure resilience requirements are being met, mishap evaluations to understand the roles of resilience strengths and gaps, and system modernization needs assessments.

Example Critical Risk: *The work system unit being assessed reliably recognizes when an agent (staff or software) they depend on is overloaded, broken, or compromised.*

Time: A minimum of 30 minutes to check for critical risks up to many hours when used by a team as a collaborative analysis worksheet. Time will be on the lower side for work system operations experts compared with analysts making assessments based on data about operations performance.

Outputs: Critical risks to the ability of a work enterprise to respond to demands or situations with resilience; critical risk ratings grouped by Tier 1 resilience enablers and compared across work units, operations, or time.

Development: Two of this chapter's authors with over 25 years each of experience studying and designing technology for use in high-consequence operations independently identified 18 critical resilience enablers among Tiers 2 and 3 TRUSTS Framework resilience enablers, the loss of which would pose a critical risk to operational performance and mission success. To develop the list, the two authors independently reviewed all Tiers 2 and 3 resilience enablers in the TRUSTS Framework and marked those that, if missing or compromised, would likely (1) cripple a work operation's ability to detect or respond to a nonroutine, extreme, or otherwise challenging demand or situation and (2) not be replaceable or compensated for by using other resilience enablers. The two then talked through each item selected and evidence underlying the selection to agree on the 18 items of the original RAD-CR worksheet.

The worksheet initially required a yes–no response and a mitigation strategy for each *no* response. Observation of worksheet use by a high-consequence operations expert with 27 years of air traffic control experience led to a switch from the yes-no response to the 5-point rating scale. Peer review of the worksheet by two resilience researchers led to the major revision of instructions, minor edits to critical risks, and the addition of two critical risks, bringing the total to 20. The two added critical risks were:

- The work system unit being assessed can switch to another resource if any critical resource becomes unavailable.
- Managerial work units can avoid interrupting frontline work units with frequent requests for situation updates during nonroutine and rapidly changing situations.

During its development, the RAD-CR worksheet was trialed by two analysis teams. One team used it to assess the resilience properties of an electronic lookout system, resulting in minor editing of one critical risk. The other team used it to assess the resilience of three work units involved in a close-call event as separate and joint units [44]. The latter team's work led to significant changes to the analysis support provided by the worksheet. The worksheet now includes restaurant and aviation operations examples to illustrate each critical risk, instructions for assessing more than one work unit, and the TRUSTS Framework resilience source represented by each critical risk. Their use of the worksheet led to simplification of the focus of analysis to always be the work system unit versus human and technology agent types and the work unit. It also led to the addition of an option to check a *Performed as Required* box in lieu of a making a rating, fields for documenting evidence used to make each rating and the final assessment, and discussion questions and topics that contributed to assessments.

11.5.7 RAD-Simulation Probes

Overview: RAD-Simulation Probes (RAD-SP) [45] is a guide for designing training and test scenarios that systemically test enablers of resilience. These include scenarios used in computer agent-based, human-in-the-loop, field, in situ, tabletop exercises, and scenario-based expert interviews. The RAD-SP guide may also be used as a checklist to ensure a given scenario includes a representative range of resilience challenges. The guide specifies for each Tier 1 resilience source of the TRUSTS Framework two to four probes that can be used directly, accompanied by two to four examples of observable responses to the probes that would be indicative of resilience. The RAD-SP ensures an underdevelopment technology is fully exercised against challenging as well as routine conditions, enables systematic comprehensive coverage of a range of resilience enablers, and helps to reveal resilience strengths and gaps. At the end of the RAD-SP guide, instructions are provided for identifying metrics and assessing responses to probes. Assessment, as described, requires familiarity with both quantitative and qualitative data analysis.

Example Probe for Guided Local Control: *Frontline participants are presented with a novel situation that requires an immediate response and for which standard procedures do not apply.*

Example Observable Response: *Participants coordinate and develop a response on the fly without asking permission. Information systems and other technologies support and do not limit coordination or improvisation.*

Time: May somewhat increase the time for exercise scenario planning.

Outputs: Scenarios featuring one or more probe challenges to systematically test a work system's resilience and impact of new technology on that resilience.

Development: The RAD-SP guide was initiated in response to interest expressed by a curriculum development project leader and to provide additional scenario development support for RAD-XP. The five Tier 1 resilience enablers of the TRUSTS Framework were used to provide a resilience-focused structure. Probes and example observable responses were developed iteratively by two chapter authors. They developed them for each resilience source based on reviews of system resilience failure case studies and interviews with operations experts. One developed an initial set of probes for military scenarios and the other for healthcare scenarios, and these were iteratively adapted until all probes were applicable to high-consequence work domain scenarios in general. The tool was reviewed by a modeling-and-simulation expert with over 20 years of experience, which led to refinement of the instructions.

11.6 Discussion

Enablers of CAS resilience have been studied and proposed by a number of researchers and theorists. Key among them are four abilities of resilient systems identified by [46]: the abilities to respond (appropriately and adaptivity), monitor (to detect changes), learn, and anticipate. Another body of influential work comes from Woods (e.g., [47–50]), who identifies system resilience essentials that include agent initiative, reciprocity among system units and agents, monitoring to improve the match between behavior and goals in light of conditions, adapting "good operating points" (where parameter settings and operator behavior patterns maintain stable effective performance) as the operational landscape changes, and managing associated trade-offs. Researchers who focus on cognitive work in complex work operations have also weighed in on the requirements of CAS resilience, taking a high-consequence work system perspective. They define these requirements in terms of *macrocognitive functions* that enable work system resilience [68], functions that include replanning, projecting, sensemaking, coordinating, and detecting problems ([53, 19, 51, 52]). Resilience in CASs occurring in nature, such as in honeybee colonies and slime mold cooperatives, suggests essential resilience enablers such as signaling, feedback loops, sensitivity to selective pressures, and role adaptivity [69, 70].

The TRUSTS Framework [64, 65, 34, 37] synthesizes these bodies of work, cases studies, observational data, and interview data to organize enablers of operations resilience around the core resilience enablers of Shared Demand and Deviation Awareness, Progressive Responding, Response Coordination, Maneuver Capacity, and Guided Local Control. The framework was organized to describe operational capabilities and functions of high-consequence work system operations that all teams should strive to maintain or strengthen as they plan or design changes for CASs. The framework is also intended to foster the recurring evaluation of impacts of technology design and acquisition decisions on resilience-enabling capabilities and functions of high-consequence work CASs. As Kauffman [9, p. 38] observed in his systems thinking primer, in a complex system, "you can never do just one thing"; any change will resonate through the interconnected and interwoven functions of the system. Considering operational resilience enablers during technology development can help avoid unintended negative resilience consequences that are not recognized until resilience becomes required, i.e., at the worst time possible.

The RAD tools described in this chapter were developed to bring system resilience concepts represented by the TRUSTS Framework into the work of modernization and technology engineering for work systems that require resilience. We suggest the tools, which reflect the more generalizable Tiers 1 and 2 resilience enablers of the framework, may be used across a variety of CASs to assess their resilience capacity, identify resilience-improvement opportunities and critical risks, and guide resilience-sensitive change and modernization.

One of our main aspirations for the RAD tools is that they be used by technology end users to communicate with the technology developers about work system capabilities and characteristics that contribute to resilience capacity. This communication, including feedback about the potential interplay between a new technology and work system operations, enables the co-evolution of the technology with the work system (e.g., [71, 36, 53]). Operations personnel can use tools such as RAD-S and RAD-CR to both reveal and communicate about resilience improvement opportunities and risks. They can use RAD-E to identify high-level resilience requirements technology developers may be asked to meet. In addition, if stakeholders specify the technology achieve a certain RAD level, the development team will be incentivized to involve and obtain feedback from operations personnel during development, and RAD tools can facilitate the capture, processing, and translation of that feedback. The RAD tools can help focus their involvement and how it shapes a technology's maturation. For example, RAD-SP and RAD-O help technologists triage feedback to information relating to operational resilience. Together, the tools give operations personnel a variety of means for communicating with developers about ways a developing IT might interact with work activities that are important and even essential to resilience and, along with it, the long-term safety and success of their operations. The tools are thus designed to help shape technologies in ways that are targeted, lightweight, and compatible with technology development activities.

We hope development teams will use the tools to ensure their new technologies and other system modernization activities are responsive to the realities of work operation demands and pressures, and not simplistically or blindly focused on efficiency or other single-dimensional goals. Often, development teams have a relatively mature product in hand before they investigate how it interacts with the operational environment. Some may never consider how the technology performs in nonroutine and challenging "edge" situations [72] and instead handoff the technology for deployment in the work system after verifying its functional requirements in controlled conditions. Other teams may evaluate the ability of a new technology to withstand limited edge situations but not its ability to support operations across challenges in general, including surprise challenges. This, at best, results in technology that is robust to specific pre-identified challenges. Our main goal is to aid teams in developing technology that performs effectively in high-consequence complex work operations in which the complexity of the environment in which challenges cannot all be predicted and resilience to the unknown is a necessity.

We designed the RAD tools to be compatible with technology development practices and methods to the extent possible to reduce the likelihood of these tools failing to be adopted by technology development teams. This is primarily manifested in the lightweight nature of the tools. For example, the RAD-X is a 90-minute exercise, the RAD-E requires almost no preparation and takes 1 to 1.5 hours, and the RAD-CR and RAD-S can be used as rapidly completed checklists (or as worksheets to drive a rigorous data-based analysis). We also invite tool users to adapt them according to their goals and resources. The tools assume the involvement of operations personnel or other users and require holistic, not piecemeal, iterative development. These are common goals of technology development efforts but perhaps because they increase the complexity and time required for development, they often fall by the wayside. Iterative development enables co-evolution by means of the gradual adjustment of a technology design and work system operation, reveals when a technology design element is not in the best interest of resilience, and helps operations personnel grow their understanding of ways to change their work practices to take the best advantage of an incoming technology.

Technologies are typically developed to improve the efficiency or reliability of a process, do something better than humans, support humans, or do something that is risky for humans. The RAD tools widen, if not open the door for using technology to benefit work systems in another important way: by strengthening their resilience. In our future work, we aim to continue improving the translation of work system resilience into technology development, including into the specification and prioritization of resilience-enabling system features. We plan to explore, for example, their criticality and the amount and variety of enablers required to have an adequate capacity for resilience. We hope to use the tools to contribute to the identification of novel technology concepts for improving resilience in high-consequence work systems as well as other types of engineered CASs, and to pursue advances that can improve technology capacity for the coordinated adaptivity that is so key to CAS resilience.

Acknowledgements: We wish to express gratitude to our many colleagues who championed this work, allowed us to try out our tools on their projects, and provided thoughtful feedback, including Maurice Azar, Stan Barr, Jeff Davis, Alison Dingwall, Craig Doescher, Heath Farris, Dhanuj Gandikota, Jessica Matthews,

Rachel Mayer Ediger, Chris Niessen, Doug Owen, Brian Parrish, Kris Rosfjord, Emilie Roth, and Jerry Vaughn. This work was funded by MITRE's Independent Research and Development Program. Approved for Public Release; Distribution Unlimited. Public Release Case Number 25-0505

References

1 C. Nemeth, R. Wears, D. Woods, E. Hollnagel, and R. Cook, Minding the gaps: Creating resilience in health care. In K. Henriksen, J. B. Battles, M. A. Keyes, M. L. Grady editors. *Advances in Patient Safety: New Directions and Alternative Approaches* (Vol. 3: Performance and Tools). Agency for Healthcare Research and Quality, Rockville, 2011.

2 Y. Bar-Yam. Introduction to the strategy and methods of complex systems. In *Handbook of Research Methods in Complexity Science*, pages 5–17. Edward Elgar Publishing, 2018.

3 J. Gleick. *Chaos: Making a New Science*. Penguin, 2008.

4 F. Heylighen. Principles of systems and cybernetics: An evolutionary perspective. *Cybernetics and Systems,* 92(1992):3–10, 1992.

5 L. Tabansky and E. Lichterman Progress: The sectoral approach to cyber resilience. *International Journal of Information Security,* 24(1):18, 2025.

6 L. von Bertalanffy. *General System Theory*. George Braziller, New York, 1968.

7 D. C. Amodeo and R. A. Francis. The role of protocol layers and macro-cognitive functions in engineered system resilience. *Reliability Engineering*, 190:106508, 2019. https://doi.org/10.1016/j.ress.2019.106508.

8 M. Ladell, N. L. Jacobson, S. C. Yale, K. L. McDermott, E. L. Papautsky, K. R. Catchpole, and M. C. Scanlon. The problem with how we view medical (and diagnostic) error in emergency medicine. *Academic Emergency Medicine*, 32:340–347, 2025, https://doi.org/10.1111/acem.15076.

9 D. L. Kauffman. *Systems 1. An Introduction to Systems Thinking.* Future Systems, Inc., Minneapolis, MN, 1980.

10 J. Burnell. How operators can create safety in an uncertain world, 2022.

11 D. Ivanov, B. Sokolov, and A. Dolgui. The ripple effect in supply chains: Trade-off 'efficiency-flexibility-resilience' in disruption management. *International Journal of Production Research,* 52(7):2154–2172, 2014.

12 R. H. Mackey. *Translating Vision into Reality: The Role of the Strategic Leader.* U.S. Army War College, Carlisle, PA, 1992.

13 A. Juarrero. Dynamics in action: Intentional behavior as a complex system. *Emergence*, 2(2):24–57, 2000.

14 S. S. Tanveer and Z. A. Rana. Prioritizing software requirements by combining the usage monitoring and user feedback data. *IEEE Access*. 2024. https://doi.org/10.1109/ACCESS.2024.3409847.

15 R. Anwar and M. B. Bashir. A systematic literature review of AI-based software requirements prioritization techniques. *IEEE Access*, 11:143815–143860, 2023. https://doi.org/10.1109/ACCESS.2023.3343252.

16 A. J. Ko, M. J. Lee, V. Ferrari, S. Ip, and C. Tran. A case study of post-deployment user feedback triage. In *Proceedings of the 4th International Workshop on Cooperative and Human Aspects of Software Engineering*, pages 1–8, 2011.

17 D. Pagano and B. Bruegge. User involvement in software evolution practice: A case study. In *2013 35th International Conference on Software Engineering (ICSE)*, pages 953–962, *IEEE*, 2013.

18 H. Sadia, S. Q. Abbas, and M. Faisal. Volatile requirement prioritization: A fuzzy based approach. *International Journal of Engineering and Advanced Technology*, 8:2467–2472, 2019.

19 S. M. Sherwood, K. J. Neville, A. L. M. T. McLean, M. M. Walwanis, and A. Bolton. Integrating new Technology into the complex system of air combat training. In: H. Handley and A. Tolk, editors. *A Framework of Human System Engineering: Applications and Case Studies*. IEEE, Washington, D.C., 2020.

20 A. A. Minai, D. Braha, and Y. Bar-Yam. Complex engineered systems: A new paradigm. In D. Braha, A. Minai, and Y. Bar-Yam, editors, *Complex Engineered Systems: Science Meets Technology*, pages 1–21, Springer, 2006. https://doi.org/10.1007/3-540-32834-3_1.

21 E. A. de Carvalho, J. O. Gomes, A. Jatobá, M. F. da Silva, and P. V. R. de Carvalho. Employing resilience engineering in eliciting software requirements for complex systems: experiments with the functional resonance analysis method (FRAM). *Cognition, Technology & Work*, 23:65–83, 2021.

22 R. Hoffman and R. Neville. Using cognitive task analysis to explore issues in the procurement of intelligent decision support systems. *Cognition, Technology and Work*, 1(11):57–70, 2009.

23 B. W. Boehm. A spiral model of software development and enhancement. In *Readings in Human–Computer Interaction*, pages 281–292, Morgan Kaufmann, 1995.

24 B. Boehm and W. Hansen. The spiral model as a tool for evolutionary acquisition. *CrossTalk*, 14(5):4–11, 2001.

25 P. Abrahamsson, O. Salo, J. Ronkainen, and J. Warsta. Agile software development methods: Review and analysis, 2017. https://doi.org/10.48550/arXiv.1709.08439.

26 The Agile Manifesto. *Manifesto for Agile Software Development*. Agilemanifesto.org, 2001.

27 R. R. Hoffman and P. A. Hancock. Measuring resilience. *Human Factors*, 59(4):564–581, 2017.

28 P. Parrend and P. Collet. A review on complex system engineering. *Journal of Systems Science and Complexity*, 33(6):1755–1784, 2020.

29 D. D. Woods. *Stella: Report from the SNAFUcatchers Workshop on Coping with Complexity.* The Ohio State University, 2017.

30 G. Klein and S. Hegde. *Tools Used in Naturalistic Decision Making (NDM) and Resilience Engineering (RE).* Zenodo,2020. https://doi.org/10.5281/zenodo.4248112.

31 A. Azadeh, V. Salehi, B. Ashjari, and M. Saberi. Performance evaluation of integrated resilience engineering factors by data envelopment analysis: The case of a petrochemical plant. *Process Safety and Environmental Protection*, 92(3):231–241, 2014.

32 E. Hollnagel. Coping with complexity: Past, present and future. *Cognition, Technology Work*, 14:199–205, 2012.

33 E. Hollnagel. *FRAM: The Functional Resonance Analysis Method: Modelling Complex Socio-Technical Systems*. CRC Press, 2017.

34 K. J. Neville, D. Owen, G. Lematta, S. L. Dorton, J. Winner, E. Barrett, and E. S. Patterson. *Support for Developing Technology for Work Operations that Require Resilience: A Conceptual Framework of Work-system Resilience Enablers* [manuscript under review]. Mclean, VA: MITRE, 2025.

35 R. R. Hoffman and M. J. McCloskey. Envisioning desirements. *IEEE Intelligent Systems*, 28(4):82–89, 2013.

36 H. Benbya and B. McKelvey. Using coevolutionary and complexity theories to improve is alignment: A multi–level approach. *Journal of Information Technology*, 21(4):284–298, 2006.

37 E. S. Neville and K. J. Patterson. Mitre resilience aware development survey (RADS) to assess the impact of a new technology, policy, or concept of operations on work system resilience, 2023.

38 Stephen L. Dorton, Emily Barrett, Theresa Fersch, Andrew Langone, and Kelly J. Neville. RAD-XP: Tabletop exercises for eliciting resilience requirements for sociotechnical systems. *Systems*, 11(10):487, 2023.

39 E. Barrett, K. J. Neville, T. Fersch, S. L. Dorton, and E. S. Patterson. Mitre resilience aware development exploration (RAD-E) exercise to identify ways a new technology, concept of operations, or policy can benefit operations resilience, 2023.

40 E. S. Patterson and K. J. Neville. Mitre resilience aware development - observation (RAD-O): Form to assess the impact of a new technology, policy, or concept of operations on work system resilience during a simulation-based activity, 2023.

41 G. Lematta, K. Neville, E. S. Patterson, S. Dorton, and B. Parrish. *Mitre Resilience-Aware Development Levels: A Technology Maturity Blueprint for Operational Readiness, Version 2.* MITRE, McLean, VA, 2023. https://trusts.mep.mitre.org/rrlevels.

42 J. C. Mankins. *Technology Readiness Levels: A White Paper*. NASA. Washington, DC, 1995.

43 E. S. Patterson and K. J. Neville. Mitre resilience aware development critical risks (RAD-CR) screening checklist, 2024.

44 G. Coville. Assessing resilience and human variability within aviation safety. In *Paper Presented at A Safer World Symposium, November 19–20, McLean, VA*, pages 19–20, 2024.

45 K. J. Neville and E. S. Patterson. Mitre resilience aware development scenario probe (RAD-SP) design guide, 2024.

46 E. Hollnagel. The four cornerstones of resilience engineering. In: C. P. Nemeth, E. Hollnagel, and S. Dekker, editors. *Resilience Engineering Perspectives, Volume 2: Preparation and Restoration*, pages 117–135. Ashgate Publishing Limited, Aldershot, UK, 2009.

47 D. D. Woods. Four concepts for resilience and the implications for the future of resilience engineering. *Reliability Engineering & System Safety* 141:5–9, 2015. https://doi.org/10.1016/j.ress.2015.03.018.

48 D. D. Woods. The theory of graceful extensibility: basic rules that govern adaptive systems. *Environment Systems and Decisions*, 38(4):433–457, 2018.

49 D. D. Woods. Essentials of resilience, revisited. In *Handbook on Resilience of Socio-technical Systems*, pages 52–65, 2019.

50 D. D. Woods. Resolving the Command–Adapt Paradox: Guided adaptability to cope with complexity. In *Compliance and Initiative in the Production of Safety: A Systems Perspective on Managing Tensions and Building Complementarity*, pages 73–87, Cham: Springer Nature Switzerland, 2024.

51 E. S. Patterson and R. R. Hoffman. Visualization framework of macrocognition functions. *Cognition, Technology & Work*, 14:221–227, 2012.

52 G. Klein, K. G. Ross, B. M. Moon, D. E. Klein, R. R. Hoffman, and E. Hollnagel. Macrocognition. *IEEE Intelligent Systems,* 18(3):81–85, 2003.

53 M. A. Khan. Understanding a co-evolution model of business and it for dynamic business process requirements. *International Journal of Advanced Computer Science and Applications,* 7(2):348–352, 2016.

54 R. R. Hoffman, G. Klein, and D. D. Woods. The theoretical foundation of macrocognition. In *Proceedings of the International Conference on Naturalistic Decision Making 2019, July 16–20,* San Francisco, CA, 2019.

55 K. Øien, S. Massaiu, R. Tinmannsvik, and F. StØrseth. Development of early warning indicators based on resilience engineering. In *Proceedings of the 10th International Conference on Probabilistic Safety Assessment and Management (PSAM)*, pages 1762–1771, 2010.

56 M. Reynolds. Making 'systems' work for evaluation. In *Rising to Challenges: How Does Evaluation Rise to the Challenge of Competing Issues, Crisis and Uncertainty?, Proceedings of the 5th October at UK Evaluation Society Annual Conference, London, UK, 3–5 October 2023,* London, UK, The Open University: Milton Keynes, UK, 2024.

57 T. A. Saurin and N. J. B. Werle. A framework for the analysis of slack in socio-technical systems. *Reliability Engineering & System Safety*, 167:439–451, 2017.

58 S. L. Dorton, E. Barrett, T. Fersch, A. Langone, and K. J. Neville. RAD-XP: Tabletop exercises for eliciting resilience requirements for sociotechnical systems. *Systems*, 11(10):487, 2023.

59 S. L. Dorton, G. Lematta, and K. J. Neville. The tough sell of resilience engineering. *IEEE Transactions on Technology and Society*, 6(1):47–53, 2024. https://doi.org/10.1109/TTS.2024.3484176.

60 M. L. Bolton and E. J. Bass. Evaluating human-automation interaction using task analytic behavior models, strategic knowledge-based erroneous human behavior generation, and model checking. In *2011 IEEE International Conference on Systems, Man, and Cybernetics,* pages 1788–1794, IEEE, 2011.

61 D. A. Morey, D. Della Vella, M. F. Rayo, D. J. Zelik, and T. B. Murphy. Joint activity testing: Towards a multi-dimensional, high-resolution evaluation method for human-machine teaming. *Proceedings of the Human Factors and Ergonomics Society Annual Meeting*, 66(1):2214–2219, 2022. https://doi.org/10.1177/1071181322661373.

62 M. Demir, N. J. Cooke, and P. G. Amazeen. A conceptual model of team dynamical behaviors and performance in human-autonomy teaming. *Cognitive Systems Research,* 52:497–507, 2018.

63 S. E. van der Merwe, R. Biggs, and R. Preiser. Sensemaking as an approach for resilience assessment in an Essential Service Organization. *Environment Systems and Decisions*, 40(1):84–106, 2020.

64 K. J. Neville, H. Rosso, and B. Pires, A systems-resilience approach to technology transition into high-consequence work systems. In *Proceedings of the 15th Conference on Naturalistic Decision Making and 9th Symposium on Resilience Engineering: Bouncing Forward from Global Crises and Challenges*, pages 72–75, Resilience Engineering Association, 2021.

65 K. J. Neville, B. Pires, P. Madhavan, M. Booth, K. Rosfjord, and E. S. Patterson. The TRUSTS work system resilience framework: A foundation for resilience-aware development and transition. *Proceedings of the Human Factors and Ergonomics Society Annual Meeting*, 66(1):820–824, 2022. https://doi.org/10.1177/1071181322661365.

66 K. J. Neville, S. L. Dorton, E. S. Patterson, E. Barrett, T. Fersch, G. Lematta, A. Langone, T. Sanders, B. Pires, X. Chau, B. Jones, E. Hua, and D. Bang. MITRE RAD (resilience-aware development) and the TRUSTS framework, The MITRE Corporation, 2023. https://trusts.mep.mitre.org/research/tos.

67 W. K. Vaneman, The system of systems engineering and integration "Vee" model. In *2016 Annual IEEE Systems Conference (SysCon)*, pages 1–7, IEEE, 2016.

68 P. A. Savage-Knepshield, C. L. Hernandez, and S. O. Sines. Exploring the synergy between human systems integration and human readiness levels: A retrospective analysis. *Ergonomics in Design*, 29(4):16–24, 2021.

69 P. Ward, J. M. Schraagen, J. Gore, and E. M. Roth, editors. *The Oxford Handbook of Expertise*. Oxford University Press, 2019.

70 C. M. Grozinger and H. M. Patch. Resilience and collapse in bee societies and communities. In M. A. Centeno, P. W. Callahan, P. A. Larcey, and T. S. Patterson, editors, *How Worlds Collapse.What History, Systems, and Complexity Can Teach Us About Our Modern World and Fragile Future*, pages 297–311, Routledge, 2023.

71 K. Rutter. Lessons from slime mold: How to survive and thrive in ever-changing organizational environments. *Bulletin of the American Society for Information Science and Technology*, 35(6):29–37, 2009.

72 F. Amarilli, B., van den Hooff, and M. van Vliet. Business-IT alignment as a coevolution process: An empirical study. *The Journal of Strategic Information Systems*, 32(2):101776, 2023.

73 E. S. Patterson, E. M. Roth, and D. D. Woods. Facets of complexity in situated work. In E.S., Patterson and J. Miller, editors, *Macrocognition Metrics and Scenarios: Design and Evaluation for Real-world Teams*, pages 221–252, Ashgate, 2010.

12

Artificial Intelligence Networks: Structure, Dynamics, and Economics

Oscar M. Granados[1] and Nicolás De la Peña[2]*

[1]Computational Finance and Digital Economics Lab, Universidad Jorge Tadeo Lozano, Bogotá, D.C., Colombia
[2]Faculty of Economics, Business, and Sustainable Development, Universidad De la Salle, Bogotá, D.C., Colombia

*Corresponding author: oscarm.granadose@utadeo.edu.co

12.1 Introduction

The complex interactions within the global ecosystem of artificial intelligence (AI) are causing it to change rapidly. Initially, data collection was an essential process. Improving algorithmic results required ever-increasing amounts of data, so the challenge was to create new methods of collection and classification [1]. This is illustrated by contests of visual pattern recognition, in which the error rate was above 25% in 2011 but fell to less than 5% in 2016 [2]. The following challenge was to increase computational power to run larger models for many actors. This way, companies, governments, and researchers capitalized on the precision of pattern recognition and predictive processes to consolidate other methods, such as generative models. These increasing capabilities then became a hardware problem: designing and producing more powerful semiconductors and computational architectures [3, 4]. This led to the graphics processing unit (GPU) boom and the rise of NVIDIA as one of the most valuable company in global markets in 2023. Nowadays, in 2025, it is an infrastructure problem because the AI global ecosystem requires extracting energy to power the projected data centers, for which the necessary data and computational capabilities are available, and the financial resources to underpin them [5, 6]. As seen, the global ecosystem of AI has a myriad of components, and it works as a complex adaptive system that is made up of a series of interacting components with several levels of feedback.

The first is economic capabilities. The global expansion of AI, driven by the aforementioned components, requires substantial economic and financial resources. The price of GPUs has been steadily increasing since 2020 [7], investment in data centers in the US increased about fourfold from 2014 to 2024 [8], and Trump's Stargate Initiative (1 GW) to invest 500 billion in AI infrastructure or Meta's Plan of 2 GW of power and housing 1.3 million GPUs from Nvidia illustrates the key role of funding [9, 10]. Financial capabilities underpin the complex array of knowledge, hardware, and energy, which require research and innovation processes. Public and private funding have been key to developing the AI global ecosystem.

Second, knowledge. The models of AI, such as large language models (LLMs) or any other generative artificial intelligence (GenAI), are based on patented innovations [11]. This constitutes the basis for AI development, as research and innovation are the sources of new techniques and improvements to existing ones. AI entangles a global collaboration network [12], as with other complex scientific enterprises. However, it is concentrated in a few research centers [13]. This reflects the well-funded national innovation strategies of leading countries [14] and the collaboration between private and public actors [15]. As well, this layer is dependent on economic capabilities and investment. For instance, the Trump administration's decision to end 1600 grants from the National Science Foundation for several topics [16], as well as the halt of new grants for foreign researchers [17], will introduce barriers to knowledge creation and diffusion at both the national and international levels.

Third, semiconductors. To effectively perform mathematical operations, many AI models require the processing of extensive datasets [1]. Thus, semiconductors, especially GPUs and TPUs, provide the physical resources for training and developing AI models [18]. The production process of AI hardware requires a global value chain that encompasses minerals, production equipment, such as EUV lithography machines, and knowledge [19]. This complexity creates an ideal scenario for weaponization, in which central states can exclude others from participation [20]. Therefore, the global data centers are powered by American microchips and located worldwide through China's and the United States' firms [21]. Nonetheless, microchip production requires a massive investment. By 2030, producing firms will invest $1 trillion around the world [22]. More specifically, in the US, the CHIPS Act of 2022 included a tax incentive for manufacturing microchips, which in 2025 generated the announcement of more than 100 projects accounting for over $540 billion [23].

Fourth, energy. The data centers where AI is processed consume vast amounts of energy for computational tasks [24]. Furthermore, LLMs require ever-increasing energy resources. For instance, the energy consumption of a ChatGPT query is approximately 10 times that of a typical Google search [25]. This has proven a decisive factor in the data center's location [26] and has led to investment plans in nuclear energy by companies such as Microsoft, Meta, and Amazon [27]. In 2025, if data centers were a country, they would constitute the 10th global consumer of energy. More strikingly, for 2030, they would be the fourth consumer, just after China, USA, and India [28]. Nonetheless, energy supply will be restricted by physical capacity and economic constraints. For example, Gartner projects that power shortages will operationally constrain approximately 40% of AI data centers by 2027, primarily due to the escalating energy demands of AI and generative AI workloads [29]. In this way, economic capacity remains the primary driver of the AI global ecosystem.

The field of AI is fundamentally concerned with the development of computational tools to address complex challenges across various sectors, including technology, business, science, and medicine. AI serves as a pivotal instrument for core tasks, including problem-solving, knowledge representation, automated reasoning, modeling, and prediction, with applications that span from specialized scientific research to daily user experiences. However, it is crucial to recognize that AI is not a monolithic entity. Instead, it is a heterogeneous collection of methods with roots in diverse fields of knowledge. This array of techniques—ranging from foundational agent-based systems to sophisticated architectures, such as convolutional neural networks (CNNs), generative adversarial networks (GANs), and LLMs—is increasingly applied to solve real-world problems [30–33].

So, in this stage of the AI age, questions arise: What is the ecosystem structure? Who can access AI hubs, and under what conditions? What parameters of the economic layer affect the ecosystem? How robust is the AI global ecosystem for economic shocks? To address these questions, we represent the elements of a complex system, and their interactions with nodes and links can help us provide insights into the system's structure, dynamics, and resilience [34]. In our case, countries are agents, characterized by sovereignty and autonomy structures that are considered self-similar and homogeneous, because they are recognized by other countries, but may exhibit specific hierarchies or role specializations [35]. However, for the AI global ecosystem, the abstraction of its organization into a single layer of nodes and links is not sufficient. For this reason, we develop a multilayer network. Hence, we aim to analyze the structure and dynamics of the AI global ecosystem driven by the economic and financial capabilities layer.

This chapter contributes to the literature on the global implications of AI in two ways. First, it empirically demonstrates the structure of the AI global ecosystem. The multilayer network approach allows us to understand the shape of each dimension (economics and finance, semiconductors, energy, and knowledge) as well as the relevance of one layer over others and the network topology of the assembled ecosystem, highlighting specific properties such as connectivity and co-evolution [36]. Second, it evidences the dynamics of the AI global ecosystem.

Here, we connect the robustness of the main layer (economics) in the presence of a specific failure. The complexity and complex adaptive system approaches help to understand the confluence of dimensions and the evolution of their relationship.

The remainder of this chapter is structured as follows. Section 12.2 proposes a multilayer network model for analyzing the structure and dynamics of AI and describes the datasets employed. Section 12.3 presents the results to highlight the unequal structure and the interrelationship among and between layers. Section 12.4 presents a discussion framed around the initial research questions. Finally, Section 12.5 concludes the chapter.

12.2 Methods and Materials

12.2.1 Methods

Countries have varying endogenous capacities to develop within the international system. These could be produced if states are interested, have the necessary capacities, and compete for links with other countries or a specific position in particular contexts or markets. To identify these differences, we propose a framework for the strength of international fitness based on differences among AI global ecosystem elements, following the analysis by Granados & De la Peña [37]. This version of the model provides information about the structure and dynamics of the AI global ecosystem, as well as the countries and interactions present in each layer of the network.

Specifically, we employ weights for different edges that represent the relevance of each variable defined above. To do so, each country is assigned a fitness parameter n_i chosen from the distribution $\rho(\eta)$ and the maximum possible fitness in the economic world system determined by $\eta_{\max}$. Namely, each new country connects to countries present in the world system, and the probability $P_{(i)}$ of this connection depends on connectivity $k_{(i)}$ influenced by preferential attachment π and fitness η_i [38]. Thus:

$$P_i = \frac{\eta_i \pi(k_i)}{\sum_j \eta_j \pi(k_j)}. \tag{12.1}$$

If the country lacks these elements, then it is likely to belong to the periphery, as not all countries succeed in developing international capabilities and performing effectively in global scenarios. They appear at different times and grow at different rates. To simplify, we assume that $\pi(k) = k$. Thus, a country will increase its connectivity k_i, at a rate that is proportional to the probability P_i that a new country will attach to it, i.e., each new country of the world economic system adds μ links to the system [39], following,

$$\frac{\eta_i k_i}{\sum_j \eta_j k_j} = \frac{\eta_i \pi(k_i)}{\sum_j \eta_j \pi(k_j)}. \tag{12.2}$$

The evolution of connectivity $k_i(t)$ established that all countries increase their connectivity in time, mathematically, $k_i(t) = \left(\frac{t}{t_1}\right)^i$, where t_i is the period at which country i has been added into the world economic system and $\beta = \left(\frac{t}{t_1}\right)$ [40]. Namely, the probability $P(k)$ that a country in the world system interacts with other countries reduces as a power law, given by $P(k)\ k^{-\gamma}$ [40].

In other words, countries habitually define their connections from their preferential attachment and other opportunities with their fitness, which means that the time dependence of a country's connectivity depends on the fitness η_i of the country, such that: $k_{\eta i}(t, t_i) = \mu\left(\frac{t}{t_1}\right)^{\beta(\eta i)}$, where, t_i is the time when the country i was included in the world system $k_\eta = k_{\eta i}(t, t_i)$; $\beta(\eta)$ is bounded between 0 and 1 because a country increases its links in t, $\beta(\eta) > 0$ and $k_i(t)$ cannot increase faster than t, i.e., $\beta(\eta) < 1$. Since $\beta(\eta) < 1$, in $t \to \infty$ limit $t^{\beta(\eta)}$, can be compared to t and obtain $\lim_{t\to\infty}(\sum_j \eta_j k_j) = C\mu t(1 + O(t^{-\epsilon}))$, where: $C = \int \delta\eta\rho(\eta)\frac{\eta}{1-\beta(\eta)}$ and which has a solution given by $\epsilon = (1 - \max_\eta \beta(\eta)) > 0$ [39]. Using these elements, Equation 12.3 can be written as:

$$\frac{\delta k_\eta}{\delta t} = \frac{\eta k_\eta}{C_t}, \tag{12.3}$$

Which has a solution given by $\eta(\eta) = \frac{\eta}{C}$. To find the connectivity distribution notated by probability $P(k)$ that in this case is a result of a weighted sum over different power laws, namely, the cumulative probability for a particular country $P(k_\eta(t)) > t = t\left(\frac{\mu}{k}\right)^{\left(\frac{C}{\eta}\right)}$ is given by:

$$P(k)\infty \int \delta\eta\rho(\eta)\frac{C}{\eta}\left(\frac{\mu}{k}\right)^{\left(\frac{C}{\eta}\right)+1}. \tag{12.4}$$

With the international fitness model, we created a multilayer network $\mathcal{M}$ of the AI global ecosystem comprising the following elements. First, we defined a single-layer network $G(V, E)$ as a set of countries V, coupled with a set of edges E describing their relations in a specific layer topic. Second, a multilayer network $\mathcal{M}$ was formed by V countries $i = 1, 2, \ldots, V$ and L layers $\alpha = L1, L2, \ldots, L$. Each layer is a specific network of the AI global ecosystem described by an adjacency matrix A^α [41, 42]. Third, we defined a multilayer network where the link does not connect the layers, but the layers share the same set of countries [43].

In the case of a multilayer network, the value $A_{ij}^{\alpha} = w_{ij}^{\alpha}$ if the directed link for country j to country i has weight w_{ij}^{α} and $A_{ij}^{\alpha} = 0$, thus, we differentiate between the in-degree $k_i^{in,\alpha}$ and the out-degree $k_i^{out,\alpha}$ of country i in layer α given by:

$$k_i^{in,\alpha} = \sum_{j=1}^{N} \theta(A_{ij}^{\alpha}),\ k_i^{out,\alpha} = \sum_{j=1}^{N} \theta(A_{ij}^{\alpha}). \tag{12.5}$$

In this way, assembling a sequence of elementary layers enables the creation of a multilayered network from the elementary layer combinations [41]. The types of participation and interactions of a country i at time$t$ are given by a matrix formed by a set of elementary layers that represent factors of the AI global ecosystem (Figure 12.1). Thus, the ecosystem was encoded by matrices $A(t)$ and represented by layers $\alpha = L_1, L_2, \ldots, L_\alpha$ with different weighted combinations $w\alpha(t)$ (Figure 12.1).

Additionally, the adjacency and Laplacian matrices properties are relevant to understanding the network features [44, 45]. If proper matrix representations are introduced in multilayer networks, we identify similar characteristics.

Given an AI global ecosystem multilayer network $\mathcal{M}$, several adjacency matrices can give information about its structure: the adjacency matrix A^{α} of each layer G_α, the adjacency matrix $\overline{A_{\mathcal{M}}}$ of the projection network proj($\mathcal{M}$), and the supra-adjacency matrix $A_{\mathcal{M}}$. The spectrum of the supra-adjacency matrix is directly related to several dynamic processes in a multilayer network. If $\lambda_1 \leq \cdots \leq \lambda_N$ is the spectrum of the supra-adjacency matrix $A_{\mathcal{M}}$ of an undirected multilayer network and $\mu_1 \leq \cdots \leq \mu_{n_\alpha}$ is the spectrum of the adjacency matrix A^{α} of layer G_α, then for every $1 \leq k \leq n_\alpha$ [34, 45–47]

$$\lambda_k \leq \mu_k \leq \lambda_{k+N-n_\alpha} \tag{12.6}$$

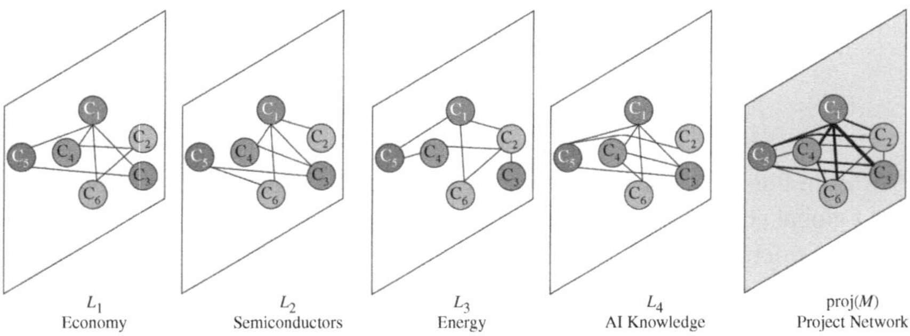

Figure 12.1 AI global ecosystem multilayer network.

A similar situation occurs for the Laplacian matrix of a multiplex network. The Laplacian matrix $\mathscr{L}_{\mathcal{M}} = \mathscr{L}$ of a multiplex $\mathcal{M}$ is defined as the $MN \times MN$ matrix of the form:

$$\mathscr{L} = \begin{pmatrix} D_1\mathbf{L^1} & 0 & \dots & 0 \\ 0 & D_2\mathbf{L}^2 & \dots & 0 \\ \vdots & \vdots & \ddots & \vdots \\ 0 & 0 & \dots & D_M\mathbf{L^M} \end{pmatrix} + \begin{pmatrix} \sum_\beta D_{1\beta}\mathbf{I} & -D_{12}\mathbf{I} & \dots & -D_{1M}\mathbf{I} \\ -D_{21}\mathbf{I} & \sum_\beta D_{2\beta}\mathbf{I} & \dots & -D_{2M}\mathbf{I} \\ \vdots & \vdots & \ddots & \vdots \\ -D_{M1}\mathbf{I} & -D_{M2}\mathbf{I} & \dots & \sum_\beta D_{M\beta}\mathbf{I} \end{pmatrix}. \quad (12.7)$$

In Equation (12.7), $\mathbf{I}$ is the $N \times N$ identity matrix and $\mathbf{L}^\alpha$ is the usual $N \times N$ Laplacian matrix of the network layer α whose elements are $L^\alpha_{ij} = s^\alpha_i \delta_{ij}-$ diffusion dynamics on a multiplex network is strongly related to the spectral properties of $\mathcal{L}$ [34]. If $\lambda_1 \leq \dots \leq \lambda_N$ is the spectrum of the Laplacian matrix $\mathcal{L}$ of an undirected multilayer network and $\mu_1 \leq \dots \leq \mu_{n_\alpha}$ is the spectrum of the Laplacian matrix $\mathbf{L}^\alpha$ of layer G_α, then for every $1 \leq k \leq n_\alpha$ [46]

$$\mu_k \leq \lambda_{k+N-n_\alpha}. \quad (12.8)$$

The model could identify several correlations among layers and the emergence of a multiplex network structure, representing the evolution of the AI global ecosystem, as well as the identification of countries that play a significant role in this ecosystem. Namely, fitness and preferential attachment define the system interactions, and the hypothesis of random distribution of links among the countries is not significant.

In both the international system and the AI global ecosystem, countries are not equivalent. Some methods in weighted networks identify the influential nodes as the degree centrality, the betweenness centrality, and closeness centrality [48–50]. However, node connectivity in these networks depends not only on its neighbors but also on the connection strengths between them, as well as the distribution of the total strengths on each edge [51].

We included resilience and percolation to understand the elements that could affect the evolution of G_{AI}. On the one hand, resilience refers to a network's ability to maintain its primary topological structure despite the presence of some form of damage. On the other hand, percolation is the process by which a fraction of nodes and their corresponding links are removed. This process can be used to study the fragility and robustness phenomena of our AI global ecosystem.

In G_{AI}, some layer nodes are interdependent on nodes in other layers. A node is interdependent on another node in another layer if it needs the other node to function correctly. When two or more networks are interdependent, a fraction of node failures in one layer can trigger a cascade of failures that propagate in the multilayer network, increasing the fragility of these structures [52, 53]. The AI global ecosystem could respond to the damage inflicted in a single layer or the interactions between specific nodes in a layer. Namely, either all the layers have a percolating cluster, or none of them have one [54]. In this case, we include the failure in the economic layer.

When a network of networks with a given superdegree sequence has layers with different numbers of interdependencies, they respond differently to failure. In the AI global ecosystem, each node does not interact directly with nodes of another layer; however, the implications for the ecosystem could be catastrophic because the damage in one layer or its interactions affects the next one due to the relevance of the first one. In this case, the layers with higher interdependencies, i.e., with higher superdegree, are more fragile than those with smaller superdegree [52]. However, the impact could affect all layers of our model, such as knowledge, semiconductors, or energy multilayers.

The robustness of multilayer networks can be evaluated by calculating the size of their mutually connected giant component (MCGC) when the damage affects a fraction $1 - p$ of the nodes in the system. The MCGC of a multilayer network is the largest component that remains after the random damage propagates back and forth in the different layers. To construct the MCGC for the case of the economic layer, where the replica nodes are interdependent, as well as in the other multilayer networks of our AI global ecosystem.

Assume that a group of nodes in layer L_1 is damaged, all the nodes in the giant component of layer L_1 are active, while the nodes that are not in the giant component are not active. All the nodes of other layers that are interdependent with nodes that are not active in layer L_1 are damaged. Next, set as active all the nodes that remain in the giant component of other layers, and set as not active all the nodes that are not in the giant component. By repeating the algorithm and alternating the analysis of layers, it is possible to characterize an avalanche of failure events that propagates from one layer to another, until the avalanche no longer propagates.

The set of active nodes at the end of the iteration is the MCGC, also known as the viable cluster of the network [55]. Between any two nodes in this mutually connected component, there are at least two paths, one in the economic layer L_1 and one in other layers, that connect both nodes and pass only through nodes that belong to the mutually connected component. Thus, for this generalized percolation problem, the algorithm is termed as the "message passing algorithm" [56, 57]. In our case, the message represents the transmission of failure.

If the node (i, α) is not damaged, i.e., $s_{i\alpha} = 1$, the message $S'_{i\alpha\to i\beta}$ will be equal to one if all the messages coming to the node (i, α) from the interdependent nodes (i, γ) different from (i, β) are equal to one, and if at least one message $\sigma_{\ell\alpha\to i\alpha}$ coming from a node (ℓ, α) neighbor of the node (i, α) in the layer α, is equal to one. If the node (i, α) is instead damaged, the message $S'_{i\alpha\to i\beta}$ will be equal to zero [34]. Therefore:

$$
\begin{aligned}
\sigma_{i\alpha\to j\alpha} &= s_{i\alpha} \prod_{\beta\in\mathcal{N}_i(\alpha)} S'_{i\beta\to i\alpha}\left[1 - \prod_{\ell\in N_\alpha(i)\setminus i} (1-\sigma_{\ell\alpha\to i\alpha})\right],\\
S'_{i\alpha\to i\beta} &= s_{i\alpha} \prod_{\gamma\in\mathcal{N}(\alpha)\setminus\beta} S'_{i\gamma\to i\alpha}\left[1 - \prod_{\ell\in N_\alpha(i)} (1-\sigma_{\ell\alpha\to i\alpha})\right]
\end{aligned}
\tag{12.9}
$$

where $N_\alpha(i)$ indicates the set of nodes (ℓ, α) which are neighbors of node i in layer α, and $\mathcal{N}_i(\alpha)$ indicates the layers β of the nodes (i, β) interdependent on the node (i, α). Finally, $S_{i\alpha} = 1$ ($S_{i\alpha} = 0$) indicates that the node (i, α) is (is not) in the mutually connected component, namely:

$$
S_{i\alpha} = s_{i\alpha} \prod_{\beta\in N_i(\alpha)} S'_{i\beta\to i\alpha}\left[1 - \prod_{\ell\in N_\alpha(i)} (1-\sigma_{\ell\alpha\to i\alpha})\right]. \tag{12.10}
$$

In other words, a node (i, α) is in the mutually connected component if it is not damaged (i.e., $s_{i\alpha} = 1$), if all the messages $S'_{i\beta\to i\alpha}$ coming from the interdependent nodes are equal to one, and if at least one message coming from a neighbor node (ℓ, α) in the layer α is equal to one. These message-passing equations can be solved for every network of networks of this type for the messages $\sigma_{i\alpha\to j\alpha}$ within each layer.

Finally, for $S_{i\alpha}$ as functions of the messages $\sigma_{i\alpha\to j\alpha}$ as

$$
s_{i\alpha} = s_{i\alpha} \prod_{\beta\in C_i(\alpha)} \left\{ s_{i\beta}\left[1 - \prod_{\ell\in N_\beta(i)} (1-\sigma_{\ell\beta\to i\beta})\right]\right\}. \tag{12.11}
$$

For this multilayer network, the mutually connected component can be defined as: A node (i, α) is in the mutually connected giant component if and only if: (a) It has at least one neighbor node (j, α) in the layer α that belongs to the mutually connected component; (b) All the nodes (ℓ, β) that can be reached from the node (i, α) by interdependent links have at least one neighbor node that belongs to the mutually connected component.

Furthermore, consider a multiplex network in which every layer α has a degree distribution $P_\alpha(k)$ and in which every node (i, α) is interdependent on all of its replica nodes. To evaluate the size of the mutually connected component, we average the messages across this ensemble of network of networks. We indicate with σ_α the average message within a layer $\sigma_{i\alpha \to j\alpha} = \sigma_\alpha$. Thus, the average messages σ_α are given in terms of the parameter p and the generating functions $G_0^\alpha(z)$ and $G_1^\alpha(z)$ defined as

$$G_0^\alpha = \sum_k P_\alpha(k) z^k$$
$$G_1^\alpha = \sum_k \frac{kP_\alpha(k)}{\langle k \rangle_\alpha} z^{k-1}, \tag{12.12}$$

where $\langle k \rangle_\alpha$ indicates the average degree in the layer α, i.e., $\langle k \rangle_\alpha = \sum_k kP_\alpha(k)$. We assume here that the damage of the nodes in the economic layer is correlated and, therefore, that a node (i, α) is damaged with probability p together with all its replica nodes. Thus,

$$P\left(\{s_{i,a}\}\right) = \prod_{i=1}^{M} \left[p \prod_{\alpha=1}^{M} \delta\left(1, s_{i,\alpha}\right) + (1-p) \prod_{\alpha=1}^{M} \delta\left(0, s_{i,\alpha}\right) \right], \tag{12.13}$$

where $\delta(x, y)$ indicates the Kronecker delta. This assumption is needed if we interpret the set of different replica nodes (i, α) in a multiplex as a single node having various types of interactions. If all layers β have the same topology, then the average messages within different layers β are all equal $\sigma_\beta = \sigma$ [55, 58, 59], and are given by:

$$\sigma = p[1 - G_0(1-\sigma)]^{M-1} [1 - G_1(1-\sigma)]. \tag{12.14}$$

Furthermore, the probability $S = \langle S_{i\alpha} \rangle$ that a node in a given layer is in the mutually connected component is given by

$$S = p[1 - G_0(1-\sigma)]^M. \tag{12.15}$$

In the case of targeted attacks toward high or low degree nodes, some kinds of failures have been considered in the economic layer. For example, a reduction of public research funds, restrictions on exports of semiconductors, or restrictions on new data centers in specific countries' regions. Like scale-free networks, interdependent networks are challenging to defend using the strategy of protecting high-degree nodes [60].

12.2.2 Materials

We categorized the data for each layer into two groups. The first group represents the node's size, and the second one represents the weight of the edge. For the economic layer (L_1), the node is given by the GDP in constant 2015 USD. The source is [61]. The links are the exports of goods, thus connecting exporters and importers. The data is collected from UN Comtrade [62] and includes all the HS codes. Previous studies validated these data sources [37, 63].

For the semiconductor layer (L_2), semiconductor patents provide the node. We took the triadic patent families for the WIPO domain of semiconductors, which includes GPUs [64]. The inventor's country of origin serves as a proxy for the origin of the patent. Exports of semiconductors provide the link. The data is collected from the BACI dataset [65]; the HS codes are 8571 and 8572.

For the energy layer (L_3), the node is given by the energy capacity. The energy supply (in petajoules) is the sum of domestic production and net imports of energy. The source is the Energy Statistics Yearbook [66]. The links are the exports of electricity, thus connecting exporters and importers. The data is collected from UN Comtrade [62], and the HS code is 271600.

For the knowledge layer (L_4), we selected the triadic patent families for AI-related technologies, which are considered their building blocks [67]. The source is [68]. We use conferences and patents as indicators of knowledge production. The current research on AI has been consolidated by industry [69]. However, academia remains an essential actor in AI research. The links are the countries of institutional affiliation of the authors of AAAI conferences, who are researchers working on the forefront developments [70]. The data is scraped from the AAAI website.

12.3 Results

Centrality measures such as closeness and betweenness centrality can be easily extended to multilayer networks once the metric and geodesic structure are defined. A different approach to measuring centrality employs the spectral properties of the adjacency matrix. The eigenvector centrality considers not only the number of links of each node but also the quality of such connections [71]. There are several ways to extend eigenvector centrality to multilayer networks, as it is used to optimize the outcome of interacting competing networks [72, 73].

Thus, the eigenvector centrality (EC) is defined as the influence of a vertex that is not only determined by the number of its neighbors but also by the influence of each neighbor. The centrality of the vertex (v_i) is proportional to the sum of neighboring centralities. The EC is defined as:

$$\mathrm{EC}(v_i) = \frac{1}{\lambda} \sum_{v_j \in V_i} \mathrm{EC}(v_j). \tag{12.16}$$

Eigenvector centrality is a measure of topological importance because it is directly related to dynamic processes occurring in networks [72]. Calculating eigenvector-like centralities in multiplex networks is to consider the eigenvector centrality $\mathbf{c}_\alpha = \left(c_1^{[\alpha]}, \dots, c_N^{[\alpha]}\right)$ in each layer $1 \leq \alpha \leq M$ separately. With this approach, for every node $i \in X$ the eigenvector centrality $\mathbf{c}_i$ is another vector:

$$\mathbf{c}_i = \left(c_i^{[1]}, \dots, c_i^{[M]}\right) \in \mathbb{R}^M, \tag{12.17}$$

where each coordinate is the centrality in the corresponding layer. Once all eigenvector centralities have been computed, the independent layer eigenvector-like centrality [73] of $\mathcal{M}$ is the matrix:

$$C = \left(\mathbf{c}_1^T \middle| \mathbf{c}_2^T \middle| \dots \middle| \mathbf{c}_M^T\right) \in \mathbb{R}^{N \times M}. \tag{12.18}$$

Notice that C is column stochastic since all components of $\mathbf{c}_\alpha$ are semipositive definite and $\|\mathbf{c}_\alpha\|_1 = 1$ for every $1 \leq \alpha \leq M$, and the centrality $\mathbf{c}_i$ of each node $i \in X$ is the ith row of C. As for the degree-type indicators, a numeric centrality measure for each node can be obtained by using an aggregation measure $f(\mathbf{c}_i)$ such as the sum, the maximum, or the ℓ_p-norm. The main limitation of this parameter is that it does not fully consider the multilevel interactions between layers and their influence on the centrality of each node [34]. However, this case is correct for the structure of the ecosystem.

The relationship between eigenvector centrality and node degree in our model reflects the characteristics of each layer, particularly the network sizes. In the economic layer, we first observe an isolated point, followed by a subsequent increase to a new set of values, with corresponding peaks in the distribution function. Eventually, the eigenvector centrality reaches saturation in the four periods, and no discontinuities are observed in the graph (Figure 12.2a).

In the energy layer, the eigenvector centrality saturates at low node degrees due to the finite possibilities of electricity exports. However, no discontinuity is observed in the graph, and it subsequently increases to a new value in 2015, representing the largest electricity exporter in Europe. For instance, in 2020, a wider dispersion of eigenvector centralities is observed for nodes with a high degree compared to earlier years. This may indicate a shift in the network's structure or the dynamics of influence propagation (Figure 12.2b). This layer, along with energy capacity, defines a significant process within the AI global ecosystem. For example, China has been altering its energy matrix, increasing its total supply from 68 833 petajoules in 2005 to 140 213 petajoules in 2020, while other countries have experienced either modest growth or reductions in their energy supply structures.

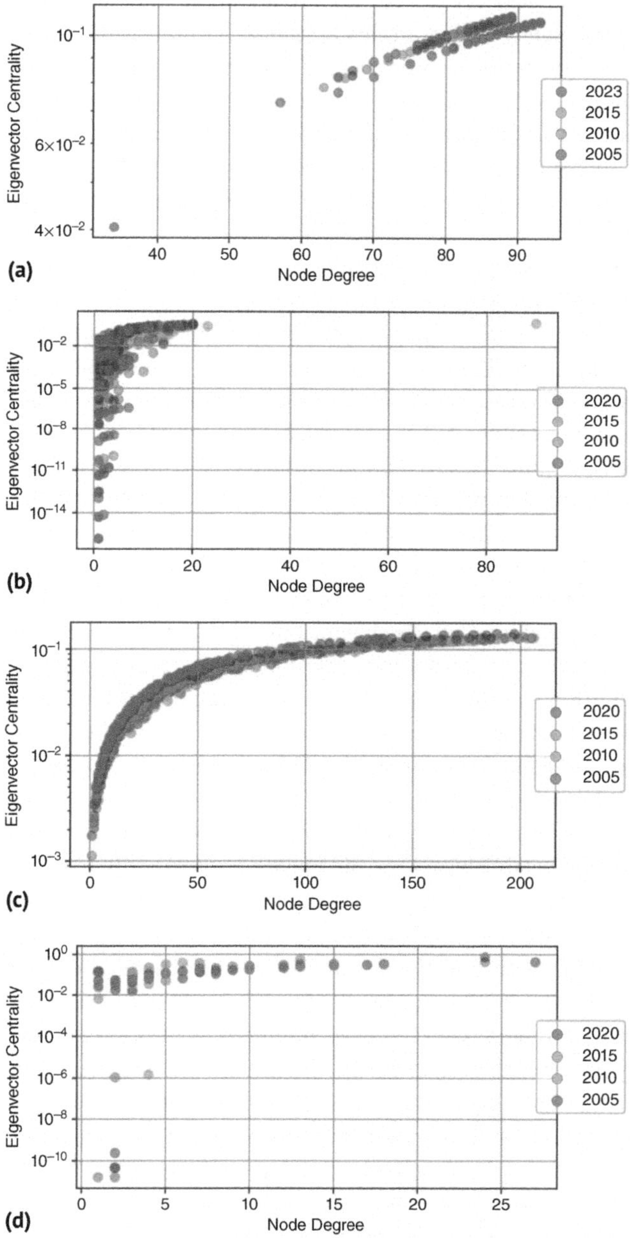

Figure 12.2 (a) Eigenvector centrality for economic layer. (b) Eigenvector centrality for energy layer. (c) Eigenvector centrality for AI knowledge layer. (d) Eigenvector centrality for semiconductors layer.

Within the semiconductor layer, the four periods exhibit similar trends (Figure 12.2c). Although the eigenvector centrality appears to saturate initially and subsequently increases to higher values with certain distances, the graph remains continuous. The definition of high eigenvector centrality indicates that a node is connected to numerous other nodes possessing high centrality. This suggests the consolidation of substantial semiconductor production for the AI global ecosystem within a limited number of countries, as shown in the top right of the figure (Figure 12.2c).

The AI knowledge layer illustrates the transformation of the scientific process, the dominance of publications from certain countries during the analysis period, and their evolution. Focusing on papers presented at the Association for the Advancement of Artificial Intelligence Conference, the AI knowledge layer demonstrates that increasing system size leads to disparities in the degree and eigenvector centralities between super-hubs, such as China and the United States, and the remaining nodes (Figure 12.2d). The prominence of Chinese and US researchers has resulted in the dominance of specific central points. As network size increases, a discontinuity emerges in the eigenvector centrality versus node degree plot as China, the United States, and, more recently, India have accelerated their research efforts and financial investments to consolidate AI advancements. In contrast, other countries such as Canada, Japan, France, Italy, and Germany have also made progress but remain in a distinct, less dominant group. Categorizing authors by their institutional affiliations highlighted the weighted interactions between authors and the significance of their nationalities. These findings indicate that China and the United States dominate the AI landscape.

To validate the argument, we utilized the results of knowledge networks in AI. In 2000, the United States took the forefront of AI research, with authors from affiliations in other countries establishing some links with authors affiliated with institutions in the United States. Nevertheless, the prevailing pattern involved interactions among authors from institutions of the same country of origin, resulting in a disconnected structure in the periphery. Specifically, the network exhibited centrality with the United States, a small cluster around it, and another group of countries at the periphery, which lacked interactions with others and occasionally only engaged with each other. By 2010, the AI network had expanded, maintaining the traditional centrality of the United States but with the addition of new sub-central nodes of lesser relevance, such as Germany, the UK, Australia, Canada, Israel, and China, countries that had escalated their resources for AI research. In addition, these nations increased their participation and interaction with authors affiliated with institutions from other countries [74].

In 2020, the network analysis revealed several features, including three structural patterns formed by China and the United States. First, self-loops indicate interactions among authors from institutions in the same country. Second, hubs of research are formed through interactions among institutions from various countries. Third, there are significant weighted links between institutions in China and the United States. Beyond the United States and China, a group of developed countries with a traditionally robust framework in research, science, knowledge, and industry, such as the UK, France, Germany, Japan, and Canada, validated the evolution of publications as outlined above.

Another pattern was the growing presence of authors with Chinese origins in institutions across the United States, Canada, Australia, and the United Kingdom, indicating China's increasing influence and consolidation of its position at the center. Throughout the snapshots, most peripheral countries exhibited minimal interaction, with a few exceptions. For instance, India and Brazil emerged as peripheral countries with commendable performance following China. Other emerging countries distanced themselves further from the center, intensifying their peripheral position, except for modest advancements observed in Indonesia, Iran, Mexico, Pakistan, and Turkey.

The big issue in the AI context is the peripheral markets. Our assessments reveal that the United States and China spearhead research, development, and deployment of AI, with India closely following. Countries like Japan, the UK, and Germany form a second-tier group. A third group includes Canada, South Korea, Australia, Italy, and France. Meanwhile, Russia, Indonesia, Brazil, Turkey, and Iran are emerging markets with notable international aspirations, and they are formulating strategies to narrow the gap with countries that might impact their regional interests or enhance their AI standing by forging partnerships with nations that have an established AI framework. This argument underscores the hypothesis, emphasizing the significance of AI as a scientific process. Countries must invest in and formulate strategies and policies in science, technology, education, and knowledge to solidify an AI framework or any other open innovation process.

In contrast, a group of countries lags significantly behind those mentioned above. In the Asia-Pacific region, countries such as Bangladesh, Vietnam, the Philippines, and Sri Lanka are notably lagging in terms of investment and research compared to the leading countries in the region. New Zealand may fall into this group based on its population levels. Within Europe, disparities are evident in various countries, such as Poland, Croatia, Slovakia, and the Czech Republic, which exhibit these lags. This is also true of developed countries, which, due to their population size, fail to generate research volumes comparable to those of the leading nations in the region. The Netherlands, Switzerland, Sweden, Norway, and Austria exemplify this situation.

Latin America, Africa, and the Middle East present a challenging scenario. In Latin America, beyond the principal economies of Brazil and Mexico, systemic lags pervade the rest of the region. Chile, Colombia, Peru, Ecuador, and Argentina form a lag group compared to regional leaders. The Caribbean, Central America, and almost all of Africa face critical challenges in science and technology investment. Low-income countries in all regions, as well as several middle-income countries, exhibit a high dependence on science and technology, coupled with structural challenges that hinder their independence in AI capabilities.

Some countries have leveraged their economic and financial capabilities to consolidate their AI capabilities. In the Middle East, several countries have started substantial investments in AI research, such as the UAE, Qatar, and Saudi Arabia, with substantial investments in various AI global ecosystem processes. Notably, those countries have established specialized institutes and are recruiting international scientists to advance AI frameworks. Those processes are part of adaptive behavior to link their countries to the core of the AI global ecosystem. However, other countries lack the economic capabilities to adapt to these processes, and they will remain in the peripheral space of the ecosystem.

Businesses from leading countries in the AI scientific process are expected to dominate the ecosystem. Intuitively, we confirm the correlation between the scientific process and business development. A similar conclusion was reached using the number of patents in other technological areas [37]. Additionally, technological companies from the United States and China compete for business across different regions, employing strategies to secure or impede business assignments.

12.4 Discussion

The structure of the AI global ecosystem presents challenges for global AI governance. First, global initiatives are challenging to coordinate due to the diverse positions on AI regulation. For instance, the United States focuses on fostering innovation, thereby prioritizing private firms and expecting them to self-regulate. China considers AI a national strategy, so the state has a relevant role in driving innovation and implementing firm regulations biased towards the government rather than companies. The European Union has a user-based approach in which companies face strict regulations and users' rights are protected [75]. Moreover, since AI is part of long-term national development and security plans, domestic interests might prevail over international ones. The deepening of these conditions would make it harder to achieve the coordination proposals between existing international institutions and develop new centralized AI global institutions [76]. Second, the asymmetry concerning most of the world's countries that do not hold central positions in the AI networks could bias the process [77]. Even if the first issue is somehow acknowledged, the striking differences in capabilities create a gap that will be used in negotiations. The global environmental governance is illustrative. The most industrialized economies, which contribute disproportionately to emissions, use their economic power to coerce others and to obtain benefits [78]. This is especially troublesome in the case of AI, given that weak institutions in many developing countries may not be well-positioned to mitigate harm, such as data privacy [79]. These elements could affect the ecosystem capacity for handling failures or errors based on interactions between agents. Although predefined mechanisms for error detection and correction are the highlighting perspective for several countries, the system complexity will be the solution to the evolution.

Our results evidence that energy capabilities are more widely distributed. It has implications for data center location, environment, competitiveness, and intellectual property. Some regions are more suitable for the location of the data center. For example, Nordic countries have an abundant energy supply, and the cold weather is favorable for refrigeration [80]. Nonetheless, the location of a data center is a relevant issue because it constitutes the physical infrastructure of AI. Therefore, it is subject to the actions and regulations of other countries regarding data. The EU, for instance, requires that data centers meet the regulatory frameworks for data collection and usage. This can confront companies to equate energy reliance and regulatory pressures. As with manufacturing outsourcing, the incentives to attract foreign investors could create a race to the bottom [81]. Industrial factories offered lower prices through environmental degradation or poor working conditions. Additionally, other electronics industry factories offered their know-how and lower prices, but they demanded the relaxation of intellectual property standards [82]. These examples highlight some consequences of the energy layer structure. Some of these issues are happening nowadays. For instance, citizens from a Uruguayan city massively protested in 2013 because the government granted permission to use water for a planned Google data center, despite the city facing water shortages. Additionally, evidence suggests that waste, deforestation, and energy shortages have affected citizens in some Mexican regions due to the construction of new data centers or the instability of electrical supply in some areas of the central United States [26].

The structure of the AI global ecosystem has geopolitical implications: power concentration, geopolitical alignment, and strategic chokepoints. States derive power from AI through economic and military applications [83]. The concentrated structure implies that a few states can obtain power, whereas others lack the necessary networks to access it. Since the states' objective is survival, they seek to increase their power while impeding rivals' access to it [84]. This derives the following two implications. On the one hand, the current and potential allies could be fostered through coercion or attraction to align with a significant AI power. Thus, countries with a central position in a network would attract more interest, although those resources would provide them with a position for negotiation. On the other hand, the concentration of power and network access in a few states creates the conditions for weaponizing such networks [85]. Hence, access to strategic resources such as knowledge, hardware, energy, or financial capacity could be restricted to rivals, as is currently the case with semiconductors in the US and rare earth minerals in China [86, 87].

The relationship between AI networks illustrates the dynamic feature of the AI global ecosystem. The knowledge layer created the basis for AI development. Data collection, AI models, and semiconductor architectures resulted from scientific discovery processes. This way, the parallel development of knowledge and hardware allowed the emergence of modern AI, in which predictive models became pervasive. The creation of GenAI, such as the LLMs, increased the demand for AI services. For example, LLMs such as Gemini, Claude, Copilot, DeepSeek, Llama, and ChatGPT offer their APIs for building applications upon them: many businesses realized that AI is affordable, increases productivity, and creates a competitive advantage. The computational power offered has proven to respond to increasing demand. Since 2018, computational resources have increased by a factor of four each year, reaching a total of four times the original amount by 2024. As the movement of electrical signals through circuits requires energy, power usage increased by 60 percent from 2020 to 2025, and it is expected to increase by more than 160 percent by 2030 compared to 2023 [88]. However, energy grids in the US are insufficient to meet future demand, and companies are under pressure to adopt sustainable operations. Thus, nuclear energy has emerged as a potential solution, and public investment plans in AI infrastructure [89]. However, it will not be feasible without the considerable funding that companies and public agencies provide.

Our approach revealed several factors that have altered the capacities and skills of numerous countries that have integrated AI into their economic strategies. One significant structural change involves certain countries incorporating the AI global ecosystem into their national economic development strategies. Additionally, technological and scientific public resources have played a pivotal role in consolidating AI research across various organizations, encompassing universities, research centers, and public institutions. Furthermore, private resources have surpassed traditional investments in science and technology, contributing to the consolidation of AI research within universities and companies. Namely, the AI ecosystem is a system where microscopic-level entities will interact, yielding macroscopic-level behaviors and structures, which in turn feed back to the microscopic-level entities. Ultimately, the relationship between structure and AI activity suggests that these structural changes can further transform the AI global ecosystem and economic power dynamics, potentially widening the gap between countries in terms of their economic, financial, and technological strengths.

12.5 Conclusion

We have several conclusions based on our research. First, the emergence of the AI race will have significant business implications, as this competence has a geopolitical perspective that affects organizations' interactions. The complexity of the AI global ecosystem drives the subsequent research, energy, and hardware strategies. Second, the AI capabilities demonstrate the consolidation of AI hubs, which are being harnessed as a tool for powerful and challenging organizations. Third, international and global perspectives will consolidate firms that understand the relevance of AI as a scientific process, which poses a significant challenge for thousands of organizations that rely on their services, platforms, and infrastructure. Fourth, some AI components will be fundamental for ecosystem dynamics. In other words, the addition of new behaviors or system components that interact with existing ones created complex networks of relationships and emergent behaviors that represent the ecosystem's change over time.

A short discussion on developing countries is worthwhile. For competitive reasons, companies and governments have deepened their use of data center services and their dependency on other technologies. However, given their lack of technological readiness, they often hire large companies and rely heavily on one or a small group of them, maintaining long-term business relationships. Will this be the same situation with AI evolution and concentration in a few organizations?

That creates a twofold risk. On the one hand, it empowers the supplier company and its host country. The panopticon and chokepoint effects are the result of surveillance through data and coercion through exclusion. On the other hand, relying on US companies could create an aversion in other countries, such as China, Russia, or those in the Middle East. In any case, developing countries are falling into a dependency that could transform into path dependence. The data would not be free, and several countries did not accept access to their population data by organizations from other countries for business purposes.

In summary, we find that the AI global ecosystem is closely tied to the robustness of central countries, as the evolution of AI capabilities requires other capabilities. Besides, this process is not autonomous because any advancement in AI will shape the development of various technological, energy, and knowledge sectors, in which economic capacity is fundamental. Finally, the billionaire resources of some organizations and governments to fund AI research could create a bubble in the ecosystem or deepen the technological periphery more when a new era of AI draws near: the combination of artificial general intelligence (AGI) and quantum computing.

References

1 Y. LeCun, Y. Bengio, and G. Hinton. Deep learning. *Nature*, 521(7553):436–444, 2015.

2 O. Russakovsky, J. Deng, H. Su, J. Krause, S. Satheesh, S. Ma, Z. Huang, A. Karpathy, A. Khosla, M. Bernstein, and A. C. Berg. Imagenet large scale visual recognition challenge. *International Journal of Computer Vision*, 115(3):211–252, 2015.

3 S. Mittal and S. Umesh. A survey on hardware accelerators and optimization techniques for RNNs. *Journal of Systems Architecture*, 112:101839, 2021. https://doi.org/10.1016/j.sysarc.2020.101839.

4 C. Silvano, D. Ielmini, F. Ferrandi, L. Fiorin, S. Curzel, L. Benini, F. Conti, A. Garofalo, C. Zambelli, E. Calore, S. Schifano, M. Palesi, G. Ascia, D. Patti, N. Petra, D. De Caro, L. Lavagno, T. Urso, V. Cardellini, G. Carlo Cardarilli, R. Birke, and S. Perri. A Survey on Deep Learning Hardware Accelerators for Heterogeneous HPC Platforms. *ACM Computing Surveys*, 57(11):286, 2025. https://doi.org/10.1145/3729215.

5 K. Crawford. *The Atlas of AI: Power, Politics, and the Planetary Costs of Artificial Intelligence*. Yale University Press, 2021.

6 K. Zeng and S. Chien. Energy-efficient AI: A survey on optimization approaches. *IEEE Access*, 10:4581–4595, 2022.

7 P. Svensson and H. Hassen. Trends in graphics processing unit prices and availability during the GPU boom: 2020–2021. *Journal of Computing and Information Technology*, 29(2):1–16, 2021.

8 International Energy Agency.What the data centre and ai boom could mean for the energy sector, 2023.https://www.iea.org/commentaries/what-the-data-centre-and-ai-boom-could-mean-for-the-energy-sector.

9 S. Holland. Trump to announce private sector ai infrastructure investment, CBS Reports, January 21 2025. https://www.reuters.com/technology/artificial-intelligence/trump-announce-private-sector-ai-infrastructure-investment-cbs-reports-2025-01-21/.Reuters.

10 Mike Isaac. Meta to increase spending to 65 billion this year in A.I. push, *NY Times*, January 27 2025.

11 B. Baker, L. Ouyang, J. Wu, D. Ziegler, R. Lowe, C. Voss, A. Radford, D. Amodei, and P. F. Christiano. Learning to summarize with human feedback. In *Advances in Neural Information Processing Systems (NeurIPS)* Vol. 33, pages 3008–3021, 2020.

12 A. Y. Ng. AI and the future of work, Harvard Business Review. July 24 2017. https://hbr.org/2017/07/ai-and-the-future-of-work.

13 Na Liu, Philip Shapira, Xiaoxu Yue, and Jiancheng Guan. Mapping technological innovation dynamics in artificial intelligence domains: Evidence from a global patent analysis. *PLoS One*, 16(12):e0262050, 2021.

14 Hamidreza Habibollahi Najaf Abadi, Zhou He, and Michael Pecht. Artificial intelligence-related research funding by the US national science foundation and the national natural science foundation of china. *IEEE Access*, 8:183448–183459, 2020.

15 Bengt-Åke Lundvall and Cecilia Rikap. China's catching-up in artificial intelligence seen as a co-evolution of corporate and national innovation systems. *Research Policy*, 51(1):104395, 2022.

16 Mary Ellen Flannery. Trump cancels federal research grants. What are the consequences?, 2025. https://www.nea.org/nea-today/all-news-articles/trump-cancels-federal-research-grants-what-are-consequences.

17 Owen Dyer. Trump halts new research grants to foreign scientists. *BMJ*, 389:r906, 2025. https://doi.org/10.1136/bmj.r906.

18 S. Venkatesh and Z. Zeng. Hardware accelerators for deep learning: A survey. *Journal of Computer Science and Technology*, 33(6):1310–1329, 2018.

19 Oscar M. Granados and Nicolas De la Peña. The supply chains of Artificial Intelligence, *arxiv*, 2025.

20 Guillaume Beaumier and Madison Cartwright. Cross-network weaponization in the semiconductor supply chain. *International Studies Quarterly*, 68(1):sqae003, 2024.

21 V. Lehdonvirta, B. Wú, and Z. Hawkins. Weaponized interdependence in a bipolar world: How economic forces and security interests shape the global reach of U.S. and Chinese cloud data centres, 2025. http://dx.doi.org/10.2139/ssrn.4670764.

22 McKinsey Company. Semiconductors have a big opportunity—but barriers to scale remain, April 2025. https://www.mckinsey.com/industries/semiconductors/our-insights/semiconductors-have-a-big-opportunity-but-barriers-to-scale-remain.

23 Semiconductor Industry Association. Semiconductor supply chain investments, March 2025. https://www.semiconductors.org/chip-supply-chain-investments.

24 E. Masanet, A. Shehabi, N. Lei, S. Smith, and J. Koomey. Recalibrating global data center energy-use estimates. *Science*, 367(6481):984–986, 2020. https://doi.org/10.1126/science.aba3758

25 Goldman Sachs. AI poised to drive 160% increase in power demand, May 14 2024. https://www.goldmansachs.com/insights/articles/AI-poised-to-drive-160-increase-in-power-demand.

26 A. Valdivia. The supply chain capitalism of AI: A call to (RE) think algorithmic harms and resistance through environmental lens. In *Information, Communication Society*, pages 1–17, 2024.

27 I. Penn and K. Weise. Hungry for energy, amazon, google and microsoft turn to nuclear power, *The New York Times*. October 18 2024. https://www.nytimes.com/2024/10/16/business/energy-environment/amazon-google-microsoft-nuclear-energy.html.

28 Christian Bogmans, Patricia Gomez-Gonzalez, Giovanni Melina, and Sneha Thube. Ai needs more abundant power supplies to keep driving economic growth, 2025. https://www.imf.org/en/Blogs/Articles/2025/05/13/ai-needs-more-abundant-power-supplies-to-keep-driving-economic-growth.

29 Gartner. Gartner predicts power shortages will restrict 40 by 2027, 2024.https://www.gartner.com/en/newsroom/press-releases/2024-11-12-gartner-predicts-power-shortages-will-restrict-40-percent-of-ai-data-centers-by-20270.

30 K. Arulkumaran, M. P. Deisenroth, M. Brundage, and A. A. Bharath. Deep reinforcement learning: A brief survey. *IEEE Signal Processing Magazine*, 34(6):26–38, 2017. https:/doi.org/10.1109/MSP.2017.2743240.

31 V. Mnih, K. Kavukcuoglu, D. Silver, A. A. Rusu, J. Veness, M. G. Bellemare, A. Graves, M. Riedmiller, A. K. Fidjeland, G. Ostrovski, S. Petersen, C. Beattie, A. Sadik, I. Antonoglou, H. King, D. Kumaran, D. Wierstra, S. Legg, and D. Hassabis. Human-level control through deep reinforcement learning. *Nature*, 518(7540):529–533, 2015. https://doi.org/10.1038/nature14236.

32 M. Wooldridge and N. R. Jennings. Intelligent agents: Theory and practice. *The Knowledge Engineering Review*, 10(2):115–152, 1995. https://doi.org/10.1017/S0269888900008122.

33 H. Naveed, A. Khan, S. Qiu, M. Saqib, S. Anwar, M. Usman, N. Akhtar, N. Barnes, and A. Mian. A comprehensive overview of large language models. https://doi.org/10.48550/arXiv.2307.06435, 2023.

34 S. Boccaletti, G. Bianconi, R. Criado, C.I. del Genio, J. Gómez-Gardeñes, M. Romance, I. Sendiña-Nadal, Z. Wang, and M. Zanin. The structure and dynamics of multilayer networks. *Physics Reports*, 544(1):1–122, 2014. https://doi.org/10.1016/j.physrep.2014.07.001.

35 Bradley Fraser, Claudia Szabo, Andrew Coyle, and Robert Hunjet. Leveraging emergent specialization in a heterogeneous multi-role swarm control architecture for positional-based uas missions. In *2020 IEEE Symposium Series on Computational Intelligence (SSCI)*, pages 2584–2591. IEEE, 2020.

36 Rob Dekkers. Complex adaptive systems. In *Applied Systems Theory*, pages 211–233. Springer, 2017.

37 Oscar M. Granados and Nicolas De la Peña. Artificial intelligence and international system structure. *Revista Brasileira de Política Internacional*, 64(1):e003, 2021.https://doi.org/10.1590/0034-7329202100103.

38 Thong Pham, Paul Sheridan, and Hidetoshi Shimodaira. Joint estimation of preferential attachment and node fitness in growing complex networks. *Scientific Reports*, 6(1):32558, 2016. https://doi.org/10.1038/srep32558.

39 G. Bianconi and A.-L. Barabási. Competition and multiscaling in evolving networks. *Europhysics Letters*, 54(4):436, 2001. https://doi.org/10.1209/epl/i2001-00260-6.

40 Albert-László Barabási and Réka Albert. Emergence of scaling in random networks. *Science*, 286(5439):509–512, 1999. https://doi.org/10.1126/science.286.5439.509.

41 Mikko Kivelä, Alex Arenas, Marc Barthelemy, James P. Gleeson, Yamir Moreno, and Mason A. Porter. Multilayer networks. *Journal of Complex Networks*, 2(3):203–271, 2014. 10.1093/comnet/cnu016.

42 Jacopo Iacovacci and Ginestra Bianconi. Extracting information from multiplex networks. *Chaos: An Interdisciplinary Journal of Nonlinear Science*, 26(6):065306, 2016. 10.1063/1.4953161.

43 Giovanni Bonaccorsi, Massimo Riccaboni, Giorgio Fagiolo, and Gianluca Santoni. Country centrality in the international multiplex network. *Applied Network Science*, 4(1):126, 2019. https://doi.org/10.1007/s41109-019-0207-3.

44 Stefano Boccaletti, Vito Latora, Yamir Moreno, Maisa Chávez, and Dong-Uk Hwang. Complex networks: Structure and dynamics. *Physics Reports*, 424:175–308, 2006. 10.1016/j.physrep.2005.10.009.

45 Andries E. Brouwer and Willem H. Haemers. *Spectra of Graphs*. Springer Science Business Media, 2011.

46 Rubén Sánchez García, Emanuele Cozzo, and Yamir Moreno. Dimensionality reduction and spectral properties of multiplex networks. *Physical Review. E, Statistical, Nonlinear, and Soft Matter Physics*, 89(5):052815, 2013.

47 Piet Van Mieghem. *Graph Spectra for Complex Networks*. Cambridge University Press, 2023.

48 A. Barrat, M. Barthélemy, R. Pastor-Satorras, and A. Vespignani. The architecture of complex weighted networks. *Proceedings of the National Academy of Sciences*, 101(11):3747–3752, 2004. https://doi.org/10.1073/pnas.0400087101.

49 M. E. J. Newman. Analysis of weighted networks. *Physical Review E*, 70:056131, 2004. 10.1103/PhysRevE.70.056131.

50 Tore Opsahl, Filip Agneessens, and John Skvoretz. Node centrality in weighted networks: Generalizing degree and shortest paths. *Social Networks*, 32(3):245–251, 2010. https://doi.org/10.1016/j.socnet.2010.03.006.

51 Ping Hu and Ting Mei. Ranking influential nodes in complex networks with structural holes. *Physica A: Statistical Mechanics and its Applications*, 490:624–631, 2018. https://doi.org/10.1016/j.physa.2017.08.049.

52 Alessandro Vespignani. The fragility of interdependency. *Nature*, 464(7291): 984–985, 2010. https://doi.org/10.1038/464984a.

53 Alexei Vázquez and Yamir Moreno. Resilience to damage of graphs with degree correlations. *Physical Review E*, 67:015101, 2003. https://doi.org/10.1103/PhysRevE.67.015101.

54 Ginestra Bianconi, Sergey N. Dorogovtsev, and José F. F. Mendes. Mutually connected component of networks of networks with replica nodes. *Physical Review E*, 91:012804, 2015. https://doi.org/10.1103/PhysRevE.91.012804.

55 G. J. Baxter, S. N. Dorogovtsev, A. V. Goltsev, and J. F. F. Mendes. Avalanche collapse of interdependent networks. *Physical Review Letters*, 109:248701, 2012. https://doi.org/10.1103/PhysRevLett.109.248701.

56 Alexander K. Hartmann and Martin Weigt. *Phase Transitions in Combinatorial Optimization Problems: Basics, Algorithms and Statistical Mechanics*. John Wiley Sons, 2006.

57 Marc Mezard and Andrea Montanari. *Information, Physics, and Computation*. Oxford University Press, 2009.

58 Sergey V. Buldyrev, Roni Parshani, Gerald Paul, H. Eugene Stanley, and Shlomo Havlin. Catastrophic cascade of failures in interdependent networks. *Nature*, 464(7291):1025–1028, 2010. https://doi.org/10.1038/nature08932.

59 Jianxi Gao, Sergey V. Buldyrev, Shlomo Havlin, and H. Eugene Stanley. Robustness of a network of networks. *Physical Review Letters*, 107:195701, 2011. https://doi.org/10.1103/PhysRevLett.107.195701.

60 Xuqing Huang, Jianxi Gao, Sergey V. Buldyrev, Shlomo Havlin, and H. Eugene Stanley. Robustness of interdependent networks under targeted attack. *Physical Review E*, 83:065101, 2011. https://doi.org/10.1103/PhysRevE.83.065101.

61 World Bank. Exports of goods and services constant 2015 USD, 2024. https://rb.gy/kylwtf (accessed June 1, 2024).

62 United Nations Statistics Division. Comtrade—united nations commodity trade statistics database, 2024. https://comtrade.un.org (accessed February 02, 2025).

63 Stephen G. Brooks and William C. Wohlforth. The rise and fall of the great powers in the twenty-first century: China's rise and the fate of America's global position. *International Security*, 40(3):7–53, 2015.

64 World Intellectual Property Organization. Patent and technology information, 2024. https://www.wipo.int/patents/en/patent-information.html (accessed June 1, 2024).

65 Guillaume Gaulier and Soledad Zignago. BACI: International trade database at the product-level. The 1994–2007 version. Working Papers 2010-23, CEPII, 2010. https://www.cepii.fr/CEPII/fr/publications/wp/abstract.asp?NoDoc=2726.

66 United Nations Statistics Division. United Nations Statistics, 2023. https://unstats.un.org/UNSDWebsite (accessed February 02, 2025).

67 Raphael Zingg. Foundational patents in artificial intelligence. In *Artificial Intelligence and Intellectual Property*, pages 75–98. Oxford University Press, 2021.

68 OECD. Patents by wipo technology domains, 2024. https://rb.gy/fi02og (accessed June 1, 2024).

69 N. Maslej, L. Fattorini, R. Perrault, V. Parli, A. Reuel, E. Brynjolfsson, J. Etchemendy, K. Ligett, T. Lyons, J. Manyika, J. C. Niebles, Y. Shoham, R. Wald, and J. Clark. Artificial intelligence index report 2024. Technical Report, Stanford University, 2024. https://aiindex.stanford.edu/report.

70 Kevin Leyton-Brown, Yatin Nandwani, Hedayat Zarkoob, Chris Cameron, Neil Newman, Dinesh Raghu. Matching papers and reviewers at large conferences. *Artificial Intelligence*, 331:104119, 2024.

71 Phillip Bonacich. Power and centrality: A family of measures. *American Journal of Sociology*, 92(5):1170–1182, 1987.

72 J. Aguirre, D. Papo, and J. M. Buldú. Successful strategies for competing networks. *Nature Physics*, 9(4):230–234, 2013. https://doi.org/10.1038/nphys2556.

73 Luis Solá, Miguel Romance, Regino Criado, Julio Flores, Alejandro García del Amo, and Stefano Boccaletti. Eigenvector centrality of nodes in multiplex networks. *Chaos: An Interdisciplinary Journal of Nonlinear Science*, 23(3):033131, 2013. https://doi.org/10.1063/1.4818544.

74 Nicolás De la Peña and Oscar M. Granados. El poder internacional y las capacidades de inteligencia artificial. *Revista Científica General José María Córdova*, 22(47):693–712, 2024. https://doi.org/10.21830/19006586.1376.

75 Teemu Birkstedt, Matti Minkkinen, Anushree Tandon, and Matti Mäntymäki. Ai governance: Themes, knowledge gaps and future agendas. *Internet Research*, 33(7):133–167, 2023.

76 Huw Roberts, Emmie Hine, Mariarosaria Taddeo, and Luciano Floridi. Global AI governance: Barriers and pathways forward. *International Affairs*, 100(3): 1275–1286, 2024.

77 Marie-Therese Png. At the tensions of south and north: Critical roles of global south stakeholders in AI governance. In *Proceedings of the 2022 ACM Conference on Fairness, Accountability, and Transparency*, pages 1434–1445, 2022.

78 Harold A. Perkins. Consent to neoliberal hegemony through coercive urban environmental governance. *International Journal of Urban and Regional Research*, 37(1):311–327, 2013.

79 Deborah G. Johnson and Mario Verdicchio. AI, agency and responsibility: The VW fraud case and beyond. *AI & Society*, 34:639–647, 2019.

80 Helen S. Adams. Norway's cool climate powers bulk data centre. *Data Centre Magazine*, 2024. https://datacentremagazine.com/articles/norways-cool-climate-powers-bulk-data-centre (accessed February 02, 2025).

81 David M. Konisky. Regulator attitudes and the environmental race to the bottom argument. *Journal of Public Administration Research and Theory*, 18(2):321–344, 2008.

82 Henry Farrell and Abraham Newman. Chained to globalization: Why it's too late to decouple. *Foreign Affairs*, 99:70, 2020.

83 Eric Schmidt. AI, great power competition national security. *Daedalus*, 151(2): 288–298, 2022.

84 John J. Mearsheimer. *The Tragedy of Great Power Politics*, 1st edition. W. W. Norton Company, New York, 2001.

85 Henry Farrell and Abraham L. Newman. Weaponized interdependence: How global economic networks shape state coercion. *International security*, 44(1):42–79, 2019.

86 Yadong Luo and Ari Van Assche. The rise of techno-geopolitical uncertainty: Implications of the United States CHIPS and Science Act. *Journal of International Business Studies*, 54(8):1423–1440, 2023.

87 Keith Bradsher, Ana Swanson, and Alexandra Stevenson. China limits exports of minerals crucial to semiconductors. *The New York Times*, December 3 2024. https://www.nytimes.com/2024/12/03/world/asia/china-minerals-semiconductors.html (accessed February 02, 2025).

88 Goldman Sachs. Is nuclear energy the answer to ai data centers' power consumption? *Goldman Sachs Insights*, 2024. https://www.goldmansachs.com/insights/articles/is-nuclear-energy-the-answer-to-ai-data-centers-power-consumption (accessed Frbruary 02, 2025).

89 Dawn Stover. AI goes nuclear. *The Bulletin of the Atomic Scientists*, 2024. https://thebulletin.org/2024/12/ai-goes-nuclear (accessed February 02, 2025).

13

Resilient Complex Adaptive Systems: A Communications Perspective

Bradley Fraser[1] and Rhett Hull[1,2]*

[1] Sensors and Effectors Division, Defence Science and Technology Group, Adelaide, SA, Australia
[2] UTS Tech Lab, University of Technology Sydney, Ultimo, NSW, Australia

*Corresponding author: bradley.fraser1@defence.gov.au

13.1 Introduction

Uncrewed aerial systems (UASs) are a perfect example of a truly disruptive technology and are now ubiquitous within our society. Although remotely controlled, UASs have existed for some time but have recently progressed to a point where non-specialist users can both afford and utilize them effectively. With relatively minor effort, these platforms have been integrated with a plethora of sensors, effectors, actuators, and computers, enabling them to address myriad use cases in nearly every industry where mobile platforms can be found, including cinematography, agriculture, mining, logistics, entertainment, research, and warfare [1].

Increasingly, we are seeing a declining amount of direct human interaction required for effective UAS operation. Indeed, mission planning software that produces automated behavior is fast becoming the norm for the control of (multi-) drone systems. A notable example of this is the spectacular light shows made of thousands of UASs that are fast replacing traditional fireworks displays [2]. While impressive, these UAS shows rely on pre-planned missions, which lack the ability to respond to environmental changes and therefore are unsuitable for dynamic environments.

For truly autonomous mission behavior, the UASs themselves must be imbued with an agency to make their own decisions to plan and respond in real time. From an implementation perspective, this requires nothing more than a UAS to be equipped with an (on-board) computer to house the decision-making algorithms.

Such approaches often outperform human operators even in especially challenging tasks such as high-speed drone racing [3]. Extending this approach to UAS teams further requires a communications channel for individuals to coordinate their actions.

Currently, the deployment of UAS teams primarily relies on wireless radio systems to communicate. However, physical layer issues such as limited, contested, or denied radio frequency (RF) spectrum, as well as link and network layer issues including routing scalability and link dynamism, practically limit the volume of information and frequency at which it can be shared with peers. Furthermore, if communication is unreliable, coordination algorithms must employ a distributed control architecture and, most importantly, make decisions without full knowledge of the network state. Where many traditional coordination approaches fail in this disrupted, denied, intermittent, and low-bandwidth (DDIL) communications environment, complex adaptive systems (CASs) are well-known to meet these requirements [4]. While plenty of attention has been given to developing CAS algorithms to exhibit specific behaviors [5], understanding CAS [6, 7], and identifying emergent behavior within them [8, 9], this chapter views CAS algorithms from the perspective of RF communications and considers the following questions:

1. How can CAS be used to enable communications in DDIL environments?
2. What happens to CAS when communications between individuals are disrupted?
3. How does explicit communication compare to implicit communication in CAS?
4. How can communicating intent between individuals assist in reducing the required communications volume?

Throughout this chapter, we examine a number of exemplar CAS algorithms in the spirit of the above questions and explore their common relationships with communications.

13.2 Enabling Communications in DDIL Environments

Mobile ad hoc networks (MANETs) are peer-to-peer wireless networks, where network nodes are free to move without regard to the physical network topology [10]. Each individual node acts as a router and forwards data on toward the destination. Specialized routing protocols adapted from those in fixed networks continually maintain routes even when peers are moving [11].

MANETs are extremely useful in search and rescue, disaster relief [12], and military scenarios when the local fixed infrastructure is either destroyed, untrusted, or otherwise unavailable. Their use in UAS swarms provides a straightforward intra-swarm communications channel for the sharing of data between individuals. However, they can also be used to augment network connectivity of other nodes. For example, for those nodes on the ground that are challenged by obstructions or interference [13].

A number of algorithms to autonomously control one or more UAS have been proposed to either reconnect disconnected components of the associated network graph [14] or physically collect and deliver network data between graph components if end-to-end connectivity is unachievable [15]. In the former *range-extension* activity, additional nodes are often referred to as repeaters, while in the latter, additional nodes are known as data ferries.

13.2.1 Range Extension

While traditional planning algorithms for this problem exist—for example, see [16], wherein an optimization approach is used to minimize graph bridges and cut (or articulation) vertices—our focus here is on CAS-style algorithms that solve the problem through self-organization. For example, Konak et al. [17] and Hunjet [18] use swarm-based principles to self-organize repeaters to maintain connectivity of other non-repeater network nodes. In the former, repeater nodes follow a heuristic algorithm based upon a variation of Reynolds' flocking rules [19], where agents follow separation, cohesion, and exploration behavior determined by nearest-neighbor positions. Graph connectivity of this approach was found to be statistically similar to a quadratic mixed-integer programming formulation solved optimally using hours of computation time. Such self-organization and emergence approaches to large-scale coordination are attractive for this reason. While not necessarily optimal, they produce a competitive solution with minimal overhead.

In a similar way, the latter work uses attractive and repulsive forces to position repeater nodes to maintain connectivity to other network nodes. These forces can act on the raw distance between neighbors or even the RF received signal strength indicator (RSSI) or signal-to-noise ratio (SNR) of the associated wireless channel of the graph edge. In this way, repeater nodes can dynamically react to barrage-jamming techniques associated with an increased RF-noise floor by decreasing intra-node distance. These virtual force techniques contain built-in platform redundancy with agents inherently reinforcing their peers without explicit signaling. This is a common trait found in resilient CAS. In contrast, an optimally solved solution would need recalculating with a change in agent numbers and initial conditions.

These range extension approaches will effectively bridge gaps between network nodes owing to the daisy-chain structures that emerge from the local interaction rules; given enough resources, the structures will bridge any-sized gap between disconnected network components. However, this cannot be guaranteed under finite resourcing. In this case, a different strategy is required where repeaters no longer form a (semi-)stable topology but move continuously between network nodes to deliver data.

13.2.2 Static Route Ferrying

In static route ferrying, we consider a static network topology with a fixed ferry path, and the problem devolves into two sub-problems. The first of them is to find an efficient fixed ferry path, which is the well-known traveling salesman problem (TSP) [20] (or traveling salesman problem with neighborhood [TSPN] when considering communications distance). Though some approaches consider finding multiple sub-TSP routes with a single ferry on each route, where the route and ferry movement is constrained such that data can be transferred between ferries when they pass each other [21], we consider a single path here. Once this path has been determined (in a pre-deployment phase), ferries can then be deployed such that their intra-node distance along the polygon perimeter is equidistant in order to maximize throughput. However, if ferries do not maintain this schedule due to, for example, speed differences caused by wind or ferry attrition or addition, then throughput will suffer.

A simple swarming-based solution [22] to the above problem is to use digital pheromones akin those laid down by certain species of foraging ants [23]. In this approach, network nodes maintain a steadily decaying digital pheromone value, which is reset to a high value when visited by a ferry. In response, a ferry will set its speed along the perimeter proportional to each pheromone value, which is described by the simple rules. (1) If the pheromone level is high, then another ferry has visited this node recently, and it should slow down to *widen* the gap between itself and the previous ferry, and (2) if the pheromone level is low, then another ferry has not visited this node recently, and it should speed up to *reduce* the gap between itself and the previous ferry. The authors model this behavior with a linearly decaying pheromone:

$$p(t) = 1 - \frac{t}{T}, \tag{13.1}$$

where $p(t)$ is the pheromone value on the interval $[0, 1]$ at time t, and the constant T is the time taken for the pheromone to decay to zero. The ferry adjusts its speed according to

$$s_i(p) = (s_{\max} - s_{\min})(1 - p) + s_{\min}, \tag{13.2}$$

where s_i is the ferry speed on the interval $[s_{\min}, s_{\max}]$, and p is the value of the pheromone of the node with which the ferry is communicating. From an implementation perspective, the node needs only to transmit the pheromone, while the ferry needs only to transmit an acknowledgment and evaluate the speed function, limiting the communication and processing requirements. This process repeats as a *stigmergic* feedback system and mimics some CAS found in nature.

Substituting Eq. (13.1) into Eq. (13.2) yields

$$s(t) = (s_{\max} - s_{\min})\left(\frac{t}{T}\right) + s_{\min}. \tag{13.3}$$

Typical CAS systems are difficult to model precisely mathematically due to the number of individuals involved; however, in this instance, the system can be modeled by considering a single node and a single ferry. Assuming the speed–time relationship solution is

$$s^* = \frac{d}{t^*}, \tag{13.4}$$

where d is the distance traveled by the ferry before returning to the node. Then the only feasible solution can be shown to be

$$s^* = \frac{s_{\min} + \sqrt{s_{\min}^2 + 4\frac{d}{T}(s_{\max} - s_{\min})}}{2}. \tag{13.5}$$

Now, the substitution of Eq. (13.4) into Eq. (13.3) yields (dropping the $*$ notation),

$$s = (s_{\max} - s_{\min})\left(\frac{d}{Ts}\right) + s_{\min}. \tag{13.6}$$

This describes a discrete, one-dimensional dynamical system through the mapping $s \mapsto f(s)$ [24]; that is,

$$s \mapsto (s_{\max} - s_{\min})\left(\frac{d}{Ts}\right) + s_{\min} := f(s). \tag{13.7}$$

In their survey of CAS, [6] notes that "one of the most mature [theories of complex systems] is dynamical systems theory, in which the behaviors of agents over time are represented individually or collectively by simple mathematical models, coupled together to represent interactions." This theory is used here in that capacity, and through the properties of dynamical systems [25], the solution converges if

$$f'(s) = \left|(s_{\max} - s_{\min})\left(-\frac{d}{Ts^2}\right)\right| < 1. \tag{13.8}$$

where $f'(s)$ is the first derivative, and therefore, a sufficient condition for convergence is that $\left(\frac{d}{T}\right) \leq s_{\min}$ holds. That is, the minimum speed ensures the pheromone does not expire completely during a full circuit. This is unsurprising since information inherent to the system dynamics would be lost in that case.

Figure 13.1a,b depicts the system plot and the *cobweb* plot, or Verhulst diagram,[1] for a toy example. In this case, the solutions are −0.5 and 1.5 and can be seen as x-axis intercepts of the parabola that corresponds to the standard quadratic form of feasible solution above and also as fixed points of the intercepts of the hyperbola that corresponds to the iterated map Eq. (13.7) and the line $y = x$.

Figure 13.1c shows the actual map-iterated solution values for four initial velocities. Three begin near the fixed-point *attractor* of the dynamical system corresponding to the derived feasible solution. For demonstration purposes, a fourth initial, infeasible, solution is shown starting near the fixed-point *repellor,* which quickly becomes unstable and converges to the attractor solution. Figure 13.1d

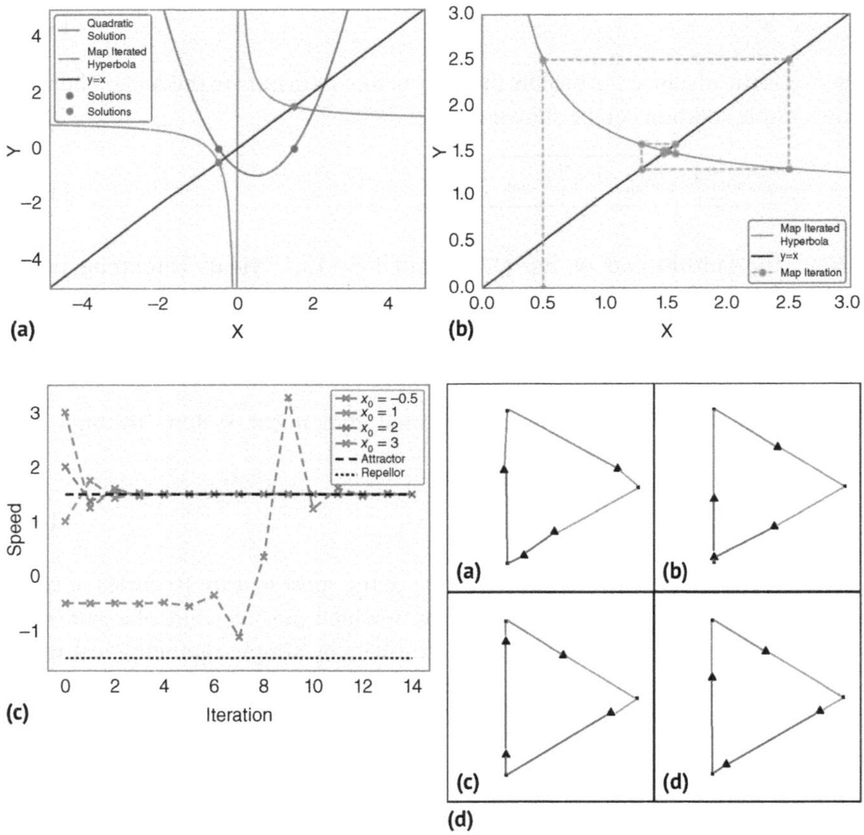

Figure 13.1 Verhulst analysis of the dynamical system. (a) System plot, (b) Verhulst diagram, (c) iterated solution, and (d) equidistant perimeter spacing.

1 Named for Pierre Verhulst, who originally published works on population growth using the continuous logistic equation [26], of which the logistic map is a discrete-time variant and closely related to the model discussed here.

shows the equidistant spacing (colored segments are of equal length) of the ferries (triangles) around the nodes (vertices) of the fixed ferry path.

This analysis can be extended to more complex systems with multiple ferries and nodes and has been demonstrated on real UAS [27, 28]. In addition, it is shown in [24] how it can be used to control the automated launch and convergence of ferries in a system of moving nodes if quality of service (QoS) agreements are breached.

CAS resiliency is of particular interest within this book, and the mathematical model of this pheromone-based coordination algorithm provides a precise explanation for such resiliency. Expressed as attractor and repellor solutions of the dynamical system, it is easy to see how the system can recover from speed perturbations experienced by the agent-based model. As such, this CAS is a particularly useful example of agent-based simulation being complemented by traditional mathematical modeling to provide insight into its collective behavior.

13.2.3 Dynamic Route Ferrying

Static route ferrying uses visit times as the metric for performance and views the network as a collection of data sources and sinks, ignoring individual data flows. At the expense of ferry predictability, the static route constraint can be removed to allow ferries to be responsive to the flow of packets through the network, where ferries are free to visit any node they desire, in any order. The focus here is no longer on positioning or timing, but rather on deciding the next hop on the dynamic path. This largely involves inspecting a ferry's packet buffer to discern traffic flows. A CAS-based approach to this problem is presented in [29, 30], wherein the ferry destination is determined by the collective packet waiting time in the ferry buffer for each destination.

This function favors nearby destination nodes that the ferry has traffic for, particularly those destinations that have not been visited for some time. This approach is assisted by geographical routing, a data routing strategy that forwards data to an intermediate node that is geographically closer to the packet's destination, if one exists.

The rules for this algorithm are then the following:

1. User[2]-to-user Interaction and Ferry-to-ferry Interaction: Exchange data in accordance with geographic routing.
2. Ferry-to-user Interaction: Ferry takes all data from user, and ferry delivers data to user if and only if the user is the destination or the user has a route to the destination or to a closer ferry.

---

2 A user node refers to a non-ferry node that produces and receives traffic.

Previously, static route ferrying used artificial pheromones to affect agent actions. Here, the data packets themselves act as pheromones in the stigmergic process that guides ferry movement and produces coordinated macroscopic-level behavior. By adding and removing data packets at user nodes, ferry nodes influence which user needs visiting at a later time; and by handing off traffic in a geographic manner, ferries tend to operate within their current local area. The authors term this as "locality," where highly local behavior (that is, very little movement) is consistent with stationary repeaters and highly non-local behavior is consistent with data ferrying. This descriptor, therefore, allows for a continuous spectrum of behavior between stationary-repeating and dynamic data ferrying and can be used to classify behavior in networks that require some degree of both strategies. To investigate such situations, consider the random geometric graph (RGG) [31].

Let $\mathcal{G}(n, r)$ represent a graph with a sequence $\mathcal{X} = \{x_1 \dots x_n\}$ of independent and uniformly distributed points on $[0, 1]^d$, where d is the space dimension. Then, given a fixed $r > 0$, connect two points if their $\mathcal{L}_p$ distance is at most r. In this chapter, we are primarily concerned with two-dimensional results ($d = 2$); thus, the distance measure is the $\mathcal{L}_2$ (Euclidean) norm. For graph connectivity analysis, it is convenient to represent the radius r as a function of n, that is, $r = r(n)$, and if D is a random variable representing the degree in $\mathcal{G}$, then the expected degree is given by [31]

$$\mathbb{E}[\mathrm{D}] = \pi(n-1)r^2 \sim \pi n r^2. \tag{13.9}$$

A connectivity threshold occurs at $\mathbb{E}[\mathrm{D}] = \ln(n)$, which gives a threshold radius of

$$r_t \sim \sqrt{\frac{\ln(n)}{\pi n}} \tag{13.10}$$

and holds *almost surely*. That is to say that the probability of this result occurring tends to one as $n \to \infty$. Figure 13.2 plots an RGG example with $r = 0.1$ and $n = 100$. The expected degree ($\mathbb{E}[\mathrm{D}] \approx 3.14$) is below the connectivity threshold of $\ln(100) \approx 4.61$ ($r_t \approx 0.12$). On inspecting the distance between components, it is clear that some components could be bridged with a repeater, while others cannot and require ferrying.

Results of the algorithm are shown in Figure 13.3, where the location histories of a 30-node RGG with three *resource*[3] nodes ($\mathbb{E}[\mathrm{D}] = 3.5$) are displayed. The resource nodes start from similar locations to find stable operating areas that connect the graph temporally.

This system again demonstrates obvious self-organization without explicit coordination between peers. The use of the RGG to model the distribution of user nodes assists in explaining the agent behavior in light of what is required by the network connectivity. However, as stated previously, a key challenge in the mathematical modeling of CAS is the complexity inherent to a system of many

3 Resource nodes are nodes that perform ferrying and/or bridging between disconnected graph components.

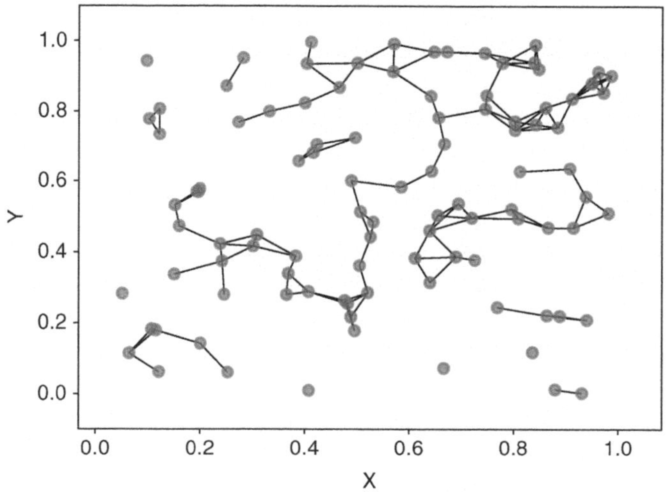

Figure 13.2 A 100 node $r = 0.1$ RGG instantiation.

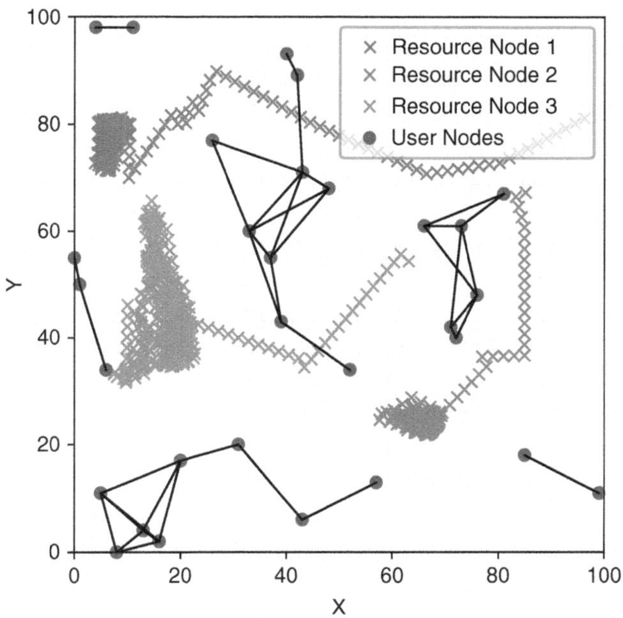

Figure 13.3 Results of the algorithm with a 30-user node RGG, $\mathbb{E}[D] = 3.5$. Crosses indicate the location histories of the three resource nodes.

interacting individuals. As such, mathematical techniques such as statistical mechanics and probability theory are often used to describe holistic behavior. Indeed, RGGs are a graph construct consisting of random variables where graph vertex degree is discussed as an expectation rather than a defined value. Further, in their model of data packet latency for the ferry system on the RGG, the authors consider ferries as atomic particles to estimate their interaction frequency and, therefore, the dissemination of data as they come into contact.

From a self-organization perspective, emergence within complex systems is thought to be from the interactions between individuals as well as between individuals and the environment. In this case, the user nodes that are being serviced by the ferries form part of the complex system's environment and are used by the system to affect its own future evolution.

13.2.4 Buffer Constraints

Another dimension to the ferrying problem is that of buffer constraints. If nodes are expected to be disconnected from the wider network for long periods of time and/or users are expected to produce large amounts of data, ferry buffers may become saturated. While this issue has been mitigated in recent years with the development and cost reduction of high-density storage chips, it may still pose an issue in remote sensor networks where edge nodes continuously collect high-resolution surveillance data.

In the dynamic route ferrying strategy discussed above, ferries work independently at the micro-level with the local rules generating cooperative macroscopic emergent behavior. Another approach is to have all ferries explicitly work together at the micro-level. This approach is presented in [32] under ferry buffer constraints. There, all ferries act and move as one. Flocking is used to keep the ferries connected (that is, create a graph with a single component), and the combined swarm moves between user node destinations according to the same rule. However, in this case, the combined connectivity footprint of the swarm is used to transfer data to and from users, and the buffer storage of the swarm is considered to be the summation of all the individual finite ferry buffers. The two approaches can be analyzed more easily in a single dimension.

Consider the problem of simplex data ferrying between two user nodes separated by distance L and having communications range r_u. The distance a ferry must travel between the users is then $L - 2r_u$. In the connected model, n ferries with buffer b, data rate R, speed v, and communications range r_f (where $r_f = r_u = r$ for simplicity) are chained together along the line joining the two users at the maximum communications range. In this formation, the reduced distance to be traversed between two user nodes is $(L - r(n+1))$. In [32], the average packet delay is shown to be

$$d_{\text{avg}}^{\text{connected}} = 2\left(\frac{L - r(n+1)}{v}\right) + \frac{3}{2}\left(\frac{nb}{R}\right). \tag{13.11}$$

Separately, [33] derive the average packet delay for the unconnected (independent ferries) case as follows:

$$d_{\text{avg}}^{\text{unconnected}} = \left(1 + \frac{1}{n}\right)\frac{L - r(n+1)}{v} + \left(n + \frac{1}{2}\right)\frac{b}{R}. \tag{13.12}$$

Henkel and Brown [33] also present another "conveyor" model wherein resources do not communicate between themselves but rather all traverse the entire distance between user nodes. This is more akin to the static route ferry approach discussed in Section 13.2.2. This conveyor model is travel-dominant with the travel time dominating the delay. On the other hand, the unconnected case is transmission-dominant. In the terminology introduced in Section 13.2.3, these two models are highly non-local and highly local, respectively. The connected model introduced in [32] is then a hybrid of the two.

The throughput and packet delay of the three models behave differently under varying buffer size constraints. Unsurprisingly, the "conveyor" model demonstrates consistently worse performance given that wireless communications will always be faster than physically moving a platform around. An interesting question then to ask is under what conditions one should employ the connected and unconnected strategies. Given that the problem is ultimately physics-based, the answer will depend on the number of ferries, the user node distribution (density and area extents), user node date generation rates, ferry speed, and, of course, buffer constraints. Holding all but the number of nodes and buffer size variables constant, Fraser [32] demonstrates empirically in two dimensions that there is a clear delineation where one strategy is better than the other in terms of packet latency.

In general, the connected approach outperforms the disconnected one for small individual buffer sizes and/or small ferry numbers, but its advantage is quickly lost as these variables increase in size. This makes sense since, as either of these variables increase, the large collective buffer size is no longer needed to fully empty user nodes encountered along the way. Intuitively, when going on a journey, having just enough luggage space prevents wastage and having to leave items behind.

The mathematical modeling derived from knowledge of the local agent rules and observation of collective simulation outcomes allows the calculation of the optimal size of the connected strategy prior to deployment [32]. This further demonstrates the interplay between agent-based and mathematical-based modeling techniques when investigating CAS. That is, agent-based modeling can be used to develop and tune behavior with mathematical modeling and then used to estimate system properties without the need to perform further simulation.

It should be noted, however, that CAS can exhibit "state changes" in their behavior as parameters are swept. Such states may exhibit remarkably different collective behavior and may not necessarily be described within the mathematical model, particularly in the case where the mathematical model was derived from observation of the simulated system.

13.3 Disrupted Communications and the Consequences for Emergence

Interaction between individuals in CASs is vital for emergent properties to form; communications are, therefore, also vital. Yet, in physical systems, the quality of communications is often overlooked. This is particularly important, as one of the promoted strengths of self-organizational and emergent systems is their ability to control very large numbers of agents, but even local communications can be disrupted due to a congested RF spectrum caused by these large systems.

Consider the case of bird flocking, where the animals use their eyes to evaluate relative peer distances.[4] What would happen if the quality of this relative distance measuring declined? For example, what if the flock flew into fog or clouds? This leads to the interesting follow-on question of what communication quality is required to maintain a particular performance of the flock. While interesting, this question and its answer are specific to the particular scenario and algorithm. A better question is "What 'amount' of emergence is required to generate a particular performance level of the emergent property?" and necessarily, "Can we measure it?" This section describes work undertaken in this endeavor. In particular, the works of [35] and [36] are in CAS interaction theory and its exploitation in measuring the breakdown of emergent behavior [37].

The importance of agent interactions within emergent systems stems from the potential state changes within the participating entities caused by the interaction. That is, it is the influence that agents have on each other that allows macroscopic emergent behavior to evolve. [35, 36] theorized that the examination of these interactions should also exhibit some sort of non-randomness within their statistics and indeed proved this to be true. By defining what a state change means for specific algorithms, a time series of state changes can be generated by recording the individual and cumulative state changes per discrete time-step of a simulation.

It was found that emergent algorithms such as the Game of Life and Boids exhibited a non-normal distribution in the cumulative state changes. Yet, a normal (or Gaussian) distribution is seen in Brownian motion or by perturbing the Boids' algorithm weights (and thereby preventing the flocking behavior). The approach

4 An approach that has been shown to be feasible on drones using modern machine learning object detection algorithms [34].

demonstrated a qualitative binary indicator of whether emergence existed within an algorithm. To extend this idea, the authors of [37] consider the auto-correlation function (ACF) of the state-change time-series statistics of a single agent. The ACF computes the similarity between observations (a value between 0 and 1) as a function of the time-series lag between them [38]. The use of the ACF comes with two assumptions:

1. The next state of a node is correlated with the current state (in some unknown way) if emergent behavior exists; and similarly
2. Emergent behavior presents as non-randomness in node interaction statistics (as shown by Chan).

This metric was applied to a static flocking algorithm that produces a regular lattice structure as nodes perform separation and cohesion movements (similar to standard flocking).

With lossless communications, the correlogram produced by the ACF of the indicator function shows a strong correlation between successive state updates, which suggests an ordered movement pattern. Naturally, graph metrics that measure the quality of the graph of the formed lattice exhibit perfect results, i.e., a fully connected, single-component graph. In the lossless case, a packet-level simulation using Wi-Fi-like modulation, encoding and propagation, log-normal path loss modeling [39], carrier sense multiple access (CSMA) , medium access control (MAC), user-level hyper text transfer protocol (HTTP) traffic generation of various rates, and non-prioritized protocol information was set up and examined for various user data rates and protocol update rates.

As expected, the algorithm performance declines as user data rates increase due to the blocking of protocol traffic and increased MAC contention. The lattice structure is less well-formed with lower graph connectivity. The lag-1 ACF value is compared for various user data rates and protocol update rates, which demonstrated a clear relationship in declining ACF value with poorer graph characteristics. This leads to a conclusion that the ACF can indeed measure some (relative) value of emergence in the system.

These approaches of finding or measuring order within systems naturally lead to the consideration of examining systems from the perspective of entropy. It has been observed that there is a strong formal similarity between the definitions of thermodynamic and information entropy. This has also led to the appropriation of entropy as a measure of macro-level disorder in multi-agent systems [40]. Given that the ACF is typically used to find non-random patterns in noisy data, it is expected that a system with correlated state changes will exhibit decreasing entropy and may be used as a metric to measure the same underlying process. The study nonetheless demonstrated the need to consider the impact of communication load and medium contention on the production of a desired emergent property within physical CASs.

13.4 Implicit vs. Explicit Communication

The allure of CAS systems lies in their potential to reduce the cost, time, and energy required to solve complex real-world problems by leveraging many low-cost platforms operating in parallel. As discussed earlier in this chapter, CAS has been applied to a range of real-world problem domains, including wireless tactical communications and intelligence, surveillance, target acquisition, and reconnaissance (ISTAR). However, these applications present significant challenges in establishing and maintaining reliable communication networks, which are often necessary for the performance coordination algorithms. For example, the Defense Advanced Research Projects Agency (DARPA) Subterranean Challenge underscores the difficulty of undertaking successful ISTAR in environments with limited communication [41, 42].

This dependence on communication poses a barrier to the widespread adoption of multi-agent systems in DDIL environments. While many coordination algorithms assume that inter-agent communication is a prerequisite, it is not always necessary. Nature offers compelling examples where effective coordination occurs without explicit communication, for instance, among ants, bees, termites, and even humans. Inspired by these biological systems, this section explores the question: How can agents coordinate and self-organize in the absence of explicit communication? In this section, we focus particularly on decentralized multi-agent systems, where each agent makes autonomous decisions based on locally available information.

While there is no universally accepted definition of implicit coordination, most studies from psychology to computer science and robotics generally refer to implicit coordination as the coordination of behavior without direct communication with other agents (though this does not preclude communication through signals in the environment). We say communication is *explicit* when agents share their intent, state, or other observations with one another and *implicit* when no direct communication occurs between agents, leaving them only to rely on their locally observed information for decision-making. We define the intent of an agent as any partial or complete solutions to the joint optimization problem that indicate the future state of the agent, for example, trajectories, waypoints, control distributions, or tasks. This section will explore some of the strategies enabling implicit and explicit coordination, emphasizing their alignment with the principles of CAS, particularly emergence and self-organization.

Bees, ants, and termites provided the inspiration for many of the early coordination algorithms, motivated by the observations of Pierre-Paul Grasse, who identified that while each of these insects or animals appears to be simple in nature and operates independently, their collective behavior is surprisingly coordinated and complex [43]. This organization occurs through an indirect coordination mechanism referred to as "stigmergy," where agents implicitly coordinate by communicating basic signals through marking the environment, thereby making local information globally accessible and impacting the local decision-making of other agents. This creates a feedback loop that allows complex behaviors such as foraging, for example, where ants will deposit pheromones on a trail leading to food, thereby reinforcing other ants to follow that particular pathway [43, 44]. This type of coordinated behavior is often referred to as "emergent" behavior, as the complexity can arise from the interactions of seemingly simple, independent agents. This feedback loop-based approach has led to the development of countless coordination algorithms for multi-agent coordination, specifically for manufacturing control systems [45–47]. However, stigmergic signal-based approaches have generally been limited to relatively simple problem domains. This is perhaps unsurprising, though, as such feedback mechanisms can fail dramatically in nature. For example, ants may reinforce ineffective behavior, leading to an "ant mill" [48], where they circle endlessly until exhaustion after losing the initial pheromone trail.

Implicit coordination is also observed in intelligent animals, ranging from flocks of birds to schools of fish. These natural phenomena have inspired the field of multi-agent swarm robotics, often referred to as "flocking" [19]. In the simplest scenarios, complex swarm behaviors such as pattern formation can emerge using only state such as position and speed of its nearest neighbors [19, 49, 50]. However, most swarming-based approaches in robotics rely on explicit communication of this state information. In contrast, animals do not explicitly share their global locations; instead, they infer this information visually by observing one another. Studies by [51, 52] demonstrate that it is possible to replicate this behavior using visual sensing, achieving remarkable similarity of behavior to schools of fish. While these swarming strategies have been successful for simple tasks, such as formation control, more advanced techniques are needed to tackle complex problems.

A more complex example of implicit coordination in everyday life is furniture moving; experienced moving crews do not necessarily rely on continuous communication with one another to move furniture through a constrained environment. A recent experiment by [53] illustrates this concept, showing that a team of ants, lacking any means of communication explicitly or via stigmergic signals, is capable of self-organization to solve a complex geometric puzzle faster than humans. Earlier research by [54–57] explored this problem domain, using a team of robots to collaboratively move objectives through a series of desired configurations. Interestingly, the approach in these works is similar to the observations in [53], where a leader robot sets the desired path of the object of interest, while the other robots use the sensed force on the object to come to a consensus on where to move the object. Both studies observed a leader–follower dynamic where one agent sets the trajectory, and others infer intent by sensing forces on the object; this closely mimics the models of ant behavior [58]. This shared understanding, facilitated by a fully observable environment and easily decomposable utility functions (like force), allows agents to reach a consensus without explicit communication.

Humans frequently engage in implicit coordination not just on simple tasks such as furniture moving but also complex problems with remarkable precision and accuracy. No example highlights this better than pedestrian and vehicle intersections [59]. Drivers, despite limited explicit communication (turn signals, brake lights, hand gestures), successfully manage to negotiate unsignaled intersections with a surprisingly low collision rate. In spite of this limited communication, cars manage to continuously and frequently pass these intersections with limited collisions to continue onto their destination. Mavrogiannis et al. [60] summarize this problem well, stating that the lack of explicit coordination provides a high degree of uncertainty about the unfolding dynamics (of other agents), which makes the decision-making problem difficult. In their works, [60, 61] address the high uncertainty by using topological braids to consider all likely future outcomes at the intersection based on the domain knowledge, for example, driving rules, the geometry of the problem, and the observed states of other agents. Their results demonstrate low collision rates even in challenging scenarios, highlighting the importance of predicting other agents' behavior for successful implicit coordination. Interestingly, as noted by [62], while agents may not explicitly seek to optimize a joint strategy, agents may adopt strategies to allow others to more readily infer the intent of their actions and thereby enabling other agents to quickly reduce the error in their predictions. In this example, the predictive reasoning based on the domain knowledge and observed states results in agents self-organizing to complete the task.

The data-ferrying communication task discussed in the previous case studies exemplifies a dual-layered coordination challenge, where most effective approaches rely on implicit coordination. In data ferrying, agents must implicitly coordinate movements to efficiently transfer the data in DDIL environments, while avoiding congestion within the network; this aligns strongly with the CAS principles of self-organization and emergence. The digital pheromone models utilized here provide basic congestion avoidance; however, being purely implicit, they fall short in dynamically prioritizing critical data sources, adapting to changing network demands as a result of their limited predictive capabilities. This limitation underscores the critical need for more sophisticated implicit coordination methods capable of proactive and adaptive decision-making, which naturally leads into more advanced frameworks such as multi-agent reinforcement learning (MARL).

To conclude our discussion, we turn to learning-based techniques, specifically MARL. Briefly put, MARL is a learning technique whereby agents learn policies that maximize a reward through repeated iterations of acting in an environment, receiving a reward and updating a policy (often represented by neural networks). For a more comprehensive description and background of MARL, readers may consult comprehensive reviews such as [63–65]. One popular paradigm for training MARL models is centralized training, decentralized execution (CTDE) [64]. CTDE is an approach whereby robots train together in a centralized manner, typically learning one policy for all agents that is distributed to each agent at execution [65]. The benefit of doing so is that CTDE allows robots to train on information not available at execution (such as the experience from other agents) to quickly reach a high-quality, robust policy. One of the useful side effects of this approach and model is that the trained policies generally only rely on locally observed state information for decision-making [65], also facilitating the need to examine state representation within these methods [66, 67]. CTDE performs exceedingly well on multi-agent problems, in particular the StarCraft multi-agent challenge (SMAC) [68–70], despite having no explicit communication of intent between agents. In SMAC, each agent acts on locally observable information (within a predefined range) such as relative position, health, and types of other entities yet can exceed human-level performance. This demonstrates that even non-communicating teams are capable of coordinating on highly complex tasks. Finally, the work by [71] provides one of the first hints about the upper bound on performance of implicit coordination methods, showing that a CTDE-based approach is unable to learn an optimal action-value function but can only estimate, based on the size of the state and action space.

13.5 Reducing Communication Volume by Communicating Intent

In DDIL communication environments, we aim to develop policies for each agent to maximize team performance under communication constraints. Communication *volume* in multi-agent systems can be measured along several axes: frequency, bandwidth, and message size. While much research has focused on determining *when* and *with which* agents to communicate, less attention has been paid to the impact of the *intent* representation on this communication volume.

In real-world multi-agent systems, communication volume is constrained by a range of factors: the physical communication layer (e.g., RF, acoustic, or laser), the network topology (e.g., peer-to-peer, mesh), and the available bandwidth. Many coordination algorithms [72–77] adopt a broadcast-like model of communications, where agents broadcast their intent to all other agents in range on the network. Intuitively, we can see that this approach is not scalable; there exists at worst quadratic scaling of the number of messages with respect to the number of agents, while the physical network capacity remains fixed. While this may not be a significant issue for small-scale systems, it can become a bottleneck as the number of agents and task complexity increases. In this section, we explore how the choice of intent representation can play a crucial role in managing this communication bottleneck, impacting both the volume of information exchanged and overall coordination performance in multi-agent systems.

Each coordination algorithm typically communicates a bespoke representation of intent depending on the specific problem, constraints, and application. To gain a better understanding of how the intent is represented and the communication volume required in these coordination algorithms, we consider a state-of-the-art planner, decentralized-Monte Carlo tree search (Dec-MCTS) [72]. In Dec-MCTS, the intent "message" is a categorical distribution over the paths with the highest expected utility extracted from the planner. Each of these intent messages contains a fixed number of paths (10 or so) and is not conditioned on the actions of other agents or changes in the environment; this approach is common across a number of algorithms [75, 77, 78]. Best et al. [72] demonstrate that this approach is effective in achieving good coordination performance on the *generalized team orienteering* problem and is robust in the case of degraded communication. Hull et al. [73] expand on this work by replacing this intent message representation with an adaptive policy in the form of a behavior tree (BT). This BT-based policy uses the same samples extracted from the planner, like Dec-MCTS, but evaluates the likely outcomes with regard to all other agents; this makes it reactive to the actions of other agents in the environment. Consequently, compared to [72], this approach demonstrates significant reductions in the number of communication rounds needed to achieve high coordination performance. The results show that this adaptive intent targets communication to the so-called *points of contention* in the coordination process, thus facilitating efficient planning. Therefore, we can see that a simple change in the intent representation to allow for an adaptive policy can result in reductions in communication volume and improved coordination performance.

If we look at the natural world, we can see that the amount of information people verbally communicate with one another depends on a range of factors, including the prior domain knowledge of each individual and the complexity of the problem. The emergence of large language models (LLMs) that has led to breakthroughs in natural language processing and generation [79–81] offers a potentially powerful tool for evolving communication in multi-agent systems. Using LLMs for communication in robotics is a promising research direction, as it provides a natural language interface that can be inspected by, interacted with, and diagnosed by the operators.

One of the first uses of LLMs for explicit planning and communication in robotics can be found in [82]. In this work, the authors compare a range of communication topologies, from round-robin to hierarchical approaches, both with and without agent feedback. Notably, the authors provide insight into the number of tokens and rounds (indicative of communication quantity) required to achieve the reported performance. As their results show, [82] demonstrate that a leader–follower system with feedback performs the best while requiring the second-least amount of information communicated, surpassed only by a leader–follower system without feedback, which has significantly poorer task performance. Further work by [83] shows that, on average, a communication-based approach requires as few or fewer planning steps than a non-communicating, centralized planner. These findings further support the evidence for the future work suggested by [84]: *multi-agent planning and communication are a co-design problem*. From these works, it is evident that communication topology and approach significantly impact the reasoning performance of the models themselves.

Our discussion here has explored the role of intent representation in multi-agent coordination, demonstrating how and what agents communicate can impact both communication volume and coordination performance. As highlighted by recent studies, careful selection of intent representation can enhance coordination algorithms while simultaneously reducing communication overhead. However, determining the optimal representation of intent remains an open question. We suggest future research should prioritize identifying and characterizing the properties of intent representations that maximize coordination performance while minimizing communication volume. A key challenge lies in developing dynamic intent representations that adapt to the constraints of the communication environment of each agent. Achieving this will pave the way for more robust, efficient, and scalable CAS.

13.6 Sensing RF Communications: A CAS Approach

Flocking-based CAS algorithms can be an effective way to deploy a swarm of sensors; the familiar lattice structure formed by virtual force-based algorithms produces a regular spacing that is ideal for spatial sampling by onboard sensors. By integrating additional local rules based on measurements from these sensors with the flocking algorithm, an autonomous and distributed sensor network can be achieved. For example, Semnani and Basir [85] demonstrate a flocking approach to surveillance to find and follow targets of interest. Individuals use a basic search strategy and flock over a target once it is found. This leverages the lattice structure to produce a combined surveillance footprint that is difficult for a target to escape once discovered.

This idea can be extended to the sensing of RF emitters. However, this problem is more difficult, as, while the emitter can likely be sensed from a greater distance than in the optical regime (dependent on emitting frequency), only magnitude can be sensed. In order to discern directional information from a single node, a mechanically driven directive antenna, multiple static directive antennas, or a digital (phased-array) antenna is needed. All of these increase both the complexity and cost of any real-world solution. With magnitude-only sensing, the typical "gradient-surfing" approach would see a single agent utilizing an isotropic antenna sampling its possible movement directions in a spiral-type pattern in order to decide which direction would move it closer to the source. Moreover, the nature of RF propagation makes the situation worse. Consider the following equation:

$$P_r(\text{dB}) = P_t + \underbrace{20\log_{10}\left(\frac{\lambda}{4\pi d_0}\right)}_{\substack{\text{Path loss due to spherical}\\ \text{point source propagation}}} - \underbrace{10\gamma\log_{10}\frac{d}{d_0}}_{\substack{\text{Combined path loss effects}\\ \text{e.g., absorption, reflection}}} + \underbrace{\chi_\sigma}_{\substack{\text{Noise due to clutter}\\ \text{e.g., fading, multi-path}\\ \text{interference}}} . \tag{13.13}$$

where P_r is receiver power in decibels referenced to one milliwatt (dBm), P_t is transmit power (dBm), λ is wavelength (m), d_0 is the reference (far-field) distance (m), γ is the unitless path loss exponent, and χ_σ is a random variable such that $\chi \sim \mathcal{N}(0, \sigma^2)$ with a standard deviation of σ dB. The signal attenuates with the logarithm of distance between transceivers and is known as the log normal path loss model (LNPLM) [39]. Looking at the model again, it is simply the combination of effects of a spherical point source propagation, attenuation due to processes such as reflection and absorption, and randomness caused by the interference of multiple propagating rays arriving at different times due to clutter in the environment.

From the equation, the spatial sampling required in order to discern the underlying gradient is dependent on (1) the distance from the source due to the logarithmic nature and (2) the random variable distribution. Thus, even a single node with sophisticated directional receivers would need to sample a significant distance to derive the power gradient.

To overcome these problems, a flocking-based gradient search CAS is developed in [86]. Using familiar attraction and repulsion rules, the agents are forced apart into the familiar lattice structure but are only attracted to neighbors who have a higher signal strength measurement than themselves. The effect of this is that the collective maintains the local structure but moves as a group in the direction of the agent(s) that are closer to the emitting source. Interestingly, this means that the agent closest to the emitting source is "pushed" closer toward it, as it does not generate any attractive force on its own.

This control scheme simply requires the transfer of position and signal strength measurements to peers, a total of four floats (ignoring protocol overhead and in two dimensions), making it a lean communications approach. Furthermore, only local, i.e., nearest-neighbor, information is required, making broadcast a suitable transport mechanism, further simplifying implementation. As it is a CAS-based algorithm, it is also robust to node failures, making it suitable for hostile environments.

The primary variables that affect the performance of the algorithm are the repulsion range between agents and the number of agents in the swarm. These variables directly map to the discussion on sampling distance, where the greater the distance between sample points, the better the localization. The effect of repulsion range in this respect is obvious; however, the swarm size is less so. Firstly, the larger the swarm, the larger the maximum sampling distance, that is, the distance between the nearest and furthest agent in the swarm and the source. This information is indirectly propagated throughout the swarm by continual local broadcasts. Secondly, the larger the swarm, the more directions there are to explore. Note that the operating frequency of the swarm must be selected such that it matches that of the emitter. If not, agents may not be able to separate enough to make useful measurements.

As with other examples within this chapter, this CAS model shows remarkable performance given its simplicity and minor communications requirements. Agents are able to self-organize to easily home in on the emitter without any explicit concept of the collective objective and, again, provide a compelling argument for the use of CAS for the coordination of large deployments of UAS.

13.7 Summary

CASs are an ideal framework for the control of large multi-agent systems. Their resilience to attrition, distributed nature, asynchronicity, local communication, and composition of relatively simple agents are all desirable characteristics for deployment in austere environments. Although challenges still remain with respect to the design and validation of CASs, they can readily be deployed in the real world. Given the unprecedented advancements in UASs hardware in recent years, they represent an exemplary target platform for real-world embodied agents. Unfortunately, however, RF spectrum is an already congested resource; wireless communications are prone to degradation due to a multitude of causes and are also an obvious threat vector for nefarious actors.

Yet, we have shown in this chapter through the collected works of CASs experts that they can still function in the absence of end-to-end communications (and we touched upon what happens to emergence when communications become completely disrupted); can enable communications to other (non-participating) agents; and can be used as sensor networks to find RF transmissions and make use of implicit information sharing to facilitate large-scale coordination.

Throughout the chapter, CASs examples were presented that are inherently robust against agent attrition and, in one mathematically provable case, resilient to perturbations of the collective system dynamics. A combination of agent-based and mathematical-based modeling was used to analyze them, with agent-based modeling finding its strength in large-scale simulation, with mathematical models being developed from observation of the simulated collective behavior. Analysis primarily focuses on the CAS's usefulness in enabling coordination for UAS -borne, wireless communications-driven, real-world (communications-based) applications.

As the enabling UASs technology is further refined, the current embodied single-agent autonomy we are beginning to see fielded will naturally extend to autonomous and cooperative teams and swarms; the understanding and application of CASs will no doubt form a significant part of this achievement.

References

1 Syed Agha Hassnain Mohsan, Nawaf Qasem Hamood Othman, Yanlong Li, Mohammed H Alsharif, and Muhammad Asghar Khan. Unmanned aerial vehicles (UAVs): Practical aspects, applications, open challenges, security issues, and future trends. *Intelligent Service Robotics,* 16(1):109–137, 2023.

2 M. Waibel, B. Keays, and F. Augugliaro. *Drone Shows: Creative Potential and Best Practices*. ETH-Zürich, 2017. https://books.google.com.au/books?id=8k5-tAEACAAJ.

3 Drew Hanover, Antonio Loquercio, Leonard Bauersfeld, Angel Romero, Robert Penicka, Yunlong Song, Giovanni Cioffi, Elia Kaufmann, and Davide Scaramuzza. Autonomous drone racing: A survey. *IEEE Transactions on Robotics* 40:3044–3067, 2024.

4 Eliza J T Middleton and Tanya Latty. Resilience in social insect infrastructure systems. *Journal of the Royal Society Interface,* 13(116):20151022, 2016.

5 Mataric Maja J. *Designing Emergent Behaviors: From Local Interactions to Collective Intelligence*. books.google.com, 1993.

6 M. E. J. Newman. Complex systems: A survey. *arXiv [cond-mat.stat-mech]*, 2011.

7 Yong Meng Teo, Ba Linh Luong, and Claudia Szabo. Formalization of emergence in multi-agent systems. In *Proceedings of the 1st ACM SIGSIM Conference on Principles of Advanced Discrete Simulation*, SIGSIM PADS '13, pages 231–240, New York, NY, USA, 2013. ACM.

8 Claudia Szabo and Yong Meng Teo. An integrated approach for the validation of emergence in component-based simulation models. In *Proceedings of the Winter Simulation Conference*, WSC '12, pages 242:1–242:12. Winter Simulation Conference, 2012.

9 Claudia Szabo and Lachlan Birdsey. Validating emergent behavior in complex systems. In Andreas Tolk, John Fowler, Guodong Shao, and Enver Yucesan, editors, *Advances in Modeling and Simulation, Simulation Foundations, Methods and Applications*, pages 47–62. Springer, Cham, 2017.

10 Toh Chai-Keong. *Ad Hoc Mobile Wireless Networks: Protocols and Systems*. 1st edition, Prentice Hall PTR, Upper Saddle River, NJ, USA, 2002.

11 Alex Hinds, Michael Ngulube, Shaoying Zhu, and Hussain Al-Aqrabi. A review of routing protocols for mobile ad-hoc NETworks (MANET). *International Journal of Information and Education Technology*, 3(1):1–5, 2013.

12 Shaik Shabana Anjum, Rafidah Md Noor, and Mohammad Hossein Anisi. Review on MANET based communication for search and rescue operations. *Wireless Personal Communications*, 94(1):31–52, 2017.

13 Laurent Reynaud and Isabelle Guérin-Lassous. Improving the performance of challenged networks with controlled mobility. In *Ad Hoc Networks, Lecture Notes of the Institute for Computer Sciences, Social Informatics and Telecommunications Engineering*, pages 205–217. Springer, Cham, 2017.

14 M. Ahmed, S. Dao, and R. Katz. Positioning range extension gateways in mobile ad hoc wireless networks to improve connectivity and throughput. In *2001 MILCOM Proceedings Communications for Network-Centric Operations: Creating the Information Force (Cat. No.01CH37277)*, volume 1, pages 331–335, 2001.

15 Wenrui Zhao and M. H. Ammar. Message ferrying: Proactive routing in highly-partitioned wireless ad hoc networks. In *The Ninth IEEE Workshop on Future Trends of Distributed Computing Systems, 2003. FTDCS 2003. Proceedings.*, pages 308–314. ieeexplore.ieee.org, 2003.

16 K. P. Hui, P. Pourbeik, P. George, D. Phillips, S. Magrath, and M. Kwiatkowski. OPAL - A survivability-oriented approach to management of tactical military networks. In *2011 - MILCOM 2011 Military Communications Conference*, pages 1127–1132, 2011.

17 Abdullah Konak, George E Buchert, and James Juro. A flocking-based approach to maintain connectivity in mobile wireless ad hoc networks. *Applied Soft Computing*, 13(2):1284–1291, 2013.

18 R. Hunjet. Autonomy and self-organisation for tactical communications and range extension. In *2015 Military Communications and Information Systems Conference (MilCIS)*, pages 1–5, 2015.

19 Reynolds Craig W. Flocks, herds and schools: A distributed behavioral model. *ACM SIGGRAPH Computer Graphics*, 21(4):25–34, 1987.

20 B. Gavish and S. C. Graves. The travelling salesman problem and related problems. *Operations Research Center, Massachusetts Institute of Technology*, pages GR–078–78, 1978.

21 W. Zhao, M. Ammar, and E. Zegura. Controlling the mobility of multiple data transport ferries in a delay-tolerant network. In *Proceedings IEEE 24th Annual Joint Conference of the IEEE Computer and Communications Societies*, volume 2, pages 1407–1418, 2005.

22 Bradley Fraser and Robert Hunjet. Data ferrying in tactical networks using swarm intelligence and stigmergic coordination. In *2016 26th International Telecommunication Networks and Applications Conference (ITNAC)*, pages 1–6. IEEE, 2016.

23 Marco Dorigo, Eric Bonabeau, and Guy Theraulaz. Ant algorithms and stigmergy. *Future Generation Computer Systems*, 16(8):851–871, 2000.

24 Bradley Fraser, Robert Hunjet, and Andrew Coyle. A swarm intelligent approach to data ferrying in sparse disconnected networks. *International Journal of Parallel, Emergent and Distributed Systems*, 36(6):630–651, 2021.

25 Holmgren Richard A. *A First Course in Discrete Dynamical Systems*. 2 edition, Universitext. Springer, New York, NY, 2000.

26 P. F. Verhulst. Recherches mathématiques sur la loi d'accroissement de la population. *Nouveaux Mémoires de l'Académie Royale des Sciences et Belles-Lettres de Bruxelles*, 18:14–54, 1845.

27 Robert Hunjet, Thomas Stevens, Mathew Elliot, Bradley Fraser, and Peter George. In *MILCOM 2017 - 2017 IEEE Military Communications Conference (MILCOM)*, pages 788–793, 2017. https://doi.org/10.1109/MILCOM.2017.8170775.

28 Robert Hunjet, Bradley Fraser, Thomas Stevens, Laura Hodges, Karina Mayen, Jan Carlo Barca, Madeleine Cochrane, Ricardo Cannizzaro, and Jennifer L. Palmer. In *2018 IEEE International Conference on Robotics and Automation (ICRA)*, pages 6381–6388, 2018. https://doi.org/10.1109/ICRA.2018.8463151.

29 Bradley Fraser, Andrew Coyle, Claudia Szabo, and Robert Hunjet. In *2019 IEEE 20th International Symposium on "A World of Wireless, Mobile and Multimedia Networks" (WoWMoM)*, pages 1–6, 2019. https://doi.org/10.1109/WoWMoM.2019.8792993.

30 B. Fraser, A. Coyle, R. Hunjet, and Claudia Szabo. An analytic latency model for a next-hop data-ferrying swarm on random geometric graphs. *IEEE Access*, 8:48929–48942, 2020.

31 Mathew Penrose. *Random Geometric Graphs*. Oxford University Press, 2003.

32 Bradley Fraser. Data ferry flocking for bulk information transfer under ferry buffer constraints. In *2018 28th International Telecommunication Networks and Applications Conference (ITNAC)*, pages 1–6, 2018. https://doi.org/10.1109/ATNAC.2018.8615303.

33 Daniel Henkel and Timothy X. Brown. On controlled node mobility in Delay-Tolerant networks of unmanned aerial vehicles. In *International Symposium on Advance Radio Technologies (ISART)*, 2006.

34 Giuseppe Laurito, Bradley Fraser, and Kent Rosser. In *2020 IEEE Symposium Series on Computational Intelligence (SSCI)*, pages 1435–1443, 2020. https://doi.org/10.1109/SSCI47803.2020.9308605.

35 W. K. V. Chan. Interaction metric of emergent behaviors in agent-based simulation. In *Proceedings of the 2011 Winter Simulation Conference (WSC), Phoenix, AZ, USA*, pages 357–368, 2011.

36 Claudia Szabo, Yong Meng Teo, and Gautam K. Chengleput. Understanding complex systems: Using interaction as a measure of emergence. In *Proceedings of the 2014 Winter Simulation Conference*, WSC '14, pages 207–218, IEEE Press, Piscataway, NJ, USA, 2014.

37 Bradley Fraser, Robert Hunjet, and Claudia Szabo. Simulating the effect of degraded wireless communications on emergent behavior. In *2017 Winter Simulation Conference (WSC)*, pages 4081–4092, 2017.

38 Gubner John A. *Probability and Random Processes for Electrical and Computer Engineers*. Cambridge University Press, Cambridge, England, 2012.

39 Rappaport Theodore S. *Wireless Communications: Principles and Practice*. 2nd edition, Prentice Hall PTR, Upper Saddle River, NJ, USA, 2002.

40 John J. Johnson, Andreas Tolk, and Andres Sousa-Poza. A theory of emergence and entropy in systems of systems. *Procedia Computer Science*, 20:283–289, 2013.

41 Graeme Best, Rohit Garg, John Keller, Geoffrey A. Hollinger, and Sebastian Scherer. Multi-robot, multi-sensor exploration of multifarious environments with full mission aerial autonomy. *The International Journal of Robotics Research*, 43(4):485–512, 2024.

42 Marco Tranzatto, Takahiro Miki, Mihir Dharmadhikari, Lukas Bernreiter, Mihir Kulkarni, Frank Mascarich, Olov Andersson, Shehryar Khattak, Marco Hutter, Roland Siegwart, and Kostas Alexis. Cerberus in the DARPA subterranean challenge. *Science Robotics*, 7(66):eabp9742, 2022

43 Guy Theraulaz and Eric Bonabeau. A brief history of stigmergy. *Artificial Life*, 5(2):97–116, 1999.

44 Francis Heylighen. Stigmergy as a universal coordination mechanism i: Definition and components. *Cognitive Systems Research*, 38:4–13, 2016. ISSN 1389-0417. Special Issue of Cognitive Systems Research – Human-Human Stigmergy.

45 Ali Vatankhah Barenji and Reza Vatankhah Barenji. Improving multi-agent manufacturing control system by indirect communication based on ant agents. *Proceedings of the Institution of Mechanical Engineers, Part I: Journal of Systems and Control Engineering*, 231(6):447–458, 2017.

46 José Barbosa, Paulo Leitão, Emmanuel Adam, and Damien Trentesaux. Dynamic self-organization in holonic multi-agent manufacturing systems: The ADACOR evolution. *Computers in Industry*, 66:99–111, 2015.

47 Wei Xiang and Heow Pueh Lee. Ant colony intelligence in multi-agent dynamic manufacturing scheduling. *Engineering Applications of Artificial Intelligence*, 21(1):73–85, 2008.

48 Delsuc Frédéric. Army ants trapped by their evolutionary history. *PLoS Biology*, 1(2), 11, 2003.

49 Michael Rubenstein, Alejandro Cornejo, and Radhika Nagpal. Programmable self-assembly in a thousand-robot swarm. *Science*, 345(6198):795–799, 2014.

50 Bayındır Levent. A review of swarm robotics tasks. *Neurocomputing*, 172:292–321, 2016. ISSN 0925-2312.

51 Florian Berlinger, Melvin Gauci, and Radhika Nagpal. Implicit coordination for 3D underwater collective behaviors in a fish-inspired robot swarm. *Science Robotics*, 6(50):eabd8668, 2021.

52 Xiaohan Wang, Fakui Wang, Zisen Nie, Yuhui Ai, and Tianjiang Hu. Optiswarm: Optical swarm robots using implicit cooperation. *IEEE Sensors Journal*, 22(24):24380–24394, 2022.

53 Tabea Dreyer, Amir Haluts, Amos Korman, Nir Gov, Ehud Fonio, and Ofer Feinerman. Comparing cooperative geometric puzzle solving in ants versus humans. *Proceedings of the National Academy of Sciences*, 122(1):e2414274121, 2025.

54 Zijian Wang and Mac Schwager. In *2015 54th IEEE Conference on Decision and Control (CDC)*, pages 380–385, 2015. https://doi.org/10.1109/CDC.2015.7402230.

55 Anastasios Tsiamis, Christos K. Verginis, Charalampos P. Bechlioulis, and Kostas J. Kyriakopoulos. Cooperative manipulation exploiting only implicit communication. In *2015 IEEE/RSJ International Conference on Intelligent Robots and Systems (IROS)*, pages 864–869, 2015.

56 Zijian Wang and Mac Schwager. In *2016 IEEE International Conference on Robotics and Automation (ICRA)*, pages 427–432, 2016. https://doi.org/10.1109/ICRA.2016.7487163.

57 Zijian Wang and Mac Schwager. Force-amplifying n-robot transport system (force-ants) for cooperative planar manipulation without communication. *The International Journal of Robotics Research*, 35(13):1564–1586, 2016.

58 Ofer Feinerman, Itai Pinkoviezky, Aviram Gelblum, Ehud Fonio, and Nir S. Gov. The physics of cooperative transport in groups of ants. *Nature Physics*, 14(7):683–693, 2018.

59 Miaomiao Liu, Guangquan Lu, Yunpeng Wang, and Zhe Zhang. Analyzing drivers' crossing decisions at unsignalized intersections in China. *Transportation Research Part F: Traffic Psychology and Behaviour*, 24:244–255, 2014. ISSN 1369-8478.

60 Christoforos Mavrogiannis, Jonathan A. DeCastro, and Siddhartha Srinivasa. Implicit multiagent coordination at uncontrolled intersections via topological braids. In Steven M. LaValle, Jason M. O'Kane, Michael Otte, Dorsa Sadigh, and Pratap Tokekar, editors, *Algorithmic Foundations of Robotics XV*, 2023.

61 Christoforos Mavrogiannis, Jonathan A. DeCastro, and Siddhartha S. Srinivasa. Abstracting road traffic via topological braids: Applications to traffic flow analysis and distributed control. *The International Journal of Robotics Research*, 43(9):1299–1321, 2024.

62 Christoforos I. Mavrogiannis and Ross A. Knepper. Multi-agent path topology in support of socially competent navigation planning. *The International Journal of Robotics Research*, 38(2–3):338–356, 2019.

63 Lucian Busoniu, Robert Babuska, and Bart De Schutter. A comprehensive survey of multiagent reinforcement learning. *IEEE Transactions on Systems, Man, and Cybernetics, Part C (Applications and Reviews)*, 38(2):156–172, 2008. https://doi.org/10.1109/TSMCC.2007.913919.

64 Kaiqing Zhang, Zhuoran Yang, and Tamer Başar. *Multi-Agent Reinforcement Learning: A Selective Overview of Theories and Algorithms*, pages 321–384. Springer International Publishing, 2021.

65 Afshin Oroojlooy and Davood Hajinezhad. A review of cooperative multi-agent deep reinforcement learning. *Applied Intelligence*, 53(11):13677–13722, 2023.

66 Maximilian Hüttenrauch, Adrian Šošić, and Gerhard Neumann. Deep reinforcement learning for swarm systems. *Journal of Machine Learning Research*, 20(1): 1966–1996, 2019. ISSN 1532-4435.

67 Bradley Fraser and Giuseppe Laurito. Train small, deploy big: Do relative world views permit swarm-safety during policy transplantation for multi-agent reinforcement learning problems? In Marcus Gallagher, Nour Moustafa, Erandi Lakshika, editors, *AI 2020: Advances in Artificial Intelligence, Lecture Notes in Computer Science*, pages 269–280. Springer International Publishing, Cham, 2020.

68 Mikayel Samvelyan, Tabish Rashid, Christian Schröder de Witt, Gregory Farquhar, Nantas Nardelli, Tim G. J. Rudner, Chia-Man Hung, Philip H. S. Torr, Jakob N. Foerster, and Shimon Whiteson. The starcraft multi-agent challenge. In *Proceedings of the 18th International Conference on Autonomous Agents and MultiAgent Systems*, pages 2186–2188, 2019.

69 Tabish Rashid, Mikayel Samvelyan, Christian Schroeder De Witt, Gregory Farquhar, Jakob Foerster, and Shimon Whiteson. Monotonic value function factorisation for deep multi-agent reinforcement learning. *Journal of Machine Learning Research*, 21(1):1–51, 2020.

70 Kyunghwan Son, Daewoo Kim, Wan Ju Kang, David Earl Hostallero, and Yung Yi. QTRAN: Learning to factorize with transformation for cooperative multi-agent reinforcement learning. In *Proceedings of the 36th International Conference on Machine Learning*, volume 97, pages 5887–5896, 2019.

71 Washim Uddin Mondal, Vaneet Aggarwal, and Satish Ukkusuri. On the near-optimality of local policies in large cooperative multi-agent reinforcement learning. *Transactions on Machine Learning Research*, 2022. ISSN 2835-8856.

72 Graeme Best, Oliver M. Cliff, Timothy Patten, Ramgopal R. Mettu, and Robert Fitch. Dec-MCTS: Decentralized planning for multi-robot active perception. *The International Journal of Robotics Research*, 38(2–3): 316–337, 2019.

73 Rhett Hull, Diluka Moratuwage, Emily Scheide, Robert Fitch, and Graeme Best. Communicating intent as behaviour trees for decentralised multi-robot coordination. In *2024 IEEE International Conference on Robotics and Automation (ICRA)*, pages 7215–7221. IEEE, 2024.

74 Aalok Patwardhan, Riku Murai, and Andrew J. Davison. Distributing collaborative multi-robot planning with gaussian belief propagation. *IEEE Robotics and Automation Letters*, 8(2):552–559, 2023.

75 Xin Zhou, Jiangchao Zhu, Hongyu Zhou, Chao Xu, and Fei Gao. Ego-swarm: A fully autonomous and decentralized quadrotor swarm system in cluttered environments. In *2021 IEEE International Conference on Robotics and Automation (ICRA)*, page 4101–4107, 2021.

76 Şenbaşlar Baskin and Gaurav S. Sukhatme. Asynchronous real-time decentralized multi-robot trajectory planning. In *2022 IEEE/RSJ International Conference on Intelligent Robots and Systems (IROS)*, pages 9972–9979, 2022.

77 Rhett Hull, Ki Myung Brian Lee, Jennifer Wakulicz, Chanyeol Yoo, James McMahon, Bryan Clarke, Stuart Anstee, Jijoong Kim, and Robert Fitch. Decentralised active perception in continuous action spaces for the coordinated escort problem. In *2023 IEEE International Conference on Robotics and Automation (ICRA)*, pages 7649–7655. IEEE, 2023.

78 Jesus Tordesillas and Jonathan P. How. Mader: Trajectory planner in multiagent and dynamic environments. *IEEE Transactions on Robotics*, 38(1):463–476, 2022.

79 Alec Radford, Jeff Wu, Rewon Child, David Luan, Dario Amodei, and Ilya Sutskever. Language models are unsupervised multitask learners. OpenAI, 2019. https://cdn.openai.com/better-language-models/language_models_are_unsupervised_multitask_learners.pdf.

80 Jacob Devlin, Ming-Wei Chang, Kenton Lee, and Kristina Toutanova. BERT: Pre-training of deep bidirectional transformers for language understanding. In *BERT: Pre-training of Deep Bidirectional Transformers for Language Understanding*, pages 4171–4186. Association for Computational Linguistics, 2019.

81 Tom B. Brown, Benjamin Mann, Nick Ryder, Melanie Subbiah, Jared Kaplan, Prafulla Dhariwal, Arvind Neelakantan, Pranav Shyam, Girish Sastry, Amanda Askell, Sandhini Agarwal, Ariel Herbert-Voss, Gretchen Krueger, Tom Henighan, Rewon Child, Aditya Ramesh, Daniel M. Ziegler, Jeffrey Wu, Clemens Winter, Christopher Hesse, Mark Chen, Eric Sigler, Mateusz Litwin, Scott Gray, Benjamin Chess, Jack Clark, Christopher Berner, Sam McCandlish, Alec Radford, Ilya Sutskever, and Dario Amodei. Language models are few-shot learners. In *Proceedings of the 34th International Conference on Neural Information Processing Systems*, NIPS '20, 2020.

82 Yongchao Chen, Jacob Arkin, Yang Zhang, Nicholas Roy, and Chuchu Fan. Scalable multi-robot collaboration with large language models: Centralized or decentralized systems? *IEEE International Conference on Robotics and Automation (ICRA)*, pages 4311–4317, 2024. https://api.semanticscholar.org/CorpusID:263143275.

83 Zhao Mandi, Shreeya Jain, and Shuran Song. RoCo: Dialectic multi-robot collaboration with large language models. In *2024 IEEE International Conference on Robotics and Automation (ICRA)*, pages 286–299, 2024.

84 Jennifer Gielis, Ajay Shankar, and Amanda Prorok. A critical review of communications in multi-robot systems. *Current Robotics Reports*, 3(4):213–225, 2022.

85 Samaneh Hosseini Semnani and Otman A. Basir. Semi-flocking algorithm for motion control of mobile sensors in large-scale surveillance systems. *IEEE Trans Cybern*, 45(1):129–137, 2015.

86 Bradley Fraser. Radio-frequency emitter localisation using a swarm of search agents. In *2018 28th International Telecommunication Networks and Applications Conference (ITNAC)*, pages 1–6, 2018. https://doi.org/10.1109/ATNAC.2018.8615450.

14

Scalable Rigidity Maintenance Strategies for Networked Multirobot Systems: Overcoming Decentralized Coordination Challenges

J. Francisco Presenza,[1,2] *Ezequiel Pecker-Marcosig,*[1,3] *Ignacio Mas,*[4]
J. Ignacio Alvarez-Hamelin,[1,2] *and Juan I. Giribet*[4]*

[1] *Facultad de Ingeniería, Universidad de Buenos Aires, Ciudad Autónoma de Buenos Aires, Argentina*
[2] *INTECIN, CONICET - Universidad de Buenos Aires, Ciudad Autónoma de Buenos Aires, Argentina*
[3] *Instituto de Ciencias de la Computación, CONICET–Universidad de Buenos Aires, Ciudad Autónoma de Buenos Aires, Argentina*
[4] *Universidad de San Andrés and CONICET, Victoria, Buenos Aires, Argentina*

*Corresponding author: jgiribet@conicet.gov.ar

14.1 Introduction

In the last decade, the coordination of networked multirobot systems (MRSs)—also denominated robotic *swarms*—has become an intensive area of research, motivated by their advantages in adaptability, coverage, robustness, and scalability across a wide range of applications. Thanks to advancements in new materials, sensor technology, and embedded control systems, mobile robots are achieving increasingly higher levels of autonomy. This has sparked growing interest in the coordinated control of MRSs across both academia and industry. The appeal lies in the ability of task distribution to enhance performance, improve fault tolerance, and enable new capabilities. Currently, these systems are employed in various activities, including search and rescue missions [1, 2]; the inspection of hazardous environments and environmental monitoring [3, 4]; object transportation and aerial manipulation [5, 6]; patrolling and surveillance purposes [7, 8]; cinematography [9, 10]; and more. Progress in the theoretical and practical study of multirobot mobile systems is reported in Verma and Ranga [11].

MRS can be viewed as a particular case of more general complex adaptive systems (CASs), where agents (robots) adapt their behavior over time through local interactions with one another and the environment, leading to emergent global behaviors that are not explicitly programmed. This type of complex collective behavior is often described as swarm intelligence [12, 13]. The three key components of a resilient CAS—presence of perturbations, mechanisms to quantify the impact of those perturbations on behavior, and controllable aspects (both inherent or aggregated) to mitigate their negative effects—can be directly mapped to networked MRSs. In this context, noise in sensor measurements, the stochastic behavior of communications, and uncertainties due to unmodeled or under-modeled behaviors are examples of perturbations. It is worth noticing the interdependence among agents causes local uncertainties to propagate and aggregate, leading to complex system-wide behavior. Moreover, local and global metrics such as the rigidity eigenvalue and tracking errors help to quantify the effect of perturbations. Finally, to counteract the negative effects of perturbations, the robots count with controllable inputs to receive control actions (such as speed commands) from the local and/or global system-wide controllers.

In the operation of MRS, the control method employed to govern the behavior of each robot is a crucial element. The design of control laws for these applications draws inspiration from various fields, including biology [14], game theory [15], and kinematic chains [16, 17]. Two typical schemes within MRS control are *formation control* [18] and *flocking* [19]. In *formation control* the global objective is that the whole team adopts a desired geometric pattern. *Flocking*, on the other hand, is inspired by the collective behavior of animals and is governed by three local rules of interaction: *separation*, *alignment*, and *cohesion*. Both objectives represent typical examples of CAS with emergent behavior from the direct application of simple rules based on local interactions.

The decentralized implementation of cooperative missions poses several challenges, particularly concerning energy, sensing, and communication constraints. These challenges are exacerbated in GPS-denied environments, where only relative information among robots is available, e.g., inter-agent displacements or distances, with the lack of a common reference frame [20]. A crucial aspect of any multirobot cooperation scheme is the exchange of information [21, 22]. Consequently, the topology of the underlying network, defined by the interactions among agents, plays a critical role. When the group consists of a few agents, the communication network structure is trivial, either by connecting all robots to each other or by linking each one to a central node acting as a centralized monitoring and control station [23]. As the number of robots increases—and consequently the volume of data exchanged—the network topology becomes a design parameter in the formation [24]. Hence, it becomes appropriate to model the multirobot network with a graph and apply results from graph theory in the study and design of control algorithms.

This chapter contributes to the broader vision outlined in the book by presenting discussions and innovative solutions to some challenges faced by multirobot coordination, firmly grounded in rigorous theoretical foundations and practical considerations. It is organized as follows. Section 14.2 introduces key concepts related to the deployment of MRSs, which will be used throughout this chapter. Section 14.3 presents a discussion of the classical control algorithms for maintaining the rigidity of a multirobot network, along with the specific models employed in our simulations. Section 14.4 presents the main subject of this chapter, which is the use of a framework decomposition into smaller groups (called subframeworks) to overcome the limitations of the previous methods. A fully scalable, gradient-based controller is proposed to maintain network rigidity through robot mobility. The validity of this controller is analyzed by simulations in Section 14.5 with two illustrative test cases. Finally, Section 14.6 concludes this chapter and outlines future lines of research.

14.2 Preliminaries

14.2.1 The Role of Simulation in Robotics Development

The growing number of applications involving robots necessitates efficient design techniques to shorten system development cycles. Heuristic design methods, such as trial-and-error approaches applied directly to the system, can be costly and dangerous [25]. Alternatively, starting with a physical-mathematical model of the system allows the use of design techniques based on analytical tools. However, given the complexity of a hybrid dynamic model with sufficiently heterogeneous components, it is not always possible to obtain a closed analytical solution to the dynamic equations. Analytical techniques are often limited by the complexity of the underlying models, requiring significant simplifications. As a result, numerical experimentation becomes essential for simulating complex systems. Particularly, in the case of mobile vehicles, the ability to experiment with simulation platforms is particularly relevant due to the costs, challenges, and risks involved in direct experimentation with real devices [26].

From a bottom-up perspective, a complex adaptive system can be intuitively modeled as an agent-based model (ABM). In the context of multirobot systems, each robot is represented as an agent that executes a local set of rules and interacts with neighboring robots for exchanging information. Depending on the model granularity, agents' models cover from the robots' physics to the controllers in charge of executing the rules, including the communication modules for the exchange of data and the mechanisms for the measurement and/or estimation of the robot state. An ABM will be introduced along with this chapter for the validation of the rigidity maintenance controller.

14.2.2 Information Processing in Multirobot Networks

It is common in multirobot networks that global information, i.e., data that depends on the state of all agents (or at least a significant subset), needs to be processed. Due to the inherent limitations of wireless networks, it cannot be assumed that all robots are able to exchange information with one another at any given time. The way that the information is processed can be classified as *centralized* or *decentralized*.

A centralized approach relies on a central computing station with the resources to simultaneously control a group of vehicles. While gathering all available information in one computing hub offers certain advantages, centralized techniques face significant drawbacks, including high computational demands on the central node, communication bottlenecks, and a lack of robustness due to a single point of failure. These limitations hinder the scheme's scalability, i.e., the ability to maintain functionality and performance as the number of robots increases. In contrast, decentralized techniques rely on agents acting solely based on their sensor-driven perception of the environment and local interactions with neighboring vehicles through point-to-point communication. This approach is better suited to handle computational and energy constraints, as well as the limited range of wireless communications, making decentralized control techniques highly promising.

14.2.3 Cooperative Localization

The tasks performed by robots are becoming increasingly sophisticated, and as a result, greater precision is required in estimating the state of the agents (position, orientation, velocity, etc.). Sensors employed in the estimation of the state of a single robot are categorized into two groups: *introceptive sensors*, such as inertial measurement units (IMUs), odometers, and optical flow sensors; and *exteroceptive sensors*, like light detection sensors (LIDARs), radio frequency (RF) sensors, global positioning systems (GPS, GLONASS, GALILEO, etc.), and scene capture systems using RGB cameras.

Introceptive sensors typically provide information on linear and angular velocity, or specific force, at a high acquisition rate. However, to estimate the vehicle's position or orientation, these observations must be integrated, leading to an accumulation of measurement errors that grow over time. This can be a problem unless high-precision inertial sensors are used, which are expensive, large, and heavy, making them unsuitable for installation on most robots. Exteroceptive sensors typically provide distance, position, or orientation information associated with another robot or an external reference; however, they have a low sampling rate and significant errors. Moreover, in many scenarios, exteroceptive sensors are not available. For example, in indoor environments, underwater settings, deep space, mine interiors, dense forests, and highly built-up cities. To overcome these limitations, the most widely used technique is *sensor fusion*, which allows for combining the capabilities of different sensors to achieve robust and accurate positioning [27].

When dealing with a multirobot formation, it is desirable for cooperative localization schemes to make the most of each agent's sensing capabilities while reducing sensor redundancy, all while maintaining some level of robustness [28, 29]. Typically, the network relies on the existence of *anchor nodes* that have access to some external positioning system, while the others (usually called *unknown nodes*) obtain relative information with respect to nearby agents, including anchors. In this chapter, we focus on teams of robots capable of measuring relative distances to their neighbors. A common way to do this is to use communication channels between agents to estimate the distance between nodes using techniques such as time of arrival (TOA) and received signal strength indicator (RSSI) [30, 31].

Relative position estimation from distance measurements is a long-standing problem; see Refs. [32, 33]. However, no efficient method with global convergence is currently known. In practice, it is typically assumed that an initial estimate of the agents' positions is available. If the *framework* is *infinitesimally rigid* (see Section 14.2.4), local convergence can be guaranteed. A gradient-based protocol for robot localization using relative distance measurements can be found in Calafiore et al. [34]. The step size used for the updates is shared by all agents and is determined through a consensus algorithm, which impacts scalability. On the other hand, a fully scalable method that requires only the exchange of estimated positions between neighbors was presented in Presenza et al. [35]. In Section 14.5, the latter approach was used to conduct the corresponding simulations.

14.2.4 Rigidity

This section provides a general overview of rigidity theory and its application to MRSs; interested readers may refer to Refs. [36, 37]. Rigidity is a fundamental property in the coordination of MRSs, as it enables the inference of relative robot positions from distance measurements between neighbors. The mathematical approach to this problem, commonly used in control applications, relies on the concept of infinitesimal rigidity of frameworks. As is common in multirobot applications, we focus on proximity frameworks and will refer to them simply as frameworks.

Consider a simple graph $G = (V, E)$ consisting of a finite set of nodes $V = \{1, \dots, n\}$ representing the robots, accompanied by a collection of edges E representing pairwise connections, i.e., $\{i, j\} \in E$ whenever robots i and j share a communication link. The neighbors of a node i, denoted $N(i)$, are the nodes that have a connection with i. The degree of node i is defined as $\deg(i) := |N(i)|$. The average degree is denoted deg(G). The distance between two nodes in a connected graph, dist(i, j), is the number of hops in a (not necessarily unique) shortest path between them. The maximum distance in a graph is called the diameter, denoted here as diam(G).

Typically, in multirobot applications, two robots are connected if and only if they lie within a prespecified distance, known as the communication range.

Definition 14.1: Let $\boldsymbol{p} := (\boldsymbol{p}_1, \dots, \boldsymbol{p}_n) \in \mathbb{R}^{dn}$ be a list of n points, and let $\Phi > 0 \in \mathbb{R}$ denote the communication range. A *proximity graph* $G(\boldsymbol{p})$ is defined by the node set V and the edge set $E(\boldsymbol{p}) := \{\{i, j\} \subseteq V \; : \; \|\boldsymbol{p}_i - \boldsymbol{p}_j\| \leq \Phi\}$.

Definition 14.2: A d-dimensional *proximity framework* is a pair $F(\boldsymbol{p}) := (G(\boldsymbol{p}), \boldsymbol{p})$, where $G(\boldsymbol{p})$ is a proximity graph and $\boldsymbol{p} := (\boldsymbol{p}_1, \dots, \boldsymbol{p}_n)$ represents a realization of the nodes in $\mathbb{R}^d$.

Definition 14.3: Let $F(\boldsymbol{p})$ be a d-dimensional framework with n nodes and m edges. The Rigidity matrix $\boldsymbol{R}(\boldsymbol{p})$ is the $m \times dn$ matrix where each row R_k with $k = 1, \dots, m$ corresponds to an edge $e_k = \{i, j\}$, such that $R_k = [R_{k,1}^{\top} \;\cdots\; R_{k,n}^{\top}]$ with $R_{k,i} = -R_{k,j} = \dfrac{(\boldsymbol{p}_i - \boldsymbol{p}_j)^{\top}}{\|\boldsymbol{p}_i - \boldsymbol{p}_j\|}$ and $R_{k,l} = 0$ if $l \notin \{i, j\}$. The Stiffness matrix is defined as $\boldsymbol{L}(\boldsymbol{p}) := \boldsymbol{R}(\boldsymbol{p})^{\top}\boldsymbol{R}(\boldsymbol{p})$.

It follows that $\boldsymbol{L}(\boldsymbol{p})$ is symmetric and positive semidefinite, so its eigenvalues are typically ordered such that $\lambda_k(\boldsymbol{L}(\boldsymbol{p})) \leq \lambda_{k'}(\boldsymbol{L}(\boldsymbol{p}))$ whenever $k \leq k'$. The number of zero eigenvalues of the Stiffness matrix determines the infinitesimal rigidity of $F(\boldsymbol{p})$.

Theorem 14.1 (See [38]): *Let $F(\boldsymbol{p})$ be a d-dimensional framework. Then, $F(\boldsymbol{p})$ is infinitesimally rigid if and only if $\lambda_{D+1}(\boldsymbol{L}(\boldsymbol{p})) > 0$, where $D := \frac{d(d+1)}{2}$.*

The value $\lambda_{D+1}(\boldsymbol{L}(\boldsymbol{p}))$ is known as the *rigidity eigenvalue* of $F(\boldsymbol{p})$ and acts as an indicator of the degree of rigidity of a framework.

14.3 Rigidity Maintenance Control

As robots move to carry out a mission, communication links may form or break. The loss of links can compromise rigidity. Therefore, to ensure that the positioning algorithm functions, rigidity maintenance must be incorporated into the control scheme's objectives. Significant efforts have been made to design control strategies that preserve or restore rigidity in networks with dynamic topologies. Proposed techniques in the literature can be categorized as either *connectivity-based*

[39, 40] or *mobility-based* [35, 38, 41, 42]. Connectivity-based approaches manage link creation and removal through combinatorial methods. Mobility-based strategies control the robots' positions, treating the loss of rigidity as an obstacle in the space to be avoided. In both cases, developing fully decentralized algorithms is challenging due to the inherently global nature of rigidity.

Connectivity-based approaches for maintaining rigidity were introduced in Williams et al. [39]. Local rules for adding and removing links between agents are used to guide the robots' movements. The combinatorial nature of this scheme enhances efficiency, as computations and information exchange are only required during topology transitions. However, the underlying theory applies solely to frameworks in two-dimensional space, restricting its applicability in real-world scenarios. A more recent work [40] presents a distributed technique to recover from the loss of rigidity caused by a link failure. A lattice of configurations is considered to locally recover the rigidity of faulty subframeworks by adding new links. Such work pioneered the use of subframeworks, although it is limited to one-hop neighbors. As the authors prove, the rigidity of all one-hop subframeworks is a sufficient (but not necessary) condition for framework rigidity, which ensures the validity of the technique adopted.

Alternatively, mobility-based strategies use the rigidity eigenvalue as a measure of a framework's rigidity. A gradient descent controller guides the robots' movements to keep the rigidity eigenvalue positive, enabling changes in the network topology. The strategy for decentralized implementation relies on consensus filters, which allow each agent to locally estimate the rigidity eigenvalue and its corresponding eigenvector—parameters needed to compute the gradient. The control's performance heavily depends on the ability to quickly estimate the rigidity eigenvalue and eigenvector, which are global properties of the entire network. However, the minimum number of iterations required for the convergence of these estimators is, at best, equal to the network's diameter [43]. This is the main challenge for scalability, as the network's diameter grows with the number of nodes, since each robot can only have a limited number of connections.

Mobility-based methods are the focus of this chapter since they (1) are not conservative in terms of the number of necessary connections, as all redundant links can be detected; and (2) consider the rigidity eigenvalue as a global metric, meaning they have an optimization criterion related to the position of agents within the formation. The main challenge of these methods, as noted previously, is the global information exchange, which compromises scalability. Section 14.4 discusses a novel adaptation mechanism to address this issue originally presented in Presenza et al. [35]. This is done by employing a framework decomposition into subframeworks of arbitrary size. This leads to a solution that maintains rigidity while (1) not requiring global information exchange; (2) optimizing the positions of the robots; and (3) being applicable to any framework.

14.3.1 Mobility-based Rigidity Maintenance

In this section, we review a decentralized mobility control based on what was presented in Zelazo et al. [41], which was experimentally validated in Zelazo et al. [38]. The control action at each time step is obtained as the solution to an optimization problem. The goal is to guide the network of mobile robots while pursuing multiple objectives, which are incorporated into the scheme using a cost function $J(\boldsymbol{p})$ that depends on the positions of all the robots.

Typically, in this scheme, two objectives are mandatory: (1) collision avoidance (CA) between robots and (2) rigidity maintenance (RM) of the framework, a necessary condition for accurate vehicle localization from distance measurements. These objectives are achieved using functions $J_{\mathrm{CA}}, J_{\mathrm{RM}}$, where their values grow unbounded as two vehicles approach each other or as rigidity deteriorates, respectively. Therefore, by steering the framework towards positions that minimize these functions, the objectives can be achieved. Closed formulas for such objectives are shown below (see Eqs. (14.3) and (14.5)). Additionally, a cost function J_{M} may be incorporated to account for mission-specific goals, such as trajectory following or the collection of objects located within the environment.

The control objective is thus defined as guiding the robots to follow a smooth trajectory $\boldsymbol{p}(t)$ with $t \in [0, T]$, minimizing the total cost $J(\boldsymbol{p}) := J_{\mathrm{CA}}(\boldsymbol{p}) + J_{\mathrm{RM}}(\boldsymbol{p}) + J_{\mathrm{M}}(\boldsymbol{p})$. The control action is based on a continuous gradient descent protocol, designed to guide the group of robots toward a local minimum of $J(\boldsymbol{p})$. If $\dot{\boldsymbol{p}}_i \in \mathbb{R}^d$ denotes the velocity of robot i, then

$$\dot{\boldsymbol{p}}_i(t) = -\frac{\partial J(\boldsymbol{p}(t))}{\partial \boldsymbol{p}_i}. \tag{14.1}$$

14.3.1.1 Robot and Network Models

The direct application of the control law Eq. (14.1) is feasible as long as the vehicles are equipped with a velocity tracking controller, a functionality typically included in commercial robotic vehicles. It is customary to consider mobile robots with such capabilities, allowing us to model the agents' position dynamics as a simple integrator $\dot{\boldsymbol{p}}_i = \boldsymbol{u}_i$, where $\boldsymbol{u}_i$ represents the control action.

Since communication ranges are limited, as the robots move, the graph experiences changes in its topology. The control nature Eq. (14.1) requires that transitions in the network topology are modeled with smooth functions. Therefore, the approach of assigning smooth weights to the graph's edges is adopted. Let $\Phi > 0$ be the communication range. The weight

$$w_{ij} := \left(1 + e^{-\beta(\gamma - \|\boldsymbol{p}_i - \boldsymbol{p}_j\|)}\right)^{-1} \tag{14.2}$$

is assigned to the edge $\{i, j\}$, where $\beta > 0$ is a parameter to adjust the rate of decay, and γ determines the curve's midpoint. Typically, $0 < \gamma < \Phi$ so that $w_{ij} \to 1$ when

$\|\boldsymbol{p}_i - \boldsymbol{p}_j\| \to 0$ and $w_{ij} \to 0$ when $\|\boldsymbol{p}_i - \boldsymbol{p}_j\| \to \Phi$. Our simulations suggest setting the parameters to $\beta = \frac{50}{\Phi}$ and $\gamma = 0.95 \cdot \Phi$ for realistic performance.

14.3.1.2 Collision Avoidance

To achieve CA between agents, it is sufficient to consider only those that are neighbors, as it is assumed that if two agents are not connected, they are at a safe distance from each other. To design the control action enabling CA, we employ the potential function:

$$J_{\text{CA}}(\boldsymbol{p}) := \sum_{\{i,j\}\in E} \|\boldsymbol{p}_i - \boldsymbol{p}_j\|^{-a}, \tag{14.3}$$

where $a > 0$. Based on our simulations, suitable values for a are within the range $[1, 1.5]$. The gradient of J_{CA} with respect to a single robot position is

$$\frac{\partial J_{\text{CA}}}{\partial \boldsymbol{p}_i} = -a \sum_{j\in N(i)} \|\boldsymbol{p}_i - \boldsymbol{p}_j\|^{-(a+2)} \left(\boldsymbol{p}_i - \boldsymbol{p}_j\right), \tag{14.4}$$

where $N(i)$ is the neighborhood of agent i.

14.3.1.3 Rigidity Maintenance

Let $\boldsymbol{L}_w(\boldsymbol{p}) := \boldsymbol{R}^{\mathsf{T}}\boldsymbol{W}\boldsymbol{R}$ be a weighted version of the stiffness matrix that includes a $m \times m$ weight matrix $\boldsymbol{W}(\boldsymbol{p})$. Diagonal entries of W are weights w_{ij} defined as in Eq. (14.2), while off-diagonal elements are zero. From this point onward, any mention of the Stiffness matrix refers to the weighted version described above. The controller must guarantee the positivity of the weighted version of the rigidity eigenvalue $\boldsymbol{\lambda}_w(\boldsymbol{p}) := \boldsymbol{\lambda}_{D+1}(\boldsymbol{L}_w(\boldsymbol{p}))$. To this end, we define the rigidity cost function as follows, which grows unbounded as the rigidity eigenvalue of the framework approaches zero,

$$J_{\text{RM}}(\boldsymbol{p}) := \boldsymbol{\lambda}_w(\boldsymbol{p})^{-b}, \tag{14.5}$$

where $b > 0$. Our simulations suggested that good choices for b lie within the interval $[0.3, 0.5]$. The gradient of J_{RM} is

$$\frac{\partial J_{\text{RM}}}{\partial \boldsymbol{p}_i} = -b\boldsymbol{\lambda}_w^{-(b+1)} \boldsymbol{\nu}^{\mathsf{T}} \frac{\partial \boldsymbol{L}_w(\boldsymbol{p})}{\partial \boldsymbol{p}_i} \boldsymbol{\nu}, \tag{14.6}$$

where $\boldsymbol{\nu}$ is a unit norm eigenvector associated with $\boldsymbol{\lambda}_w$. This expression was derived using the formula for the derivative of the eigenvalues of a symmetric matrix, as detailed in [41, 42]. Note that both the rigidity eigenvalue and its eigenvector are global information, as they depend on the entire network's configuration. Consequently, according to this model, each robot requires a network-wide information exchange to compute its own actions.

14.3.2 Further Comments

It is worth noticing that the cost functions Eqs. (14.3) and (14.5) depend on the set of edges E that are present at time t. However, due to the proximity graph model used, this set of edges is a function of $\boldsymbol{p}$, that is $E = E(\boldsymbol{p}(t))$. Therefore, the cost functions (and the associated gradients) may experience discontinuities during edge transitions. These discontinuities, however, are bounded, and they should not cause major problems.

14.4 Subframework-based Rigidity

In this section, we present a novel formulation that enables the decomposition of a framework in terms of a set of subframeworks. This approach is valuable for two main reasons: (1) it provides a new perspective on rigidity by expressing it as a distributed property across the network; and (2) it lays the foundation for designing multirobot rigidity control schemes that rely solely on local information.

14.4.1 Framework Decomposition

As stated earlier, one of the drawbacks of the rigidity theory of frameworks is the global nature of this property. That is, rigidity depends on the entire framework: it is a function of the positions of all nodes and all existing edges. Therefore, the simple removal of a single edge can make a previously rigid framework flexible. This makes the direct application of the property in multirobot networks impractical, as it requires robots to access a collective state—an increasingly challenging task as the network scales. For this reason, it is desirable to adapt the study of rigidity to subsets of nearby robots that form rigid *subframeworks*.

Definition 14.4: Let $F(\boldsymbol{p})$ be a framework. A subset of nodes $V_{(i,h)}$ is defined by two parameters: a node $i \in V$ called the *center*, and an integer $h > 0$ called the *hop limit*, such that $V_{(i,h)} := \{j \in V : \text{dist}(i, j) \leq h\}$. These nodes induce a subframework $F(\boldsymbol{p}_{(i,h)})$, where $\boldsymbol{p}_{(i,h)}$ is the list of positions restricted to $V_{(i,h)}$.

When the positions are clear from the context, we will use the notation $F_{(i,h)} := F(\boldsymbol{p}_{(i,h)})$. Figure 14.1 shows a framework and two subframeworks with hop limits $h_1 = 1$ and $h_5 = 2$. It is possible to establish a necessary and sufficient condition for the rigidity of the entire framework based on the rigidity of its subframeworks, see Theorem 14.2.

Lemma 14.1: *Let $F(\boldsymbol{p})$ be a framework and $\boldsymbol{h} = (h_1, \dots, h_n)$ a list of hop limits. Then, h defines a decomposition into n subframeworks $F_{(1,h_1)}, \dots, F_{(n,h_n)}$ such that their union equals F.*

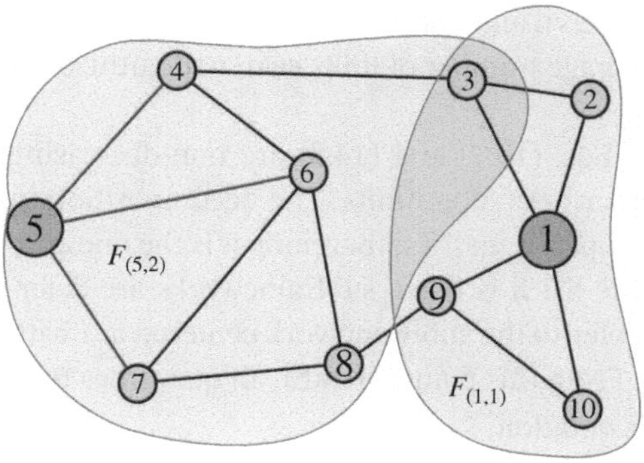

Figure 14.1 A framework and two subframeworks (shaded areas) with centers at $i = 1$ and 5. Centers are highlighted using bigger markers. *Source*: Presenza, J. Francisco.

Theorem 14.2 (*See* [35]): *Let $F(\boldsymbol{p})$ be a framework such that the underlying graph is connected. Then, F is infinitesimally rigid if and only if there exists a decomposition such that all subframeworks are infinitesimally rigid.*

14.4.1.1 Optimal Decomposition

To compare different decompositions of the same framework and select the most suitable $\boldsymbol{h}$, it is necessary to define metrics that evaluate specific desired behaviors. A decomposition that optimizes these metrics is then selected. Firstly, to implement a rigidity algorithm in a decentralized manner, it is natural for all nodes within the same subframework to communicate with each other by routing messages through the center, resulting in paths of length no greater than twice the subframework's hop limit. Therefore, to quantify the worst-case round-trip delay (RTD), we define the following metric:

$$\mathrm{RTD}(\boldsymbol{h}) := 2 \max_{i \in V} h_i. \tag{14.7}$$

Secondly, the bidirectional exchange of information between the nodes of a subframework and its center entails a certain level of communication channel utilization. As the subframework's hop limit increases, the center must transmit messages to a larger number of nodes, resulting in an increased communication load (CL). To measure this, we propose the following metric:

$$\mathrm{CL}(G, \boldsymbol{h}) := \frac{1}{|V|} \sum_{i \in V} c_i \cdot \deg(i), \quad c_i := |\{k \in V \ : \ \mathrm{dist}(i, k) < h_k\}|. \tag{14.8}$$

The load coefficient $c_i \geq 1$ quantifies the amount of information that the i-th node must transmit by counting the number of subframeworks in which it is contained, excluding those where it sits on the boundary. It can be checked that

$\mathrm{CL}(G, \boldsymbol{h}) \geq \deg(G)$, and that equality holds if and only if $h_i = 1$ for all $i \in V$. Hence, Eq. (14.8) can be interpreted as the average number of links each node utilizes in one communication round.

Note that both metrics defined in Eqs. (14.7) and (14.8) are non-decreasing functions with respect to the subframeworks' hop limits. The decomposition in Definition 14.5 below optimizes the proposed metrics; therefore, it is the one chosen from this point forward. When the list $\boldsymbol{h}$ is fixed, subframeworks are determined by their center; therefore, we refer to the subframework centered at i with hop limit $h_i^\star$ simply as $F_{(i)} := F_{(i,h_i^\star)}$. From this point forward, all quantities relative to $F_{(i)}$ will be indicated using this notation.

Definition 14.5: Let F be an d-dimensional framework that is infinitesimally rigid. Then, the *optimal decomposition* is given by $\boldsymbol{h}^\star = (h_1^\star, \dots, h_n^\star)$ where

$$h_i^\star := \min\{\eta \ : \ F_{(i,\eta)} \text{ is infinitesimally rigid in } \mathbb{R}^d\}. \tag{14.9}$$

14.4.1.2 Random Framework Model

To gain deeper insight into the trade-offs introduced by the framework decomposition strategy and to compare the performance of optimal decomposition with classical methods, we propose the following model for the creation of rigid frameworks.

Definition 14.6: Let $\mathcal{R} \subset \mathbb{R}^d$ be open, bounded, and connected, $n \in \mathbb{N}$, and $\alpha \in [0, 1] \subset \mathbb{R}$. Let

$$\boldsymbol{p}_i \sim \mathbf{U}(\mathcal{R}) \text{ for all } 1 \leq i \leq n, \tag{14.10}$$

where $\mathbf{U}$ denotes the uniform random distribution, and

$$\Phi_0 := \min\{\Phi \ : \ F(\boldsymbol{p}) \text{ is infinitesimally rigid}\}, \tag{14.11}$$

$$\Phi_1 := \max\{\|\boldsymbol{p}_i - \boldsymbol{p}_j\| \ : \ i, j \in V\}. \tag{14.12}$$

Then, the random framework model $(\mathcal{R}, n, \alpha)$ produces frameworks with n nodes, positions in $\mathcal{R}$, and a communication range of $\Phi(\alpha) := (1 - \alpha)\Phi_0 + \alpha\Phi_1$.

By definition, all frameworks produced with this random model are infinitesimally rigid. The parameter α is used to control the number of edges in the graph. When $\alpha = 0$, the graph has the fewest possible number of edges that ensures rigidity for a given position $\boldsymbol{p}$. In contrast, $\alpha = 1$ results in a complete graph.

Using this random framework model, in Figure 14.2, the obtained values for $\mathrm{RTD}(\boldsymbol{h}^\star)$ (with $\boldsymbol{h}^\star$ from Eq. (14.9)) and the graph diameter $\mathrm{diam}(G)$ are compared for different numbers of nodes n and three different α values. This comparison is appropriate, as the graph diameter represents a lower bound for the delay in information exchange via consensus. It is observed that for $\alpha = 0$ (associated

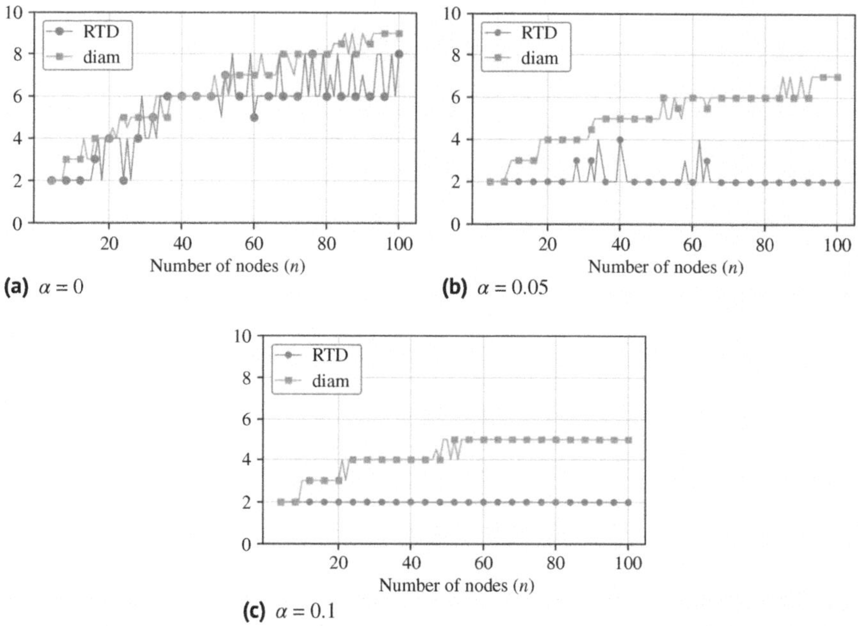

Figure 14.2 Median values for the round-trip delay RTD($\boldsymbol{h}^{\star}$) and the diameter of 50 random frameworks evaluated for each value of n. *Source*: Presenza, J. Francisco.

with fewer edges), RTD($\boldsymbol{h}^{\star}$) closely follows the graph diameter, slightly diverging below it for values of $n \approx 60$ nodes. This indicates that for $\alpha = 0$, the subframework with the highest hop limit has a delay on the order of the diameter, suggesting a virtual equivalence with existing techniques. However, for α values as small as 0.05 and 0.1, RTD($\boldsymbol{h}^{\star}$) decreases more significantly than the diameter, causing both curves to diverge. It can be observed that RTD($\boldsymbol{h}^{\star}$) (and therefore the subframeworks' hop limits) remain nearly constant with n, suggesting that the subframework decomposition is a scalable technique.

In Figure 14.3, we observe the evolution of CL(G, $\boldsymbol{h}^{\star}$) and the average degree with the number of nodes n, along with three different α values. The latter represents the average channel usage in a scheme where each robot exchanges data with its one-hop neighbors only. It can be seen that for $\alpha = 0$, the subframework approach entails a higher load than a one-hop messaging scheme, particularly as the number of nodes grows. This could pose a challenge to the scalability of subframework decomposition. However, for $\alpha = 0.05$, the load closely follows the growth of the average degree, remaining only slightly above it. Furthermore, it becomes evident (Figure **14.3c**) that for $\alpha = 0.1$, both parameters are virtually identical.

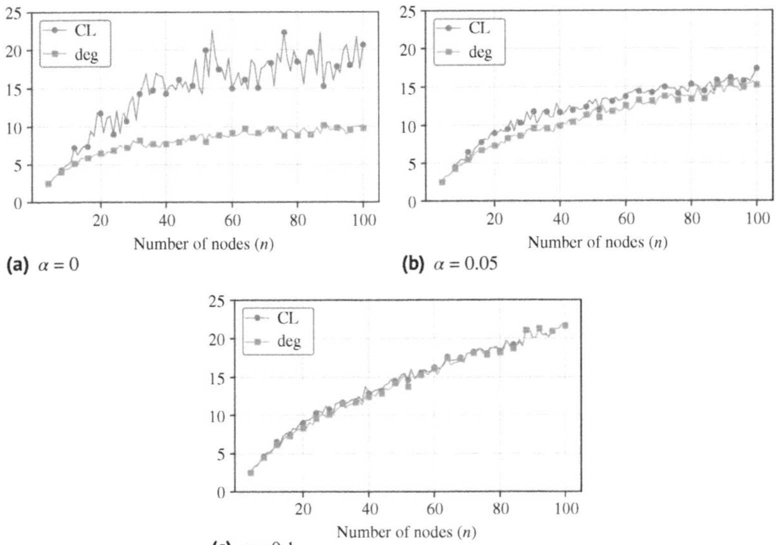

(a) $\alpha = 0$ (b) $\alpha = 0.05$ (c) $\alpha = 0.1$

Figure 14.3 Median values for the communication load $\mathrm{CL}(G, \boldsymbol{h}^\star)$ and the average degree of 50 random frameworks evaluated for each value of n. *Source*: Presenza, J. Francisco.

14.4.2 Subframework-based Control

The primary drawback of the classical approach (Section 14.3) is that it requires consensus algorithms which, as discussed, hinder scalability. The innovation in the control technique described here, from Presenza et al. [35, 44], lies in applying the aforementioned control scheme to individual subframeworks rather than to the entire framework. This scheme allows expressing the rigidity of the framework in local terms through the rigidity of its subframeworks. The rigidity eigenvalues associated with them naturally arise as local rigidity metrics that can be computed and controlled by robots without network-wide exchange of information.

14.4.2.1 Adapted Rigidity Control

To maintain the rigidity of each subframework, we employ the associated stiffness matrix $\boldsymbol{L}_w(\boldsymbol{p}_{(i)})$, where $\boldsymbol{p}_{(i)}$ is the list of positions restricted to the i-th subframework $F_{(i)}$. Therefore, the controller must guarantee the positivity of each rigidity eigenvalue $\boldsymbol{\lambda}_w(\boldsymbol{p}_{(i)}) := \boldsymbol{\lambda}_{D+1}(\boldsymbol{L}_w(\boldsymbol{p}_{(i)}))$. To this end, we define the rigidity cost function as follows, which grows unbounded as the rigidity eigenvalue of any subframework approaches zero,

$$J_{\mathrm{RM}}(\boldsymbol{p}) := \sum_{i \in V} \lambda_w(\boldsymbol{p}_{(i)})^{-b}. \tag{14.13}$$

The gradient of J_{RM} is

$$\frac{\partial J_{\mathrm{RM}}}{\partial \boldsymbol{p}_i} = \sum_{j:\ i \in V_{(j)}} v_{ji}, \quad v_{ji} := -b\lambda_w(\boldsymbol{p}_{(j)})^{-(b+1)} \boldsymbol{\nu}_{(j)}^{\mathsf{T}} \frac{\partial \boldsymbol{L}_w(\boldsymbol{p}_{(j)})}{\partial \boldsymbol{p}_i} \boldsymbol{\nu}_{(j)}, \tag{14.14}$$

where, for each $i \in V$, $\boldsymbol{\nu}_{(j)}$ is a unit norm eigenvector associated with $\lambda_w(\boldsymbol{p}_{(j)})$. Note that, unlike Eq. (14.6), expression Eq. (14.14) includes only information related to the subframeworks that contain node i (j such that $i \in V_{(j)}$), and not the whole framework. This means that the ith robot can apply the rigidity maintenance control input with information that is confined within a few hops (determined by the near subframeworks' hop limits).

14.4.2.2 Communication Load

As analyzed in Section 14.4.1, the information exchange required by a subframework-based control scheme may result in an increased CL on the links. To maintain the load at appropriate levels through mobility control, the scheme may also include active CL minimization. To this end, we propose the incorporation of the following cost:

$$J_{\mathrm{CL}}(\boldsymbol{p}) := \sum_{i \in V} \left(c_i \sum_{j \in N(i)} w_{ij} \right), \tag{14.15}$$

which is the weighted version of Eq. (14.8). The gradient results:

$$\frac{\partial J_{\mathrm{CL}}}{\partial \boldsymbol{p}_i} = \sum_{j \in N(i)} (c_i + c_j) \frac{\partial w_{ij}}{\partial \boldsymbol{p}_i}. \tag{14.16}$$

Note that $c_i + c_j$ assigns a relative importance to edges such that following the negative gradient induces the removal of the most heavily loaded links.

14.4.3 Decentralized Implementation

A message exchange scheme must be implemented to allow robots to access the necessary information for applying Eq. (14.1) in a decentralized manner. It is assumed that the entire framework is rigid at the initial instant.

14.4.3.1 Initialization Phase

The protocol starts with a process in which the robots remain stationary and exchange messages. The goal of this process is for robots to obtain two parameters: their own hop limit $h_i^\star$ and the maximum hop limit among subframeworks that contain it $h_i^\dagger := \max\{h_j^\star \ :\ i \in V_{(j)}\}$. To achieve this, each robot i sends an initial position message (IPM) once to its neighbors, containing information about its position,

$$\text{IPM}_i = (i, \eta, \hat{p}_i), \tag{14.17}$$

where η is the amount of hops that the message must traverse, and $\hat{p}_i$ is its estimated position. Initially, if the graph's diameter is known, set η to this value. Otherwise, estimate it as $\lceil \frac{n-1}{d} \rceil$, which serves as an upper bound for the diameter in rigid frameworks within $\mathbb{R}^d$ [45]. Each time IPM_i is retransmitted, the value of η is updated to $\eta - 1$ until it eventually reaches $\eta = 0$. As robot i receives positions from other robots, it constructs the sequence of subframeworks $\{F_{(i,\delta)} \;:\; \delta = 1, 2, ...\}$. It stops at the first δ for which the resulting subframework is infinitesimally rigid and sets $h_i^\star = \delta$.

Once this process is completed for all (in η steps), each robot sends a one-time message called the initial control message (ICM),

$$\text{ICM}_i = (i, h_i^\star). \tag{14.18}$$

This message travels $h_i^\star$ hops to all robots in the i-th subframework. Based on the ICMs received, a robot i determines the set of subframeworks it is contained in, thus it can compute $h_i^\dagger$ and the load coefficient c_i needed to compute Eq. (14.16).

14.4.3.2 Control Phase

In this phase, each robot emits position messages (PMs) at regular time steps:

$$\text{PM}_i(t) = (i, t, h_i^\dagger, \hat{p}_i), \tag{14.19}$$

which must be retransmitted until it reaches all the centers whose subframeworks contain i. Additionally, robot i collects the positions of those in its subframework and sends control messages (CMs):

$$\text{CM}_{ij}(t) = (i, t, h_i^\star, v_{ij}), \tag{14.20}$$

for each $j \in V_{(i)} \setminus \{i\}$, where v_{ij} is the part of the control action that j must follow to ensure the rigidity of $F_{(i)}$. This scheme facilitates the necessary exchange of data for each agent to apply Eq. (14.14).

14.4.4 Further Comments

In the subframework-based control scheme, the situation discussed in Section 14.3.2 might be more compromising since the discontinuities in the cost Eq. (14.13) might be unbounded. To see this, consider a subframework $F_{(i)}$. The problem resides in the dynamic nature of the subset of nodes $V_{(i)}$. Consider a node $k \notin V_{(i)}$. Then $F_{(i)}$ loses its rigidity if either of the following two situations occurs. Firstly, if a new connection between k and $j \in V_{(i)}$ is established, such that after the connection $\text{dist}(i, k) \leq h_i^\star$ and k do not have enough neighbors in $V_{(i)}$ (less than d). Secondly, if a disconnection that preserves the rigidity of the original subframework (therefore allowed by the controller) expels the node k out of $V_{(i)}$, but the resultant subframework $(F_{(i)} \setminus k)$ is not rigid.

It is noteworthy that only the subframeworks are vulnerable to this type of rigidity loss; the rigidity of the framework is not compromised. These compromising scenarios were not observed during our simulations, suggesting that they are uncommon. In any case, they can be solved by finding a new hop limit for the compromised subframework. However, we acknowledge that this solution can be challenging to implement during the ongoing mission. Research aimed at solving this problem efficiently is being conducted.

14.5 Simulation Results

In this section, we conduct a series of simulation experiments to validate the presented multirobot control strategy. Given the simplicity of the physical models considered—namely, the robots' dynamics and the proximity graph—we opted to develop a custom simulator implemented in Python.

Notice that the three main components of the model—the continuous robot dynamics, the control logic, and the event-driven interactions between robots—exhibit different mathematical natures: namely, ordinary differential equations, discrete-time systems, and discrete-event systems, respectively. Therefore, they were mapped to a common discrete-time representation with a minimum time step able to cover all the meaningful effects. To determine an appropriate value for this time step in terms of accuracy and required computational resources, we varied its value, considering 1, 25, and 100 ms for the simulation of a model including an inter-robot communication delay of $\tau = 300$ms. While a time step of 100ms is infeasible due to the lack of accuracy (which impacts the simulation results), 1 and 25ms produce similar results (despite the difference in time discretization). The upcoming simulation results were conducted with a time step of 25ms, due to the fewer computational resources required.

For the sake of simplicity and to keep the representation straightforward, we simulated the multirobot network in $\mathbb{R}^2$. However, this scheme can be seamlessly applied to $\mathbb{R}^3$ without any modification. In the following examples, the controllers use the estimated positions of the robots as inputs, obtained through the localization algorithm outlined in Section 14.2.3. This shows that the control-estimation loop is effectively closed.

14.5.1 Multiple Objectives Under Time Delays

In this case study, we aim to evaluate the control scheme's ability to meet multiple objectives in the presence of communication delays. This is important since the rigidity control action $\boldsymbol{v}_{ij}(t)$ applied by node $j \in V$ at step t, see Eq. (14.14), is computed using the estimated positions $\{\hat{p}_k(t - T_{ik} - T_{ij}) \ : \ k \in V_{(i)}\}$, where $T_{ij} := \tau \cdot \text{dist}(i, j)$ assumes a fixed delay τ associated with one-hop communications. Although total delays $T_{ik} + T_{ij}$ are not uniform for different k, j, they are bounded by $\tau \cdot \text{RTD}(\boldsymbol{h}^\star)$.

To evaluate the achievement of multiple objectives, we propose a cooperative task involving the collection of stationary objects that are randomly placed within the same region. This introduces a force that tends to separate the robots, stressing the rigidity condition due to the dynamic graph topology. This results in a trade-off between two objectives that the controller must manage. For this purpose, the following cost function $J_{\text{OC}}(\boldsymbol{p})$ is employed, whose minimization acts as an attractive potential toward the objects to be collected. Namely,

$$J_{\text{OC}}(\boldsymbol{p}) := \sum_{i \in V} \|\boldsymbol{p}_i - \boldsymbol{r}_i\|^c, \quad c > 0, \tag{14.21}$$

where $\boldsymbol{r}_i$ represents the position of the closest non-collected object to the i-th robot. The manually adjusted value of $c = 0.5$ gave good results.

To this end, we performed simulations with a network of $n = 10$ agents randomly positioned within $\mathcal{R} = [-40\,\text{m}, 40\,\text{m}] \times [-40\,\text{m}, 40\,\text{m}]$, with a communication range of $\Phi = 8.5$m. We conducted four simulations under the same initial conditions but with different one-hop delays: $\tau = 100, 200, 300$ and 400ms. This example has $\text{RTD}(\boldsymbol{h}^\star) = 6$, so the delays reach up to 2.4s. In this test, the distances between robots are small (on the order of 10 meters), creating a challenging scenario for the controller. This is because a small delay can cause position control errors of just a few meters, which are sufficient to break links necessary for maintaining rigidity.

Figure 14.4 shows the state of the robot collective and the objects to be collected. Uncollected objects are represented by square markers, while collected objects are marked with crosses. In the subsequent figures, the progress of object collection is shown for the case without delay. The trajectories corresponding to cases with $\tau > 0$ are omitted, as there are no significant differences in these illustrations. It is important to observe that, although the rigidity eigenvalue of the framework is not actively controlled and is only guaranteed to remain positive (due to Theorem 14.2), its time evolution closely mirrors that of the subframeworks' eigenvalues.

Also, it is noteworthy that, as the mission progresses, a regular pattern resembling a triangular lattice emerges, which tends to preserve inter-agent distances, resembling a *formation control*. This is a result of the control effort to maximize

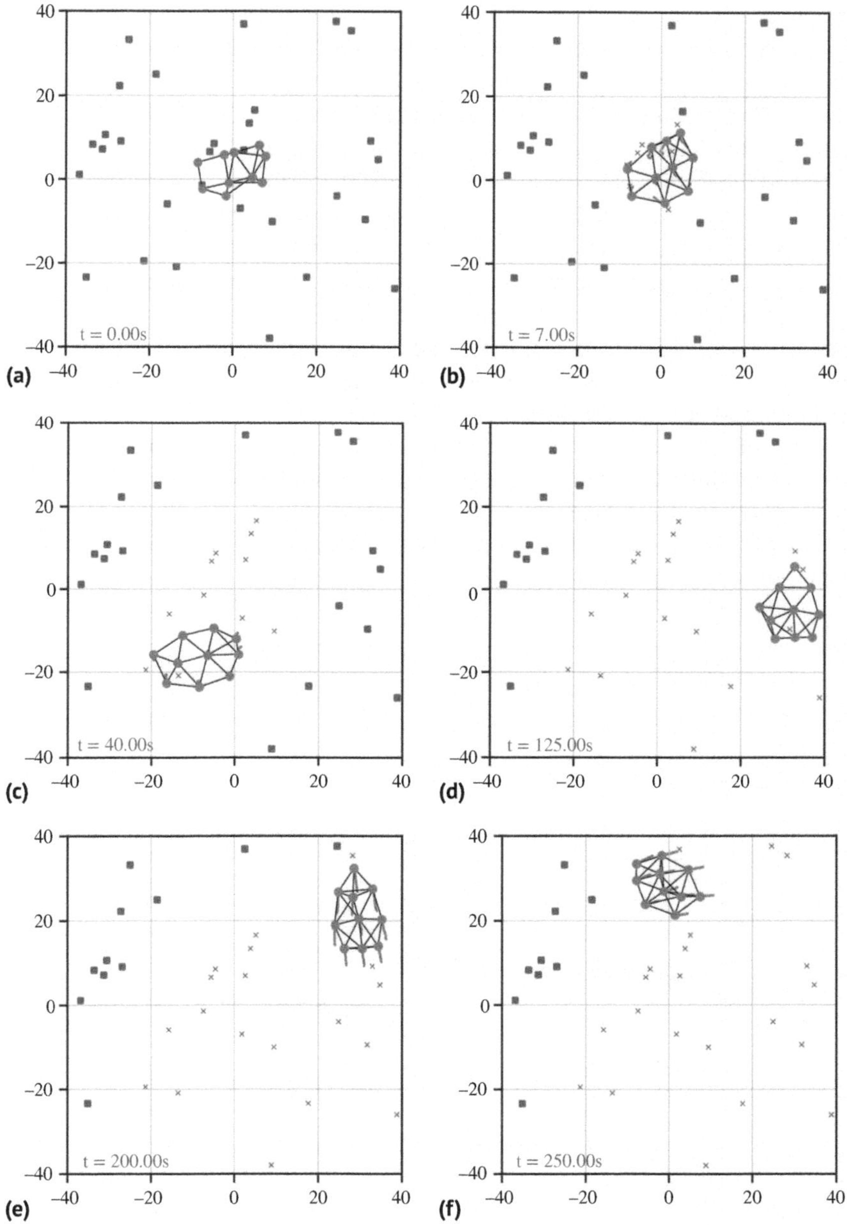

Figure 14.4 Framework snapshots at 0s, 7s, 40s, 125s, 200s, and 250s. Square markers represent uncollected objects, while the collected ones are marked with crosses. Circular markers represent the robots' positions, while the lines represent communication links. Tails are added to help visualize the trajectory followed by the robots. *Source*: Presenza, J. Francisco.

the rigidity eigenvalues, which prioritize regular structures. Moreover, in order to preserve cohesion, robots tend to move with a common velocity, yielding a form of *flocking*. It is interesting to note that these behaviors are not postulated as goals; these results show the upsurge of an *emergent property* from the robot collective.

On the other hand, Figure 14.5 shows the rigidity eigenvalues of the subframeworks and the framework for the four evaluated scenarios involving different values of τ. In all cases, the rigidity eigenvalues remain greater than zero at all times, indicating the robustness of the controller against time delays. However, it can be observed that the controller faces greater difficulty in maintaining these values stable as τ increases. This is especially evident in the initial stage, where the rapid movement of the agents during initialization, combined with the time delay, causes the rigidity of the subframeworks to become more compromised, reaching very low values for $\tau = 300, 400$ms.

Finally, the effect of latency can be observed by analyzing the positions of the vehicles over time, as shown in Figure 14.6. The evolution of the x-coordinate for the 10 robots is shown between 0 and 25s, where the first 5s are employed for running the initialization phase. It reveals the emergence of oscillations characteristic of the time delays. These oscillations increase in amplitude and period as

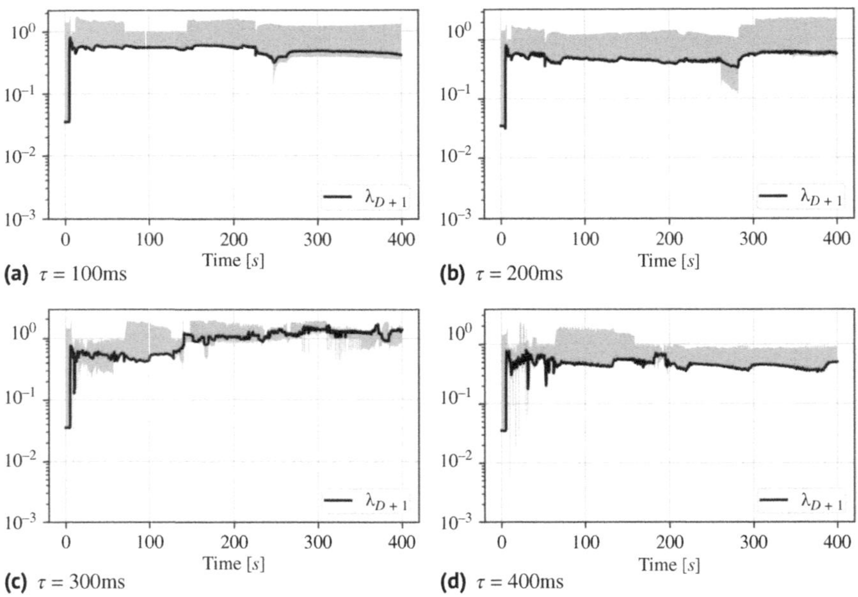

(a) $\tau = 100$ms **(b)** $\tau = 200$ms **(c)** $\tau = 300$ms **(d)** $\tau = 400$ms

Figure 14.5 Rigidity metrics evolution for different values of delay τ. The line represents the framework's rigidity eigenvalue, while the shadowed area indicates the range between the minimum and maximum rigidity eigenvalues of the subframeworks. *Source*: Presenza, J. Francisco.

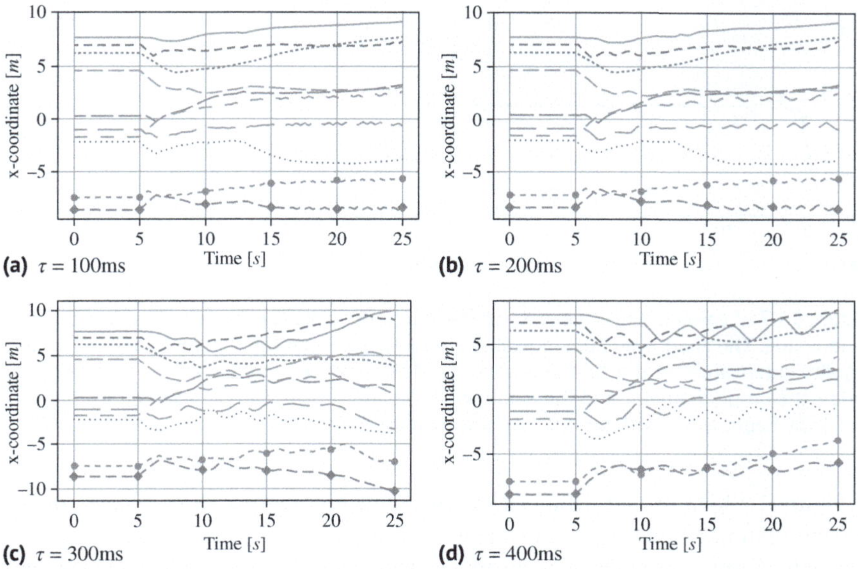

Figure 14.6 Performance degradation as a consequence of time delays. Oscillations in the trajectories followed by the n robots can be observed between 0 and 25s. *Source*: Presenza, J. Francisco.

a function of τ, and although the controller is able to maintain the rigidity of the framework, the performance is degraded.

14.5.2 Scalability Test

In this case, we aim to evaluate the scalability of the proposed strategy by employing a framework $F(\mathbf{p}(t))$ with $n = 60$ robots. In this validation scenario, the MRS is tasked exclusively with minimizing the collective CL. This simulation assumed instant communications between robots; that is, the delay associated with one-hop communication is set to $\tau = 0$. Despite no object collection being considered, the CL objective also acts as a repulsive force, moving the robots apart in such a way as to eliminate the most loaded links. This opposes the maintenance of the framework's rigidity, which must be guaranteed by the controller.

The initial positions are $\mathbf{p}_i(0) \sim \mathbf{U}(\mathcal{R})$, where $\mathcal{R} = [-100\,\text{m}, 100\,\text{m}] \times [-100\,\text{m}, 100\,\text{m}]$, and the communication range is $\Phi = 40\,\text{m}$. It can be checked that, with these parameters, $F(\mathbf{p}(0))$ results in infinitesimally rigid. The optimal decomposition corresponding to the framework is $\boldsymbol{h}^\star \equiv \{1^{(53)}, 2^{(1)}, 3^{(3)}, 4^{(3)}\}$, where the superscript denotes the number of subframeworks with such a hop limit.

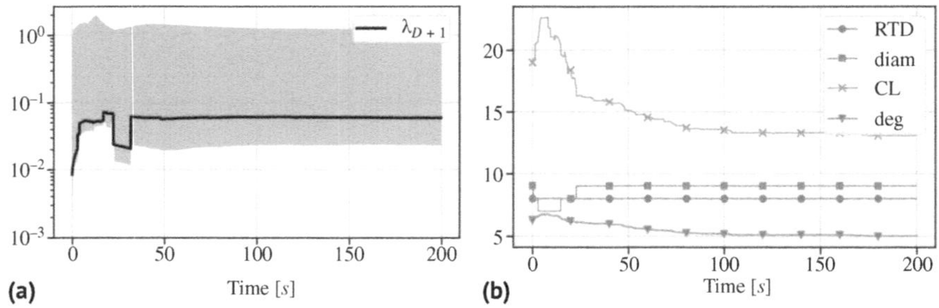

Figure 14.7 Scalability test metrics. (a) The line represents the framework's rigidity eigenvalue. The shadowed area indicates the range between the minimum and maximum rigidity eigenvalues of the subframeworks. (b) The communication load, the round-trip delay, the graph diameter and the average degree are shown. *Source:* Adapted from Ref. [35]. Reproduced with permission of the American Automatic Control Council (AACC).

Figure 14.7 shows performance metrics, and Figure 14.8 shows snapshots of the simulation results. Figure 14.7a shows the evolution of the rigidity eigenvalues, which are kept positive at all times. The shadowed area is bounded by the minimum and maximum rigidity eigenvalues of the subframeworks, while the line represents the rigidity of the framework. Figure 14.7b shows the CL, the RTD, the graph diameter (diam), and the average degree (deg). It can be observed that in the first 25s, the controllers manage to increase the rigidity eigenvalue of the framework by optimizing the robots' positions. This control effort has the effect of creating new links, which leads to a corresponding increase in the CL. From that point onward, the minimization of the cost J_{CL} begins to take effect.

In our second scalability test, we incorporated a dynamic approach to adapt the optimal decomposition $h^\star$ as the network topology changes. This scheme allows robots to reduce the size of subframeworks as the application progresses, which helps reduce both the RTD and the CL. To do this, each robot periodically evaluates the rigidity eigenvalue of its reduced subframework, i.e., the one obtained by reducing the hop limit to $h_i - 1$. If it is above a small threshold (in our simulation, we set it to 0.05), then the value h_i is reduced by one. After running 200s of simulation, the framework decomposition had $\boldsymbol{h}^\star \equiv \{1^{(54)}, 2^{(3)}, 3^{(1)}, 4^{(2)}\}$, which is an improvement from the initial conditions. Results are shown in Figure 14.9, where CL is greatly reduced. However, RTD is not improved in this case, since the maximum hop limit is not reduced.

14.5.3 Further Comments

The discrete-time approach chosen to develop the simulations faces scalability constraints, particularly regarding the number of robots (and, by extension, the number of events), as well as the ability to incorporate more complex dynamics.

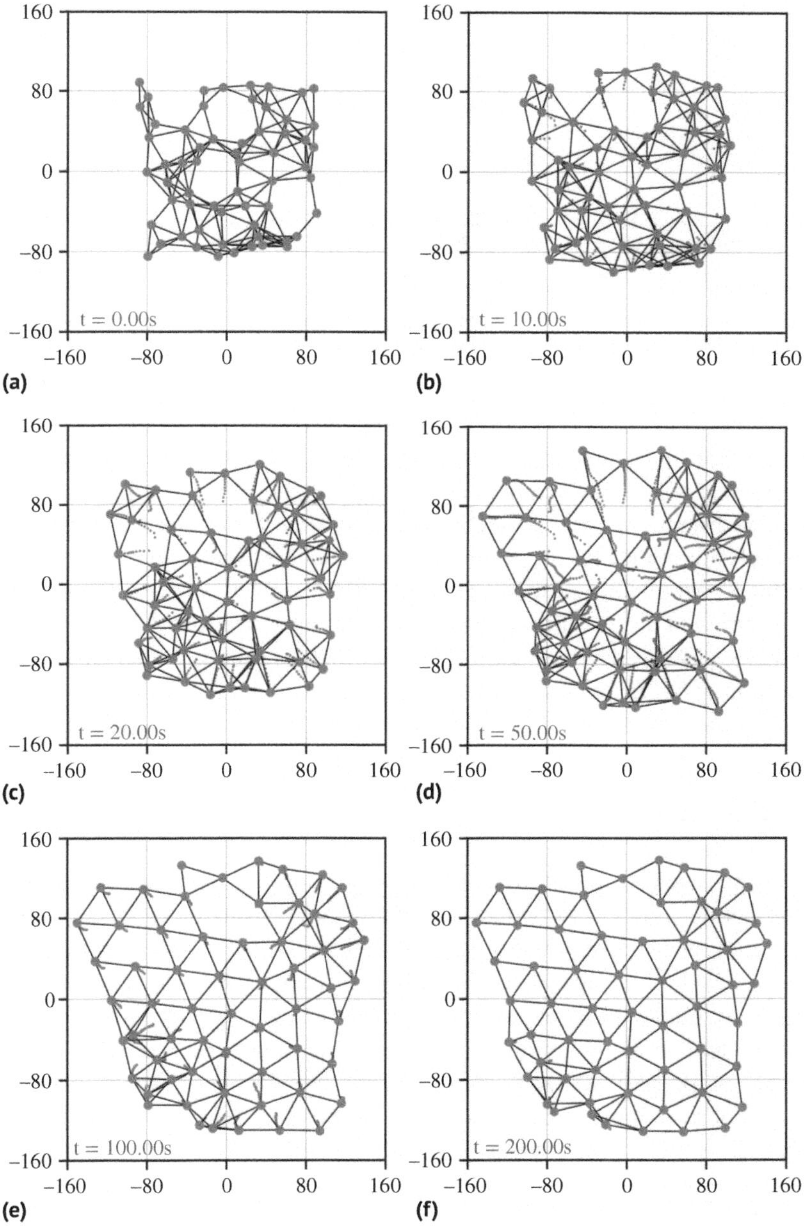

Figure 14.8 Framework snapshots at 0, 10, 20, 50, 100, and 200s . The big dots represent the agents' position, while the lines represent communication links. The smaller dots tails help to visualize the trajectory followed by the robots. *Source:* Adapted from Ref. [35]. Reproduced with permission of the American Automatic Control Council (AACC).

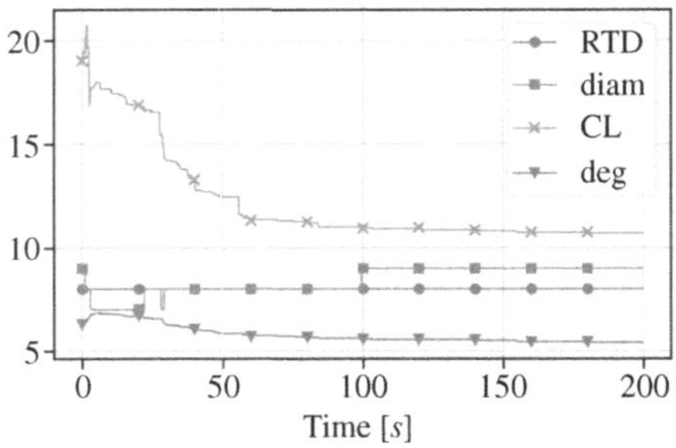

Figure 14.9 Scalability test metrics with the adaptive mechanism to dynamically adjust the hop limits $h^{\star}$. *Source:* Presenza, J. Francisco.

To address these limitations, a more efficient approach would be the use of a mathematical formalism capable of representing such hybrid dynamical systems and that strictly separates models from the simulation engine, which is common to all models. The discrete event system (DEVS) specification formalism has proven to be a valuable alternative for the representation of hybrid dynamical systems [46]. It is the most general formalism for describing discrete-event systems, which can naturally represent discrete-time and event systems and approximate continuous-time systems with any degree of desired accuracy. This formalism offers the additional advantage of a strict separation between the model and the simulator, relieving the modeler of handling the discrete events generated during simulation.

14.6 Conclusions

In this chapter, we discussed fundamental problems in the decentralized control of multirobot networks, a particular case of a CAS. Focusing on robots that can measure only relative distances to their neighbors (a long-standing problem in MRS), we studied rigidity maintenance control—a desired collective behavior that enables the estimation of robot positions from distance measurements.

First, we highlighted the limitations of expressing rigidity as a property of the entire framework, particularly as its size increases. We then formally defined the concept of a subframework and introduced theorems that enable rigidity to be expressed in terms of subframeworks. Subsequently, we discussed metrics to assess the performance associated with a given decomposition. These metrics consider indicators such as delay and CL, which are critical in the design of control schemes and communication protocols. In accordance with the metrics proposed, we defined the optimal decomposition and then performed simulations to evaluate its features. The results indicated that, typically, the size of the subframeworks is smaller than the graph diameter, resulting in reduced information exchange delays. Moreover, we observed that the size of subframeworks tends to remain constant regardless of the number of nodes, unlike the diameter. This result is highly promising for the scalability of the subframework scheme.

Subframework rigidity theory was then applied for the coordinated control of rigidity maintenance in a framework. To achieve this, we proposed an optimization-based multirobot control scheme and developed cost functions tailored to meet the multiple objectives outlined. The proposed scheme does not require global information exchange, thereby avoiding high delays, which is particularly valuable in large-scale networks. However, network load must be monitored to ensure it remains at acceptable levels. We presented two approaches to limit its value: (1) by incorporating CL reduction as an extra control objective and (2) through the dynamic reduction of subframework hop limits.

To validate the proposed scheme, we conducted a series of simulations aimed at evaluating its scalability and robustness under network latency—conditions that challenge control performance. To this end, we relied on an ABM of the networked MRS. We observed that the controller effectively balances multiple objectives in both challenging scenarios, demonstrating a critical property for applications of this nature.

The simulation results show that, while the robot system is guided by basic rules derived from high-level objective—such as rigidity maintenance—collective patterns emerge that tend to preserve inter-agent distances and to induce common velocities, resembling *formation control* and *flocking* schemes, respectively. These cohesive patterns are not explicitly defined, allowing the controller to dissolve them if necessary to achieve its objectives.

The topics discussed in this chapter pave the way for different avenues to explore. Among the potential research directions arising from the work developed in this chapter, we highlight the following. Firstly, we will endow the control scheme with the ability to dynamically augment the subframework hop limits to adapt to changing scenarios. This would solve the subframework rigidity loss discussed in Section 14.4.4. Secondly, the relaxation of the presented framework decomposition will be analyzed, allowing for the possibility that not every agent has an associated subframework, which leads to a sparser decomposition. This sparsity could significantly reduce the CL. This relaxation and the corresponding adaptation of Theorem 14.2 are currently under development.

Finally, it is worth validating the subframework-based control scheme in simulation environments specifically targeted to MRSs, such as [47–50]. Another approach, perhaps better suited for large networks, is the use of simulator frameworks built specifically for complex adaptive systems. In this avenue, we are currently developing a simulation testbed for MRS that takes advantage of a DEVS-based simulation framework specifically targeted at complex adaptive systems [51]. Indeed, the emergent behavior DEVS (EB-DEVS) formalism [52] exploits a microscopic–macroscopic communication mechanism offered by the simulator for local agents to interact via global states, which helps reduce the number of connections and, therefore, the number of exchanged events. We expect that simulation performance in this case will be significantly improved in both accuracy and performance.

References

1 Francisco Javier Pérez-Grau, Ricardo Ragel, Fernando Caballero, Antidio Viguria, and Aníbal Ollero. Semi-autonomous teleoperation of UAVs in search and rescue scenarios. In *2017 International Conference on Unmanned Aircraft Systems (ICUAS)*, pages 1066–1074, 2017.

2 Zendai Kashino, Goldie Nejat, and Beno Benhabib. Aerial wilderness search and rescue with ground support. *Journal of Intelligent & Robotic Systems*, 99:147–163, 2020.

3 Michael Mugnai, Massimo Teppati Losè, Massimo Satler, and Carlo Avizzano. Towards autonomous firefighting UAVs: Online planners for obstacle avoidance and payload delivery. *Journal of Intelligent & Robotic Systems*, 110:10, 2024.

4 Huy Xuan Pham, Hung Manh La, David Feil-Seifer, and Matthew C. Deans. A distributed control framework of multiple unmanned aerial vehicles for dynamic wildfire tracking. *IEEE Transactions on Systems, Man, and Cybernetics: Systems*, 50(4):1537–1548, 2020.

5 Bruno Gabrich, David Saldaña, Vijay R. Kumar, and Mark H. Yim. A flying gripper based on cuboid modular robots. In *2018 IEEE International Conference on Robotics and Automation (ICRA)*, pages 7024–7030, 2018.

6 Francisco Guido Rossomando, Claudio Darío Rosales, Javier Gimenez, Lucio Rafael Salinas, Carlos Miguel Soria, Mário Sarcinelli Filho, and Ricardo O. Carelli. Aerial load transportation with multiple quadrotors based on a kinematic controller and a neural SMC dynamic compensation. *Journal of Intelligent & Robotic Systems*, 100:519–530, 2020.

7 Ezequiel Pecker-Marcosig, Sebastián Zudaire, Rodrigo Castro, and Sebastián Uchitel. Correct and efficient UAV missions based on temporal planning and in-flight hybrid simulations. *Robotics and Autonomous Systems*, 164:104404, 2023.

8 Sayyed Majid Esmailifar and Fariborz Saghafi. Cooperative localization of marine targets by UAVs. *Mechanical Systems and Signal Processing*, 87:23–42, 2017.

9 Rogerio Bonatti, Wenshan Wang, Cherie Ho, Aayush Ahuja, Mirko Gschwindt, Efe Camci, Erdal Kayacan, Sanjiban Choudhury, and Sebastian Scherer. Autonomous aerial cinematography in unstructured environments with learned artistic decision-making. *Journal of Field Robotics*, 37(4):606–641, 2020.

10 Patricio Moreno, J. Francisco Presenza, Ignacio Mas, and Juan I. Giribet. Aerial multi-camera robotic JIB crane. *IEEE Robotics and Automation Letters*, 6(2):4103–4108, 2021.

11 Janardan Kumar Verma and Virender Ranga. Multi-robot coordination analysis, taxonomy, challenges and future scope. *Journal of Intelligent & Robotic Systems*, 102:10, 2021.

12 Soon-Jo Chung, Aditya Avinash Paranjape, Philip Dames, Shaojie Shen, and Vijay Kumar. A survey on aerial swarm robotics. *IEEE Transactions on Robotics*, 34(4):837–855, 2018.

13 Mohamed Abdelkader, Samet Guler, Hassan Jaleel, and Jeff Shamma. Aerial swarms: Recent applications and challenges. *Current Robotics Reports*, 2:309–320, 2021.

14 Mohammad Jafari, Hao Xu, and Luis Rodolfo Garcia Carrillo. A biologically-inspired reinforcement learning based intelligent distributed flocking control for multi-agent systems in presence of uncertain system and dynamic environment. *IFAC Journal of Systems and Control*, 13:100096, 2020.

15 Julian Barreiro-Gomez, Ignacio Mas, Juan I. Giribet, Patricio Moreno, Carlos Ocampo-Martinez, Ricardo Sánchez-Peña, and Nicanor Quijano. Distributed data-driven UAV formation control via evolutionary games: Experimental results. *Journal of the Franklin Institute*, 358(10):5334–5352, 2021.

16 Juan I. Giribet, Leonardo J. Colombo, Patricio Moreno, Ignacio Mas, and Dimos V. Dimarogonas. Dual quaternion cluster-space formation control. *IEEE Robotics and Automation Letters*, 6(4):6789–6796, 2021.

17 Ignacio Mas, Patricio Moreno, Juan I. Giribet, and Diego Barzi. Formation control for multi-domain autonomous vehicles based on dual quaternions. In *International Conference on Unmanned Aircraft Systems—ICUAS*, pages 723–730, 2017.

18 Yefeng Liu, Jingjing Liu, Zengpeng He, Zhenhong Li, Qichun Zhang, and Zhengtao Ding. A survey of multi-agent systems on distributed formation control. *Unmanned Systems*, 12(05):913–926, 2024.

19 G. Vásárhelyi, Cs. Virágh, G. Somorjai, N. Tarcai, T. Szörenyi, T. Nepusz, and T. Vicsek. Outdoor flocking and formation flight with autonomous aerial robots. In *2014 IEEE/RSJ International Conference on Intelligent Robots and Systems*, pages 3866–3873, 2014.

20 Martin Saska, Tomás B'aca, Justin R. Thomas, Jan Chudoba, Libor Preucil, Tomás Krajník, Jan Faigl, Giuseppe Loianno, and Vijay R. Kumar. System for deployment of groups of unmanned micro aerial vehicles in GPS-denied environments using onboard visual relative localization. *Autonomous Robots*, 41:919–944, 2017.

21 Wei Ren, Randal W. Beard, and Ella M. Atkins. Information consensus in multivehicle cooperative control. *IEEE Control Systems Magazine*, 27(2):71–82, 2007.

22 Sonia Martínez, Jorge Cortés, and Francesco Bullo. Motion coordination with distributed information. *IEEE Control Systems*, 27:75–88, 2007.

23 Samira Hayat, Evsen Yanmaz, and Raheeb Muzaffar. Survey on unmanned aerial vehicle networks for civil applications: A communications viewpoint. *IEEE Communications Surveys & Tutorials*, 18:2624–2661, 2016.

24 Torsten Andre, Karin Anna Hummel, Angela P. Schoellig, Evsen Yanmaz, Mahdi Asadpour, Christian Bettstetter, Pasquale Grippa, Hermann Hellwagner, Stephan Sand, and Siwei Zhang. Application-driven design of aerial communication networks. *IEEE Communications Magazine*, 52(5):129–137, 2014.

25 Aicha Idriss Hentati, Lobna Krichen, Mohamed Fourati, and Lamia Chaari Fourati. Simulation tools, environments and frameworks for UAV systems performance analysis. In *Proceedings of the 2018 14th International Wireless Communications Mobile Computing Conference (IWCMC)*, pages 1495–1500. IEEE, 2018.

26 Tom Erez, Yuval Tassa, and Emanuel Todorov. Simulation tools for model-based robotics: Comparison of Bullet, Havok, MuJoCo, ODE and PhysX. In *Proceedings of the 2015 IEEE International Conference on Robotics and Automation (ICRA)*, pages 4397–4404, 2015.

27 Wangyan Li, Zidong Wang, Guoliang Wei, Lifeng Ma, Jun Hu, and Derui Ding. A survey on multisensor fusion and consensus filtering for sensor networks. *Discrete Dynamics in Nature and Society*, 2015:683–701, 2015.

28 Marco Cognetti, Giuseppe Oriolo, Pietro Peliti, Lorenzo Rosa, and Paolo Stegagno. Cooperative control of a heterogeneous multi-robot system based on relative localization. In *2014 IEEE/RSJ International Conference on Intelligent Robots and Systems*, pages 350–356, 2014.

29 Salil Goel, Allison Kealy, Vassilis Gikas, Guenther Retscher, Charles K. Toth, Dorota Brzezinska, and Bharat Lohani. Cooperative localization of unmanned aerial vehicles using GNSS, MEMS inertial, and UWB sensors. *ASCE Journal of Surveying Engineering*, 143:04017007, 2017.

30 Guangwei Zhu and Jianghai Hu. A distributed continuous-time algorithm for network localization using angle-of-arrival information. *Automatica*, 50(1):53–63, 2014.

31 Youngchul Bae. Robust localization for robot and IoT using RSSI. *Energies*, 12(11):2212, 2019.

32 J. Aspnes, T. Eren, D. K. Goldenberg, A. S. Morse, W. Whiteley, Y. R. Yang, B. D. O. Anderson, and P. N. Belhumeur. A theory of network localization. *IEEE Transactions on Mobile Computing*, 5(12):1663–1678, 2006.

33 Zheng Yang and Yunhao Liu. Understanding node localizability of wireless ad-hoc networks. In *2010 Proceedings IEEE INFOCOM*, pages 1249–1260, 2010.

34 Giuseppe C. Calafiore, Luca Carlone, and Mingzhu Wei. A distributed gradient method for localization of formations using relative range measurements. In *2010 IEEE International Symposium on Computer-Aided Control System Design*, pages 1146–1151, 2010.

35 Juan F. Presenza, J. Ignacio Alvarez-Hamelin, Ignacio Mas, and Juan I. Giribet. Subframework-based rigidity control in multirobot networks. In *2022 American Control Conference (ACC)*, pages 3431–3436, 2022.

36 Brian D.O. Anderson, Changbin Yu, Baris Fidan, and Julien M. Hendrickx. Rigid graph control architectures for autonomous formations. *IEEE Control Systems Magazine*, 28(6):48–63, 2008.

37 Guangwei Zhu and Jianghai Hu. Stiffness matrix and quantitative measure of formation rigidity. In *Proceedings of the 48h IEEE Conference on Decision and Control (CDC) Held Jointly with 2009 28th Chinese Control Conference*, pages 3057–3062, 2009.

38 Daniel Zelazo, Antonio Franchi, Heinrich Bülthoff, and Paolo Robuffo Giordano. Decentralized rigidity maintenance control with range measurements for multi-robot systems. *The International Journal of Robotics Research*, 34(1):105–128, 2015.

39 Ryan K. Williams, Andrea Gasparri, Attilio Priolo, and Gaurav S. Sukhatme. Distributed combinatorial rigidity control in multi-agent networks. In *52nd IEEE Conference on Decision and Control*, pages 6061–6066, 2013.

40 Ali Moradi Amani, Guanrong Chen, Mahdi Jalili, Xinghuo Yu, and Lewi Stone. Distributed rigidity recovery in distance-based formations using configuration lattice. *IEEE Transactions on Control of Network Systems*, 7(3):1547–1558, 2020.

41 Daniel Zelazo, Antonio Franchi, P. Allgöwer, Heinrich Bülthoff, and Paolo R. Giordano. Rigidity maintenance control for multi-robot systems. In Nicholas Roy, Paul Newman, and Siddhartha Srinivasa, editors, *Robotics: Science and Systems VIII*, pages 473–480. MIT Press, 2013.

42 Zhiyong Sun, Changbin Yu, and Brian D. O. Anderson. Distributed optimization on proximity network rigidity via robotic movements. In *2015 34th Chinese Control Conference (CCC)*, pages 6954–6960, 2015.

43 Julien M. Hendrickx, Raphael M. Jungers, Alexander Olshevsky, and Guillaume Vankeerberghen. Graph diameter, eigenvalues, and minimum-time consensus. *Automatica*, 50(2):635–640, 2014.

44 J. Francisco Presenza, Juan I. Giribet, J. Ignacio Alvarez-Hamelin, and Ignacio Mas. Control de rigidez por subframeworks en sistemas multirobot. In *XI Jornadas Argentinas de Robótica*, 2022.

45 J. Ignacio Alvarez-Hamelin, Juan I. Giribet, Ignacio Mas, and J. Francisco Presenza. Extremal properties and bounds for the generalized algebraic connectivity of graphs in Euclidean spaces. *Linear Algebra and its Applications*, 727:24–36, 2025.

46 Ezequiel Pecker-Marcosig, Juan I. Giribet, and Rodrigo D. Castro. Hybrid adaptive control for UAV data collection: A simulation-based design to trade-off resources between stability and communication. In *2017 Winter Simulation Conference (WSC)*, pages 1704–1715, 2017.

47 Srikrishna Acharya, Mukunda Bharatheesha, Yogesh Simmhan, and Bharadwaj Amrutur. A co-simulation framework for communication and control in autonomous multi-robot systems. In *2023 IEEE/RSJ International Conference on Intelligent Robots and Systems (IROS)*, pages 11087–11094, 2023.

48 Miguel Calvo-Fullana, Daniel Mox, Alexander Pyattaev, Jonathan Fink, Vijay Kumar, and Alejandro Ribeiro. ROS-NetSim: A framework for the integration of robotic and network simulators. *IEEE Robotics and Automation Letters*, 6(2):1120–1127, 2021.

49 Harish Anand, Stephen A. Rees, Zhiang Chen, Ashwin Jose Poruthukaran, Sarah Bearman, Lakshmi Gana Prasad Antervedi, and Jnaneshwar Das. OpenUAV cloud testbed: A collaborative design studio for field robotics. In *2021 IEEE 17th International Conference on Automation Science and Engineering (CASE)*, pages 724–731, 2021.

50 Andrea Testa, Andrea Camisa, and Giuseppe Notarstefano. ChoiRbot: A ROS 2 toolbox for cooperative robotics. *IEEE Robotics and Automation Letters*, 6(2):2714–2720, 2021.

51 Ezequiel Pecker Marcosig, Juan F. Presenza, Ignacio Mas, J. Ignacio Alvarez-Hamelin, Juan I. Giribet, and Rodrigo Castro. Simulation of decentralized coordination strategies for networked multi-robot systems with emergent behavior-DEVS. In *Proceedings of the 2025 Winter Simulation Conference*, 2025. doi: https://doi.org/10.1109/WSC68292.2025.11338929

52 Daniel Foguelman, Philipp Henning, Adelinde Uhrmacher, and Rodrigo Castro. EB-DEVS: A formal framework for modeling and simulation of emergent behavior in dynamic complex systems. *Journal of Computational Science*, 53:101387, 2021.

Part IV

Digital Twins and Socio-ecological Applications

15

Immersive and Cognitive Urban Digital Twins for Sustainable and Resilient Smart Territories

Mamadou K. Traoré, Mama Diakité, and Zeeshan Ali*

IMS UMR 5218, University of Bordeaux, CNRS, Bordeaux INP, Talence, France

*Corresponding author: mamadou-kaba.traore@u-bordeaux.fr

15.1 Urban Areas Are Complex Adaptive Systems

Urbanization can be traced back to around 3500 BC in the Middle and Near East (today known as Iraq), later spreading to Mexico, China, India, and Mesoamerica. In the developing world, industrialization concept triggered rapid urbanization in the nineteenth and twentieth centuries. By 1950, a quarter of the world's population was living in urban areas; during the late twentieth century, the trend accelerated in developing countries; by 2008, over half of the world's population was urban; and projections suggest an increase in the share of the world's urban population to 75% by 2050, with most occurring in developing countries [1]. Undoubtedly, urbanization has paved the way for numerous benefits for human lives. Urbanization not only supports social well-being but also contributes to the commerce and trade supporting economic activities. Urban areas have better access to basic services and necessities of life such as transportation, hospitals, education, electricity, water, and waste management. These benefits push industries, enterprises, start-ups, and businesses to start their functions in urban areas, which consequently increases the employment opportunities. These benefits make it clear why the world is moving to urban territories, but in the meanwhile, urbanization brings many complex challenges that are multidimensional, interlinked, and interdependent.

The escalation of urbanization is catalyzing a concurrent expansion of urban challenges. As identified by Biljecki [2], the consistent increase in many challenges is directly and indirectly concerned with the urbanization, including air pollution (as particles of 2.5 microns or less in diameter, nitrogen oxides, carbon oxides, greenhouse gases, etc.), soil pollution, urban heat island (UHI), poor waste disposal, water, minerals, and fossil fuel consumption, habitat fragmentation and degradation, mobility, energy utilization, etc.

These challenges are not just affecting the urban ecosystem or the environment alone, but peri-urban and urban communities are also highly affected by the urbanization. The increase in soil pollution diminishes the soil fertility in many peripheral areas; similar to air pollution, UHI not only increases the local temperature, but also the surroundings and peripheral; habitat fragmentation and degradation decline the local and peripheral biodiversity; and the lack of poor waste disposal has highlighted the significant concentrations of water contaminants, affecting both human and aquatic life in surrounding and even nearby cities.

The population growth in cities has escalated the house prices while decreasing their availability. People with low incomes prefer to live in nearby peri-urban areas while working in urban areas. This phenomenon greatly affects the multiple dimensions of urban and peri-urban areas, such as area expansion, land use, mobility, etc. As stated earlier that urban challenges are interlinked and interdependent also; for instance, mobility alone inherits several challenges directly and indirectly that an urban ecosystem faces regularly, such as mobility infrastructure, resource optimization, traffic congestion, air contamination, CO_2 emission, noise pollution, limited transit options for disabled groups, etc.

Similar to the peri-urban areas, there is a surge in challenges at the community level also. Huge buildings are built to entertain a large amount of people; as the community population grows, consequently their needs also grow, which raises the challenges of parking management, resource management, electricity demand and supply, lack of green environment, more water utilization, increase in waste production, etc. All these urban, peri-urban, and urban community challenges hinder the cities' ability to provide sustainable urban services and emphasize the urgent need for innovative strategies to enhance the overall urban framework.

According to future projections established by experts in the field, urban areas will continue to expand, the population will increase [3], and, consequently, the challenges related to the urban ecosystem will intensify, highlighting the urgent need for sustainable and adaptive mitigation strategies. However, there is no simple answer to such urban issues, as they have cascading effects on each other. Even the sustainability metrics that are widely used in urban contexts do not provide holistic insights, and some focus narrowly on specific facets [2]. As such, urban areas are inherently complex adaptive systems (CASs), characterized by nonlinearity, feedback loops, emergent behavior, and adaptability.

To sum up, urbanization is occurring worldwide and urban territories are facing complex major challenges that are multidimensional, interlinked, and interconnected, which make the urban territories not only a CAS but also a complex adaptive system of systems (CASoSs). Therefore, there is a need for holistic approaches, rather than the fragmented approaches that are limited to certain aspects of urban phenomena.

15.2 Resilience and Sustainability Needs

Urban migration and city internal challenges are not the sole reasons that are affecting the urban territories. They are also greatly influenced by many external and natural factors. External factors (such as political issues, trade, conflicts, war, etc.) largely affect the urban territories (e.g., the Russia–Ukraine conflict resulted in a significant increase of oil, gas, and energy prices, and consequently a stress on urban services). Natural hazards and pandemics also have the ability to disrupt urban services. Natural events (such as flooding, earthquakes, pandemics, etc.) can be traced from long back human history in different eras. As these issues are likely to repeat more and more frequently in the future, urban territories need to be resilient, i.e., be adaptive to maintain a satisfying level of service. This is particularly challenging because internal and external factors often lead to emergent behaviors that are very hard to predict due to the complexity of the different components of the CASoS.

Besides resilience, urban territories also need to be sustainable in order to meet present needs without compromising future generations. The United Nations (UN) defined sustainability in terms of a balance between social equality, economic growth, environmental preservation, and the key to the problems that cities face [4]. The Millennium Development Goals (MDGs 2000–2015) have been defined, based on eight key principles, aiming to reduce global poverty and to improve basic human needs and well-being, with a special focus on developing countries and poverty-related issues [5]. The Sustainable Development Goals (SDGs 2016–2030) expanded the MDGs, with 17 key principles and 169 targets, integrating other dimensions of global collaboration, social inclusion, economic growth, and environmental protection [5]. SDGs assist in developing healthy societies aligned with collective well-being and outline key issues that cities and governments should prioritize when allocating resources to address contemporary problems [4].

Many sustainability initiatives are being taken; for instance, United Nations for Smart Sustainable Cities (U4SSCs) and the International Telecommunication

Union-Telecommunication (ITU-T) standardization [6] are among few initiatives that promote smart sustainable urban development based on the SDGs. The U4SSC key performance indicators (KPIs) serve as a mapping standard and comprehensive framework that helps cities and communities worldwide to assess and improve their levels of smartness and sustainability in compliance with SDGs. The U4SSC KPIs are adopted by over 200 cities globally, empowering them to become more sustainable, innovative, responsive, and people-centric smart cities for the local and global well-being.

There is a global context of resilience and sustainability, for which the modern world is striving to achieve. Sustainability and resilience have the potential to transform urban territories into more adaptable, responsive, eco-friendly, resource-enriched, connected, and smart territories.

15.3 Urban Digital Twins

To address urban challenges, there is a need for innovative model-based approaches that can capture and reflect the complexity of urban behavior.

A model is a representation of a selected part of the system of interest, and modeling helps to understand and simplify its reality through abstraction. Model-based system engineering (MBSE) formalizes the application of modeling principles, languages, methods, and tools to the entire lifecycle of large, complex, interdisciplinary, sociotechnical systems [7]. Thus, to engineer the large-scale complex urban phenomena, it is necessary to align the design, functionality, and governance of models with the principles and dynamics of CASoS that are capable of learning, evolving, and adapting under stress or disruption. The collaborative world teams must speak the same language and must work on the same system model, and the communication mechanisms must be supported by flexible, standardized, and friendly modeling languages, for which MBSE is leading the way.

Urban territories are multiscale and multiperspective. Urban issues are widespread from individual to territorial scale, encompassing building, district, city, and regional scales. Furthermore, each of these urban scales incorporates several challenges that could be dependent or independent of each other, making them multiperspective. In fact, MBSE approaches have already recognized their importance long ago, but traditional methods are not sufficient to address urban challenges, as they suffer two major issues: (1) the lack of holistic approach, i.e., the integration of multiscale and multiperspective urban horizons; and (2) the lack of integration of the high variability nature of complex urban territories and their operating context.

Addressing these complex urban horizons requires a **multiperspective modeling and holistic simulation** approach that can integrate various urban phenomena and at various scales. Also, to address the high variability nature of urban territories, there is a need to **dynamically enrich the models** with data collected on urban territories and their operating context.

Both challenges can be addressed with the urban digital twin (UDT) technology. In order to respond (1), UDTs have the potential to address the different urban challenges at different scales. Figure 15.1 represents the maturity of UDT in terms of multiscale and multiperspective horizons, where *x*-axis distributes the urban phenomena (order could be random), while *y*-axis represents the urban ecosystem in ascending order. In response to (2), UDT is capable of dynamic representation of the physical as well as dynamic enrichment of the model with the urban data.

However, the UDT technology is still in its infancy. The lack of standards and the challenge of integrating fragmented solutions complicate UDTs' adoption and implementation. The next section introduces a framework in an attempt to overcome these limitations.

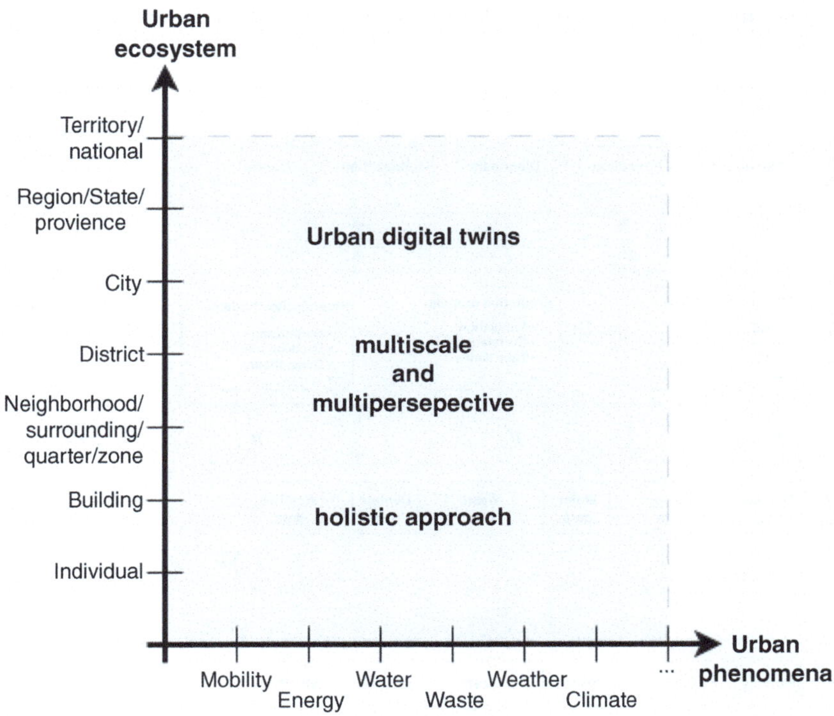

Figure 15.1 UDT maturity level in terms of multiscale and multiperspective urban horizons.

15.4 UDT Framework

The proposed UDT framework consists of data, models, and services (DMS framework), the three pillars of a digital twin (DT) technology. The framework description is given in terms of a conceptual **UDT value chain** accompanied by a formal **UDT specification**.

Our conceptual UDT value chain provides a comprehensive view of the constituent elements of the UDT ecosystem by providing the internal composition of each element, as well as the internal and external interactions. As illustrated in Figure 15.2, this ecosystem is based on three main components: the physical system or twin of interest, its virtual representation (UDT), and the stakeholders.

Physical smart city is referred to as the twin of interest (ToI). It may involve different urban concerns to be managed, such as mobility, energy use, water distribution, waste, etc.

Stakeholders are the actors involved in the UDT, who can intervene at different levels and for various purposes. Their role is fundamental in the functioning of smart cities [8]. Users, citizens, etc. can interact with the models, thus enabling a “human-in-the-loop” approach. This interaction facilitates community training

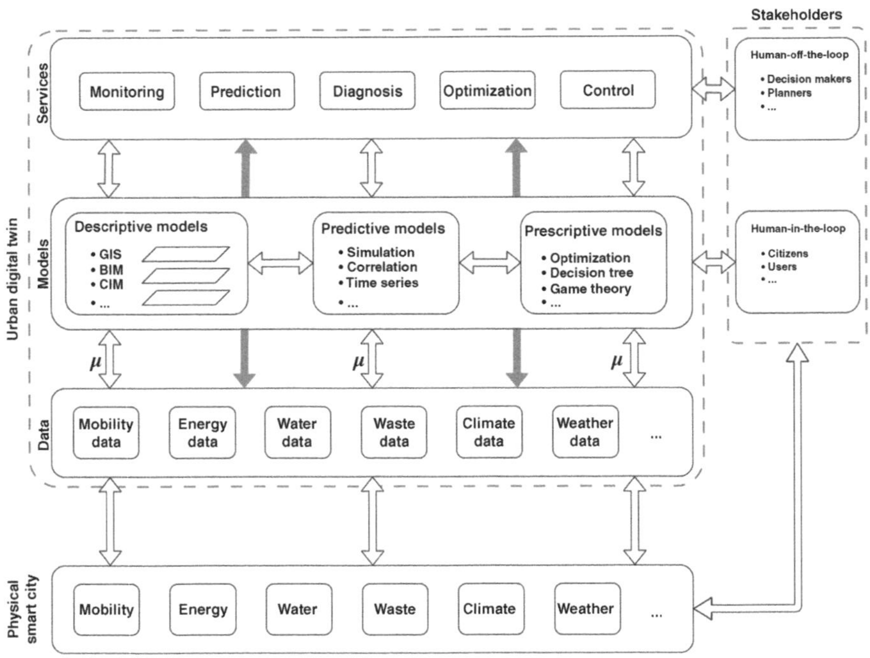

Figure 15.2 Conceptual UDT DMS framework.

and improves the observability of socio-environmental transitions in the city. On the other hand, decision-makers and urban planners interact with the services provided by the UDT to analyze the results and to develop policies. These policies aim to frame and accelerate the socio-environmental transition, thus contributing to the construction of sustainable and resilient territories.

UDT represents a virtual model synchronized with a ToI through data collected. This is an abstraction that reflects several perspectives of the ToI (static, dynamic, functional, etc.). The UDT architecture is based on three main nodes: DMS, as well as interaction between them, among which μ is the synchronization function.

Data node is crucial; it gathers all the relevant information collected to provide information on the state of the ToI. This information can come not only from real-time sensors but also from various existing sources and data from other services. Ramos et al. [9] emphasize the importance of distinguishing different levels of information according to their origin, including context data, system data, and interface data.

Models node contains the collection of models representing different aspects of the urban ToI. These models are updated according to the data collected on the ToI. The models can be descriptive models (such as three-dimensional (3D) models, geographical information systems, building information models, city information models, etc.), predictive models (such as simulation models, correlation equations, etc.), or prescriptive models (such as optimization algorithms, decision trees, learning models, etc.).

Services node depicts the services provided by the UDT. These services are driven by both the models and collected data, thus enabling informed decision-making. Two types of services can be distinguished:

- Services Derived Directly from Raw Data: These services enable structured access to historical and real-time data, mainly for monitoring operations. They offer stakeholders immediate visualization of information without requiring complex modeling structures (illustrated by the two-way arrow between services and data in Figure 15.2).
- Model-based Services: These services leverage and combine digital models to provide forecasts, diagnostics, and optimization solutions. By strategically aligning these models with specific services, the UDT maximizes its decision-making capabilities and improves the efficiency of urban systems (illustrated by the two-way arrow between services and models in Figure 15.2).

Synchronization function (μ) is a fundamental element of DTs. It ensures that the virtual representation faithfully reflects the state of the real system by integrating the collected data. This process, often called model inference, is illustrated by the two-way arrow between models and data in Figure 15.2. Several methods can ensure this synchronization, including:

- Classical updating techniques, such as mapping and recalibration algorithms.
- Artificial intelligence (AI) and machine learning (ML) techniques, capable of improving the prediction of future states of the urban system.
- Statistical methods, such as data fitting, Kalman filters, or Bayesian inference, are used to optimize the integration of information and improve the accuracy of the models.

The formal UDT specification that we propose is based on the interpretation of our conceptual UDT value chain presented in Figure 15.2. Its objective is to allow the symbolic manipulation of UDT concept, thus facilitating a better understanding and a rigorous analysis of the urban territories.

Let $\mathcal{M}$ be the set of dynamical systems. We formulate this specification as follows:

UDT = $\langle \mathrm{D}, \mathrm{M}, \mathrm{S}, \mu \rangle$, where:

- D represents the type of collected data from ToI (abstract set);
- $M \in \mathcal{M}$ is the set of models of the ToI;
- $S \in \mathcal{M}$ is the set of UDT services, parameterized by M and D;
- $\mu : \Omega_D \times \mathcal{M} \rightarrow \mathcal{M}$ is the data assimilation function, where Ω_D is the set of admissible data segments.

Let $\omega \in \Omega_D$ represent a measured data segment, then: $\omega : T_U \rightarrow \mathrm{D}$,
where T_U represents universal time, $Dom(\omega) \subset T_U$, and $Im(\omega) \subset D$.

15.5 Cognitive UDT

Integrating cognition into UDTs improves their ability to adapt, learn, and optimize the management of urban territories. By combining dynamic models and AI approaches, including ML, etc., it becomes possible to develop UDTs capable of reasoning, predicting, and adjusting their representations in real time. This approach makes it possible to:

- Improve the adaptability of UDTs by incorporating autonomous update mechanisms based on the evolution of urban data.
- Optimize decision-making through a more detailed understanding of trends and systemic behaviors.
- Create intelligent interactions with stakeholders, facilitating better exploration of scenarios and more realistic simulation of urban developments.

UDT models can integrate cognitive capabilities as multiagent systems, e.g., by coupling dynamic models with artificial neural networks (ANNs). For example, a recurrent neural network (RNN) can be used to monitor the evolution of urban traffic, while convolutional networks (CNNs) can process satellite images to analyze spatial dynamics. Thanks to these learning methods, cognitive UDTs become capable of providing more accurate predictions, adapting their simulations in real time, and interacting intelligently with urban stakeholders to support strategic decision-making.

Also, the inference function can integrate cognitive capabilities to ensure the proper alignment of the DT with the ToI. Such a capability has to be developed in our framework between the data layer and the models layer.

15.6 Cityverse: An Immersive UDT

The word metaverse is the combination of "meta" and "universe." The concept is defined as *an integrative ecosystem of virtual worlds offering immersive experiences to users that modify pre-existing and create new value from economic, environmental, social, and cultural perspectives* [10]. Bridging the metaverse with UDT is what we call "cityverse."

The metaverse technology enables humans to experience immersive interactions in 3D virtual space via virtual reality (VR) or augmented reality (AR) devices, through avatars representing their digital identities. In this regard, cityverse facilitates digital urbanism through the integration and augmentation of urban ecosystems, offering immersive interaction experiences to stakeholders.

The cityverse approach is depicted by Figure 15.3. Smart territories operate on large, heterogeneous, and dynamic data often coming from various sensors and conveyed through the internet of things (IoT) to data lakes, managing data sources at various computational levels, from centralized cloud level to decentralized fog level and to distributed edge levels. This enables the design of highly detailed and data-rich digital models that capture urban processes, using hybrid models. These

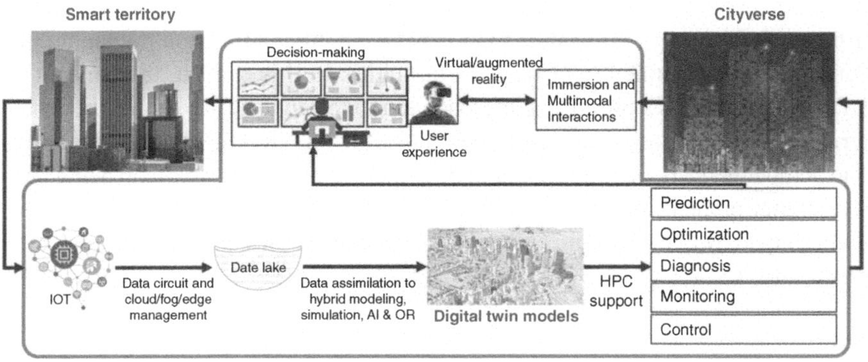

Figure 15.3 Cityverse value chain.

diverse UDT data and models enable the provision of advanced services of monitoring, prediction, diagnosis, "what-if" scenario exploration, optimization, and feedback control.

The presented UDT value chain recognizes the role of its stakeholders that can interact and communicate with UDT to explore the reality about the real world in depth. This concept can be transformed into the cityverse by providing the user with an immersive experience thanks to the technology of extended reality (XR), consisting of VR, AR, and mixed reality (MR), each providing its own range of environment spectrum from fully virtual to fully real with varying degrees of immersion. VR creates a fully immersive experience using the digital environment that replaces the real world, while AR overlays digital elements on the user's view of the real world, and MR blends both VR and AR, allowing the real and virtual elements to interact in real time.

To fully implement cityverse, there is a need for 3D descriptive models, such as geographical information system (GIS), building information model (BIM), city information model (CIM), 3D photogrammetry, drone 3D imagery, etc. These 3D city models have different so-called levels of detail (LoD), the basic ones ranging from LoD1 to LoD4 models, and the more complex ones going up to LoD9 and even more depending on the level of detail required [11]. The higher the LoD, the more satisfying the immersive experience.

UDTs can leverage detailed descriptive models to realize the immersive experience using VR glasses for human-in-the-loop by creating the stakeholders' "avatar" through which they can interact, communicate, and explore scenarios and interact with other avatars. Cityverse adds value to territories and their communities, as well as enhances citizen engagement.

15.7 Conclusion and the Way Forward

The rapid transformation of urban spaces poses major challenges in infrastructure management, crisis resilience, and environmental sustainability. Increasing urbanization requires innovative tools capable of anticipating and managing the complexity of systemic interactions within cities. In this context, the concept of UDT is emerging as a promising approach to model, analyze, and optimize urban territories through an advanced integration of DMS. UDT helps understanding and achieving resilience and sustainability in CAS because models stop being passive digital replicas and become living, learning, evolving systems, informed by data, shaped by participation, and guided by the principles of resilience and complexity science.

However, despite its potential, UDT remains an emerging technology with challenges related to its integration and the lack of standards, thus hindering its large-scale adoption. The DMS framework proposed in this chapter provides a clear structuring of UDT based on three essential pillars: DMS. This framework allows not only a rigorous conceptualization of UDT but also its symbolic manipulation and a more efficient exploitation of digital resources for urban management. The integration of cognitive capabilities into UDTs offers a major asset by giving UDTs the capacity for dynamic adaptation, prediction, and optimization. By combining these capabilities with immersive visualization via the cityverse, a metaverse environment dedicated to smart territories, it becomes possible to strengthen stakeholder engagement and propose interactive and accessible solutions for urban planning. Cognitive and immersive UDTs are enablers of advanced strategy for the intelligent management of cities and territories of the future in the era of Society 5.0. Their standardized development, and verification and validation (V&V) will be key to ensuring their effective adoption for improved sustainability, resilience, and quality of life in urban areas.

Future research directions include (but are not limited to) the following:

- A holistic framework for multiperspective and multiscale UDT modeling is critical for addressing the inherent complexity of urban systems. Tackling urban challenges across different spatial and temporal scales necessitates both (1) a detailed understanding of underlying phenomena, requiring the formulation of models at a microscopic level of abstraction; and (2) decision-support models typically developed at a macroscopic level of abstraction. A fundamental challenge, therefore, lies in enabling consistent and meaningful information transfer across these scales. This challenge gives rise to several core research questions: How do high-level policy or planning decisions influence low-level urban dynamics? Conversely, how can insights from fine-grained phenomena inform strategic decision-making processes? Finally, how can these complementary perspectives be coherently integrated within a unified and interoperable UDT modeling framework?
- End-to-end engineering and deployment of UDTs extend beyond the development of the core model within the DMS framework, requiring integration and interoperability with smart city infrastructures, human users, and other UDTs. Interactions with the smart city involve uni/bi-directional communication between physical and digital layers, particularly through sensor and actuator networks. These interactions raise key research questions concerning how data characteristics (such as frequency, volume, latency, and completeness) affect the reliability and accuracy of UDT-enabled services. They also highlight the need for robust mechanisms to integrate heterogeneous and

geographically distributed data sources at both technical (e.g., IoT architectures) and semantic (e.g., ontology-based representations) levels. Interactions with human users encompass both human-out-of-the-loop contexts, where users consume UDT services via standalone, web-based, or mobile interfaces, and human-in-the-loop contexts, where users actively participate in simulations through immersive environments. These interactions prompt inquiries into the design of standardized architectures for digital–physical communication, including the representation of human agents through avatars and the use of multimodal interfaces leveraging eye tracking, gesture recognition, and other sensory modalities. Finally, interactions among multiple UDTs introduce the overarching challenge of interoperability, underscoring the need to conceptualize UDTs as interdependent components within a larger "system of systems" paradigm.

- The inherent complexity of UDTs necessitates the establishment of comprehensive V&V frameworks. Existing approaches to UDT V&V are largely confined to the assessment of simulation models, offering only a partial evaluation of system integrity. To ensure reliability and trustworthiness, V&V must extend beyond simulation to encompass the entire UDT lifecycle. This broader scope should explicitly address data integration processes, model fidelity, and service-level functionality, thereby enabling a holistic assessment of UDT performance and consistency within real-world operational contexts.

References

1 Xizhe Peng, Xiangming Chen, and Yuan Cheng. *Urbanization and Its Consequences*, Vol. 2, Eolss Publishers, Paris, France, 2011.
2 B. Bera, S. Chinta, D. A. Mahajan, A. Sailaja, and R. Mahajan. Urbanization and its impact on environmental sustainability: A comprehensive review. *Journal of Harbin Engineering University*, 44(8):1310–1318, 2023.
3 UN Habitat. World cities report: Envisaging the future of cities. In *United Nations Human Settlements Programme (UN-Habitat)*, Nairobi, Kenya, 2022.
4 Pierre Major, Guoyuan Li, Hans Petter Hildre, and Houxiang Zhang. The use of a data-driven digital twin of a smart city: A case study of Ålesund, Norway. *IEEE Instrumentation & Measurement Magazine*, 24(7):39–49, 2021.
5 Lucien Georgeson and Mark Maslin. Putting the united nations sustainable development goals into practice: A review of implementation, monitoring, and finance. *Geo: Geography and Environment*, 5(1):e00049, 2018.
6 Ziqin Sang and Keng Li. ITU-T standardisation activities on smart sustainable cities. *IET Smart Cities*, 1(1):3–9, 2019.

7 Ana Luísa Ramos, José Vasconcelos Ferreira, and Jaume Barceló. Model-based systems engineering: An emerging approach for modern systems. *IEEE Transactions on Systems, Man, and Cybernetics, Part C (Applications and Reviews)*, 42(1):101–111, 2011.

8 Gary White, Anna Zink, Lara Codecá, and Siobhán Clarke. A digital twin smart city for citizen feedback. *Cities*, 110:103064, 2021.

9 Mama Diakité and Mamadou Kaba Traoré. Formalizing a framework of inference capabilities for digital twin engineering. *SIMULATION*, 100(9):887–902, 2024. https://doi.org/10.1177/00375497241228281.

10 ITU. ITU focus group on metaverse (FG-MV): Overview of metaverse. Technical Report FGMV-32, International Telecommunication Union (ITU), 2024.

11 Filip Biljecki. Level of detail in 3D city models. PhD thesis, 2017.

16

The Role of Digital Twins in Improving River Water Quality

Souvik Barat, Shankar Kausley, Puneet Singh Thakur, Surabhi Shrivastava, Vinay Kulkarni, and Beena Rai*

Tata Consultancy Services Research, Tata Consultancy Services, Pune, Maharashtra, India

*Corresponding author: souvik.barat@tcs.com

16.1 Introduction

Rivers are essential freshwater sources for supporting diverse ecosystems, including human societies. They serve as primary sources for drinking water, irrigation for agriculture, industrial usage, environments for aquatic life, and also support recreation, tourism and transportation. Preserving water quality across their entire length is, therefore, not only crucial for ensuring public health but also for sustaining the ecosystem and ensuring biodiversity [1]. Any degradation of river water quality poses severe risks to both human health and the environment in several ways. For example, contaminated water can carry pathogens and toxic substances that lead to a range of diseases and health issues, such as cholera and dysentery. Additionally, compromised water quality can disrupt aquatic ecosystems, leading to the loss of biodiversity and affecting fish population, thus adversely impacting the economy. The environmental impacts of poor water quality extend beyond rivers themselves; they affect surrounding wetlands, floodplains, and even groundwater supplies.

In spite of self-cleaning ability [2], the river water quality often degrades along the stretch from source to sea due to a wide range of environmental, socio-economic, and human-induced factors. These factors alter the physical, chemical, and biological properties of river water [3]. One of the primary difficulties in maintaining river water quality lies in spatiotemporal changes involving natural phenomena, such as seasonal shifts and climatic variations, and hydraulic modifications due to heterogeneity in geographic characteristics in addition to human activities that worsen water quality.

Climate change, for instance, exacerbates water pollution by altering rainfall patterns, leading to increased runoff and the spread of pollutants into rivers. More intense storms can overwhelm sewage systems, resulting in raw sewage being discharged into rivers. While prolonged droughts can concentrate pollutants in lower water volumes, resulting in high pH levels and low dissolved oxygen (DO). Furthermore, the natural processes impacting water pollution include erosion, leaching of minerals from soil into water and the decomposition of organic matter contributes to the organic load and chemical imbalance of the river.

Moreover, sedimentation caused by habitat destruction and the alteration of natural river flow through the construction of dams and weirs further disrupts the delicate balance of river ecosystems. Rivers are often subjected to point-source pollution from industrial discharges, agricultural runoff containing fertilizers and pesticides, and untreated sewage and stormwater runoff from urban areas. Non-point source pollution, such as atmospheric deposition and runoff from urban landscapes, diffuses, and spreads over a wide area, further complicating efforts to control water quality. Human-led activities, such as urbanization, industrialization, and agricultural intensification also lead to increased water consumption, thus indirectly leading to water pollutants in river systems. Discharges from housing societies, overflows from wastewater treatment plants (WWTPs), and contamination from industries contribute to the degradation of water quality. Other contributing factors include mining activities that release heavy metals and chemicals into the water, deforestation that increases sedimentation rates and land-use changes that increase nutrient loading into rivers from agricultural runoff.

Current practices for maintaining water quality involve a combination of monitoring, regulation, and pollution control measures. For example, regulatory bodies periodically monitor river water quality by measuring parameters, such as pH, conductivity, total dissolved solids (TDS), dissolved oxygen (DO), biochemical oxygen demand (BOD), chemical oxygen demand (COD) and turbidity, at various locations. While current methods, including manual or remote sensing-based monitoring, can detect specific indicators of pollution, they fail to explain the underlying causes of water quality degradation [4, 5]. Therefore, the key challenges for localized decision-making for improving water quality remain in the following aspects: what interventions should be considered, where, and why? Is there a need for long-term sustained interventions in terms of policies? If so, who should be impacted, and how? How can we balance economic development with environmental protection? What role can community engagement play in improving water quality?

These questions require a precise understanding of the various factors that are negatively influencing water quality, their properties, and dynamics with other factors, such as land use, climate change, and industrial activities. Additionally, understanding the social and economic factors that drive pollution is essential for crafting effective solutions.

Given the complexity of river water quality dynamics, which are influenced by hydraulic flows, atmospheric factors, and human-induced disruptions, a more comprehensive and integrated approach is imperative. Achieving a holistic understanding of the various factors that affect water quality requires an in-depth understanding of the dynamics involving the interplay between river characteristics, such as flow velocity, gradient, and the heterogeneous nature of riverbeds, along with environmental elements like temperature, rainfall, and humidity, and human activities, including industrial discharges, agricultural runoff, and untreated sewage that contribute significantly to the degradation of water quality. To make informed and effective decisions, it is essential to understand these multidimensional aspects and their nonlinear interactions across both time and space. Traditional monitoring methods, however, often fall short of capturing the full complexity of these dynamics as they tend to focus on individual factors without considering how they interact across different locations and time periods [6, 7].

We visualize the river ecosystem as a complex and dynamic *system of systems* and leverage a system modeling and simulation (M&S) approach by representing the river ecosystem as a digital twin to understand the dynamics of different aspects and their influences towards emergent system behavior. Moreover, it serves as an in silico environment for exploring the efficacy of different interventions and policies under various scenarios prior to their implementation. In particular, this chapter presents a simulatable river ecosystem digital twin [8] to represent and analyze relevant aspects of rivers, their environment, and possible human-made disruptions. The core of the digital twin is our simulatable agent model, where segments and their auxiliary ecosystems are represented as interacting agents with their own characteristics, behaviors, and interactions with neighboring segments. We seamlessly integrate the agent modeling paradigm with a physics-based model within a simulatable framework to capture river flow, quality, and their interplay accurately. By conforming to a fine-grained bottom-up modeling paradigm, the agent abstraction helps capture the inherent heterogeneity of various segments and entities associated with the river ecosystem at microlevel. The physics-based models, in contrast, are used to capture the natural phenomena of hydrology and water quality dynamics using mass-balance equations [9] at the type or mesolevel. Our River Digital Twin, which seamlessly integrates agent-based modeling with first-principles physics-based models, offers transformative aid for introducing quantitative rigor in decision-making for river water quality management. By simulating various river management strategies and predicting the emerging outcomes of environmental uncertainties and potential mitigation alternatives (interventions) of unanticipated situations, the proposed approach enables more effective and sustainable management practices. Combining cutting-edge sensing technology, it can provide a comprehensive framework to safeguard the long-term health of river ecosystems and the communities that rely on them.

The remainder of the chapter is structured as follows: Section 16.2 describes water quality dynamics and highlights the inherent complexities for understanding the underlying causes of water quality degradation. Section 16.3 presents an overview of state-of-the-art quality assessment techniques. Section 16.4 discusses our key hypothesis for conceptualizing a novel river ecosystem digital twin as an environment for risk-free experimentation of interventions and strategies. In Section 16.5, we outline our approach. Section 16.6 showcases an application of our approach. Finally, Section 16.7 concludes the chapter by discussing our early experiences, limitations, and avenues for future research.

16.2 Systemic Complexity in Understanding River Quality

From a system theory perspective, a river ecosystem that includes the river under consideration, associated water bodies and entities influencing water quality can be visualized as a dynamic system of systems. Each system, representing a segment of the river or connected water bodies, may exhibit unique characteristics defined by various factors across four interconnected dimensions: river characteristics, environmental changes, human-made disruptions, and control measures as shown in Figure 16.1. These factors interact in a nonlinear manner and exhibit inherent uncertainty, dynamism, and emergent behavior. For example, an upstream river segment that flows through a rural or forested area may feature fast-flowing water, steep gradients, and minimal human activity. Those river segments are, typically, highly sensitive to changes in rainfall or temperature but relatively unaffected by human disruptions like industrial effluence. In contrast, a downstream urban segment may display slower flow, shallower depth and higher pollution levels from urban runoff, sewage, and agricultural discharge.

Conventional modeling approaches often overlook the inherent segment-specific heterogeneities within river systems, as highlighted in Shanafield et al. [10]. Consequently, these approaches tend to disregard localized variations when assessing the water quality of the entire river. In this context, we visualize the water quality of the river system along four aspects, described as follows:

River Characteristics River characteristics, such as flow velocity, hydrodynamics, gradient, and riverbed composition, are fundamental drivers of river water

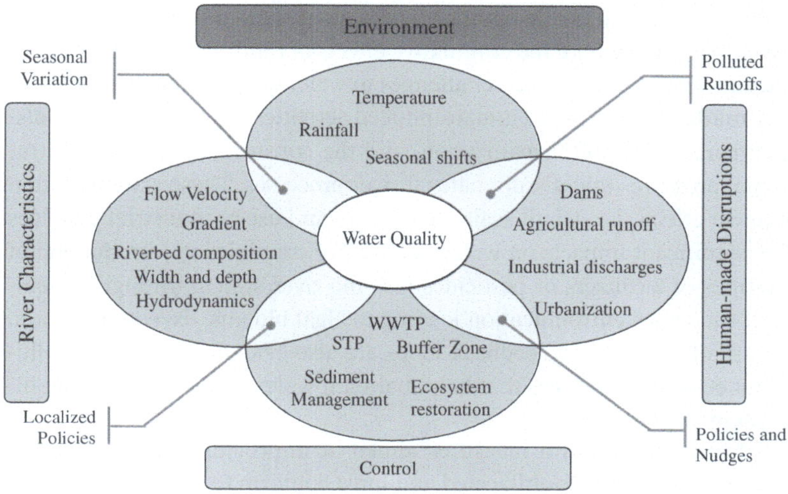

Figure 16.1 Systemic complexities.

and its quality. For example, changes in river flow can influence sediment transport, which in turn alters the riverbed structure and affects nutrient cycling. The self-healing capacity of the riverbed may change if pollution levels exceed its ability to recover. These characteristics also exhibit nonlinearity, i.e., small changes in flow velocity, for instance, can significantly alter the dispersion and dilution of pollutants, causing unpredictable changes in water quality. Furthermore, river characteristics interact with environmental factors and human-made changes, leading to feedback loops that further influence water quality. It is important to note that these characteristics are not homogeneous across the river length. They change over time and along the length of the river. This segment-specific variation further complicates the prediction and management of water quality.

Environmental Changes Environmental factors, such as temperature, rainfall patterns, and climate change significantly influence river water quality. For example, seasonal shifts or long-term climate trends can amplify the effects of pollution. Uncertainty arises as these environmental conditions are subject to variability, and their interactions with other aspects may not always be well understood. A higher temperature can increase the rate of microbial activity, leading to faster breakdown of organic material, which in turn alters BOD levels in the river. Similarly, changes in precipitation and runoff can introduce large volumes of water that may positively impact water quality, but can also introduce significant quantities of pollutants, such as nutrients and toxins, exacerbating water quality degradation in an unpredictable manner. These environmental changes can create emergent behavior, where the cumulative effects of multiple shifts in environmental factors result in unforeseen challenges to water quality management.

Human-made Disruptions Human-induced activities, such as industrial discharges, agricultural runoff, urbanization, and the construction of dams, introduce external pressures that disrupt natural river processes. These activities often have nonlinear effects, i.e., small-scale changes in land use or industrial practices can lead to significant impacts on water quality. For example, an introduction of large quantities of fertilizers or pesticides into the river system through agricultural runoff can cause eutrophication leading to algal blooms, oxygen depletion, and the death of aquatic life. Feedback loops are also evident when these pollutants disrupt ecosystem services that help maintain water quality, such as the filtration capacity of wetlands.

Control Measures Control measures aimed at improving and maintaining river water quality must be multifaceted and must conform to all other aspects of interest. These measures typically include regulations, pollution reduction strategies, and ecosystem restoration efforts—all of which need to be contextual and adaptable to spatiotemporal characteristics of the river. Moreover, the heterogeneity of river characteristics across different segments calls for localized control measures tailored to specific conditions, rather than a one-size-fits-all approach. For example, implementation of buffer zones in rural areas to prevent agricultural runoff and installation of wastewater treatment facilities in urban settings to remove household and industrial pollutants before they enter water bodies could be useful for ensuring water quality.

These four aspects need to be correlated meaningfully for comprehending and controlling water quality dynamics—precisely, controlling river water quality requires effective management of human-made disruptions in cognition with the river characteristics along the river length as well as environmental variations. Therefore, recognizing the interconnected, dynamic, and nonlinear nature of these elements and comprehending the feedback mechanisms, uncertainty, and emergent behaviors that arise from the complex interactions of these factors is essential. Moreover, the segment-specific heterogeneity of the river ecosystem necessitates that control measures be localized to address the unique characteristics of each river segment.

16.3 State-of-the-art Approaches and Technologies

Traditionally, river water quality assessment has relied on periodic sampling and laboratory analysis. While these methods provide valuable snapshots of water conditions, they are labor-intensive, expensive, and time-consuming. Often, these methods fail to capture temporal variations or sudden pollution events [11] resulting in inadequate insight for effective control measures in a timely manner.

To address limitations of traditional methods, several computational approaches have emerged as pragmatic alternatives for estimating and predicting water quality and its spatiotemporal trends. Various modeling techniques, including physics-based models, data-driven models, and more recently, digital twins, have been used for river water quality monitoring and control. Physics-based models simulate river systems by using mathematical representations of physical processes, such as flow dynamics, sediment transport, and chemical reactions. These models are critical for understanding the physical and chemical properties of water.

On the other hand, data-driven models have revolutionized river water quality prediction by analyzing large datasets and identifying patterns. Artificial intelligence (AI) and machine learning (ML) techniques such as artificial neural networks (ANNs) [12], recurrent neural networks (RNNs) [13], support vector machines (SVMs) [14], neuro-fuzzy (NF) models [15], and genetic programming (GP) [16] are used extensively to predict water quality indicators like DO levels, salinity, and BOD. These computational and AI-driven models excel at learning from historical data and adapting to changing conditions. Notably, ANNs have been widely applied due to their ability to handle non-linear relationships and provide continuous updates and refined predictions.

More complex models like long-short-term memory (LSTM) and bidirectional-LSTM (Bi-LSTM) [9] have been explored to analyze time-series water quality data, while hybrid approaches such as combining DL-based models along with other ensemble methods, such as bootstrap aggregating (BA) and alternating model trees (AMT), have demonstrated improved predictive accuracy [17, 18]. Table 16.1 presents a comparative overview of state-of-the-art river water quality control approaches.

Despite these advancements, existing data-driven approaches are chiefly constrained by limited sample sizes, region-specific methodologies, and the need for recalibration to generalize across diverse systems. They face multiple challenges in terms of data acquisition, integration, accuracy, and requirements of computational resources. Management and modeling of high-dimensional data, such as water discharge, temperature, and pollutant concentrations, can be difficult and require optimization of input configuration. These methods fail to predict situations that have not been observed in the past. As a result, their effectiveness is reduced with increasing urbanization, climate change, and changes in land use and land cover. For example, AI-driven models may struggle to predict water quality outcomes in river segments where the urban landscape is rapidly changing as new types of pollutants or land modifications have not been part of the historical data. Similarly, these models may be less effective in forecasting the impact of extreme weather events, such as unprecedented floods or droughts, that are

Table 16.1 Comparison of different models predicting river water quality.

S. No.	Case study	Objectives	Algorithms and data used	Predictions and interventions	Limitations
1	ML-based forecasting model for Danube River water temperature, Romania [19]).	Develop a robust prediction tool for water temperature to enhance operational decision-making in water resource management.	Artificial neural network (ANN). Data: air temperature, water temperature, and water discharge from 2008 to 2015.	The model predicts temperature with high accuracy, forecasting one-week ahead.	Limited generalization due to inconsistent historical data. Variations in air temperature and discharge may affect accuracy.
2	ML approach for optimized monitoring and prediction of water quality for the Santiago River, Mexico [20].	Identify optimal data subsets to enhance ML model training, improve prediction accuracy, and optimize monitoring networks.	ANFIS, ANN, SVM, CA, TSA. Data: 42 water quality parameters from 13 monitoring stations.	Cluster analysis reduced the network to six stations; ANFIS achieved high accuracy ($R^2 = 0.92$).	Region-specific; depends on long-term historical data and focused on physicochemical parameters, neglecting biological indicators.
3	Deep learning for water quality prediction in Clackamas River, USA [17].	Develop hybrid deep learning models to predict turbidity and DO for water quality; optimize input configurations.	Bi-LSTM, Bootstrap Aggregating (BA) with Alternating Model Tree (BA_AMT). Data: Daily records of discharge, temperature, conductance, and pH.	Bi-LSTM effectively predicted turbidity and DO with low RMSE, supports real-time monitoring.	Needs optimization of hyperparameters for Bi-LSTM; limited generalizability across other river systems.

4	Deep learning-based water pollution and water quality assessment in Mahanadi River, Odisha, India [18].	Evaluate surface water quality based on physicochemical parameters and assess seasonal water quality variations.	ANN, inverse distance weighted interpolation. Data: Physicochemical parameters such as pH, DO, BOD, TSS, EC, TDS, and ion concentrations.	ANN achieved high accuracy (0.96), identified pollutants and water quality variations.	Small sample size and seasonal limitations may impact generalizability.
5	Water quality assessment and forecasting for sustainable management in Liffey (Ireland) and Andarax (Spain) rivers [21].	Assess water quality in contrasting river systems using advanced learning algorithms for sustainable water resource management.	Multilayer Perceptron and synthetic data augmentation via Gaussian processes. Data: DO, pH, EC, TDS, turbidity, and nutrients.	DL model achieved high predictive accuracy (R^2 0.87–0.98), predicted significant quality changes under different conditions.	The small dataset required synthetic augmentation for effective model training.

becoming more common due to climate change. Additionally, shifts in agricultural practices, such as the introduction of new fertilizers or pesticides, could further challenge the accuracy of predictions based on historical data. Furthermore, these models typically focus on correlational patterns within datasets without identifying the root causes of deviations in water quality [22]. This limitation hinders their applicability in addressing emerging water quality issues to a large extent.

16.3.1 Tools and Technologies

Several software tools, such as MIKE21 [23] and CORIWAQ [24], are used for river modeling. These tools primarily focus on hydrodynamics, flow dynamics, and steady-state conditions. Therefore, they provide valuable insights into river flow patterns and help simulate various environmental conditions with limited support for water quality modeling. Moreover, these tools typically rely on predefined sets of factors and can analyze only a limited number of key performance indicators (KPIs), such as BOD, DO, and COD. They often overlook other important water quality indicators, including total suspended solids (TSS), TDS, and turbidity.

Additionally, tools like QUAL2E [25], WEAP [26], and RS MINERVE [27] designed to predict river flow and flooding conditions, face challenges in accurately incorporating water quality models. QUAL2E, for example, is focused on simulating water quality in steady-state conditions but struggles to capture the complexities of dynamic and nonlinear interactions in river ecosystems. Similarly, WEAP and RS MINERVE are useful for water resource planning but can be limited in their ability to predict the real-time impacts of changing land use or climate patterns on water quality.

In contrast, more advanced tools like the SWAT (Soil and Water Assessment Tool) [28] model and the HEC-RAS (Hydrologic Engineering Center's River Analysis System) [29] offer more comprehensive approaches by integrating both hydrodynamics and water quality modeling. SWAT, for example, is particularly useful for large-scale watershed modeling and can simulate the effects of land use changes, climate variations, and best management practices on water quality. However, its complex setup and the need for extensive calibration data can be a limitation. HEC-RAS, which models flow and sediment transport, is powerful for riverine floodplain modeling but can face challenges in accurately simulating water quality dynamics, particularly when complex pollutant interactions are involved.

These limitations critically highlight the need for more adaptable, dynamic, and integrated water quality models that can account for the full spectrum of river ecosystem complexities along river characteristics, environmental factors, and human-related disruptions.

16.4 Our Hypothesis

State-of-the-art river modeling approaches, including historical data-centric AI models and physics-based models, often rely on simplified assumptions that represent a river system or a segment as a homogeneous and monolithic unit. These coarse-grained approaches fail to capture the inherent spatiotemporal heterogeneities present across river segments and other water bodies within river ecosystems. As discussed earlier, these ecosystems exhibit significant heterogeneity across multiple dimensions. We argue that modeling a river as a digital twin consisting of a sequence of heterogeneous segments aligned with the system-of-systems nature can provide more accurate and context-specific insights into how various factors interact and influence water quality.

We further hypothesize that a fine-grained *Agent* abstraction is well-suited for representing river segments, water bodies, and other relevant entities. By treating each river segment as an individual agent, an agent-based digital twin can capture localized physical characteristics, environmental conditions, and human-related disruptions at the segment level. This bottom-up approach allows for more granular insights into how changes affect pollutant levels within specific river segments. It also facilitates both decomposition and composition mechanisms supporting complex composition and emergentism. Furthermore, the use of agent abstraction while conceptualizing a digital twin helps to support a useful multi-modeling technique. We consider a combination of rule-based agent specifications, first-principle models representing laws of physics and chemistry, and data-centric approaches to represent three aspects discussed earlier. For instance, rule-based models, such as Event-Condition-Action (ECA) rules, can represent agent behavior when the dynamics are known with some degree of certainty but lack a first-principle model. These rules allow for the simulation of complex interactions based on specific events or conditions. ECA specification can further be augmented with nondeterministic specification for capturing inherent uncertainties. On the other hand, physics-based models (e.g., hydrodynamic models) can represent physical characteristics like flow velocity, water depth, and segment geometry, along with chemical reactions related to BOD and DO composition. Additionally, AI techniques and ML can be applied to model agent behaviors that are not yet well understood and well-studied fields.

Chiefly, our hypothesis is to use the concept of a digital twin with a fine-grained agent-based representation to account for the system of systems nature, modularity, and spatiotemporal heterogeneity. It also advocates a multi-modeling technique, where we argue that the combination of agent-based ECA specification augmented with nondeterminism, physics-based modeling, and data-driven approaches will provide the necessary support to represent ecosystem complexities. Finally, we consider simulation to predict water quality outcomes under different situations, i.e., what-if scenario, more effectively, thus enabling stakeholders to make informed decisions and optimize water quality management strategies.

16.5 Approach

We adopt a digital twin-based approach for precise understanding of the dynamics and evaluating the effectiveness of potential interventions. The proposed approach is outlined in Figure 16.2. As shown, our method comprises four key steps: (1) constructing and validating a digital representation of the river ecosystem, (2) synchronizing the constructed model with real-world data, (3) conducting simulation-based what-if experiments, and (4) selecting the most effective interventions based on these simulations.

In our approach, we apply a simulatable agent abstraction combined with a multi-modeling technique that integrates fine-grained agent modeling and physics-based models to comprehensively represent the complexities of river ecosystems discussed in Section 16.2. To achieve this, we leverage a river ecosystem metamodel proposed by Shrivastava et al. [8], where the elements of this metamodel, shown in Figure 16.3, represent the associated entities within the river ecosystem. Conceptually, these elements are extended forms of simulatable Agents/Actors exhibiting stochastic behavior [30]. This agent-based conceptualization helps to specify distinct and fine-grained characteristics, behaviors, and unique interaction patterns, enabling a bottom-up modeling paradigm. Additionally, the simulatable nature of this approach helps in understanding the nonlinear and emergent behaviors of the complex system. Next, we conceptually validate the constructed model where we engage domain experts to examine the model

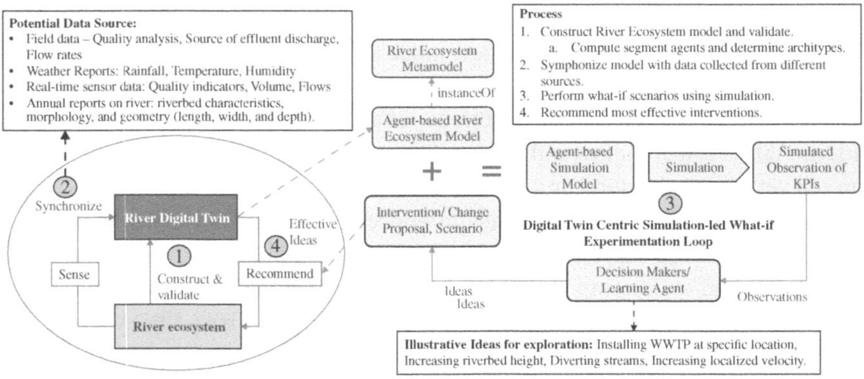

Figure 16.2 An overview of proposed approach.

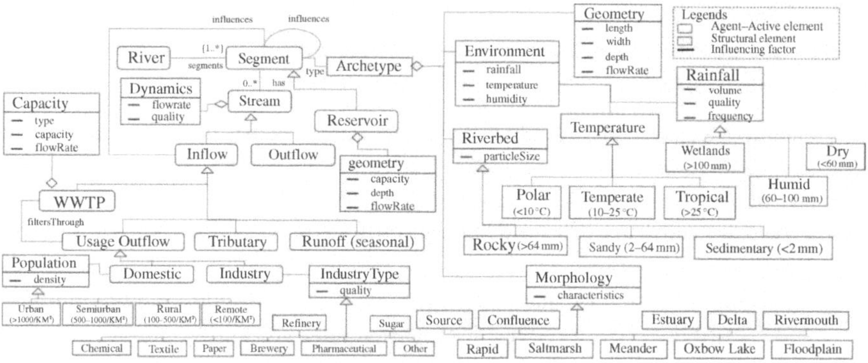

Figure 16.3 River metamodel with core concepts and relationships.

structure and assumptions to ensure that the model sufficiently mimics the river ecosystem. This step adheres to the principles of conceptual validation outlined in canonical simulation research by Sargent [31].

In the synchronization phase, we populate the model parameters with real-world data collected from various sensors and sources distributed across the river ecosystem. The synchronized model effectively becomes the digital twin of the river ecosystem, which is then utilized for simulations to explore strategies or interventions aimed at improving water quality, such as installing (WWTPs or deploying river cleaning techniques like sediment traps.

Prior to experimenting with different scenarios, we simulated the model using known conditions and compared the simulation results with real-world observations to establish operational validity as recommended by Sargent [31]. Iterative simulations of the validated model provide critical insights into potential interventions, aiding in the selection of optimal solutions while accommodating trade-offs. The remaining sections elaborate on our modeling approach, the schema for data synchronization, and the technical implementation of a river ecosystem digital twin.

16.5.1 River Ecosystem Model Construction

The key physical element of a river ecosystem is a long stretch of river with heterogeneous characteristics along its path, and it is augmented with a range of streams that flow into and out of it. Conforming to the concept of a system of systems, we conceptualize a river as a composite river agent. As shown in metamodel depicted Figure 16.3, a river agent comprises a sequence of interconnected segments. Each of these segments represents localized characteristics and unique dynamics in terms of incoming and outgoing streams, volumetric flow and water quality, such as BOD and DO. We introduce segment archetypes to characterize the variations in flow and quality dynamics at mesolevel, where we aim to define a finite set of segment archetypes for a river ecosystem based on factors, including:

- Geometric characteristics: Such as length, width, depth, and flow rate.
- Riverbed composition: For instance, rocky, sandy, or sedimentary formations.
- Morphological features: As depicted in Figure 16.3.
- Types and numbers of inflowing and outflowing streams: Including tributaries or runoff sources.
- Environmental factors: Such as precipitation, humidity levels (wet, humid, and dry), and temperature profiles (polar, temperate, and tropical).

These river segments often comprise different forms of water bodies or streams. Metamodel supports different types of inflows and outflows with unique characteristics with varying capacities. Inflow streams, which directly influence flow dynamics and nonlinearly impact water quality, are categorized into four types: tributaries joining the river, runoff from local areas (due to rainfall and flooding), direct outflows from domestic and industrial sources, and treated water discharge from WWTPs. The dynamics of water usage often emerge as a result of complex domestic and industrial behavior patterns. These can be categorized based on population density and industrial types, as shown in Figure 16.3. Archetypes for these patterns are defined to represent different domestic and industrial usage scenarios at the mesolevel. For example:

- Domestic archetypes: Based on population density, which directly affects wastewater quality and volume. High-density areas with low compliance, such as suburban regions lacking adequate treatment facilities, may discharge highly polluted water.
- Industrial archetypes: Classified by industry type and treatment methods, influencing effluent quality and environmental impact.

Additional archetypes can be developed under existing categories or new ones to account for advanced systems. For instance, agricultural water usage could also be incorporated to represent runoff and irrigation discharge.

16.5.2 Formation of River Segments

In our formulation, a river is visualized as a sequence of segments characterized by segment archetypes. While segment archetypes are derived based on geometric, morphological, weather, terrain, and stream topology variations as shown in Figure 16.3, discretizing a river into segments is a challenge. In reality, these characteristics gradually change along the course of a river, necessitating potentially infinite number of segments to accurately reflect the segmental heterogeneity of the river. Moreover, they can be dynamic, i.e., change over the time. To address this complexity in a pragmatic way, an approximation algorithm, illustrated in Figure 16.4a, is considered to derive a manageable set of river segments. The primary goal of this segmentation algorithm is to optimize the number of river segments while ensuring that heterogeneity is not compromised beyond a range, i.e., threshold difference value ($\delta_{\text{threshold}}$). This threshold may vary between rivers and can be heuristically correlated with noticeable variations in characteristics along specific stretches of the river.

In our algorithm, as depicted in Figure 16.4a, we first calculate the cumulative characteristic difference (δ) between the weighted sums of all characteristic differences from the beginning to the end of a predefined stretch (let's say X km) of a river path. It is important to note that the value of X also may vary depending on the river and its inherent heterogeneity. Characteristic values at the start and end of the stretch are derived, monitored, or sensed from environmental and river-specific data sources, such as literature, databases, and other resources. The weights represent the inverse of average values of each parameter across the stretch. These cumulative characteristic differences (δ) are then compared against a threshold value, $\delta_{\text{threshold}}$, to determine whether the subsequent segment (i.e., next X km) should be part of the current segment (if $\delta \leq \delta_{\text{threshold}}$) or whether it should constitute a new segment. A new segment is classified if the δ value exceeds the threshold, i.e., exceeding the noticeable/tolerable difference. Next, an archetype is assigned based on the average values of all associated characteristics (to associate mesolevel quality equations discussed in Section 16.5.2.1). At the end of this process, a river model is conceptualized as a sequence of interconnected segments with segment-specific archetypes. In this formulation, consecutive segments of a river interact with each other according to various sets of rules and interactions along hydraulic flows and quality-specific chemical reactions as discussed below.

16.5.2.1 Physics-based Equation for River Dynamics

As illustrated in Figure 16.4b, each segment, represented as Agent with unique characteristics, exhibits a set of quality parameters (e.g., BOD and DO), where the dynamics of these quality parameters are primarily governed by the law of physics and hydrology (at archetype or mesolevel). We define a set of mass balance equations to represent the dynamics of these quality parameters, specifically for BOD and DO. Primarily, it is a weighted average of mixing streams (Eqs. 16.1–16.3), ensuring precise representation of mixing effects on water quality.

Equation (16.1) represents the mixing of different streams at the start of an ith river agent, where Q_{li} is an inlet from the previous agent of river/reservoir and Q_{2_i} to Q_{li} is inlet from external agents (industries, WWTP, rainfall), Q_E is volume

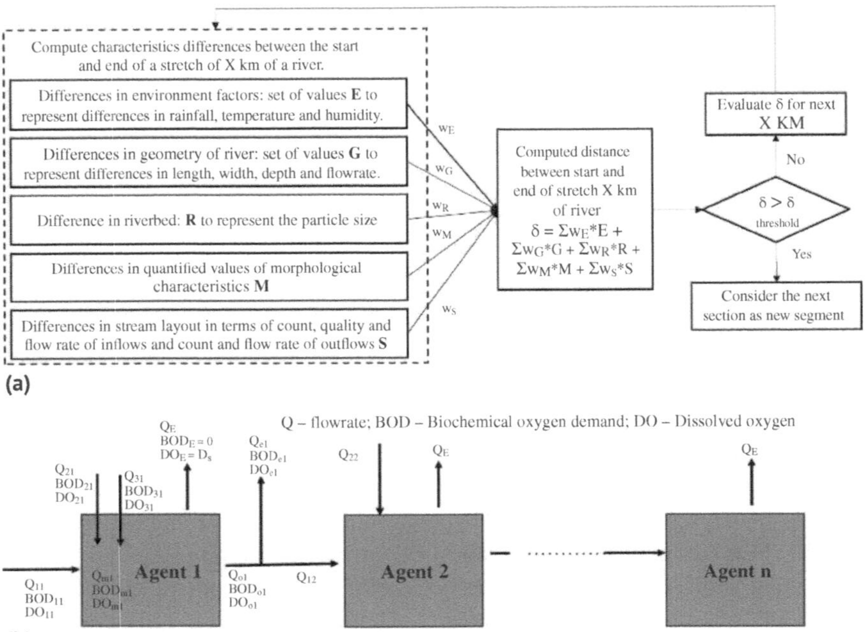

Figure 16.4 Methodology for developing a model to predict the river water quality (agent, agent segment, and physical-based behavior). (a) Algorithm to calculate number of river segments. (b) Illustration of agent behavior and interactions.

loss by evaporation. Quality parameters, i.e., BOD and DO change due to mixing is represented by Eqs. (16.2) and (16.3).

$$Q_{mi} = Q_{1i} + Q_{2i} + Q_{3i} + \cdots + Q_{li} - Q_E \tag{16.1}$$

$$\mathrm{BOD}_{mi} = \frac{Q_{1i} \times \mathrm{BOD}_{1i} + Q_{2i} \times \mathrm{BOD}_{2i} + Q_{3i} \times \mathrm{BOD}_{3i} + \cdots + Q_{li} \times \mathrm{BOD}_{li}}{Q_{mi}} \tag{16.2}$$

$$\mathrm{DO}_{mi} = \frac{Q_{1i} \times \mathrm{DO}_{1i} + Q_{2i} \times \mathrm{DO}_{2i} + Q_{3i} \times \mathrm{DO}_{3i} + \cdots + Q_{li} \times \mathrm{DO}_{li}}{Q_{mi}} \tag{16.3}$$

The evaporation from the river surface can be estimated using the modified Penman-Monteith equation, suggested by Valiantzas [32]. The evaporation is calculated in mm/day and can be converted to m^3/day by multiplying evaporation by the length (L) and width (W) of the river of that agent. The BOD and DO of the river agent are instantiated by the mass balance equation in (16.2) and (16.3), but it changes as it leaves the agent to the other agents according to Eqs. (16.4) and (16.5).

$$\mathrm{BOD}_{\mathrm{oi}} = \frac{\mathrm{BOD}_{\mathrm{mi}}}{1 + k_{\mathrm{di}}\tau_{\mathrm{i}}} \tag{16.4}$$

$$\mathrm{DO}_{oi} = \frac{\mathrm{DO}_{mi} + k_{ri}D_s\tau_i - k_{di}\left(\frac{\mathrm{BOD}_{mi}}{1+k_{di}\tau_i}\right)\tau_i}{1 + k_{ri}\tau_i} \tag{16.5}$$

where k_{di} and k_{ri} are the BOD degradation coefficient and reaeration coefficient of the river section/agent, which depends on the temperature, velocity, and riverbed properties. BOD_{mi} and DO_{mi} are the mixed BOD and DO values, respectively, calculated from mass balance equations at the inlet of any river agent. Ds is the saturation DO, which is dependent on the temperature of the river section. τ_i is calculated by dividing the volume of the river section (L × W × D) with the total flow in the river section Q_{mi}. The pictorial representation of the interaction of different agents is given in Figure 16.4b. The reaeration rate (k_r) is a function of the velocity and bed height of the river section and deoxygenation rate (k_d) is a function of temperature. Similarly, any other physics of the system can be added in the river class to include more dynamics or behavior in the river system, such as algal growth or erosion of the riverbed, etc.

16.5.2.2 Correlation Between Agents

Effluents from industry, streams, households, and WWTP are added or taken up by the river. The reservoir may discharge or take up water based on location and factors like height. Rainfall adds water directly to the river, with its volume determined by Eq. (16.6) [33].

$$\begin{aligned}\ln(V_{\mathrm{rain}}) &= 4.856 + 0.035 \times P - 0.07698T + 1.137 \times 10^{-6}PT \\ &= 2.296 \times 10^{-4}P^2 + 1.098 \times 10^{-3}T^2 \times 1.152 \times 10^{-5}P^2T \\ &\quad -5.931 \times 10^{-5}PT^2\end{aligned} \tag{16.6}$$

where P is the monthly average rainfall in mm and V_{rain} is water influx in river due to rainfall m^3/day. Equation (16.6) can be converted to daily rainfall data for finding daily influx of water in the river due to rainfall.

In certain scenarios, the household and industry agents may also directly interact with the WWTP. Depending on the situation, they may discharge water to the WWTP, which subsequently releases treated water back into the river. The WWTP, in turn, receives input from industry or household agents. Additionally, the rainfall contributes to the water influx in the WWTP. The model accounts for the influx of water in the WWTP due to rainfall using Eq. (16.7) [34]. This equation enables the accurate representation of the impact of rainfall on the WWTP's water inflow.

$$V_{\text{rain}}\text{WWTP} = 1.48P + 935.35 \times 1000 \tag{16.7}$$

16.5.2.3 Illustration of the Proposed Concepts

To illustrate the concepts and methodology of our digital twin formulation, we use a simplified example of a unidirectional river spanning 48 km as shown in Figure 16.5 [8]. The river in this scenario has a one-way flow, i.e., free from any backflow. Its ecosystem includes a rural area that channels all its wastewater to a WWTP, which then releases treated water into the river at a point 18 km from the start. Furthermore, an external tributary joins the flow of the river at the 23-km mark. We also consider two industries along the river. The first is a brewery situated at the 28-km mark, which withdraws water from the river. The second is an automotive industry located at the 41-km mark. The environmental characteristics of the river, such as its riverbed, morphology, and surrounding weather conditions, gradually vary as it flows downstream. Specifically, the first 24-km experience temperate and humid conditions, followed by a tropical and dry climate for the next 12 km, and finally transitioning to tropical and humid conditions for the remaining stretch.

A schematic of the representative river with segments and segment archetypes is depicted in Figure 16.5. In this case, we considered X value to be 1 km and $\delta_{\text{threshold}} = 1$. Based on the proposed algorithm, we derived 10 segments and 9 segment archetypes, as shown in Figure 16.5. The segment 1 (Archetype A) and segment 2 (Archetype B) are different due to riverbed characteristics, segments 2 and 3 (Archetype C) are different due to WWTP inflow dynamics, segments 2 and 4 are same Archetype, segment 5 differs from segments 2 and 3 due to tributary inflow. Segments 6–8 are different from earlier segments as the weather is different, and they have different inflow characteristics. Here, segments 1 and 7 are morphologically similar but belong to different archetypes due to weather. The weather of segments 9 and 10 differs from earlier. Segment 9 differs from segment 10 in terms of riverbed and morphology.

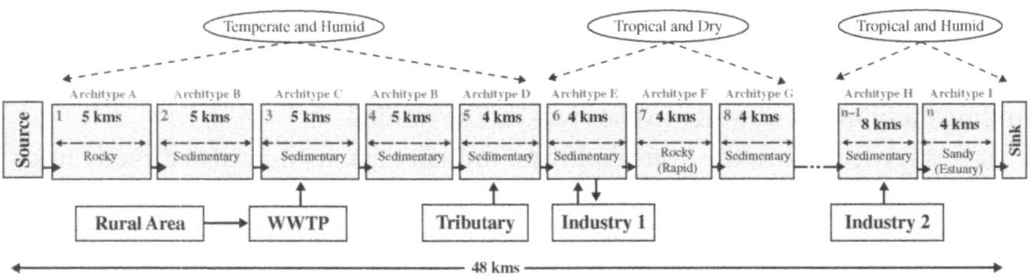

Figure 16.5 Illustration of a representative river ecosystem.

16.5.3 Data Synchronization

We employ a systematic process, as depicted in Figure 16.2, to construct a digital twin of a river ecosystem. Initially, we model the river or a section of it by utilizing the metamodel outlined in Figure 16.3. This entails replicating the river's geometry, riverbed, morphology, and topological features, including input and output streams along with their respective characteristics. Subsequently, we determine the required number of segment agents, formulate agents, and define their archetypes using the algorithm presented in Figure 16.4a.

Real-world data, obtained from the system of records as highlighted in Table 16.2, is then integrated into the constructed model to synchronize the model of the digital twin with the real system. This data may originate from various sources and adhere to the schema highlighted in Figure 16.2. Data for the river properties such as riverbed characteristics, morphology, and geometry (length, width, and depth), are sourced from different reports published by pollution control bodies, NGOs, and government agencies. These properties are generally static and do not change frequently, especially for perennial rivers. However, for rain-fed rivers, there can be substantial seasonal variations due to the changes in rainfall and waterflow, leading to fluctuations in these properties. River characteristics data is typically collected through geological surveys and riverbed and water column sampling, updated annually. Sediment load, and soil composition, essential for understanding riverbed dynamics and sedimentation, are also gathered through monthly riverbed and water column sampling, along with geological surveys. This helps track erosion or deposition processes that affect water flow and quality over time. Vegetation (riparian) data, which can affect river ecosystems and water filtration, is collected through remote sensing and field surveys, usually updated monthly.

Table 16.2 Data required for developing a river digital twin, its source and update frequency.

Sr. No.	Data type	Data source	Update frequency
1	Temperature, humidity	Meteorological data	Hourly
2	Rainfall	Meteorological data	Daily
3	Flow volumes and quality	Sensory data and testing	Daily
4	River-bed characteristics	Reports	Annually
	Population demographic	Reports	Quarterly
5	Water chemistry (pH, DO, TDS, etc.)	In situ sensors, lab testing	Hourly/daily/weekly
6	Water quality indices (BOD, COD)	Water testing laboratories	Weekly/monthly

(Continued)

Table 16.2 (Continued)

Sr. No.	Data type	Data source	Update frequency
7	Sediment load	Riverbed and water column sampling	Annually
8	Soil and sediment composition	Geological surveys	Annually
9	Groundwater levels	Monitoring wells	Weekly/monthly
10	Pollution load (e.g., nutrient concentration)	Environmental monitoring stations	Weekly
11	Land use and cover	Satellite imagery, land surveys	Annually
12	Vegetation (Riparian)	Remote sensing, Field surveys	Monthly

Flow volumes and water quality are monitored using sensory data and testing stations, which track water volume, quality, and inflows/outflows to assess point-source pollution. This data, updated daily, helps identify pollution from industrial or agricultural runoffs. Pollution load data, such as nutrient concentration, is gathered weekly from environmental monitoring stations for real-time contaminant tracking. Field sampling points further validate water quality and detect any deviations. Effluent volume and quality indicators, such as BOD and DO, are monitored with real-time sensors along the river, with updates weekly or monthly. Data from water testing labs validates these measurements to ensure accurate pollution tracking. The data on river water chemistry, including pH, DO, total dissolved solids (TDS), and others, is captured through in situ sensors, validated with regular lab testing, and typically updated on an hourly, daily, or weekly basis. This data provides insights into the river's health and its susceptibility to pollution.

The information on environmental parameters, such as rainfall, temperature, and humidity, is collected from meteorological data. This data is generally updated in real-time, with hourly updates for temperature and humidity and daily updates for rainfall. It is essential for understanding the environmental conditions that affect water quality and river dynamics. The data about type of household, population demographics, and type of industries present in the river's catchment area is also gathered from published reports, often available online. This information can be crucial for understanding anthropogenic pressures on the river system and is typically updated quarterly or annually, depending on the source.

16.5.4 Validation

We validate the constructed model in two phases. The first phase focuses on conceptual validity, ensuring the integrity of modeling concepts and assumptions. This is a manual process where models and assumptions are presented to stakeholders and collected feedback is incorporated using the methodology proposed by Sargent [31]. This validation guarantees that the structural and conceptual foundation of the model is grounded with domain and contextual knowledge.

The second validation phase involves operational validity, which ensures the behavioral aspects and dynamic accuracy of the simulation. This is achieved by simulating historical scenarios and comparing the results against observed data collected from the actual system. Operational validation examines emergent behaviors, collective dynamics, and responses to varying historical inputs to ensure that the digital twin accurately represents the evolving nature of the real-world system it emulates.

16.5.5 What-if Experimentations

A fully functional digital twin synchronized with real data can then be simulated for what-if experimentations such as effect of different interventions, changing scenarios in the case of new industries, population explosion, and stringent guidelines by the pollution control bodies. Different interventions can be checked on the digital twin to identify the intervention best suited to the current situation to ensure sustainable practices in river ecosystems and allowing stakeholders to adjust their strategies based on immediate environmental changes. At the beginning of the simulation, initial values of the river can be taken from the river water quality reports or a particular value can be assumed.

We simulate instantiated, synchronized, and validated river model, i.e., river digital twin, to perform what-if experimentations. The agent-based simulation is implemented using the MESA framework package available in Python [35]. Different agents are defined using classes, which allow for the encoding of their respective behaviors. The simulation progresses in discrete time steps, with each step equivalent to one day. To ensure this, all effluent discharge volumes, rainfall volumes, and other quantities are expressed in metric cubic meters per day.

16.6 Application: Simulations for Predicting Water Quality and Interventions

To evaluate the applicability and coherence of the digital twin model and simulation capabilities, we considered a 480 km stretch of the Ganga River spanning between Kanpur and Varanasi, out of its total river length of 2525 km [36, 37].

First, the physical layout of the stretch of the river under consideration, as shown in Figure 16.6a, is derived from the CPCB website [38]. Our key assumption in modeling is that the main river stream and major water sources are considered. All small streams, which are not contributing significantly, are either ignored or they are aggregated with the nearest major stream. Table 16.3 shows the details of

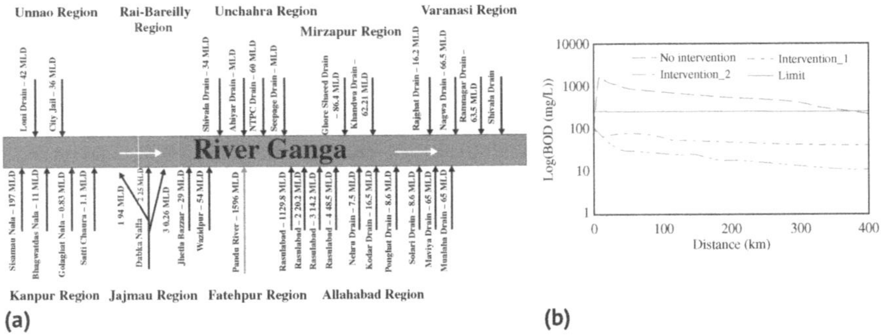

Figure 16.6 Stretch of Ganga from Kanpur to Varanasi. (a) River Ganga ecosystem from Kanpur to Varanasi. (b) Simulation of river stretch with and without interventions.

31 major streams that we considered in our model. As can be found in Table 16.3 and Figure 16.6a, most of the streams that are considered are the major "nallas" or drainages of towns and cities of that area. The drains contain the wastewater of all the nearby municipal area as well as small industries. As can be seen from the data (Figure 16.6a), the first drain, which is Sisamau Nalla has the maximum BOD load (product of BOD and volume discharge), followed by Wazidpur Nalla and Dabka Nalla. In a stretch of 480 km, approximately 30 streams with high BOD load are getting added in the main river at regular intervals. A major wastewater stream from NTPC plant is getting added 200 km downstream of the start of the river and high volumes of water from Pandu River (tributary of River Ganga) and Yamuna River get added to the main river stream daily. Coal fields are situated on the Pandu River leading to generally high BOD levels of the river.

Considering $\delta_{\text{threshold}}$ value to be 5, X to be 10 km, and the algorithm described in Figure 16.4a, the 480 km stretch of Ganga from Kanpur to Varanasi is divided into 31 segments and 9 unique archetypes.

The different archetypes are chiefly derived on the basis on the number and quality of inflows in different sections, since other parameters such as riverbed composition, environmental factors, and morphology are almost similar in this stretch. The model is synchronized with location-specific environmental data (temperature, humidity, and rainfall) collected from weather site as discussed in Table 16.2. The initial values of the BOD and DO of the stretch are taken from the quality values published in reports for different sampling points.

Following the construction of the digital twin for the stretch, the digital twin was simulated for a large period (100 days) to understand the long-term effect on the river water quality. The log of quality values of each section was plotted against the distance of the section from the start of the stretch. For comparison purposes, CPCB limit for the quality is also plotted on the same graph to check the sections, which are crossing the set limit value of 50 mg/L. As can be seen from Figure 16.6b,

Table 16.3 External streams adding to the River Ganga stretch considered in our study.

Point source	Flow (MLD)	BOD load (kg/day)	BOD (mg/L)	Distance (km)
Sisamau Nala	197	544 980	2766	13
Loni drain	42	4860	116	25
Bhagwatdas nala	11	1144	104	38
City Jail drain	36	7208	201	50
Golaghat nala	1	114	137	63
Permiya nala	186	11 485	62	75
Satti Chaura	1	97	88	88
Dabka Nalla-1 (Kachha nala)	94	15 792	168	113
Dabka Nalla-2 (Pakka nala)	25	3475	139	125
Dabka Nalla-3 (Pakka nala)	0	10	38	138
Shetla Bazar (Kachha nala)	29	12 296	424	150
Wazidpur Nalla	54	45 522	843	163
Pandu River	1396	34 900	25	188
Arihari drain	34	127	4	188
NTPC drain	60	1121	19	200
Rasulabad-1 (Pakka nala)	30	20 264	680	213
Rasulabad-2 (Pakka nala)	20	5656	280	225
Rasulabad-3 (kachha nala)	14	1320	93	238
Rasulabad-4 (Kachha nala)	49	2376	49	250
Ghore Saheed drain	86	4121	48	263
Nehru drain	7	61	9	275
Khandwa drain	62	5350	86	288
Kodar drain	20	1040	52	300
Pongaghat drain	8	161	20	313
Solari drain	35	1087	31	325
Rajghat drain	16	808	50	338
Maviya drain	65	3380	52	350
Mugalaha drain	46	598	13	363
Nagwa drain	66	4060	61	363
Ramnagar drain	24	963	41	375

in the current condition of quality and quantity of streams added, the overall quality of the river Ganga stretch is way above the specified limits. Hence, necessary actions and interventions need to be taken to improve the quality and keep it under check. To explore the effectiveness of interventions, two different interventions or actions were taken to limit the quality within the desired limits. For the first intervention, 100 mg/L BOD limit for all the streams was imposed. No stream can discharge water greater than 100 mg/L. This can be achieved by at-source treatment or treatment before they enter the river stretch. The second intervention is when a limit of 50 mg/L was imposed for all incoming streams. The system was simulated with two different interventions to highlight the importance of changing norms and regulations and in making the right choice. Figure 16.6b shows the variation of BOD in logarithmic scale with two interventions, without intervention. The second intervention showed better performance as the incoming BOD limit was decreased for all the incoming streams from intervention 1, whereas very high-BOD values can be seen when there is no limit on the BOD value of the effluent stream discharged in the river. Other similar interventions like creating sewage treatment plants (STPs) on the course of the river, increasing riverbed height, diverting streams to different locations can be simulated on the digital twin to choose the best possible intervention in a given scenario for a particular river stretch. The interventions studied can be related to different initiatives launched by the government of India to improve the quality of the Ganga River, under the Namami Gange program. One such initiative is the treatment of sewage carrying drains joining the River Ganga. The sewage handling projects are primarily being taken up in towns along the banks of the river Ganga, through construction of STPs. However, these projects require a long duration for commissioning, are capital intensive and entail higher operating and maintenance costs. Recently, innovative technologies such as Modular STP, Phytorid, Geo-Tube, etc. have been tried which require short commissioning time and are less expensive. Additionally, in situ treatments have been tried where naturally occurring microbial consortia/nutrient media are used in the treatment process or activated to degrade sewage in flowing conditions without diverting the flow. The in situ methods are found to be simple, easy-to-operate, and do not require major modification of drains [39]. Additionally, the digital twin can be simulated for different environmental conditions to understand its effects on the overall quality and helpful in deciding preventive actions to maintain the quality within the desired limits.

16.7 Conclusion

This chapter considers river ecosystems complex systems and emphasizes the necessity of conceptualizing such systems as multifaceted system of systems governed by a network of interrelated physical, chemical, and anthropogenic factors.

By integrating spatially explicit environmental datasets, segment-specific characteristics, and socio-environmental dynamics, our approach demonstrates the efficacy of our modeling and simulation approach. The proposed digital twin approach conceptually combines the agent modeling paradigm with physics-based modeling techniques and adopts a hybrid methodology that supports fine-grained and coarse-grained modeling methodologies. Additionally, it uses the pragmatic assumptions of mesolevel modeling with context-specific archetypes. They collectively demonstrate how the synergistic use of existing modeling techniques with methodological rigor can help model and comprehend complex water quality dynamics in large river ecosystems.

By utilizing sophisticated computational models, our approach facilitates the simulation of current conditions as well as hypothetical interventions. The case application on the Ganga River has validated the utility of our digital-twin-based simulation—it helps to identify critical degradation zones and assess the viability of diverse intervention measures, such as imposing BOD limits and deploying sewage treatment technologies. The successful simulation of interventions on the River Ganga highlights the potential of this approach to support water quality management efforts specifically and to comprehend complex systems in general.

However, it is crucial to acknowledge certain limitations, including the ongoing need for calibration and validation to ensure the accuracy and reliability of the model, particularly regarding the definition of segments and segment archetypes. Future research will explore how the constructed model and digital twin may be made adaptive, allowing them to evolve automatically as the river and its environment change due to climate change and other environmental factors.

References

1 N. T. Anh, N. T. Nhan, B. Schmalz, and T. L. Luu. Influences of key factors on river water quality in urban and rural areas: A review. *Case Studies in Chemical and Environmental Engineering*, 8:100424, 2023.

2 Y. Wen, G. Schoups, and N. Van De Giesen. Organic pollution of rivers: Combined threats of urbanization, livestock farming and global climate change. *Scientific Reports*, 7(1):43289, 2017.

3 S. J. Interlandi and C. S. Crockett. Recent water quality trends in the Schuylkill River, Pennsylvania, USA: a preliminary assessment of the relative influences of climate, river discharge and suburban development. *Water Research*, 37(8):1737–1748, 2003.

4 A. T. Demetillo, M. V. Japitana, and E. B. Taboada. A system for monitoring water quality in a large aquatic area using wireless sensor network technology. *Sustainable Environment Research*, 29:1–9, 2019.

5 J. Huang, Y. Zhang, H. Bing, J. Peng, F. Dong, J. Gao, and G. B. Arhonditsis. Characterizing the river water quality in China: recent progress and on-going challenges. *Water Research*, 201:117309, 2021.
6 A. D. Sutadian, N. Muttil, A. G. Yilmaz, and B. J. C. Perera. Development of river water quality indices—a review. *Environmental Monitoring and Assessment*, 188:1–29, 2016.
7 S. H. Ewaid, S. A. Abed, and S. A. Kadhum. Predicting the Tigris River water quality within Baghdad, Iraq by using water quality index and regression analysis. *Environmental Technology & Innovation*, 11:390–398, 2018.
8 S. Shrivastava, S. Barat, S. Kausley, V. Kulkarni, and B. Rai. River digital twin for water quality prediction. In *2024 Winter Simulation Conference (WSC)*, pages 608–619. IEEE, December 2024.
9 S. Chen, J. Huang, P. Wang, X. Tang, and Z. Zhang. A coupled model to improve river water quality prediction towards addressing non-stationarity and data limitation. *Water Research*, 248:120895, 2023.
10 Margaret Shanafield, Sarah A. Bourke, Margaret A. Zimmer, and Katie H. Costigan. An overview of the hydrology of non-perennial rivers and streams. *Wiley Interdisciplinary Reviews: Water*, 8(2):e1504, 2021.
11 J. Y. Ho, H. A. Afan, A. H. El-Shafie, S. B. Koting, N. S. Mohd, W. Z. B. Jaafar, H. Sai Lai, M. A. Malek, A. N. Ahmed, W. H. M. W. Mohtar, A. Elshorbagy, and A. El-Shafie. Towards a time and cost effective approach to water quality index class prediction. *Journal of Hydrology*, 575:148–165, 2019. doi: 10.1016/j.jhydrol.2019.05.01.
12 K. P. Singh, A. Basant, A. Malik, and G. Jain. Artificial neural network modeling of the river water quality—a case study. *Ecological Modelling*, 220(6):888–895, 2009.
13 Larry R. Medsker and Lakhmi Jain. Recurrent neural networks. *Design and Applications*, 5(64-67):2, 2001.
14 W. C. Leong, A. Bahadori, J. Zhang, and Z. Ahmad. Prediction of water quality index (WQI) using support vector machine (SVM) and least square-support vector machine (LS-SVM). *International Journal of River Basin Management*, 19(2):149–156, 2021.
15 A. Azad, H. Karami, S. Farzin, S. F. Mousavi, and O. Kisi. Modeling river water quality parameters using modified adaptive neuro fuzzy inference system. *Water Science and Engineering*, 12(1):45–54, 2019.
16 H. Jafari, T. Rajaee, and O. Kisi. Improved water quality prediction with hybrid wavelet-genetic programming model and Shannon entropy. *Natural Resources Research*, 29:3819–3840, 2020.
17 K. Khosravi, A. A. Farooque, M. Karbasi, M. Ali, S. Heddam, A. Faghfouri, and S. Abolfathi. Enhanced water quality prediction model using advanced hybridized resampling alternating tree-based and deep learning algorithms. *Environmental Science and Pollution Research*, 32:1–20. 2025.

18 A. Das. Water pollution and water quality assessment and application of criterion impact loss (CILOS), geographical information system (GIS), artificial neural network (ANN) and decision-learning technique in river water quality management: An experiment on the Mahanadi catchment, Odisha, India. *Desalination and Water Treatment*, 321:100969, 2025.

19 C. S. Ionescu, I. Opriş, D. E. G. Nistoran, and C. A. Baciu. Forecasting model for Danuberiver water temperature using artificial neural networks. *Hydrology*, 12(2), 21, 2025.

20 A. F. del Castillo, M. V. Garibay, D. Díaz-Vázquez, C. Yebra-Montes, L. E. Brown, A. Johnson, A. Garcia-Gonzalez, and M. S. Gradilla-Hernández. Improving river water quality prediction with hybrid machine learning and temporal analysis. *Ecological Informatics*, 82:102655, 2024.

21 E. Abushandi. Water quality assessment and forecasting along the liffey and andarax rivers by artificial neural network techniques toward sustainable water resources management. *Water*, 17(3):453, 2025.

22 S. B. H. S. Asadollah, A. Sharafati, D. Motta, and Z. M. Yaseen. River water quality index prediction anduncertainty analysis: A comparative study of machine learning models. *Journal of Environmental Chemical Engineering*, 9(1), 104599, 2021.

23 I. R. Warren and H. Bach. Mike 21: a modelling system for estuaries, coastal waters and seas. *Environmental Software*, 7(4):229–240, 1992.

24 T. T. Nguyen, I. Keupers, and P. Willems. Conceptual river water quality model with flexible model structure. *Environmental Modelling & Software*, 104:102–117, 2018.

25 R. Paliwal, P. Sharma, and A. Kansal. Water quality modelling of the river Yamuna (India) using QUAL2E-UNCAS. *Journal of Environmental Management*, 83(2): 131–144, 2007.

26 J. Sieber. *WEAP: Water Evaluation and Planning System*. Software, 2006.

27 A. Foehn, J. García Hernández, B. Roquier, J. Fluixá-Sanmartín, T. Brauchli, J. Paredes Arquiola, and G. De Cesare. RS MINERVE – User manual (v2.15). Centre de recherche sur l'environnement alpin (CREALP), 2020. https://www.crealp.ch/en/rs-minerve-software/.

28 P. W., Gassman, M. R., Reyes, C. H. Green, and J. G. Arnold. The soil and water assessment tool: historical development, applications, and future research directions. *Transactions of the ASABE*, 50(4), 1211–1250, 2007.

29 G. W. Brunner. HEC-RAS river analysis system. Hydraulic reference manual. Version 1.0., 1995.

30 T. Clark, V. Kulkarni, S. Barat, and B. Barn. ESL: an actor-based platform fordeveloping emergent behaviour organisation simulations. In *International Conference on Practical Applications of Agents and Multi-agent Systems,* pages 311–315, Cham: Springer International Publishing, 2017.

31 R. G. Sargent. Verification and validation of simulation models: an advanced tutorial. In *2020 Winter Simulation Conference (WSC)*, pages 16–29. IEEE, December 2020.

32 J. D. Valiantzas. Simplified versions for the Penman evaporation equation using routineweather data. *Journal of Hydrology*, 331(3-4), 690–702, 2006.

33 S. Kostić, M. Stojković, S. Prohaska, and N. Vasović. Modeling of river flow rate as a function of rainfall and temperature using response surface methodology based on historical time series. *Journal of Hydroinformatics*, 18(4), 651–665, 2016.

34 A. Mesdaghinia, S. Nasseri, A.H. Mahvi, H. R. Tashauoei, and M. Hadi. The estimation of per capita loadings of domestic wastewater in Tehran. *Journal of Environmental Health Science and Engineering*, 13(1), 25, 2015.

35 David Masad and Jackie L. Kazil. Mesa: an agent-based modeling framework. In *SciPy*, pages 51–58, 2015.

36 M. Singh, I. B. Singh, and G. Müller. Sediment characteristics and transportation dynamics of the Ganga River. *Geomorphology*, 86(1-2), 144–175, 2007.

37 I. B. Singh. The Ganga River. *Large Rivers: Geomorphology and Management, Second Edition*, 521–550, 2022.

38 Central Pollution Control Board. Pollution assessment: River Ganga. Ministry of Environment, Forest and Climate Change, Government of India, 2013. https://cpcb.nic.in/wqm/pollution-assessment-ganga-2013.pdf

39 National Mission for Clean Ganga. Ministry of Jal Shakti, Government of India, n.d. https://nmcg.nic.in/csr/biodrains.aspx.

17

Integrating Simulation and Real-world Models for Harmful Algal and Cyanobacterial Blooms Management

José L. Risco-Martín,[1] Roberto Domínguez,[1] Román Cárdenas,[2] Eva Besada-Portas,[1] and Segundo Esteban[1]*

[1] *Department of Computer Architecture and Automation, Universidad Complutense de Madrid, Madrid, Spain*
[2] *Department of Electronics Engineering, Universidad Politécnica de Madrid, Madrid, Spain*

*Corresponding author: jlrisco@ucm.es

17.1 Introduction and Related Work

Harmful algal and cyanobacterial blooms (HACB) pose significant threats to ecosystems, drinking water, and recreational water use. These blooms occur when cyanobacteria, a group of photosynthetic prokaryotes, grow massively under favorable conditions, often exacerbated by global changes such as warming and eutrophication. The presence of cyanobacteria in water bodies can lead to the production of cyanotoxins, which are harmful to both aquatic life and humans. These toxins can cause severe health issues, including renal, hepatic, and neurological damage, and have been linked to an increased risk of primary liver cancer in some regions [1, 2]. Additionally, cyanobacterial blooms can alter the organoleptic properties of water, affecting its color, odor, and taste and can lead to the accumulation of scum on water surfaces, which disrupts both ecosystem and recreational activities [3]. This is a widespread environmental challenge, with significant, recurring blooms documented in major water systems globally, such as the River Murray in South Australia, which poses an ongoing threat to water security [4]. The complexity of these blooms, driven by factors such as biological growth, competition, and environmental conditions, makes them difficult to predict and manage effectively [5].

Traditional methods for monitoring HACB include (1) manually collecting water samples for laboratory analysis to identify toxins and (2) using multiparametric probes on buoys near water treatment facilities or in recreational areas [6, 7]. Current technological approaches involve (3) analyzing multispectral satellite images of water surfaces [8–10]; (4) employing predictive models to forecast HACB occurrences and severity [11]; and (5) implementing early warning systems to inform managers, authorities, and the public about water quality [12]. Each method has its limitations: (1) is restricted to occasional sampling at specific locations; (2) only detects conditions conducive to HACB at the buoy; (3) is ineffective under poor visibility; (4) lacks precision in predicting HACB location and status; and (5) is not well-suited for HACB. Moreover, these methods do not provide detailed information on water species, and their time resolution is insufficient for timely management responses. Consequently, only reactive measures, such as restricting water use, sourcing water from different levels, or employing costly treatments [13], are feasible. This situation underscores the need for a proactive, efficient, cost-effective, timely, and comprehensive solution for HACB monitoring, early warning, and management. Such a solution should integrate the strengths and address the limitations of existing methods while taking advantage of new emerging technologies in related sciences like biology, ecology, and toxicology.

To efficiently manage HACB, we have recently proposed a novel Model-based systems engineering (MBSE) driven paradigm. This approach involves modeling the entire management infrastructure, including sensors, water bodies, operators, unmanned surface vehicles (USVs), fog services, and cloud services, to create a comprehensive framework for proactive HACB management. This paradigm is detailed in the IA-GES-BLOOM project [14], which aims to develop an integral system for the alert and management of HACB in continental waters. The project emphasizes the use of internet of things (IoT) technologies, artificial intelligence (AI), and modeling and simulation (M&S) techniques to enhance the characterization and prediction of HACB, thereby enabling timely and effective management responses. Additionally, the SMART-BLOOMS project [15] extends this concept by incorporating digital twins (DTs) and edge computing to further optimize the deployment and operation of the management system, as discussed in our recent work [16]. These efforts collectively aim to transform the traditional reactive approaches into a proactive, efficient, and sustainable management strategy for HACB.

To address the challenges and limitations of existing methods for managing HACB, we have developed DEVS-BLOOM, a comprehensive tool that embodies the concepts and approaches discussed earlier [17]. DEVS-BLOOM is designed to facilitate real-time monitoring and management of these blooms by integrating MBSE with cyber-physical system (CPS) infrastructure through an IoT architecture. This tool utilizes the discrete event system specification (DEVS) formalism to provide a robust framework for simulating and controlling the dynamics of harmful blooms. It offers a scalable and efficient solution for predicting bloom occurrences and optimizing management strategies.

The next phase in the development of DEVS-BLOOM involves demonstrating the feasibility of deploying the model in a real-world environment. This transition is expected to be seamless through the use of the capabilities of MBSE and the DEVS formalism. In this chapter, we validate this hypothesis by executing a controlled deployment of the DEVS-BLOOM model in a real-world scenario. This involves utilizing a small-scale aquatic environment and an USV equipped with sensors and actuators, designed to mimic the operational conditions of larger systems. The deployment is facilitated by our DEVS-RT framework, which ensures real-time interaction and data processing. This setup allows us to test the integration of simulated models with physical components, thereby providing a comprehensive evaluation of the system's performance in monitoring and managing HACB.

The chapter is organized as follows. Section 17.2 provides an overview of the foundational technologies that underpin our approach, with a focus on the DEVS formalism and the real-time simulation capabilities of the xDEVS framework. In Section 17.3, we detail the integration of models and the deployment of the system, highlighting the steps taken to ensure smooth operation in a real-world environment. Section 17.4 presents the results and validation of the system's performance, demonstrating its effectiveness in monitoring and managing HACB. Finally, Section 17.5 discusses the significance of our findings and outlines potential directions for future research.

17.2 Foundational Technologies

17.2.1 The DEVS Formalism

In DEVS, systems are modeled using atomic and coupled models [18]. Atomic models describe autonomous behavior through state transitions and reactions to external events. They are formally defined by the following tuple:

$$\mathrm{AM} = \langle X, Y, S, \delta_{\mathrm{int}}, \delta_{\mathrm{ext}}, \delta_{\mathrm{con}}, \lambda, ta \rangle , \tag{17.1}$$

where X and Y are input and output sets, S is the set of states, and ta is the time advance function. The functions δ_{int}, δ_{ext}, and δ_{con} govern state transitions, while λ defines output events.

Alternatively, coupled models represent systems as interconnected DEVS submodels, defined by:

$$\mathrm{CM} = \langle X, Y, C, \mathrm{EIC}, \mathrm{EOC}, \mathrm{IC} \rangle , \tag{17.2}$$

where C is the set of components, and EIC, EOC, and IC define the coupling relations between inputs, outputs, and internal connections of the components, respectively.

A common approach in DEVS modeling is to use ports to classify input and output events [19]. The input set X is divided into subsets, each assigned to a port, forming the input port set P_X. Each input event $x \in X$ is expressed as a tuple $\langle p, x_p \rangle$, where $p \in P_X$ is the port and x_p is the event value:

$$X = \{\langle p, x_p \rangle, p \in P_X\}. \tag{17.3}$$

Similarly, the input set X_p for a port p is defined as the set of values for events assigned to that port:

$$X_p = \{x_p | \langle p, x_p \rangle \in X\}. \tag{17.4}$$

This scheme is also applied to the output set Y and facilitates the definition of coupled models by structuring event propagation through port couplings.

17.2.2 Real-time DEVS Simulation in xDEVS

The xDEVS framework [20] has been updated to support real-time (RT) simulation, allowing DEVS models to work with external systems while keeping in sync with real-world time. This improvement is important for applications that need to match physical processes, like hardware-in-the-loop (HIL) simulations and DTs. Figure 17.1 shows a comparison between virtual-time (VT) and RT simulations using the xDEVS engine.

In the RT simulation setup shown in Figure 17.1b, the xDEVS framework adds three main parts: the RT coordinator, the input handler, and the output handler. These parts work together to control event flow and keep the simulation within RT limits. The **RT coordinator** is an improved version of the usual DEVS root

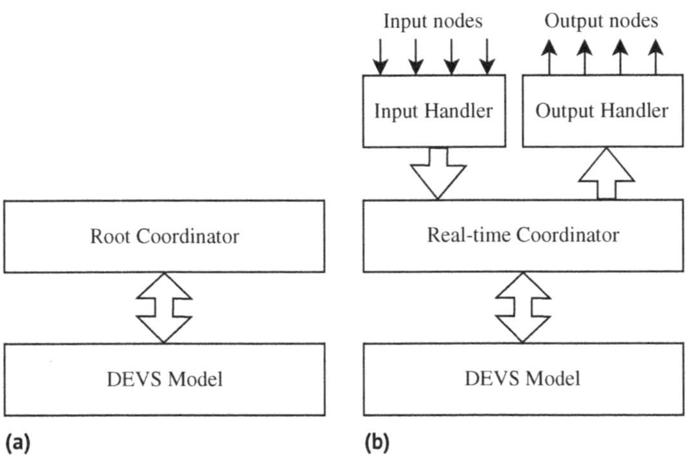

Figure 17.1 Comparison between (a) VT simulation and (b) RT simulation on xDEVS.

coordinator from Figure 17.1a, made to handle RT tasks. It connects the DEVS models with the outside world and runs the DEVS state transitions while working with the input and output handlers. Its job is to match the simulation's virtual time with real-world time, making sure they stay in sync.

The input and output handlers help DEVS models communicate with external systems. The **input handler** takes events from one or more external sources and sends them into the simulation. In xDEVS, input events are added using a multiple producer single consumer (MPSC) queue, where the input handler takes events from the queue while external sources add them. On the other hand, the **output handler** collects events from the simulation and sends them to external systems. In xDEVS, output events use a broadcast queue, where the output handler sends events, and output nodes receive them and pass them on to external systems.

Figure 17.2 illustrates the RT simulation process, showing how the RT coordinator works with the input and output handlers to handle events while keeping

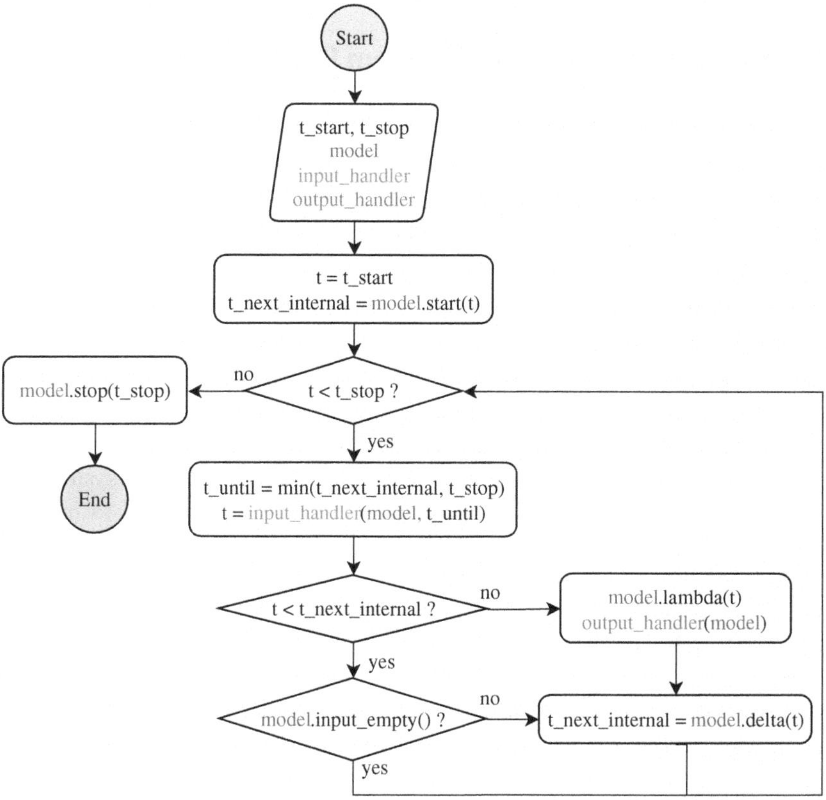

Figure 17.2 Workflow of RT simulation in xDEVS.

the simulation time aligned with the wall-clock time. The output handler runs every time after the output function *lambda*. Also, when the state transition function *delta* is executed, the RT coordinator uses the input handler to wait for the required wall-clock time for any external input event.

In xDEVS, RT simulations run on a completely isolated environment, and the only way to interact with the model under simulation is through the input and output handlers. These handlers can act as gateways between the models and numerous external entities, such as Transmission Control Protocol (TCP) sockets, Message Queuing Telemetry Transport (MQTT) brokers, or even hardware peripherals connected to the processing system running the simulation (e.g., general purpose input/output [GPIO] or inter-integrated circuit [I2C] master nodes). This approach has proven to solve previous issues with other proposals (e.g., costly modifications to the original DEVS simulation algorithm or requiring modification of the DEVS model under study), and it is being adopted in other DEVS-compliant simulators [21].

17.3 Model Integration and System Deployment

17.3.1 Overview of DEVS-BLOOM

DEVS-BLOOM is a comprehensive framework designed for the real-time monitoring and management of HACB. Its primary purpose is to provide a robust system that integrates both simulated and real-world models to enhance the detection and management of HACB in aquatic environments. The framework is built upon the DEVS formalism, which allows for modular and hierarchical modeling of complex systems [17]. The core components of DEVS-BLOOM include USVs equipped with various sensors to collect data from water bodies, simulated and real sensors used to gather data on water quality parameters such as temperature, oxygen levels, and nutrient concentrations, inference models to predict the occurrence and movement of HACB based on the data collected by them, and control systems to manage the operation of USVs and ensure that they navigate effectively to areas of interest.

A significant aspect of DEVS-BLOOM is its ability to integrate simulated models with real-world components. This integration is achieved through the use of the DEVS-RT architecture described in Section 17.2, and the implementation of DTs, which serve as virtual representations of physical entities. DTs facilitate the seamless transition between simulation and real-world deployment, allowing for the testing and validation of models in a controlled environment before actual implementation. The integration of simulated and real models is crucial for several reasons: (1) validation and testing of different strategies and models in a risk-free environment; (2) scalability and flexibility as the framework can easily scale to include additional sensors or USVs, and it can adapt to different water bodies; and (3) real-time decision-making, since the integration ensures that real-time data from sensors can be used to make informed decisions quickly, crucial for the timely management of HACB, where delays can lead to significant ecological and economic impacts.

17.3.2 System Architecture

Figure 17.3 shows the architecture of DEVS-BLOOM. It is structured into three primary layers: Edge, Fog, and Cloud. Each layer plays a distinct role in the system, ensuring efficient data processing, analysis, and decision-making for the management of HACB.

The Edge layer is the closest to the physical environment and includes all devices directly interacting with the water body. This layer comprises sensors and

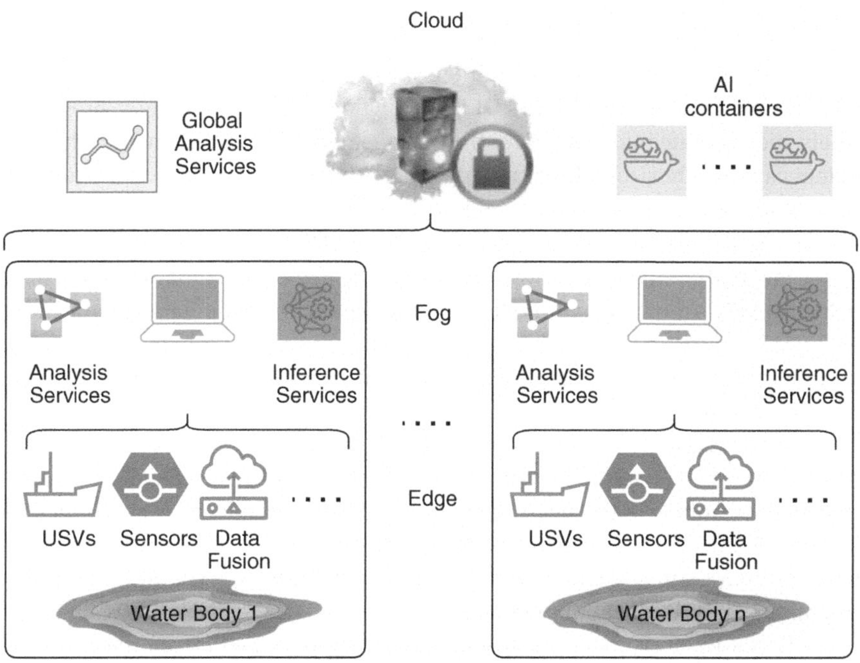

Figure 17.3 DEVS-BLOOM IoT architecture.

USVs equipped with various instruments to collect real-time data on water quality parameters such as temperature, oxygen levels, and nutrient concentrations. The primary function of the Edge layer is to gather data and perform initial processing to reduce the amount of data sent to higher layers. The Fog layer acts as an intermediary between the Edge and Cloud layers. It includes devices with more substantial computing power than those at the Edge, such as local servers or workstations at ground control stations (GCSs). The Fog layer is responsible for aggregating data from multiple Edge devices, performing more complex data processing, and running predictive models to infer the likelihood and potential movement of HACB. This layer supports real-time decision-making by providing processed data and insights to operators and domain experts, who can then plan USV trajectories and other management actions. The Fog layer also ensures that only relevant and processed data is sent to the Cloud layer, optimizing bandwidth and storage usage. Finally, the Cloud layer is the topmost layer, providing extensive computational resources and storage capabilities. It is responsible for long-term data storage, complex data analytics, and machine learning tasks that require significant processing power. The Cloud layer supports strategic decision-making by providing a comprehensive view of multiple water bodies and enabling authorities to compare reports and make high-level decisions. It also facilitates the training and updating of predictive models used in the Fog layer, ensuring that the system adapts to new data and conditions over time.

This layered architecture allows DEVS-BLOOM to operate efficiently, providing timely and accurate information for the management of HACB. It ensures that the system can scale horizontally by adding more sensors and USVs and vertically by integrating more complex models and analytics.

Figure 17.4 illustrates a segment of the DEVS-BLOOM model. It integrates both virtual and real components, allowing it to operate in different modes: virtual, real, or hybrid [17]. In the virtual mode, the model uses synthetic data generated by the EEMS software [22], which provides comprehensive water body simulations, including all necessary measurements and algal or cyanobacterial concentrations for validation purposes.

At the end, each sensor within the DEVS-BLOOM framework is implemented as a DT. These DTs can either simulate the behavior of real sensors or directly process data from actual sensors using our DEVS-RT architecture, as detailed in Section 17.2. This dual capability allows the system to seamlessly transition between simulation and real-world operations, facilitating the testing and calibration of sensors and other components.

The DEVS-BLOOM model is structured to support scalability and flexibility. It can be expanded to include additional sensors or USVs and can adapt to different water bodies. This is achieved through the modular nature of the DEVS formalism, which allows for the hierarchical and compositional modeling of complex

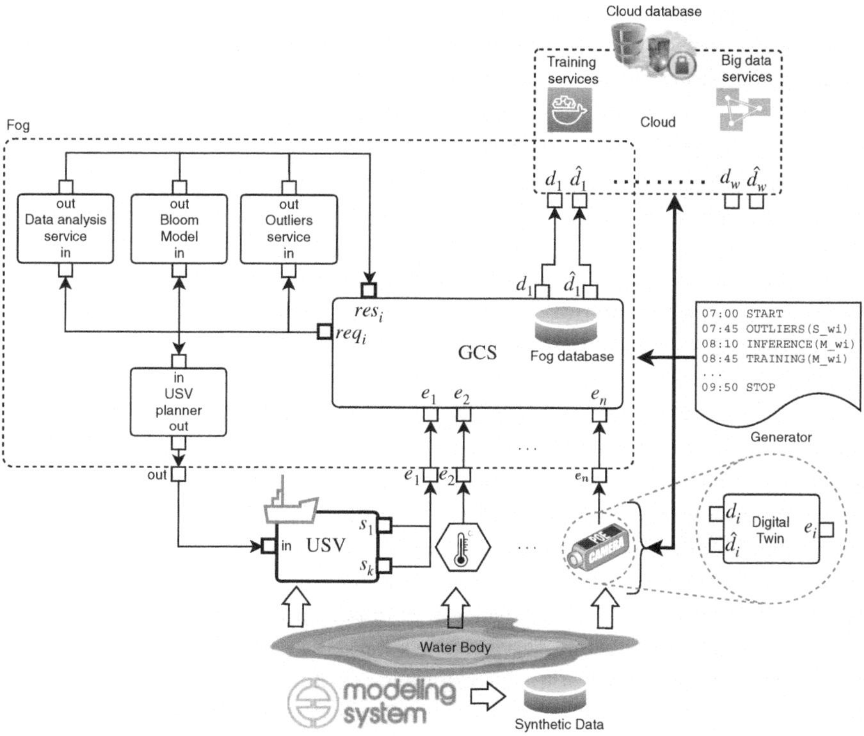

Figure 17.4 DEVS-BLOOM model.

systems [17]. The public implementation of DEVS-BLOOM is available for further exploration and use [23].

17.3.3 Model Adaptation

The DEVS-BLOOM virtual model has undergone extensive verification across various simulation scenarios, as documented in previous studies [16, 17, 24]. These simulations have been conducted on multiple (virtual) lakes, each with distinct conditions, demonstrating the model's robustness and adaptability. The results have shown that the three-layer architecture of DEVS-BLOOM functions effectively, with successful implementation of alarms, dashboards, and other critical features. A significant advantage of this approach is that the simulation model can directly serve as the deployment model. This eliminates the need to re-implement components such as USV planning software and real-time analysis tools, which are already integrated into the DEVS-BLOOM framework.

The next step involves deploying the model in a controlled real-world scenario to validate the proposed conceptual model. This deployment will test the model's ability to transition from a simulated environment to a real-world application seamlessly. In the following, we detail the straightforward adaptation of the DEVS-BLOOM model for real-world testing.

The DEVS-BLOOM model has been adapted as shown in Figure 17.5 to validate bloom detection and USV navigation and control. The deployment involves a scaled-down, 3D-printed boat that simulates a USV and is equipped to follow bloom predictions, collect data, and trigger alarms if necessary. The test will take place in a small lake, with two human operators on-site. One operator will manage a laptop, while the other oversees the 3D-printed USV, which contains a Raspberry Pi for processing.

In this setup, DEVS-BLOOM operates as a distributed DEVS model. The cloud-coupled model (in labeled "Cloud" in Figure 17.5) runs on the laptop, while the Raspberry Pi handles the ground control station (GCS) coupled model. The Raspberry Pi also runs the *Generator model* to start and stop the test, the *Bloom* atomic model for bloom prediction, as well as the *USV model* and *USV planner* for navigation and control (located in the "Fog" coupled model in Figure 17.5). Other atomic models have been disconnected for simplicity, like the Data Analysis service or the Outliers Detection service. The small lake environment has been pre-simulated with the EEMS software to provide synthetic data and replace some sensors and provide validation over the prediction. The EEMS simulation results file is stored in the Raspberry Pi, ensuring that the system can operate independently in the field.

In DEVS-BLOOM, there are two groups of sensors. The first group, the bottom block of sensors in Figure 17.5, feeds the bloom model, and the second group is used to rule the navigation system, the middle block of sensors in Figure 17.5.

The first group acts as DTs that process data from the EEMS simulation and provide information to the bloom model. These sensors represent key environmental parameters relevant to the bloom model. Each sensor model has specific characteristics, including measurement range, precision, response delay, and noise properties, which contribute to the accuracy and reliability of the simulation:

- ALG – Bloom Detector Sensor (mg/L): This sensor provides simulated bloom detection data for validation purposes. It measures algal concentration in the water column.
- BTH – Bathymetry Sensor (m): The bathymetry sensor captures underwater topography data, representing the depth profile of the water body (in meters).
- DOX – Dissolved Oxygen Sensor (mg/L): This sensor monitors dissolved oxygen levels in the water, which is critical for aquatic life.

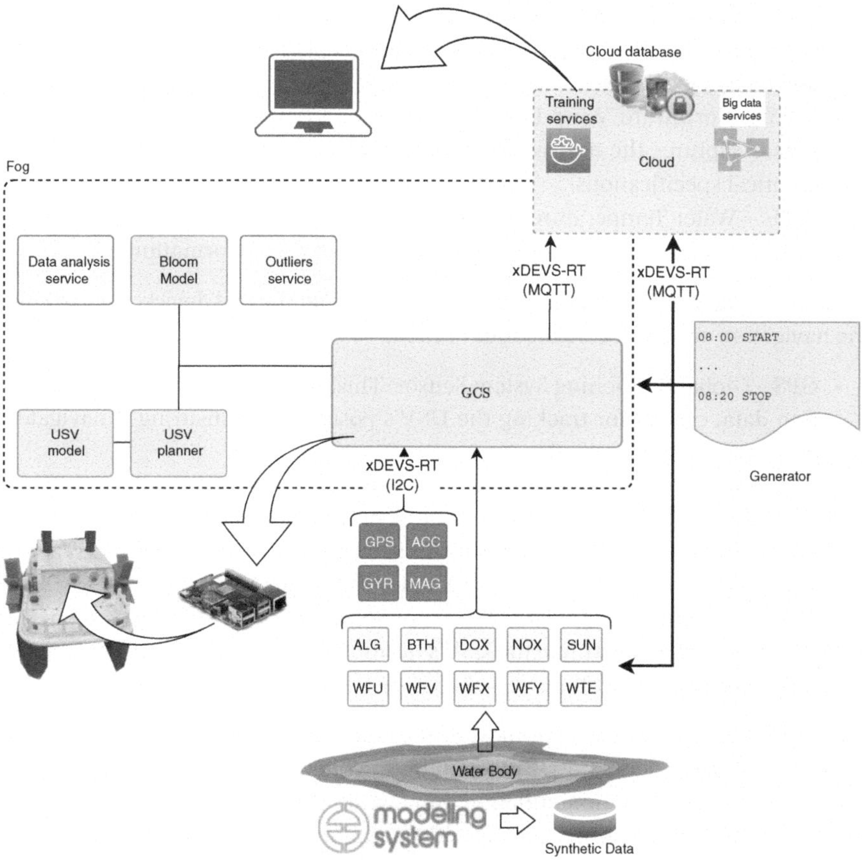

Figure 17.5 DEVS-BLOOM deployed model.

- NOX – Nitrogen Concentration Sensor (mg/L): The nitrogen sensor measures nitrogen compounds in the water, essential for nutrient cycling analysis.
- SUN – Solar Radiation Sensor (normalized to 1): This sensor records solar radiation, which influences photosynthetic activity and water temperature.
- WFU – Eastward Water Flow Sensor (m/s): This sensor measures the eastward component (x-direction) of water velocity, which affects pollutant dispersion and ecosystem dynamics.
- WFV – Northward Water Flow Sensor (m/s): Similar to WFU, this sensor captures the northward component (y-direction) of water flow, with identical properties.

- WFX – Eastward Wind Flow Sensor (m/s): This sensor monitors the eastward component (x-direction) of wind flow, influencing surface water movement and energy exchange.
- WFY – Northward Wind Flow Sensor (m/s): Complementary to WFX, this sensor captures the northward component (y-direction) of wind flow, with identical specifications.
- WTE – Water Temperature Sensor (°C): This sensor records water temperature, a crucial factor in biological processes and bloom formation.

In the second group, sensors act as propagators from the real device, used to rule the navigation system. There are four of them:

- GPS – Global Positioning System Sensor: This sensor provides real-time location data, crucial for tracking the USV's position and ensuring it navigates accurately to the target areas.
- ACC – Accelerometer: This sensor measures the acceleration of the USV in three dimensions, helping to monitor its movement dynamics and stability.
- GYR – Gyroscope: This sensor measures the angular velocity of the USV, providing data on its orientation and rotational movements.
- MAG – Magnetometer: This sensor measures the magnetic field intensity, which is used to determine the USV's heading relative to the Earth's magnetic field, aiding in navigation.

In the DEVS-BLOOM model, most of the connections depicted in Figure 17.5 are managed using standard DEVS couplings, which handle the communication between models within the simulation framework. However, for real-time deployment, certain connections require specific communication protocols to facilitate interaction with external systems.

The connection between the GCS and the Generator models with the Cloud model is established using the MQTT protocol. MQTT is a lightweight messaging protocol designed for low-bandwidth, high-latency, or unreliable networks, making it ideal for IoT applications. It operates on a publish-subscribe model, where the GCS and Generator models publish messages to specific topics, and the Cloud model subscribes to these topics to receive updates.

For the connection between real sensors and their corresponding atomic models, the I2C protocol is used. I2C is a synchronous, multi-master, multi-slave, packet-switched, single-ended, serial communication bus widely used for attaching lower-speed peripheral devices to processors and microcontrollers. In this setup, the real sensors act as I2C slaves, providing data to the atomic models, which function as I2C masters, requesting and processing the sensor data. The xDEVS-RT implementation, as described in Section 17.2, ensures that these communication protocols are seamlessly integrated into the DEVS-BLOOM framework.

17.3.4 Deployment Strategy

Deploying DEVS-BLOOM in a real-world environment involves several critical steps to ensure the system operates effectively and efficiently. The deployment process begins with a thorough assessment of the target aquatic environment to understand its specific characteristics and constraints. This includes evaluating the size, depth, and water quality parameters of the water body, as well as identifying potential sources of interference, such as nearby industrial activities or natural obstacles.

The next step involves the physical installation of the USVs and associated sensors. The USVs must be equipped with the necessary hardware, including global positioning system (GPS), accelerometers, gyroscopes, and magnetometers, to ensure accurate navigation and data collection. The sensors must be calibrated to account for local environmental conditions, such as variations in magnetic fields or water salinity, which could affect sensor accuracy.

Once the hardware is in place, the DEVS-BLOOM software must be configured to integrate with the physical components. This involves setting up communication protocols, such as MQTT and I2C, to facilitate data exchange between the USVs, sensors, and the central control system. The software must be tailored to the specific requirements of the deployment site, including adjusting the parameters of the inference models to reflect local conditions.

A critical aspect of the deployment is the establishment of a reliable communication network. This network must support real-time data transmission between the USVs, the GCS, and the cloud infrastructure. The network should be robust enough to handle potential disruptions, such as signal loss due to environmental factors or hardware malfunctions. In our case, we have used Wi-Fi.

During deployment, it is essential to conduct a series of controlled test runs to validate the system's performance. These tests should cover various scenarios, including different weather conditions and water quality states, to ensure the system can adapt to changing environments. The test runs also provide an opportunity to fine-tune the system's parameters and address any issues that arise.

Another important consideration is ensuring the system's resilience to external factors, such as power outages or hardware failures. To address this, the deployment strategy should include contingency plans, such as backup power supplies and redundant communication links, to maintain system functionality in the event of a failure. In our case, we implemented a remote control system using an Xbox controller to manually operate the boat if needed, bypassing the DEVS-BLOOM framework. This approach was added as a precautionary measure and is not part of the core DEVS-BLOOM development.

17.4 Results and System Validation

In this section, we present the results and validation of the DEVS-BLOOM system, focusing on its ability to integrate simulated and real models effectively. The primary goal was to validate the system's capability to operate in a real-world environment while maintaining the integrity of the simulated models.

The DEVS-BLOOM system was deployed in a controlled aquatic environment at La Alhóndiga Lake in Getafe, Madrid, Spain. This deployment aimed to test its real-time operation and integration of real sensor data. A small-scale USV equipped with sensors for GPS, acceleration, angular velocity, and magnetic field intensity was used. These sensors provided real-time data to the DEVS-BLOOM system, enabling accurate monitoring and management of HACB.

The mission involved the USV following the bloom predicted by the bloom predictive model and validating the prediction with the measurements stored in the EEMS simulation. The objective was to assess the accuracy of the predictive loop and the effectiveness of the navigation system.

Figure 17.6 shows the USV actively navigating toward the predicted bloom location, which is situated close to the shoreline. The USV is equipped with a protective casing that houses the Raspberry Pi and the necessary sensors, ensuring they remain secure and functional during the operation.

In the field test, the USV demonstrated its ability to navigate toward predicted bloom locations effectively. The process begins with the USV reaching the current bloom location, as shown in Figure 17.7a. Upon arrival, the USV validates the bloom's presence against the EEMS simulation data, which helps confirm whether the prediction was accurate (true positive) or not (false positive). Once validated, the bloom predictive model generates the next target point, as illustrated in Figure 17.7b.

Figure 17.6 USV navigating toward the predicted bloom location.

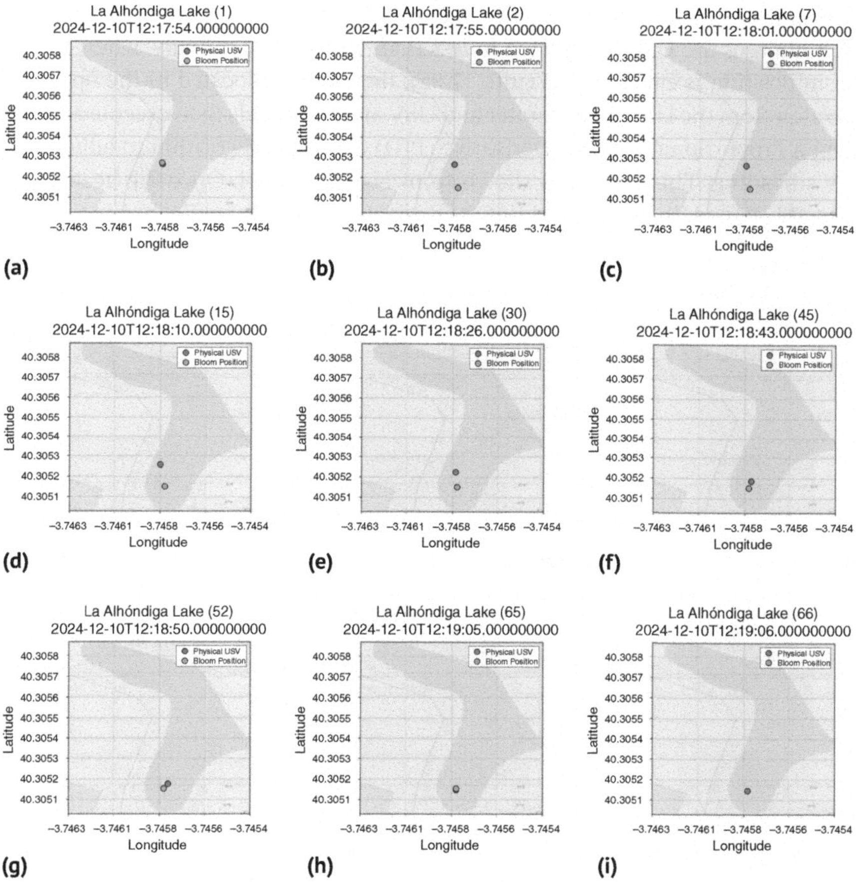

Figure 17.7 USV navigating toward the predicted bloom locations. (a) USV at the bloom; (b) New bloom point; (c–h) USV approximating; (i) USV at the bloom.

The USV then navigates toward this new target point, as depicted in Figure 17.7c–h. The navigation process involves continuous adjustments to ensure the USV remains on course toward the predicted location. Upon reaching the target, as shown in Figure 17.7i, the USV again validates the bloom's presence. If the prediction is missed or the USV takes too long to reach the bloom, the predictive model recalibrates and generates a new target point to ensure the USV can continue its monitoring mission effectively.

This iterative process of prediction, navigation, and validation allows the USV to adapt dynamically to the changing conditions of the aquatic environment, ensuring accurate and timely monitoring of harmful algal blooms.

Continuing with maneuverability, Figure 17.8 illustrates how the heading or orientation of the USV is adjusted in less than 15 seconds. Initially, when a new bloom position is calculated (Figure 17.8a), the bloom is located in the opposite direction from the USV. The Planner and USV atomic models then execute actions using a Proportional Integral Derivative (PID) controller algorithm to adjust the USV's heading. This process is shown from Figure 17.8b–i. Once the heading is correctly aligned, the USV continues its navigation toward the bloom.

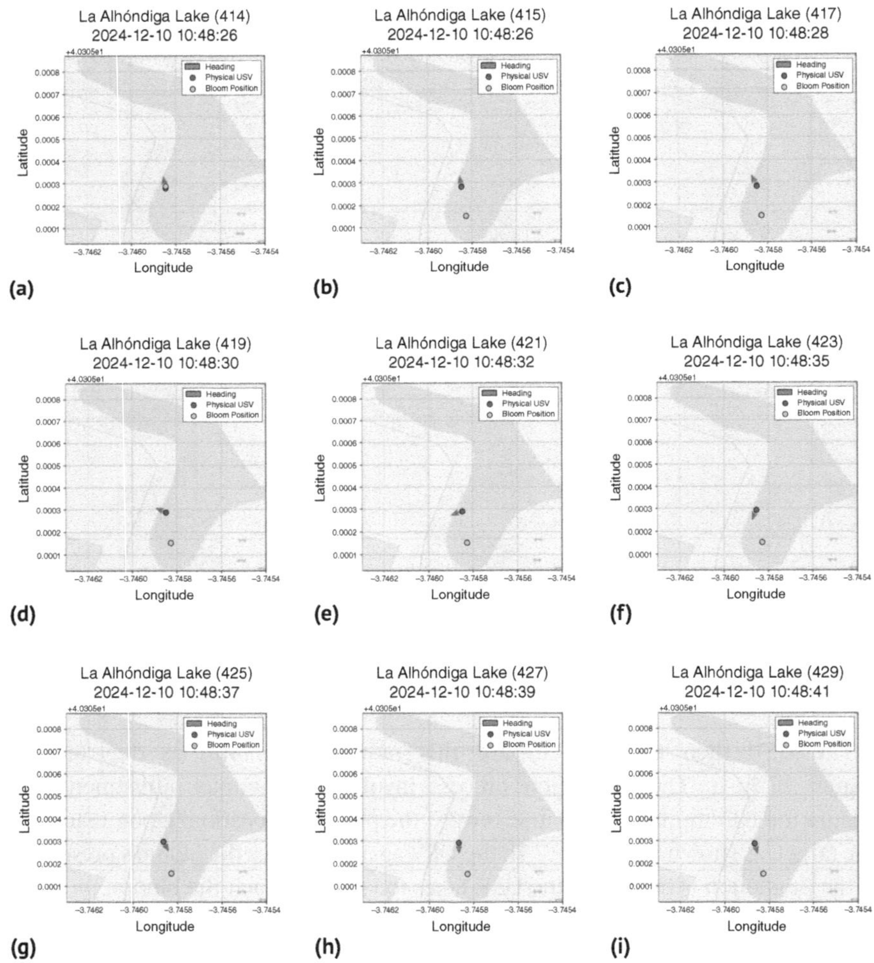

Figure 17.8 USV's reorientation process. (a) New bloom location; (b) Bloom at opposite direction; (c–h) Calibrating...; (i) USV aligned.

This data allows us to easily validate the controller algorithm used or compare it with others in a straightforward manner by simply replacing the controller algorithm in the DEVS-BLOOM model.

In Figure 17.9, we present an overlay of all the positions recorded by the USV (depicted as circles) alongside the predicted bloom locations (depicted as triangles). This visualization clearly demonstrates the effectiveness of the navigation system, as the USV consistently follows the predicted bloom path without much deviation. The close alignment between the USV's path and the predicted bloom locations validates the accuracy and reliability of the navigation system in this controlled scenario.

In the final validation of the DEVS-BLOOM predictive system, we compared the measured and inferred algae density against the original data.[1] The results are shown in Figure 17.10. The lines marked with diamonds and squares represent the simulated algae concentration (in μg/L) during the virtual simulation and the

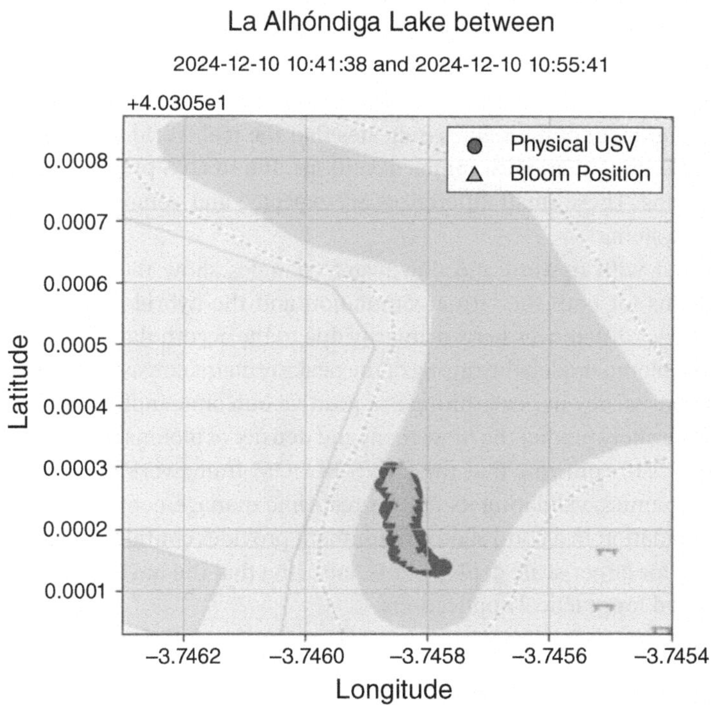

Figure 17.9 Path followed by the USV during the field test, illustrating the system's navigation capabilities.

1 EEMS allows the simulation of both harmful algal and cyanobacterial concentrations; in this case, we used the results for harmful algal concentrations.

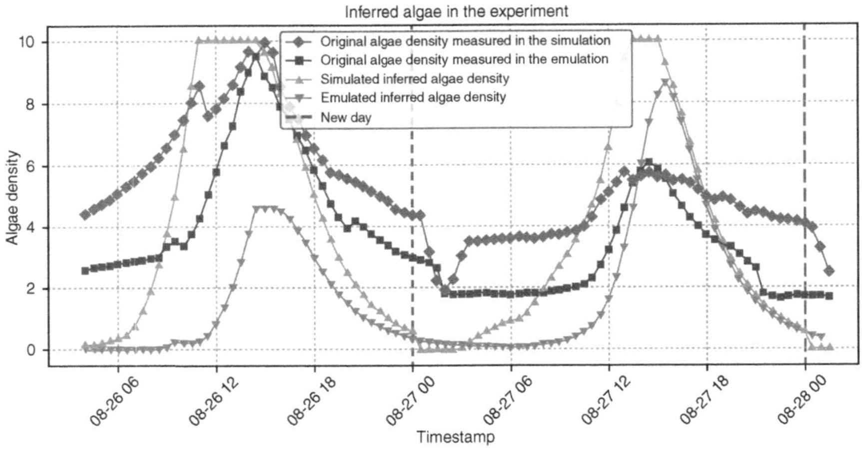

Figure 17.10 Comparison of algae density in virtual and hybrid experiments, showing the real-time system's accuracy.

hybrid emulation (conducted in the lake with the actual USV and sensors), respectively. The close alignment of these lines indicates that the real-world test closely mirrors the virtual simulation, with minor deviations due to GPS precision and real-world conditions. These small differences are expected and demonstrate the robustness of the system.

The lines marked with upward and downward triangles show the predicted algae concentrations for both the virtual simulation and the hybrid emulation. While there are larger differences here, primarily due to the accumulation of precision errors in the bloom model, the timing of the peaks remains consistent across both lines. This consistency in peak timing is a positive outcome, indicating that the system can accurately predict the movement and density of blooms over time.

Overall, Figure 17.10 confirms that the DEVS-BLOOM framework effectively predicts bloom dynamics, validating its use for real-time management of HACB. This successful validation in a controlled environment provides confidence in the system's readiness for larger-scale deployments, ensuring that the navigation system is well-prepared for practical applications.

17.5 Conclusion and Future Work

In this chapter, we presented DEVS-BLOOM, a comprehensive framework designed for the real-time monitoring and management of HACB using the DEVS formalism. The system effectively integrates both simulated and real-world models, allowing for seamless transitions between virtual simulations and real-world deployments. Our deployment strategy, which includes a three-layer architecture of Edge, Fog, and Cloud, has proven effective in managing data processing and decision-making tasks efficiently.

The real-world deployment of DEVS-BLOOM in a controlled environment demonstrated the system's capability to accurately predict and manage blooms. The integration of DTs and real-time simulation using the xDEVS framework allowed for effective navigation and control of USVs, validating the system's design and functionality.

For future work, we aim to scale the DEVS-BLOOM system for larger deployments, which will involve integrating more simultaneous USVs to enhance its capabilities. We also plan to explore the incorporation of more advanced machine learning techniques to improve the accuracy of bloom predictions. Additionally, we will focus on enhancing the system's resilience to environmental and operational challenges, ensuring it can operate effectively in diverse and dynamic aquatic environments. We hope to introduce a new paradigm in the proactive management of HACB and contribute to the sustainability of aquatic ecosystems.

Acknowledgments

This work has been supported by the Research Projects SMART-BLOOMS (TED2021-130123B-I00) by MCIN/AEI/10.13039/501100011033 and the European Union NextGenerationEU/PRTR, and INSERTION (PID2021-127648OB-C33) by the Knowledge Generation program of the Spanish Ministry of Science and Innovation.

References

1 Igor Mrdjen, Jiyoung Lee, Christopher M. Weghorst, and Thomas J. Knobloch. Impact of cyanotoxin ingestion on liver cancer development using an at-risk two-staged model of mouse hepatocarcinogenesis. *Toxins*, 14(7):484, 2022. https://doi.org/10.3390/toxins14070484.

2 Shen Gu, Mingxuemei Jiang, and Bo Zhang. Microcystin-LR in primary liver cancers: An overview. *Toxins*, 14(10):715, 2022. https://doi.org/10.3390/toxins14100715.

3 Susan B. Watson, Paul Monis, Peter Baker, and Steven Giglio. Biochemistry and genetics of taste- and odor-producing cyanobacteria. *Harmful Algae*, 54:112–127, 2016. https://doi.org/10.1016/j.hal.2015.11.008.

4 Sara G. Beavis, Vanessa N. L. Wong, Luke M. Mosley, Darren S. Baldwin, James O. Latimer, Patrick Lane, and Aparna Lal. Water quality risks in the murray-darling basin. *Australasian Journal of Water Resources*, 27(1):85–102, 2023. https://doi.org/10.1080/13241583.2022.2163475.

5 Bastiaan W. Ibelings, Rainer Kurmayer, Sandra M. F. O. Azevedo, Ingrid Chorus Susanna A. Wood, and Martin Welker. Understanding the occurrence of cyanobacteria and cyanotoxins. In *Toxic Cyanobacteria in Water*, 2 edition. CRC Press, 2021.

6 Todd R. Miller, Wilson Tarpey, Jeffrey Nuese, and Matthew Smith. Real-time monitoring of cyanobacterial harmful algal blooms with the panther buoy. *ACS ES&T Water*, 2(6):1099–1110, 2022. https://doi.org/10.1021/acsestwater.2c00072.

7 Justin D. Chaffin, Douglas D. Kane, and Alex Johnson. Effectiveness of a fixed-depth sensor deployed from a buoy to estimate water-column cyanobacterial biomass depends on wind speed. *Journal of Environmental Sciences*, 93:23–29, 2020. https://doi.org/10.1016/j.jes.2020.03.003.

8 Sachidananda Mishra, Richard P. Stumpf, Blake A. Schaeffer, P. Jeremy Werdell, Keith A. Loftin, and Andrew Meredith. Measurement of cyanobacterial bloom magnitude using satellite remote sensing. *Scientific Reports*, 9(1):18310, 2019. https://doi.org/10.1038/s41598-019-54453-y.

9 Megan M. Coffer, Blake A. Schaeffer, Wilson B. Salls, Erin Urquhart, Keith A. Loftin, Richard P. Stumpf, P. Jeremy Werdell, and John A. Darling. Satellite remote sensing to assess cyanobacterial bloom frequency across the United States at multiple spatial scales. *Ecological Indicators*, 128:1–107822, 2021. https://doi.org/10.1016/j.ecolind.2021.107822.

10 Michael Sayers, Gary L. Fahnenstiel, Robert A. Shuchman, and Matthew Whitley. Cyanobacteria blooms in three eutrophic basins of the Great Lakes: a comparative analysis using satellite remote sensing. *International Journal of Remote Sensing*, 37(17):4148–4171, 2016. https://doi.org/10.1080/01431161.2016.1207265.

11 Benny Zuse Rousso, Edoardo Bertone, Rodney Stewart, and David P. Hamilton. A systematic literature review of forecasting and predictive models for cyanobacteria blooms in freshwater lakes. *Water Research*, 182:115959, 2020. https://doi.org/10.1016/j.watres.2020.115959.

12 Husein Almuhtaram, Faith A. Kibuye, Suraj Ajjampur, Caitlin M. Glover, Ron Hofmann, Virginie Gaget, Christine Owen, Eric C. Wert, and Arash Zamyadi. State of knowledge on early warning tools for cyanobacteria detection. *Ecological Indicators*, 133:108442, 2021. https://doi.org/10.1016/j.ecolind.2021.108442.

13 Ingrid Chorus and Matthias Zessner. Assessing and controlling the risk of cyanobacterial blooms: Nutrient loads from the catchment. In *Toxic Cyanobacteria in Water*, pages 433–503. CRC Press, 2021.

14 Eva Besada-Portas and Antonio Quesada. Towards a comprehensive system for the alert and management of cyanobacterial blooms in continental waters (IA-GES-BLOOM-CM). *Research Project*, 2021. Y2020/TCS-6420, funded by the Synergic program of the Comunidad Autónoma de Madrid.

15 Antonio Quesada and José L. Risco-Martín. Beyond the use of digital technologies in cyanobacteria blooms: Smart management of cyanobacteria exploiting digital twins and edge computing (SMART-BLOOMS). *Research Project*, 2022. TED2021-130123B-I00, funded by MCIN/AEI/10.13039/501100011033 and the European Union NextGenerationEU/PRTR.

16 Giordy A. Andrade, Segundo Esteban, José L. Risco-Martín, Jesús Chacón, and Eva Besada-Portas. Modeling- and simulation-driven methodology for the deployment of an inland water monitoring system. *Information*, 15(5):1–22, 2024. https://doi.org/10.3390/info15050267.

17 José L. Risco-Martín, Segundo Esteban, Jesús Chacón, Gonzalo Carazo-Barbero, Eva Besada-Portas, and José A. López-Orozco. Simulation-driven engineering for the management of harmful algal and cyanobacterial blooms. *SIMULATION*, 99(10):1041–1055, 2023. https://doi.org/10.1177/00375497231184246.

18 Bernard P. Zeigler, Alexandre Muzy, and Ernesto Kofman. *Theory of Modeling and Simulation: Discrete Event & Iterative System Computational Foundations*. Academic Press, 2018.

19 Román Cárdenas. Integrative modeling, simulation, and optimization techniques for efficient data-intensive applications in edge computing infrastructures. PhD thesis. Universidad Politécnica de Madrid: Carleton University, 2024.

20 José L. Risco-Martín, Saurabh Mittal, Kevin Henares, Román Cardenas, and Patricia Arroba. xDEVS: A toolkit for interoperable modeling and simulation of formal discrete event systems. *Software: Practice and Experience*, 53(3):748–789, 2023.

21 Sasisekhar Govind and Gabriel Wainer. Handling Asynchronous Inputs in DEVS Based Real-Time Kernels. In *2024 Winter Simulation Conference (WSC)*, pages 2277–2288, 2024. https://doi.org/10.1109/WSC63780.2024.10838981.

22 DSI. EE Modeling System, 2025. https://www.eemodelingsystem.com; https://www.eemodelingsystem.com/user-center/downloads.

23 ISCAR. DEVS-BLOOM, 2024. https://github.com/iscar-ucm/devs-bloom-public.

24 Jesús Chacón, Giordy A. Andrade, José L. Risco-Martín, and Segundo Esteban. A bleeding edge web application for early detection of cyanobacterial blooms. *Electronics*, 13(5):1–22, 2024. https://doi.org/10.3390/electronics13050942.

18

Optimal Strategies to Contain and Prevent the Spread of Diseases in Group-structured Populations

*Varun K. Rao,[1] Siddharth Patwardhan,[2] and Filippo Radicchi[1]**

[1] Center for Complex Networks and Systems Research, Luddy School of Informatics, Computing, and Engineering, Indiana University, Bloomington, IN, USA
[2] Center for Science of Science and Innovation, Kellogg School of Management, Northwestern University, Evanston, IL, USA

*Corresponding author: filiradi@iu.edu

18.1 Introduction

Once an epidemic process begins, the organization of the social network in which it unfolds strongly influences its outcome [1–4]. In some real-world networks, for example, those formed by college students and school-going children, the vast majority of contacts that contribute to the spread of diseases are due to group activities/settings [4–6]. Individuals often interact with multiple groups and are exposed to different types of group interactions; consequently, realistic models of epidemic spreading must account for these kinds of group interactions that individuals participate [7]. These kinds of interactions are also accounted for in theoretical models of epidemic spreading [4–6, 8–10]. In the above papers, the existence of multiple types of group interactions is considered as an ingredient of the epidemic models. This perspective can be reframed by treating the exposure of individuals to multiple types of group interactions as a free parameter in a model of epidemic spreading aimed at understanding how group structure affects disease propagation dynamics. In this chapter, we summarize the main finding of our recent publication on the topic [11]. The focus is on a susceptible–infected–recovered (SIR) epidemic process defined on multiplex networks where the layers represent different settings for social interactions. We specifically consider synthetic multiplex networks with built-in community structure, where we can vary the within-layer strength and the cross-layer correlation of the community structure [12]. The most important finding is that increasing the strength and/or correlation of the community structure can (1) decrease the severity of epidemic outbreaks by leading to weaker and shorter-lived epidemic outbreaks and (2) make disease control interventions considerably more effective. Our work provides an example of how microscopic changes in a complex network of social contacts may fundamentally impact the macroscopic outcome of an epidemic process. This work is therefore highly relevant to making complex adaptive systems (CAS) more resilient to disruptions from epidemics.

18.2 Methods

18.2.1 Multiplex Networks

Multiplex networks generalize simple networks to represent relational data consisting of multiple types of interactions between the same entities [13]. They are essentially a collection of simple networks sharing the same set of nodes, each imagined on a distinct layer representing a distinct type of interactions. By capturing the multifaceted nature of real-world systems, multiplex networks provide a meaningful and convenient framework for modeling real complex systems. In social networks, multiplex networks can model multiple types of interactions, like friendships, professional connections, and shared interests.

For a multiplex network with L layers, the graph associated with layer $\ell = 1, 2, ..., L$ can be represented as $\mathcal{G}^{(\ell)} = (\mathcal{N}, \mathcal{E}^{(\ell)})$. Graphs corresponding to all the layers have the same set of nodes $\mathcal{N}$, with size $N = |\mathcal{N}|$, and have L different layers of interaction given by the edge sets $\mathcal{E}^{(1)}, \mathcal{E}^{(2)}, ..., \mathcal{E}^{(L)}$. An illustration of a multiplex network with three layers is shown in Figure 18.1.

A multiplex network can be represented using a set of L symmetric square matrices $\{A^{(1)}, A^{(2)}, ..., A^{(L)}\}$ of size $N \times N$. The element $A^{(\ell)}_{i,j}$ of the adjacency matrix represents the relationship between nodes i and j in layer ℓ. $A^{(\ell)}_{i,j} = A^{(\ell)}_{j,i} = 1$ if there exists an edge connecting i to j in layer ℓ and $A^{(\ell)}_{i,j} = A^{(\ell)}_{j,i} = 0$ otherwise.

18.2.1.1 Synthetic Multiplex Networks with Tunable Community Structure

In this work, we consider synthetic multiplex networks composed of $L = 2$ layers. Each layer has preimposed block/community structure being generated following the prescriptions of a stochastic block model where both degrees and block sizes are homogeneous [15]. Specifically, nodes' degrees in layer ℓ are random variates extracted from a Poisson distribution with average $\langle k^{(\ell)} \rangle$. Also, each individual i is associated to a group $\sigma^{(\ell)}_i = 1, ..., Q^{(\ell)}$ in layer ℓ, with all groups having the same size $q^{(\ell)}$. The mixing parameter μ quantifies the fraction of connections that each node has towards other nodes outside its own group. We use the same value of the mixing parameter for both layers. We tune the correlation among

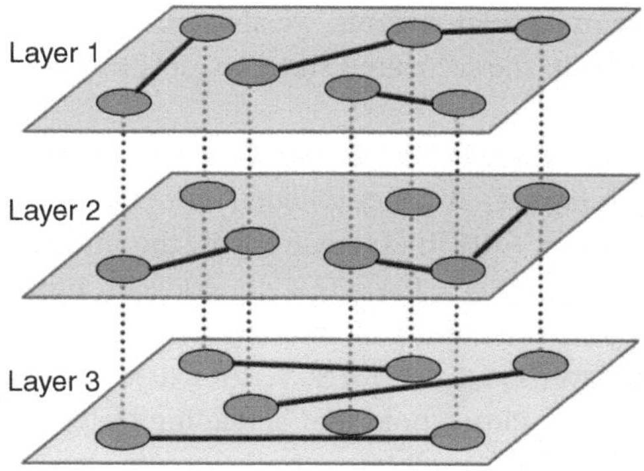

Figure 18.1 Visualization of a multiplex network composed of $N = 7$ nodes and $L = 3$ layers of interactions. *Source:* Adapted from [14].

the community structure of the layers as follows. We start from a configuration where $\sigma_i^{(1)} = \sigma_i^{(2)}$ $\forall i = 1, \ldots, N$, meaning that the layer-wise community partitions are perfectly correlated. Then, with probability r, we swap the label of each node i in layer $\ell = 1$ with that of another randomly selected node j. For $r = 0$, the two partitions are maximally correlated by construction; for $r = 1$, the two partitions are instead maximally uncorrelated. Correlation of the layer-wise community structures is quantified using normalized mutual information (NMI) [16]. Values of the NMI metric range from 0 for uncorrelated block structures to 1 for perfectly aligned communities.

In some experiments, we also consider multiplex networks with heterogeneous degrees and community sizes. There, the degree k of a node is a random variate extracted from the power-law distribution $P(k) \sim k^{-\lambda}$ for $k \in [k_{\min}, k_{\max}]$ and $P(k) = 0$ otherwise; similarly, the size q of a community is a random variate extracted the power-law distribution $P(q) \sim q^{-\tau}$ for $q \in [q_{\min}, q_{\max}]$ and $P(q) = 0$ otherwise.

18.2.2 Epidemic Spreading Process

We consider continuous-time SIR dynamics on a multiplex network [13, 17]. In the SIR model, nodes can be in three different states: S, I, or R. Susceptible nodes can become infected if in contact with a node in the infected state; infected nodes recover spontaneously; once in the recovered state, nodes do not longer participate in the spreading dynamics [1].

The rate of recovery is set to 1, without loss of generality. This is a usual practice as the properties of SIR models are determined by the ratio of the rates of spreading and recovery. We indicate with $\rho_i(t) = S, I$, or R the state of node i at time t. Our initial conditions are generally such that all nodes are in the S state, except for a randomly selected node v that is in state I, i.e., $\rho_i(t = 0) = S$ for $i = 1, \ldots, N$

and $i \neq v$, while $\rho_v(t=0) = I$. In some of our experiments, we also impose that a fraction f of randomly selected nodes is in the recovered state, i.e., $\rho_i(t=0) = R$ with probability f for all $i = 1, \ldots, N$.

Starting from an initial condition, the rules of SIR dynamics are as follows. Denote with $\beta^{(\ell)}$ the spreading rate in layer ℓ. A generic node i can change its state in the infinitesimal time interval $[t, t + \mathrm{d}t]$ if: (1) node i is in the infected state and recovers with probability $\mathrm{d}t$, i.e., $\rho_i(t) = I \rightarrow \rho_i(t + \mathrm{d}t) = R$ with probability dt; (2) node i is in the susceptible state and gets infected with probability $\beta^{(\ell)}\mathrm{d}t$ by its infected neighbor j on layer ℓ, i.e., $\rho_i(t) = S \rightarrow \rho_i(t + \mathrm{d}t) = I$ with probability $\beta^{(\ell)}\mathrm{d}t$ if $\rho_j(t) = I$ and $A_{ij}^{(\ell)} = 1$. Please note that spreading events of type of (2) can happen for all individual edges in all layers in the graph; thus, if node i is in the S state at time t, the overall probability to get infected in the time interval $[t, t + \mathrm{d}t]$ is Prob.$(\rho_i(t) = S \rightarrow \rho_i(t + \mathrm{d}t) = I) = 1 - \prod_{\ell=1}^{L}(1 - \beta^{(\ell)}\,\mathrm{d}t)^{g_i^{(\ell)}}$, where $g_i^{(\ell)} = \sum_j A_{ij}^{(\ell)}\delta(\rho_j(t), I)$ is the number of infected neighbors that node i has in layer ℓ at time t with $\delta(x, y) = 1$ if $x = y$ and $\delta(x, y) = 0$ otherwise. An absorbing dynamical state is reached when no infected nodes are present. SIR dynamics is simulated using Gillespie's algorithm [18].

18.2.2.1 Metrics of Epidemic Severity

We quantify the severity of a simulated epidemic process using the following metrics:

1. The outbreak size, i.e., the relative number of nodes in the recovered state at the end of the process.
2. The peak of infection, i.e., the maximum value of the relative number of nodes that are infected at the same time.
3. The epidemic duration, i.e., the minimum time value when infected nodes are no longer present in the network.

To run the experiments in the next sections, we assume a multiplex network and the parameters $\beta^{(\ell)}$ of the epidemic model are given. In particular, we consider only cases where $\beta^{(\ell)} = \beta$ for $\ell = 1, 2$. In all experiments, we simulate the SIR dynamics multiple times. Using the epidemic severity metrics defined above, we report the sample averages in the results of the chapter.

18.3 Results

18.3.1 Illustrative Example

Figure 18.2 provides an illustrative example of the modeling framework. All model parameters, except for the correlation of the community structure among the layers, are fixed. We generate a small multiplex network in which each layer exhibits

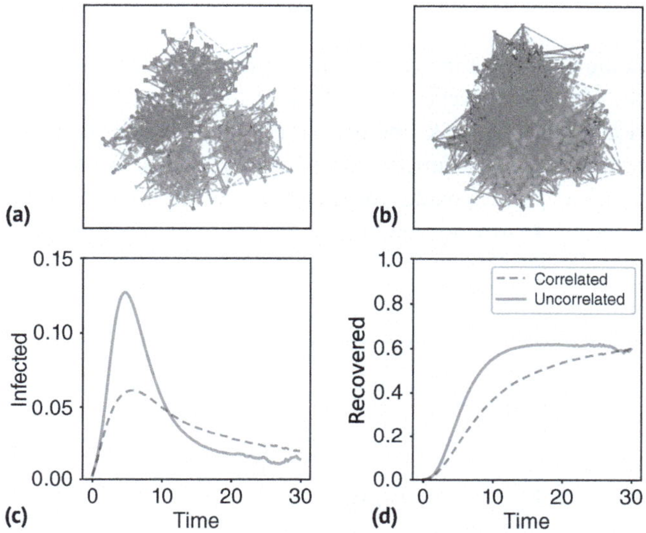

Figure 18.2 (a) To illustrate the modeling framework, we consider a multiplex network composed of $N = 400$ nodes and two layers of interactions. In both layers ($\ell = 1$ and $\ell = 2$), nodes are partitioned into four communities of size $q^{(\ell)} = 100$. Node colors (orange, purple, blue, and green) represent community membership in layer $\ell = 1$, while node shapes (circle, square, triangle, and diamond) indicate community membership in layer $\ell = 2$. In this configuration, the community structures of the two layers are maximally correlated (NMI $= 1.0$), implying a one-to-one correspondence between colors and shapes. Edges are generated such that the average degree in each layer is $\langle k^{(\ell)} \rangle = 3$, with a fraction $\mu = 0.025$ of edges connecting nodes in different communities. Edges in layer $\ell = 1$ are shown as solid gray lines, while edges in layer $\ell = 2$ are shown as dashed lines. (b) Same as panel (a), but with uncorrelated community structure between layers (NMI $= 0.0$), obtained by assigning colors and shapes to nodes independently at random. (c) We perform $V = 1\,000\,000$ simulations of the SIR model on the networks shown in panels (a) and (b), using a spreading rate $\beta = 0.38$. The fraction of infected nodes in the population is shown as a function of time. Time series are binned using intervals of width 0.25, and averages are reported only when at least 50 simulations survive within each bin. (d) Same as panel (c), but showing the fraction of recovered nodes in the population as a function of time. For color references, please refer to the online version.

a clear block structure where $\mu = 0.025$. In Figure 18.2a, the two structures are maximally correlated (NMI = 1.0), whereas in Figure 18.2b, they are completely uncorrelated (NMI = 0.0). Despite yielding comparable final outbreak sizes, the difference in group structures produces different SIR spreading dynamics. Correlated community structures lead to slower outbreaks with smaller peak intensity than in the case of uncorrelated community structures. This comparison highlights that correlations among layer-wise community structures primarily affect the temporal evolution of the process rather than the overall size of the outbreak.

18.3.2 Group Size

In Figure 18.3, we vary the group size in synthetic multiplex networks and show its effect on the metrics of epidemic severity. We look at the outbreak size, the fraction of the population infected at peak, and the duration in the correlated and uncorrelated multiplex networks as a function of the group size q. We set the community sizes $q^{(1)} = q^{(2)} = q$ and consider three different values of q. We consider graphs with size $N = 10\,000$, mixing parameter $\mu = 0.025$, and average degree $\langle k^{(1)} \rangle = \langle k^{(2)} \rangle = 6$.

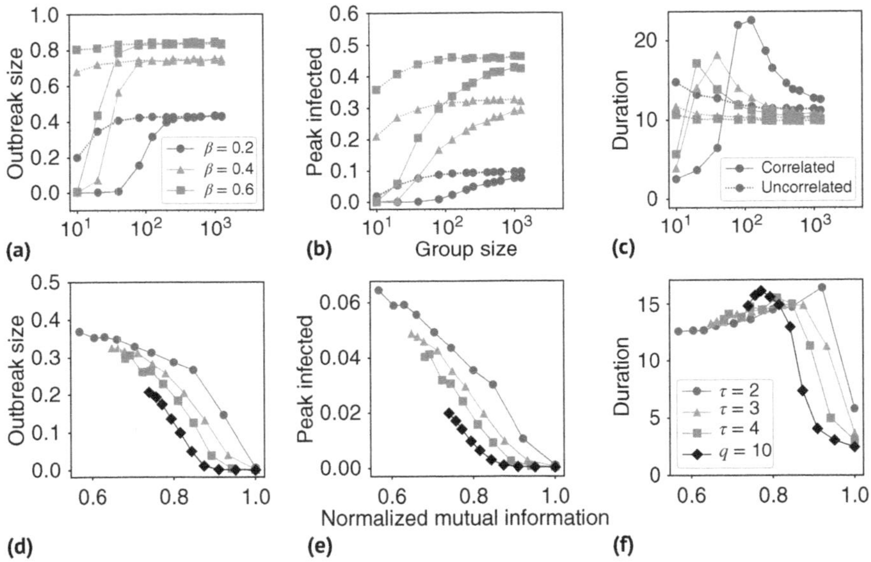

Figure 18.3 We consider multiplex networks with $N = 10\,000$. Graphs with $\langle k^{(1)} \rangle = \langle k^{(2)} \rangle = 6$ and $\mu = 0.025$ are considered. We set the community sizes $q^{(1)} = q^{(2)} = q$, and vary q from 10 to 1250. The solid and dotted curves show the indicated epidemic properties as a function of the group size q for multiplex networks with correlated and uncorrelated community structures, respectively. Different colors refer to different values of spreading rate β in panels a, b, and c. We run the model for $V = 5000$ simulations of the SIR dynamics and display average values in the plot. (a) Outbreak size, (b) fraction of the infected population at the peak, and (c) total duration of the outbreak are plotted as functions of the group size. Furthermore, we fix the spreading $\beta = 0.4$ and choose the community sizes q from the power-law distribution $P(q) \sim q^{-\tau}$ for $q \in [10, 100]$ and $P(q) = 0$ otherwise. We consider exponent values $\tau = 2, 3$, and 4. As a term of comparison, we also display results valid when communities have all identical sizes $q = 10$. Different colors refer to the different community size distributions in panels d, e, and f. (d) Outbreak size, (e) fraction of the infected population at the peak, and (f) total duration of the outbreak are plotted as functions of the NMI between the layer-wise community structures. *Source:* Adapted from [11]. For color references, please refer to the online version.

We also explore the effect of the heterogeneous community sizes on the outbreak. We consider multiplex networks with $N = 10\,000$ nodes, average degrees $\langle k^{(1)}\rangle = \langle k^{(2)}\rangle = 6$, and $\mu = 0.025$. We choose the community sizes q from the power-law distribution $P(q) \sim q^{-\tau}$ for $q \in [10, 100]$ and $P(q) = 0$ otherwise. We consider exponent values $\tau = 2$, 3, and 4. Community sizes $q_1, q_2, \ldots, q_Q$ are generated from the above power-law distribution and used in both layers to define their block structure. As a term of comparison, we also display results valid when communities have all identical sizes $q = 10$. Following the procedure outlined in Section 18.2, we vary the correlation of a layer's group structure by reassigning the community labels of a randomly selected fraction of nodes. We run $V = 5000$ simulations of the SIR dynamics and display average values in Figure 18.3.

18.3.3 Group Correlation and Strength

We systematically investigate how correlations between the community structures of multiplex network layers affect SIR spreading. Results shown in Figure 18.4 are obtained using the network construction algorithm described in Section 18.2, with $N = 10\,000$ nodes and parameters $\langle k^{(1)}\rangle = 3$, $\langle k^{(2)}\rangle = 10$, $q^{(1)} = 5$, $q^{(2)} = 25$, and $\mu = 0.025$. We consider three transmission rates, $\beta = 0.2$, 0.4, and 0.6, all corresponding to the endemic regime. Increasing the community correlation within each layer influences epidemic severity in distinct ways. The final outbreak size remains largely unchanged except when community structures approach maximal correlation, whereas reductions in epidemic intensity and increases in epidemic duration are observed over a broader range of correlation values. Overall, stronger community correlation slows the spreading process, leading to lower peak of infections and longer epidemic durations.

These observations about the effect correlation has on the severity outcomes are true when communities within each layer have sufficiently strong modular organization. In Figure 18.4d–f, we demonstrate the impact how weakening the network modular structure by increasing μ has different outcomes depending on whether the group structure is correlated or uncorrelated. When communities across layers are uncorrelated, varying the strength of community structure has a negligible effect on epidemic severity, while, when groups are correlated, progressively weakening the modular structure leads to increases in all epidemic severity metrics.

18.3.4 Immunization

In Figure 18.5, we examine how there are varying outcomes for random immunization between uncorrelated and correlated networks. Immunization is implemented by modifying the initial conditions of the dynamics: a fraction of nodes is

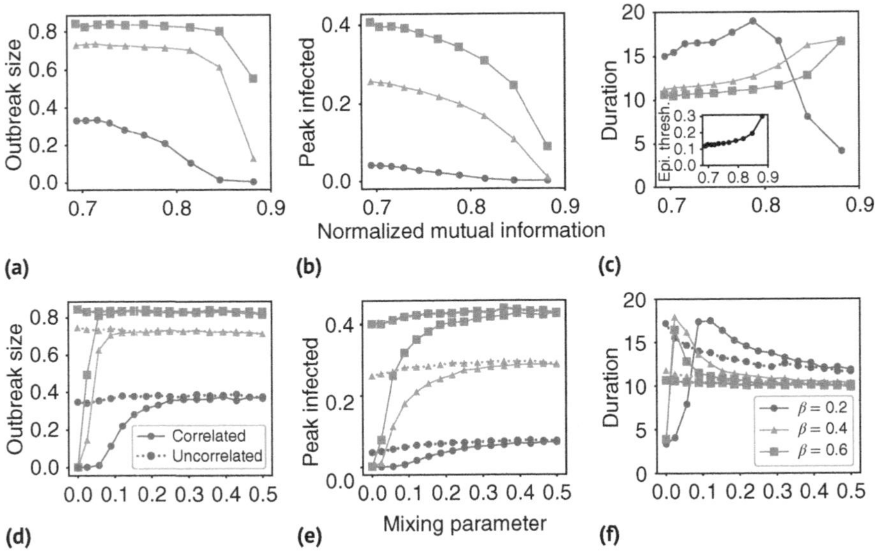

Figure 18.4 (a) We analyze multiplex networks composed of $N = 10\,000$ nodes with parameters $\langle k^{(1)} \rangle = 3$, $q^{(1)} = 5$, $\langle k^{(2)} \rangle = 10$, $q^{(2)} = 25$, and mixing parameter $\mu = 0.025$. The correlation among the community structures of the layers is systematically varied by randomly reassigning a fraction of node community labels, following the procedure described in Section 18.2. For each configuration, we simulate SIR dynamics and measure the final outbreak size, which we display as a function of the NMI between the layers' partitions. Results are averaged over $V = 5000$ realizations. Different colors and symbols correspond to transmission rates $\beta = 0.2$, 0.4, and 0.6, all in the supercritical regime (see inset in panel (c)), with corresponding reproduction numbers $R_0 = 2.6$, 5.2, and 7.8, respectively. (b) Same as panel (a), but for the peak fraction of infected nodes. (c) Same as panel (a), but for the epidemic duration. (d) We focus on configurations corresponding to maximum (solid curves, filled symbols) and minimum (dashed curves, transparent symbols) correlation among the community structures of the layers and generate networks with varying values of the mixing parameter μ. The final outbreak size is shown as a function of μ, averaged over $V = 5000$ realizations. (e) Same as (d), but for the peak fraction of infected nodes. (f) Same as (d), but for the average epidemic duration. *Source:* Adapted from [11]/American Physical Society/CC BY 4.0.

randomly assigned to the recovered state, while all remaining nodes are susceptible except for a single infected seed. Under this setup, networks with a correlated community structure exhibit substantially stronger suppression of epidemic spread than uncorrelated networks, as observed in Figure 18.5c. For identical transmission rates, correlated communities produce reductions of up to a factor of five in both final outbreak size and peak infected. As a consequence, epidemic dynamics observed at $\beta = 0.6$ with correlated communities closely resemble those obtained at $\beta = 0.2$ when community structure is uncorrelated.

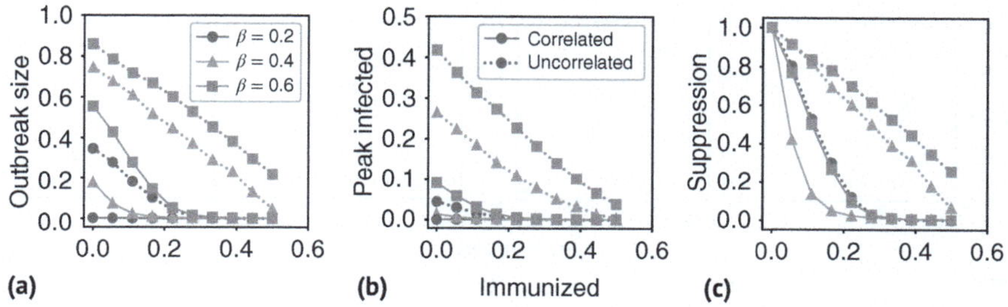

Figure 18.5 In all panels, we use the same experimental setting as in the bottom row of Figure 18.4, but, modify the initial conditions of the dynamics by immunizing a random fraction of nodes. In the first two panels, we plot the effect this has on two specific epidemic severity outcomes: (a) the outbreak size and in (b) the peak fraction of infected nodes is shown. (c) We show a novel metric, suppression, by rescaling the abscissa values of panel (a) by the outbreak size that is observed when a null fraction of nodes is immunized. All results are obtained by averaging the outcome of $V = 5000$ repetitions of the epidemic process. Source: Adapted from [11]/American Physical Society/CC BY 4.0.

In addition to immunizing nodes based on a random immunization scheme, we test how immunizing nodes based on their centrality measures would affect the size of outbreaks in correlated and uncorrelated networks. In contrast to Figure 18.5, we use a heterogeneous degree distribution $P(k) \sim k^{-2}$ (Figure 18.6). We utilized a heterogeneous degree distribution to analyze the effects of targeted immunization on the outbreak size to better match real-world systems. We test the effects of immunizing the nodes with the largest degree, adaptive degree [20], and collective influence [19]. We find that vaccinating nodes based on centrality ranking has a larger effect on decreasing outbreak size than vaccinating nodes randomly. The effect is more pronounced in correlated networks compared to uncorrelated networks. Immunizing 20% of the nodes by degree, collective influence, or adaptive degree makes the outbreak size more than 10 times smaller than immunizing the same fraction of nodes randomly in correlated networks.

18.4 Discussion

In this chapter, we reported on results concerning SIR spreading in multiplex networks with built-in community structure. In each layer, community organization is controlled by two parameters: the mixing parameter μ and the community size $q^{(\ell)}$. Correlation among the layer-wise community structures is indicative of the likelihood that two nodes belong to the same community in both layers.

We constructed synthetic multiplex networks in which key parameters are adjustable, allowing us to isolate the role of group structure in shaping spreading dynamics. Using extensive numerical simulations, we find that the degree of

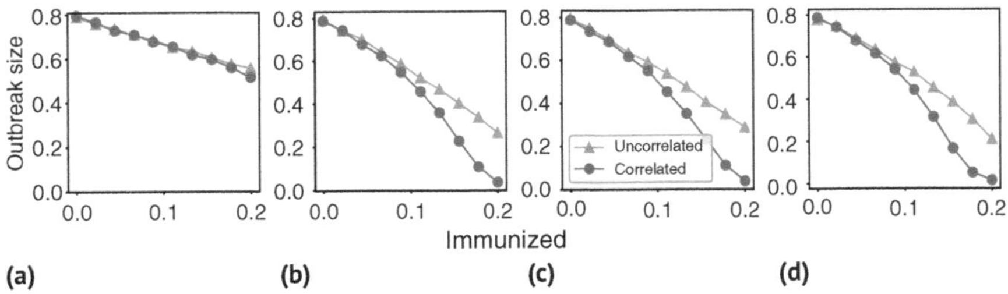

Figure 18.6 We consider a similar experimental setup, to Figure 18.5, with some key differences. In this experiment, we set the spreading rate $\beta = 1$ and utilized a heterogeneous degree distribution. Nodes in the layers are organized in communities of size $q = 100$ and their degrees are chosen from a power-law degree distribution given by $P(k) \sim k^{-2}$ for $k \in [2, \sqrt{q}]$ and $P(k) = 0$ otherwise. The mixing parameter is $\mu = 0.025$. We consider the cases when the layer-wise community structures are maximally correlated (circle) and when they are uncorrelated (triangle). Similar to Figure 18.5, we immunize a fraction of nodes randomly in (a). Then in (b), (c), and (d), we perform a targeted immunization of the top-ranked nodes based on three different centrality measures: (b) degree, (c) collective influence (with parameter $\ell = 2$) [19], and (d) adaptive degree [20]. We then plot the size of the outbreak as a function of the fraction of immunized nodes. Results are obtained by averaging the outcome of $V = 5000$ repetitions of the epidemic process.

correlation between layers strongly influences SIR dynamics in these networks. When communities are highly modular and are correlated across layers, spreading processes tend to be less severe by persisting for longer durations while having a smaller epidemic peak. In contrast, networks that are not sufficiently modular or have poorly correlated community structures give rise to spreading processes that are shorter in duration but have larger peak infected values. The level of correlation strongly affects control strategies, with both random and targeted immunization schemes performing substantially better in networks whose community structures are correlated across layers than in those that are uncorrelated.

These findings can be understood by considering how correlations among layer-wise community structures impact spreading processes. When communities are correlated between layers, interactions between nodes become increasingly redundant. For this reason, spreading is more likely to occur between already connected groups, leading to longer outbreak durations and smaller epidemic peaks. Correlating groups across layers may thus represent an effective strategy for reducing the severity of epidemic outbreaks. This seems closely related to the mitigation strategy proposed by Meidan based on an alternating lockdown policy applied to half of the population at a time [21].

The root of this work is based on analyzing how a contact network can be made more resilient to epidemics through changing its structure. By correlating the community structure between layers, systems experience longer, less severe outbreak while systems that are uncorrelated experience shorter more intense outbreaks. These results highlight how modifying local structural features in a multilayered network can lead to significant changes in the outcomes of an epidemic process. More broadly, since epidemic spread is an emergent property occurring from the interactions of the network, the results of this chapter show that small structural changes to a CAS can have a large impact on how the system is globally affected by a dynamical process. Our work supports the idea that microscopic changes to interactions within a CAS can have large impact on its resilience to disruption.

For this work, we made a few key assumptions about the spreading dynamics of realistic diseases and the social contact networks that sustain the spreading dynamics. Still, our results reveal something fundamental about how spreading occurs in multiplex systems and demonstrate that slight changes to how a system is organized could help decreasing the severity of outbreaks. Managers of large organizations, such as schools or colleges, can take under consideration some of the easy-to-implement principles highlighted in our study when planning group activities.

References

1 Romualdo Pastor-Satorras, Claudio Castellano, Piet Van Mieghem, and Alessandro Vespignani. Epidemic processes in complex networks. *Reviews of Modern Physics*, 87(3):925–979, 2015. https://doi.org/10.1103/RevModPhys.87.925.

2 Alessandro Vespignani, Huaiyu Tian, Christopher Dye, James O. Lloyd-Smith, Rosalind M. Eggo, Munik Shrestha, Samuel V. Scarpino, Bernardo Gutierrez, Moritz U. G. Kraemer, Joseph Wu, Kathy Leung, and Gabriel M. Leung.Modelling Covid-19. *Nature Reviews Physics*, 2(6):279–281, 2020.

3 Quan-Hui Liu, Marco Ajelli, Alberto Aleta, Stefano Merler, Yamir Moreno, and Alessandro Vespignani. Measurability of the epidemic reproduction number in data-driven contact networks. *Proceedings of the National Academy of Sciences*, 115(50):12680–12685, 2018.

4 David Soriano-Paños, L. Lotero, Alex Arenas, and Jesús Gómez-Gardeñes. Spreading processes in multiplex metapopulations containing different mobility networks. *Physical Review X*, 8(3):031039, 2018.

5 Clara Granell and Peter J. Mucha. Epidemic spreading in localized environments with recurrent mobility patterns. *Physical Review E*, 97(5):052302, 2018.

6 Jesús Gómez-Gardeñes, David Soriano-Panos, and Alex Arenas. Critical regimes driven by recurrent mobility patterns of reaction–diffusion processes in networks. *Nature Physics*, 14(4):391–395, 2018.

7 Wouter Van den Broeck, Corrado Gioannini, Bruno Gonçalves, Marco Quaggiotto, Vittoria Colizza, and Alessandro Vespignani. The gleamviz computational tool, a publicly available software to explore realistic epidemic spreading scenarios at the global scale. *BMC Infectious Diseases*, 11(1):1–14, 2011.

8 Clara Granell, Sergio Gómez, and Alex Arenas. Dynamical interplay between awareness and epidemic spreading in multiplex networks. *Physical Review Letters*, 111(12):1–5, 2013. https://doi.org/10.1103/PhysRevLett.111.128701.

9 S. Gómez, A. Díaz-Guilera, J. Gómez-Gardeñes, C. J. Pérez-Vicente, Y. Moreno, and A. Arenas. Diffusion dynamics on multiplex networks. *Physical Review Letters*, 110(2):1–5, 2013. https://doi.org/10.1103/PhysRevLett.110.028701.

10 Clara Granell, Sergio Gómez, and Alex Arenas. Competing spreading processes on multiplex networks: awareness and epidemics. *Physical Review E*, 90(1):012808, 2014.

11 Siddharth Patwardhan, Varun K. Rao, Santo Fortunato, and Filippo Radicchi. Epidemic spreading in group-structured populations. *Physical Review X*, 13:041054, 2023. https://doi.org/10.1103/PhysRevX.13.041054.

12 Santo Fortunato. Community detection in graphs. *Physics Reports*, 486(3–5):75–174, 2010.

13 Ginestra Bianconi. *Multilayer Networks: Structure and Function*. Oxford University Press, 2018.

14 Siddharth Patwardhan. *Optimization Problems in Multiplex Networks*. PhD Thesis, Indiana University, 2024.

15 Paul W. Holland, Kathryn Blackmond Laskey, and Samuel Leinhardt. Stochastic blockmodels: First steps. *Social Networks*, 5(2):109–137, 1983.

16 Leon Danon, Albert Diaz-Guilera, Jordi Duch, and Alex Arenas. Comparing community structure identification. *Journal of Statistical Mechanics: Theory and Experiment*, 2005(09):P09008, 2005.

17 Dongmei Fan, Guo-Ping Jiang, Yu-Rong Song, and Xu Zhang. Influence of geometric correlations on epidemic spreading in multiplex networks. *Physica A: Statistical Mechanics and Its Applications*, 533:122028, 2019.

18 Daniel T. Gillespie. Exact stochastic simulation of coupled chemical reactions. *The Journal of Physical Chemistry*, 81(25):2340–2361, 1977.

19 Flaviano Morone and Hernán A. Makse. Influence maximization in complex networks through optimal percolation. *Nature*, 524(7563):65–68, 2015.

20 Wei Chen, Yajun Wang, and Siyu Yang. Efficient influence maximization in social networks. In *Proceedings of the 15th ACM SIGKDD International Conference on Knowledge Discovery and Data Mining*, pages 199–208, 2009.

21 D. Meidan, N. Schulamann, Reuven Cohen, S. Haber, E. Yaniv, R. Sarid, and B. Barzel. Alternating quarantine for sustainable mitigation of Covid-19. *Nature Communications*, 12:220, 2021.

Part V

Next Steps

19

Research Agenda for Resilient Complex Adaptive Systems

Claudia Szabo,[1] Rodrigo Castro,[2] Saurabh Mittal,[3] and Viktoriya Semeshenko[4]*

[1] *School of Computer and Mathematical Sciences, The University of Adelaide, Adelaide, Australia*
[2] *Facultad de Cs. Exactas y Naturales and Instituto de Ciencias de la Computación, Universidad de Buenos Aires and CONICET, Ciudad Autónoma de Buenos Aires, Argentina*
[3] *Defense Science Division, Modern Technology Solutions, Inc., Dayton, OH, USA*
[4] *Facultad de Cs. Económicas and Instituto Interdisciplinario de Economía Política, Universidad de Buenos Aires and CONICET, Ciudad Autónoma de Buenos Aires, Argentina*

*Corresponding Author: Rodrigo Castro; rcastro@dc.uba.ar

19.1 RCAS Applications, Formalisms, and Tools

We present in this chapter a summary of observed themes, lessons learned, research gaps, and future research agenda. A recurring theme running through several chapters is the focus on using advanced modeling and simulation (M&S) techniques. Various chapters emphasize the importance of high-granularity data and its integration into dynamic models. For instance, the use of administrative labor flow data in Argentina allows for fine-grained analysis of labor market resilience during the COVID-19 pandemic. Similarly, urban digital twins (UDTs) are proposed as holistic, multilayered frameworks that integrate real-time data, predictive models, and cognitive capabilities—such as machine learning—to support adaptive urban management and decision-making.

Given the deep uncertainty inherent in RCAS, several authors advocate for ensemble-based approaches and multiobjective optimization to address RCAS challenges. Instead of relying on single models or objectives, researchers use sets of simulations executed through design of experiments to explore diverse futures and identify robust strategies. Visualization and immersive decision-support systems are also highlighted as key tools for communicating complex trade-offs to policymakers.

Other methods are rooted in agent-based modeling and empirical validation protocols. Frameworks are proposed to bridge simulated dynamics with real-world economic data, introducing a composite similarity index that integrates multiple statistical and spectral measures for robust model selection and validation. The goal is to demonstrate how complex adaptive systems (CAS) approaches can be grounded in empirical regularities without sacrificing their ability to capture emergent properties.

The DEVS (discrete event system specification) formalism and its extensions are recognized as robust tools for building modular, hierarchical, and reusable models. These formalisms enable the representation of complex interactions across multiple scales—from individual agents to entire systems—while supporting rigorous verification and validation. The use of DEVS tends to facilitate the study of emergent behaviors, adaptivity [1], and resilience in systems ranging from urban infrastructures to social networks and economics [2]. The integration of simulation formalisms with cyber-physical systems is also explored, for instance, with a DEVS-based management platform for harmful algal blooms that integrates DEVS with real-time digital twins and IoT-based sensing infrastructures. The goal is to demonstrate how CAS approaches can seamlessly transition from virtual simulation to field deployment, enabling proactive environmental management through layered edge–fog–cloud architectures.

Environmental resilience is explored through innovative digital twin approaches. Authors propose a river ecosystem digital twin that integrates agent-based models with physics-based hydrological simulations, enabling multiscale exploration of pollution dynamics, climate impacts, and human interventions. These works underscore the growing role of digital twins as resilience laboratories, where scenarios and interventions can be safely tested before real-world deployment.

In the domain of robotics and engineering of distributed systems, the decentralized control and coordination in multirobot systems is demonstrated by applying rigidity theory to dynamic communication graphs. The introduction of subframework-based controllers provides a scalable method to maintain network coherence under local information constraints, exemplifying how mathematical theory, simulation, and control engineering converge to enhance resilience in adaptive ensembles.

Other groups examine the structure and resilience of global artificial intelligence (AI) ecosystems through multilayer network analysis, integrating economics, knowledge, hardware, and energy dimensions. Their empirical demonstration highlights how systemic dependencies shape robustness and vulnerability in a rapidly evolving technological landscape, offering both methodological innovation and strategic insights for governance.

From a socioeconomic perspective, a hybrid framework aims at combining agent-based computational economics (ACE) with multi-agent reinforcement learning (MARL) to study macrocrisis phenomena. The goal is to integrate heterogeneous agents to adaptively learn strategies under uncertainty while capturing systemic fragilities such as coordination failures, bubbles, and cascades. It highlights how learning dynamics and policy interventions can be tested *in silico* to design more resilient economic structures. Complementing this, other groups highlight the role of resilience assessment frameworks that combine simulation, risk modeling, and participatory decision-making to manage systemic uncertainty across sectors.

Finally, network theory plays a crucial role in understanding RCAS. Methods like disparity filtering and community detection (e.g., Leiden algorithm) are used to extract meaningful structures from dense labor flow networks, revealing core-periphery patterns and modular communities that highlight systemic vulnerabilities and resilience mechanisms. These techniques help identify key sectors, persistent interactions, and the impact of shocks on system cohesion. Multilayer network analysis of AI ecosystems demonstrates how structural tools from Complexity Science can map interdependencies among economic, technological, and energy domains, thereby revealing systemic vulnerabilities and resilience pathways. This aligns with other authors' emphasis on validation and model selection techniques—including composite similarity indices—that strengthen the credibility of complex simulations when applied to policy-relevant contexts.

Across these diverse applications, several unifying methodological lessons emerge. First, multidimensional validation and integration with data are essential to ensure the relevance and trustworthiness of CAS models. Second, cyber-physical coupling via digital twins and IoT architectures allows simulations to inform real-time interventions. Third, scalable decentralized control strategies provide robustness in distributed adaptive systems, whether of robots or economic agents. Fourth, lack of formal approaches, with the exception of DEVS-based approaches, introduces risks in complexity management and model development that ultimately impacts the computational M&S infrastructure with lack of extensibility, scalability, and faster than real-time simulation. Finally, network-based analytical tools remain indispensable for capturing structural interdependencies in socio-technical systems [3].

19.2 Lessons Learned

The editors asked the chapter authors in this book to identify a few lessons learned from their research into RCAS. The lessons learned focused on (1) resilience, (2) modeling, simulation, analysis, and visualization, and (3) real-life deployments. We summarize them below.

19.2.1 Resilience

The inherent complexity of RCAS brings significant challenges in managing them, particularly in harsh and adversarial environments. A deep understanding of resilience in CAS requires analysis of the nonlinear interconnections and emergent behaviors among system components (i.e., system of systems) under uncertain conditions, at both micro- and macrolevels, from multiple viewpoints and different stakeholders' perspectives. In some cases, especially when looking at resource-constrained systems, a realistic approach involves satisfying to get partial or suboptimal solutions that are good enough as opposed to giving no solution at all. An effective adaptation/resilience loop requires meaningful observations and effective feedback to the dynamic environment within which the RCAS operates. Lastly, some authors have observed that resilience is not a linear, elastic process, but an evolutionary one, with no straightforward return to a previous state but rather a process of evolution and adaptation, supported by the study of mesostructures, i.e., structures that are intermediate between the topmost layers and the bottom-most layers. This informs the study of emergent behaviors not just at a macrolevel but a layered endeavor where intermediate part-emergent behaviors affect the overall SoS behavior.

19.2.2 Modeling, Simulation, Analysis and Visualization

The role that modeling and simulation play in the analysis of RCAS cannot be understated as they offer valuable epistemological insights by enabling the exploration of known and unknown system behaviors through theory-rooted and reality-grounded models, as validated by the subject matter experts. In fact, M&S-enabled test beds are the only tools we have to explore synthetic emergent behaviors for CAS and RCAS. These test beds afford us deep insights into the interconnections between individual components and their environment (albeit, synthetic), which is particularly effective for studying ecological problems. The design, implementation, and deployment of effective RCAS simulations require an interdisciplinary approach that is built on an in-depth understanding of the limitations of simulations, namely, embedded assumptions, approximations, and that simulations inherently cannot produce new knowledge beyond their input and algorithms: a belief that had existed a decade ago. Today, the amalgamation of algorithms and the heuristics using AI/machine learning (ML) and generative AI technologies expand the art and science to a capability where simulation-based discovery is possible, which affords detecting new patterns across different time scales, dimensions, and fidelity layers.

19.2.3 Real-world Deployment and Integration

The importance of integrating RCAS M&S infrastructure with real-world applications cannot be understated, as it allows for the integration of models from other fields within the RCAS model. Furthermore, the integration of RCAS simulation models with real-world physical systems allows for the testing and refinement of management strategies in the RCAS M&S environment, before they are tested in the real-life system, with reduced transfer time and costs. In specific domains such as systems operating at the Edge, RCAS provides an ideal framework for the coordination of large-scale deployments of real-world, embodied agents due to their reduced, asynchronous communications requirements that do not overburden the increasingly congested wireless spectrum. The RCAS' self-organizing properties allow RCAS agents to be more robust to packet loss than traditional highly coupled coordination methods. Within these domains, viewing the RCAS as a dynamical system can yield direct performance implications for communications, sensing, and other applications. A deep focus on the applied use of RCAS frameworks demonstrates RCAS cannot display a single unified mode of coordination, but rather a co-design between implicit coordination methods that rely on limited communication with emergent behavior to explicit methods that rely on explicit, frequent communication. Exploiting the strengths of each approach can lead to significant improvements in communication efficiency and coordination performance of RCAS.

19.3 Research Gaps

The editors also asked the authors to identify a few research gaps, either specific to RCAS or to the specific application domain where they were deploying specific RCAS frameworks.

19.3.1 Modeling, Analysis, and Visualization

The first research gap identified is focused on advancing the modeling and analysis techniques that would then allow deeper insight into RCAS behaviors, and strengthen their explainability and thus trust in simulation results and analysis.

There is insufficient support for multimodeling and co-simulation techniques to comprehensively represent and analyze CAS. For example, domains with few universally accepted theories about fundamental concepts have significant uncertainty and often conflicting conceptualizations or scenarios. There is a need for a multimodeling and ensemble modeling would allow for appropriate integration and resolution, also considering the effects that accuracy and uncertainty of one model may have on other connected models. This includes exploring robust decision-making techniques that go beyond single-model or single-objective optimizations toward multiobjective, sensitivity-informed solutions.

This idea of ensemble modeling expands to blended coordination, where there is a need to understand, model, and apply blended models of implicit and explicit coordination as discussed above. This has not been well studied for either AS or resilient systems or both together in RCAS, but has the potential to enable significantly more robust, scalable, and flexible RCAS.

There is also a critical need to better understand and characterize the limits of computational modeling and simulation in capturing emergent behavior, understanding the limits of modeling of fundamentally new phenomena. Research is needed into approaches that incorporate aspects of emergence beyond simply measuring and identifying it, considering also the impact on reliability and the interpretation of simulation results in practice. This uncertainty in identifying M&S limits impacts the underlying computational hardware/software infrastructure required to deliver verifiable synthetic emergent behavior. Many times, adding more hardware does not yield credible results when the issues deal with a federated simulation architecture with parallel distributed federates engaged in both asynchronous and synchronous data exchange. This has further implications in the LVC architecture for RCAS that brings together disparate systems in an SoS setting for simulation-based exploration and experimentation. V&V of RCAS is a well-acknowledged research gap.

The other identified gaps refer to the analysis and interpretation of RCAS results, and specifically in the identification of patterns and structures, connections between macro- and microlevels, and in the generalization of results across heterogeneous scenarios. Firstly, research suggests that within RCAS there exist mesostructures, such as communities and patterns that appear "at the edge" of a RCAS structure, but we lack analysis techniques to identify and extract lessons learned from them. In addition, there is a need to connect macrolevel outcomes to microlevel conditions and connect agent-level models to network-level outcomes to better analyze effects at the micro level. Similarly, RCAS, though deployed in very specific use-cases and scenarios, offer a high potential of transferability of lessons learned, either from a control, structure, or emergent behavior perspective, but the analysis, interpretation, and visualization methods to facilitate this knowledge transfer are still in their infancy.

Lastly, a critical aspect of modeling, simulation, analysis, and visualization of RCAS is that it allows researchers to effectively connect with policymakers and other relevant stakeholders. There is a need to explicitly connect specific RCAS dynamics, supported by clear, explainable results, to concrete policy outcomes which would provide evidence-based guidance across a variety of domains.

19.3.2 RCAS Metrics

The need for more sophisticated analysis techniques identified above is closely tied to the need for specific RCAS metrics that capture the dynamics, causality, and temporal dependencies inherent in such systems. Traditional metrics focusing only on instantaneous effects or system states that are insufficient to understand cascading effects and interdependencies over time and space. There is a need for community-validated metrics that connect delayed effects across decision cycles and more importantly, for capturing the indirect effect that RCAS adaptation has on measurable outcomes. This would facilitate explainability and stakeholder engagement, but also allow for the refinement of RCAS adaptation approaches, control of recognized emergent behavior, and improve qualitative analysis.

In addition, there is a need for further research into establishing metrics of resilience that capture all effects as interpreted from valid emergent behaviors. This set of metrics and associated emergent behaviors calls for a comparative research and benchmarking. The metrics for resilience, some of which are outlined in Chapter 2, need to consider that resilience is not necessarily linearly elastic (i.e., the system comes back to the exact same state), but can lead to evolved states with improved outcomes. Metrics to capture this evolution as well as the underlying supporting theory are required to better understand and explain resilience in RCAS.

19.4 Agenda for Next-generation Research

Based on the lessons learned through this collaborative endeavor and identified research gaps, we are in a position to propose a research agenda for RCAS, as a specialized class of SoS, studied and experimented through a virtual M&S-enabled test bed. We describe the agenda through the following broad categories:

1. Engineering process for RCAS
2. Modeling, simulation, and visualization tools and approaches
3. Measurement, instrumentation, and control
4. Strong emergent behavior (SEB)

19.4.1 Engineering Processes for RCAS

RCAS is a specialized class of CAS with added features/constraints of resilient behaviors. There are three broad classes of RCAS we articulated in Mittal et al. [4]: (1) Class I: the RCAS as a system that exists in the real world, (2) Class II: the RCAS test bed built using the latest in M&S technology, and (3) Class III: the artificial RCAS which we intend to engineer using the RCAS test bed.

Class I: The first class of RCAS inspires us to identify resilience in CAS, various interconnections between the perceived players/agents in a contextual environment. It gives us an observation medium to study natural CAS, where examples include financial markets, brain, ant colony, population stability, ecological systems, power grid and many others. In our book, we did not focus on this type of RCAS but included them as RCAS that our case studies used to build their artificial RCAS.

Class II: The natural RCAS is then brought into the M&S laboratory within a dedicated computational environment geared to reproduce the natural RCAS behaviors within a controlled setting in a computational way. This required levering modeling, simulation, analytics, visualization, verification, validation, and engineering technology to create a test bed that can "accurately" reproduce the observed natural RCAS behavior. The book has extensively focused on the challenges, methodology, metrics, and engineering approaches to construct such a test bed, as described in Chapter 2. Through various examples by the authors, we looked at such test beds built from authors' experience supporting the domain of interest. However, we could not find a common reference architecture for RCAS that guides the development of RCAS test beds for increased efficiencies across the board. Chapter 2 described one RCAS architecture and engineering process to build RCAS test bed to obtain synthetic emergence in a reproducible, repeatable, and controllable manner.

Class III: The third RCAS is an artificial RCAS that is an engineered one, which replicates some of the characteristics of natural CAS, reproducible in an RCAS test bed. This version of RCAS is an live, virtual and constructive (LVC) system [5], an open system, that leverages the virtual constructive test bed to inform, influence and control the live systems, which are active participants in the RCAS. Such RCAS incorporates digital twins into the RCAS architecture and integrates a robust computational infrastructure of RCAS test bed for real-time experimentation and heuristics evaluation.

The next-generation research agenda for RCAS engineering caters to all the three RCAS classes:

- **Class I:** There are three main areas for further exploration for Class I RCAS: (1) the data acquisition/collection mechanisms (2) data storage infrastructure (3) leveraging of AI/ML algorithms for emergent behavior pattern discovery. The data acquisition and collection mechanisms will be guided by the sampling required to collect data. Without the right kind of sampling, continuous real-world behavior once discretized, may or may not be sufficient to reproduce the behavior in a digital quantized environment after discretization. More research is needed to identify the correct sampling rate for specific behaviors across different domains. This data will be in large volume and may require storage of the order of terabytes or petabytes depending upon the domain. While the storage may be available (at a cost), making the data stores accessible by the RCAS test bed in an efficient manner requires further research. Do such architecture need to be cloud-enabled? Lastly, as described ahead, focused research on finding emergent behavior patterns using AI/ML algorithms and generative AI technologies is a critical enabler.

- **Class II:** This class is about developing a computational RCAS test bed and the associated engineering process, similar to one described in Chapter 2, for enabling verifiable synthetic emergence in a controllable manner for a plethora of measures (e.g., measures of entropy, measures of performance, measures of effectiveness, and the synthesized measures of resilience). Further research is needed to understand these measures in a qualitative manner for specific socio-technical domain and quantify the measured ranges for generating synthetic emergent behaviors within the RCAS test bed. Verification and validation methodologies for synthetic emergent behaviors require further research as well.
- **Class III:** This class is about developing an RCAS with live systems and/or digital twins, and the associated engineering processes. This is by far the most complex undertaking, which not only takes digital twins to the next level of integration and interoperability, but subject them to complex ecosystems for emergent behaviors. To clarify, developing a test bed for a digital twin exclusively is not a Class III RCAS. However, bringing a digital twin in a Class II RCAS makes a Class III RCAS. The critical difference is that the Class II RCAS is constructed to manifest verifiable synthetic emergent behavior. Further research is needed on Class III RCAS infrastructure when a higher fidelity digital twin is brought into the Class II architecture for accuracy, correctness, and subsequent verifiability of the synthetic emergent behavior.

19.4.2 Modeling, Simulation, and Visualization

The confluence of computational modeling and complex systems analysis has reached an inflection point, driven by the emergence of new AI paradigms [6]. Traditionally, modeling the resilience of complex systems has been hampered by a lack of an unambiguous definition of resilience [7] and a persistent "micro-macro dualism" where causation operates in both directions across different scales [8].

Classical approaches, such as agent-based modeling (ABM) and statistical methods, have proven insufficient to capture these nonlinear, multiscale phenomena.

We are witnessing a revolution where new generations of AI and machine learning models (including large language models (LLMs), graph neural networks (GNNs), and generative AI) are fundamentally changing this landscape. They represent a new epistemology, moving the field from a traditional engineering discipline toward a complex systems science [9]. Generative AI and LLMs are transforming ABM by enabling more nuanced, realistic, and adaptive agent behaviors, moving beyond simplistic, predefined rules [10]. These agents are endowed with human-like "minds" and can simulate complex social dynamics, creating "realistic" simulations of society to study emergent phenomena such as the spread of inflammatory messages or the effects of social policies [11, 12].

Graph neural networks (GNNs) provide a powerful framework for analyzing the structure and dynamics of interconnected systems, particularly in critical infrastructure such as transportation and power networks [13]. By integrating with techniques, such as deep reinforcement learning, GNNs enable dynamic, proactive network management and cyber-physical attack detection [14].

The very nature of these new AI models is that they are complex, emergent systems themselves, exhibiting behaviors that are not trivially predictable from their parts. This forces a shift in research from a mechanistic inquiry of "why" a model behaves in a certain way to a high-level, behavioral understanding of "what" it is doing [15]. The unpredictability of these systems is analogous to phase transitions in physical systems and makes them susceptible to even minor changes in input, which can lead to drastically different outputs. This phenomenon, while debated as a potential "mirage," still poses significant challenges for governance and risk management [7].

This paradigm shift requires a new research agenda for next-generation resilience. Below are suggested lines for the future:

- **Hybrid Multiscale Modeling:** The future lies in seamlessly integrating data-driven AI with physics-informed simulation methods [16]. The core idea is to integrate different modeling techniques to capture system behavior from fine-grained, microlevel details to coarse-grained, macrolevel behaviors. This approach is essential for bridging the micro- and macro gap by enabling bidirectional causation between fine-grained details and emergent system-level behaviors across scales.
- **Digital Twin for Proactive Control:** The ultimate framework for this is the digital twin, transformed from a static representation into an active, autonomous system for proactive resilience that is continuously synchronized with the physical counterpart. This requires developing autonomous feedback and control architectures, such as the "Sense, Infer, Control" (SIC)

model, which leverages generative AI to continuously sense system states, infer risks, and generate real-time, optimal control policies [17].

- **AI Model Management and Governance:** As these complex, emergent systems are integrated into critical infrastructure, new "meta-level" challenges arise. The unpredictability and data dependency of these models require new frameworks for automated model management and continuous monitoring [18, 19]. Future research should address ethical and societal challenges, including data bias and cascading failures, by advocating for adaptive and inclusive governance models.

In summary, the integration of new AI and machine learning models with traditional simulation methods is not merely an incremental improvement; it is a fundamental transformation of the field. This new approach should allow for the modeling of systems that are not just resilient to specific shocks but are inherently adaptive, autonomous, and capable of real-time optimization [20]. The unfinished work is to build the ethical and governance frameworks necessary to manage the unpredictable and powerful nature of these new systems, ensuring that they contribute to a more resilient and sustainable future.

Ultimately, given the overview of the agenda for modeling and simulation for RCAS, a core component of the research agenda should be the development of advanced visualization frameworks. Such tools should be designed as a practical "micro-macro scaler," which enables to traverse different levels of abstraction and transform opaque, "black-box" models into transparent, interpretable systems. The agenda must prioritize creating visualization tools that can display cross-scale dynamics, allowing to drill down from a high-level overview to specific, low-level details which are important for analyzing the micro-macro dualism of a complex system. This is essential for visually representing the causal relationships behind emergent behaviors, providing both a *global* explanation of a model's overall function and a *local* explanation for individual decisions, thereby ensuring comprehension and empowering informed decision-making.

19.4.3 Measurement and Control

Many perspectives can be used to classify RCAS metrics, and Chapter 2 looks at a taxonomy of metrics for emergent behavior in RCAS, considering measures of entropy, measures of performance, and measures of effectiveness. Little research to date has considered metrics of the resilience and adaptation of complex adaptive systems, and few if any comparative analyses of effectiveness of adaptation have been performed [21].

Generic, quantitative metrics of resilience are particularly hard to define as resilience is an application domain and context-specific property [2, 22, 23]. Furthermore, resilience is not a linear property, in that the state that the system has 'returned' to is not necessarily exactly the same as the previous one, so qualitative and quantitative metrics and associated analyses need to be defined. In addition, RCAS is defined by a plethora of attributes and their values at both micro and macro levels. This data has high dimensionality, is multimodal and multi-source, and defining measuring and analysis methods to capture and describe the most relevant is not trivial. Machine learning shows significant potential in this area. However, while important component attributes and their values can be identified, their importance is yet to be explained. In addition, measurement and analysis techniques often are a posteriori [24], when disturbances, failures, vulnerabilities have already been exposed. Identifying these as they happen remains an open challenge.

Chapters 2 [4] and 3 [25] discuss emergent behavior in detail and highlight both the need and potential avenues for measuring emergent behaviors but also the challenges in modeling and simulation of truly SEB, as discussed below. Emergent behavior also leads to significant control challenges as the effect of interactions and decisions are not easily predicted [26, 27]. In addition, RCAS are often characterized by distributed agency, where adaptation and decision-making are distributed among agents [20]. This makes for resilient and self-organizing systems, but also makes it very difficult to trace the control flow. Lastly, the nature of RCAS deployment implies that satisficing the local temporal requirements will often be the best strategy [28]. However, satisficing is domain and scenario dependent, and there is a need for modeling and analysis theories that are able to support this approach and be sufficiently generalizable to allow for comparative analyses and benchmarking. Considering mesostructures and patterns that appear within the agent community can offer additional perspectives on the interaction between agents [2]. Other control paradigms include bringing together implicit and explicit communication and control [28]; however, this requires sophisticated modeling paradigms. Similarly, fast and frugal architectures, inspired from Cognitive Psychology, may lead to activity-based systems [29] that may offer better embodied resilience in complex adaptive environments.

19.4.4 Strong Emergent Behavior

After exploring the diverse applications of modeling and simulation throughout this volume, it seems crucial to return to the fundamental philosophical questions that underpin our field. One of the most significant in RCAS is the very nature of emergence. To fully comprehend this concept, the distinction between strong emergence and weak emergence must be stressed.

Strong emergence is a central ontological claim in the philosophy of science. It posits that the novel properties of a system are *genuinely new* and cannot, in principle, be reduced to or predicted from the properties and interactions of its constituent parts. This view contrasts with weak emergence, which is primarily an epistemological position. The latter holds that a system's new properties are merely computationally or practically difficult to predict, but are ultimately grounded in the lower-level mechanisms.

In this context, the philosopher Mark Bedau has argued that strong emergence is "uncomfortably like magic," as it implies that an emergent phenomenon possesses an irreducible causal power that cannot be explained by the aggregation of microlevel potentialities, a notion he finds to be scientifically irrelevant [30]. By contrast, thinkers like David Chalmers champion the concept [31]. Chalmers famously defines a strongly emergent phenomenon as one where "truths concerning that phenomenon are not deducible even in principle from truths in the low-level domain," arguing that consciousness is the clearest example because its subjective "what-it's-like-ness" is not logically deducible from the physical facts alone. This position presents a profound challenge to the basic tenets of reductionism.

As we have seen in the contributions to this book, strong emergence is not a widely dealt-with concept or form of analysis, most likely because of its properties of irreducibility. Instead, the approaches tend to treat emergence as a result of the inherent complexity of systems and use models to manage and/or mitigate the effects of this complexity. This aligns with an epistemological view of emergence, where the main objective is to understand and control the phenomenon rather than to explore its more philosophical underlying nature. Given the limitations and insights explored throughout this volume, we propose the following research agenda to guide future work on strong emergence in RCAS.

19.4.4.1 Modeling Approaches for SEB

Current modeling paradigms—including agent-based models (ABMs), system dynamics, and hybrid approaches—are well-suited for weak emergence but struggle to represent strong emergence without ad hoc or post hoc adjustments. Future research should develop advanced multiparadigm modeling frameworks that integrate mechanistic, statistical, and AI-based components to capture irreducible emergent novelty, while also formalizing causal emergence within modeling languages (e.g., extended DEVS, ontological frameworks) [32, 33] to distinguish between weak and strong emergence during—not after—model execution.

19.4.4.2 Simulation Methods for SEB

Simulating SEB requires computational power and epistemological rigor. Current simulations often fail to generate genuinely novel behaviors without the external injection of information or *ex ante* semantic labeling. We propose investing in real-time and accelerated simulation platforms [34–36] that can handle cross-scale feedback and nonlinear tipping points at a fine-grained level of detail (which is typically extremely computationally demanding). We also advocate the development of simulation-based generative experiments that employ interactive human-in-the-loop or AI-in-the-loop approaches to label and learn from emergent novelties as they occur.

19.4.4.3 AI/ML and Explainability for SEB

While AI and machine learning offer powerful tools for pattern detection and system control, their application to SEB is still nascent. The future will see new ways of using machine learning to identify and explain new structures that emerge, without using labels that are already defined. It will also see models that are self-aware, and can detect and justify SEB using explanatory coherence and counterfactual reasoning.

19.4.4.4 Entropy and Metrics for SEB

Entropy and information-theoretic measures have proven useful for quantifying uncertainty and complexity in RCAS. However, these are not yet enough to capture SEB [27]. For future research, we should use multiscale entropy metrics to distinguish between patterns that are easy to predict and those that are new and unexpected. This should include developing measures of entropy that quantify the flow of information between the micro and macro levels, especially during phase transitions or systemic reconfigurations.

19.4.4.5 Visualization and Human Understanding for SEB

Strong emergence often cannot be visualized using conventional techniques, which are typically designed for steady-state or cyclic behaviors. We foster the development of interactive, multiperspective visual analytics that allow users to explore emergent structures across temporal and spatial scales, including narrative-based visualization to contextualize novel behaviors within domain-specific stories [37], aiding sense- and decision-making under uncertainty [22].

19.4.4.6 Validation and Governance for SEB

Strong emergence poses challenges for validation, governance, and ethics. Research is needed into dynamic validation frameworks that can adapt to evolving real system behaviors and redefine "ground truth" during experimentation, along with new governance models for responsible innovation in RCAS, especially when emergence leads to unintended or irreversible consequences. Efficient verification mechanisms need to be developed that cater to not just timing and synchronization instrumentation in a distributed landscape but ones that are focused on the tracking emergent behavior duration in context with the quality and quantity of underlying data.

As we close this book, one lesson stands clear: resilience in complex adaptive systems is not a property to be installed but a capability to be continuously expanded upon and optimized. Across the chapters, we have moved from foundations, to modeling frameworks and formalisms, to engineering processes to practical domains. The authors linked microlevel interactions to macrolevel outcomes across a variety of application domains using various approaches and engaging different stakeholders. The agenda ahead is unmistakably transdisciplinary: advance metrics that capture dynamics and cascading effects; design control processes that are robust and resilient to deep uncertainty and partitions; develop modeling, simulation, and visualization approaches that consider multiple perspectives and deep stakeholder engagement. Resilience work is slow, iterative, and often invisible, until unfortunately, its importance is noted, sometimes after the fact. There is a critical need for research to facilitate benchmarking and comparative analyses to allow us to safeguard the numerous application domains where RCAS are embedded.

Disclaimer: Dr. Mittal's affiliation with MTSI is provided for identification purposes only and is not intended to convey or imply concurrence with, or support for, the positions, opinions, or viewpoints expressed by this author.

References

1 Saurabh Mittal. Emergence in stigmergic and complex adaptive systems: A formal discrete event systems perspective. *Cognitive Systems Research*, 21:22–39, 2013.

2 V. Shemeshenko and S. A. De Raco. Intersectoral labor mobility in Argentina: Evolution and resilience. In C. Szabo, R. Castro, S. Mittal, and V. Shemeshenko, editors, *Resilient Complex Adaptive Systems: Modeling, Simulation, Visualization and Engineering Approaches*. John Wiley & Sons, Hoboken, NJ, 2025.

3 Claudia Szabo, Rodrigo Castro, Joachim Denil, and Susan Sanchez. Resilience and complexity in socio-cyber-physical systems. In *2023 Winter Simulation Conference (WSC)*, pages 660–670. IEEE, 2023. https://ieeexplore.ieee.org/abstract/document/10408660.

4 S. Mittal, C. Szabo, R. Castro, and V. Shemeshenko. Emergent behavior in resilient complex adaptive systems: Challenges, metrics and engineering approach. In Claudia Szabo, Rodrigo Castro, Saurabh Mittal, and Viktor Shemeshenko, editors, *Resilient Complex Adaptive Systems: Modeling, Simulation, Visualization and Engineering Approaches*. John Wiley & Sons, Hoboken, NJ, 2025.

5 Saurabh Mittal. *New Frontiers in Modeling and Simulation in Complex Systems Engineering: The Case of Synthetic Emergence*, pages 173–194. Springer International Publishing, Cham, 2019. https://doi.org/10.1007/978-3-030-17164-3_9.

6 Ahmad Muhammad Ayyaz, George Baryannis, and Richard Hill. Defining complex adaptive systems: An algorithmic approach. *Systems*, 12(2):45, 2024. https://doi.org/10.3390/systems12020045.

7 Noam Kolt, Michal Shur-Ofry, and Reuven Cohen. Lessons from complex systems science for AI governance. *Patterns*, 6(8):101341, 2025. https://doi.org/10.1016/j.patter.2025.101341.

8 Massimo Marchiori and Lino Possamai. Micro-macro analysis of complex networks. *PLoS One*, 10(1):1–27, 2015. https://doi.org/10.1371/journal.pone.0116670.

9 Ari Holtzman, Peter West, and Luke Zettlemoyer. Generative models as a complex systems science: How can we make sense of large language model behavior? *Journal of Social Computing*, 6(2):75–94, 2025. https://doi.org/10.23919/JSC.2025.0009.

10 Chen Gao, Xiaochong Lan, Nian Li, Yuan Yuan, Jingtao Ding, Zhilun Zhou, Fengli Xu, and Yong Li. Large language models empowered agent-based modeling and simulation: A survey and perspectives. *Humanities and Social Sciences Communications*, 11(1):1259, 2024. https://doi.org/10.1057/s41599-024-03611-3.

11 Jinghua Piao, Yuwei Yan, Jun Zhang, Nian Li, Junbo Yan, Xiaochong Lan, Zhihong Lu, Zhiheng Zheng, Jing Yi Wang, Di Zhou, Chen Gao, Fengli Xu, Fang Zhang, Ke Rong, Jun Su, and Yong Li. Agentsociety: Large-scale simulation of LLM-driven generative agents advances understanding of human behaviors and society, 2025. https://arxiv.org/abs/2502.08691.

12 Peter Sunehag and Joel Z. Leibo. Simulation streams: A programming paradigm for controlling large language models and building complex systems with generative AI, 2025. https://arxiv.org/abs/2501.18668.

13 Aldo Ramirez-Arellano and Juan Bory-Reyes. The resilience of complex network: An approach for relevant nodes extraction. *FRACTALS (fractals)*, 29(01): 1–14, 2021. https://doi.org/10.1142/S0218348X21500092.

14 Xuzeng Li, Tao Zhang, Jian Wang, Zhen Han, Jiqiang Liu, Jiawen Kang, Dusit Niyato, and Abbas Jamalipour. Achieving network resilience through graph neural network-enabled deep reinforcement learning, 2025. https://arxiv.org/abs/2501.11074.

15 Herbert A. Simon. The architecture of complexity. *Proceedings of the American Philosophical Society*, 106(6):467–482, 1962.http://www.jstor.org/stable/985254.

16 Sabine Attinger and Petros Koumoutsakos. *Multiscale Modelling and Simulation*. Lecture Notes in Computational Science and Engineering. Springer, 2004. https://doi.org/10.1007/978-3-642-18756-8.

17 Mohammad Reza Boskabadi, Yudong Cao, Behnam Khadem, William Clements, Z. Nevin Gerek, Eric Reuthe, Abhishek Sivaram, Christopher J. Savoie, and Seyed Soheil Mansouri. Industrial agentic ai and generative modeling in complex systems. *Current Opinion in Chemical Engineering*, 48:101150, 2025. https://doi.org/10.1016/j.coche.2025.101150.

18 Tala Talaei Khoei and Naima Kaabouch. Machine learning: Models, challenges, and research directions. *Future Internet*, 15(10):332, 2023.https://doi.org/10.3390/fi15100332.

19 Radwa Elshawi, Mohamed Maher, and Sherif Sakr. Automated machine learning: State-of-the-art and open challenges, 2019. https://arxiv.org/abs/1906.02287.

20 R. Castro, V. Shemeshenko, and P. Mira. A novel integrated agent-based computational economics and multi-agent reinforcement learning framework: Application to macrocrisis phenomena. In C. Szabo, R. Castro, S. Mittal, and V. Shemeshenko, editors, *Resilient Complex Adaptive Systems: Modeling, Simulation, Visualization and Engineering Approaches*. John Wiley & Sons, Hoboken, NJ, 2025.

21 Claudia Szabo, Robin Baker, Glen Pearce, Eyoel Teffera, and Anthony Perry. Overview and challenges of distributed decision making in resource contested and dynamic environments. *ACM Computing Surveys*, 57(9):1–34, 2025.

22 L. Yilmaz. Self-aware models with explanatory capabilities: Learning explanatory models from simulation experiments. In C. Szabo, R. Castro, S. Mittal, and V. Shemeshenko, editors, *Resilient Complex Adaptive Systems: Modeling, Simulation, Visualization and Engineering Approaches*. John Wiley & Sons, Hoboken, NJ, 2025.

23 Seyedmohsen Hosseini, Kash Barker, and Jose E. Ramirez-Marquez. A review of definitions and measures of system resilience. *Reliability Engineering & System Safety*, 145:47–61, 2016.

24 Claudia Szabo and Yong Meng Teo. Post-mortem analysis of emergent behavior in complex simulation models. In *Proceedings of the 1st ACM SIGSIM Conference on Principles of Advanced Discrete Simulation*, pages 241–252, 2013.

25 A. Tolk. Epistemology of resilient complex adaptive system simulations. In C. Szabo, R. Castro, S. Mittal, and V. Shemeshenko, editors, *Resilient Complex Adaptive Systems: Modeling, Simulation, Visualization and Engineering Approaches*. John Wiley & Sons, Hoboken, NJ, 2025.

26 Saurabh Mittal and Larry Rainey. Harnessing emergence: The control and design of emergent behavior in system of systems engineering. In *Proceedings of the Conference on Summer Computer Simulation*, pages 1–10, 2015.

27 Claudia Szabo, Joshua Groot, Thomas McAtee, and Christo Pyromallis. Who did this: Identifying causes of emergent behavior in complex systems. In *2019 IEEE Symposium Series on Computational Intelligence (SSCI)*, pages 262–270. IEEE, 2019.

28 K. Chan and F. T. Johnsen. A framework for characterizing, understanding and predicting behavior in complex edge computing systems. In C. Szabo, R. Castro, S. Mittal, and V. Shemeshenko, editors, *Resilient Complex Adaptive Systems: Modeling, Simulation, Visualization and Engineering Approaches*. John Wiley & Sons, Hoboken, NJ, 2025.

29 Saurabh Mittal. Model engineering for cyber complex adaptive systems. In *Proceedings of the European Modeling and Simulation Symposium*, 2014.

30 Mark A. Bedau. Is weak emergence just in the mind? *Minds and Machines*, 18(4):443–459, 2008. https://doi.org/10.1007/s11023-008-9122-6.

31 Chalmers. David J. Strong and weak emergence. *The Re-Emergence of Emergence*, 675:244–254, 2006.

32 Daniel Foguelman, Philipp Henning, Adelinde Uhrmacher, and Rodrigo Castro. EB-DEVS: A formal framework for modeling and simulation of emergent behavior in dynamic complex systems. *Journal of Computational Science*, 53:101387, 2021.

33 Saurabh Mittal. *Netcentric Complex Adaptive Systems,* Chapter 23, pages 639–661. CRC Press, 2014.

34 Saurabh Mittal and José L. Risco-Martín. *Simulation-Based Complex Adaptive Systems*, pages 127–150. Springer International Publishing, Cham, 2017. https://doi.org/10.1007/978-3-319-61264-5_6.

35 J. L. Risco-Martin, R. Dominguez, R. Cárdenas, E. Besada-Portas, and S. Esteban. Integrating simulation and real-world models for harmful algal and cyanobacterial blooms management. In C. Szabo, R. Castro, S. Mittal, and V. Shemeshenko, editors, *Resilient Complex Adaptive Systems: Modeling, Simulation, Visualization and Engineering Approaches*, Hoboken, NJ, 2025.

36 B. Fraser and R. Hull. Resilient complex adaptive systems: A communications perspective. In C. Szabo, R. Castro, S. Mittal, and V. Shemeshenko, editors, *Resilient Complex Adaptive Systems: Modeling, Simulation, Visualization and Engineering Approaches*. John Wiley & Sons, Hoboken, NJ, 2025.

37 I. Honhaga, V.A Do, E. Teffera, A. Perry, R. Baker, and C. Szabo. Multi-agent reinforcement learning for decision making in contested environments. In C. Szabo, R. Castro, S. Mittal, and V. Shemeshenko, editors, *Resilient Complex Adaptive Systems: Modeling, Simulation, Visualization and Engineering Approaches*. John Wiley & Sons, Hoboken, NJ, 2025.

Acronyms

ACF	Auto-correlation Function, 14, 15
BT	Behavior Tree, 19
CAS	Complex Adaptive Systems, 1–6, 8, 9, 13, 15, 16, 18, 20–23
CSMA	Carrier Sense Multiple Access, 14
CTDE	Centralized Training, Decentralized Execution, 18
DARPA	Defense Advanced Research Projects Agency, 15
DDIL	Disrupted, Denied, Intermittent, and Low-bandwidth, 3, 15, 18, 19
Dec-MCTS	Decentralized-Monte Carlo Tree Search, 19
HTTP	Hyper Text Transfer Protocol, 14
ISTAR	Intelligence, Surveillance, Target Acquisition, and Reconnaissance, 2, 15
LLM	Large Language Model, 20
LNPLM	Log Normal Path Loss Model, 20
MAC	Medium Access Control, 14
MANET	Mobile Ad hoc Network, 3
MARL	Multi-agent Reinforcement Learning, 18
QoS	Quality of Service, 8
RF	Radio Frequency, 1–4, 13, 19, 21, 23
RGG	Random Geometric Graph, 9, 10
RSSI	Received Signal Strength Indicator, 4
SMAC	StarCraft Multi-agent Challenge, 18
SNR	Signal-to-noise Ratio, 4
TSP	Travelling Salesman Problem, 5
TSPN	Traveling Salesman Problem with Neighborhood, 5
UAS	Uncrewed Aerial System, 2, 3, 8, 23

Index

Note: Page numbers in *italics* and **bold** refers to figures and tables respectively.

c

d

e